Lecture Notes in Electrical Engineering

Volume 767

The book series *Lecture Notes in Electrical Engineering* (LNEE) publishes the latest developments in Electrical Engineering - quickly, informally and in high quality. While original research reported in proceedings and monographs has traditionally formed the core of LNEE, we also encourage authors to submit books devoted to supporting student education and professional training in the various fields and applications areas of electrical engineering. The series cover classical and emerging topics concerning:

- Communication Engineering, Information Theory and Networks
- Electronics Engineering and Microelectronics
- Signal, Image and Speech Processing
- Wireless and Mobile Communication
- Circuits and Systems
- Energy Systems, Power Electronics and Electrical Machines
- Electro-optical Engineering
- Instrumentation Engineering
- Avionics Engineering
- Control Systems
- Internet-of-Things and Cybersecurity
- Biomedical Devices, MEMS and NEMS

For general information about this book series, comments or suggestions, please contact leontina.dicecco@springer.com.

To submit a proposal or request further information, please contact the Publishing Editor in your country:

China

Jasmine Dou, Editor (jasmine.dou@springer.com)

India, Japan, Rest of Asia

Swati Meherishi, Editorial Director (Swati.Meherishi@springer.com)

Southeast Asia, Australia, New Zealand

Ramesh Nath Premnath, Editor (ramesh.premnath@springernature.com)

USA, Canada:

Michael Luby, Senior Editor (michael.luby@springer.com)

All other Countries:

Leontina Di Cecco, Senior Editor (leontina.dicecco@springer.com)

**** This series is indexed by EI Compendex and Scopus databases. ****

More information about this series at http://www.springer.com/series/7818

Sanjeevikumar P. · Nagesh Prabhu ·
Suryanarayana K.
Editors

Advances in Renewable Energy and Electric Vehicles

Select Proceedings of AREEV 2020

Springer

Editors
Sanjeevikumar P.
CTIF Global Capsule (CGC) Laboratory
Department of Business Development
and Technology
Aarhus University
Herning, Denmark

Nagesh Prabhu
Department of Electrical and Electronics
Engineering
NMAM Institute of Technology
Karkala, Karnataka, India

Suryanarayana K.
Department of Electrical and Electronics
Engineering
NMAM Institute of Technology
Karkala, Karnataka, India

ISSN 1876-1100 ISSN 1876-1119 (electronic)
Lecture Notes in Electrical Engineering
ISBN 978-981-16-1641-9 ISBN 978-981-16-1642-6 (eBook)
https://doi.org/10.1007/978-981-16-1642-6

This Springer imprint is published by the registered company Springer Nature Singapore Pte Ltd.
The registered company address is: 152 Beach Road, #21-01/04 Gateway East, Singapore 189721, Singapore

Contents

About the Editors

Sanjeevikumar Padmanaban (Member'12-Senior Member'15, IEEE) received a Ph.D. degree in electrical engineering from the University of Bologna, Bologna, Italy 2012. He was an Associate Professor at VIT University from 2012 to 2013. In 2013, he joined the National Institute of Technology, India, as a Faculty Member. In 2014, he was invited as a Visiting Researcher at the Department of Electrical Engineering, Qatar University, Doha, Qatar, funded by the Qatar National Research Foundation (Government of Qatar). He continued his research activities with the Dublin Institute of Technology, Dublin, Ireland, in 2014.

Further, he served as an Associate Professor with the Department of Electrical and Electronics Engineering, University of Johannesburg, Johannesburg, South Africa, from 2016 to 2018. From March 2018 to February 2021, he has been a Faculty Member with the Department of Energy Technology, Aalborg University, Esbjerg, Denmark. Since March 2021, he has been with the CTIF Global Capsule (CGC) Laboratory, Department of Business Development and Technology, Aarhus University, Herning, Denmark.

S. Padmanaban has authored over 300 scientific papers and was the recipient of the Best Paper cum Most Excellence Research Paper Award from IET-SEISCON'13, IET-CEAT'16, IEEE-EECSI'19, IEEE-CENCON'19 and five best paper awards from ETAEERE'16 sponsored Lecture Notes in Electrical Engineering, Springer book. He is a Fellow of the Institution of Engineers, India, the Institution of Electronics and Telecommunication Engineers, India, and the Institution of Engineering and Technology, U.K. He is an Editor/Associate Editor/Editorial Board for refereed journals, in particular the IEEE SYSTEMS JOURNAL, IEEE Transaction on Industry Applications, IEEE ACCESS, *IET Power Electronics, IET Electronics Letters*, and Wiley-*International Transactions on Electrical Energy Systems*, Subject Editorial Board Member-*Energy Sources-Energies Journal*, MDPI, and the Subject Editor for the *IET Renewable Power Generation, IET Generation, Transmission and Distribution*, and *FACETS* journal (Canada).

Dr. Nagesh Prabhu is currently a Professor and Head at the Department of Electrical and Electronics Engineering, NMAM Institute of Technology, Nitte, Karnataka, India. He obtained his AMIE Electrical Engineering from Institution of Engineers, Kolkatta, M.Tech. from Karnataka Regional Engineering College, Surathkal, and Ph.D. in Power Systems and Power Electronics from Indian Institute of Science (IISc) Bangalore. His major areas of research interest include Power System Dynamics and control HVDC, FACTS, Power Quality & Custom Power Controllers for Smart Grid Applications. He has published over 60 papers in International Journals, conferences. He is a Member of Indian Society for Technical Education, Senior Member of IEEE, USA, Fellow of Indian Society for Lighting Engineers. He has delivered invited lectures at several international conferences and has been a resource person for several FDPs & workshops. He has guided four research scholars for their Ph.D. degree and two research scholars are currently pursuing their Ph.D. under VTU Belagavi. India. He is presently serving as an Associate Editor for *IEEE ACCESS Multidisciplinary Journal* and is a regular reviewer for *IEEE Transactions on Power Delivery*, *IEEE Transactions on Power Systems*, *International Journal of Power and Energy Systems* (Elsevier), *International Journal of Electric Power System Research* (Elsevier), *Ain Shams Engineering Journal* (Elsevier), *International Journal of IET—Generation Transmission & Distribution*, *Journal of Power Electronics*, Korea, *International Journal of Control*, Singapore and *International Journal of Mechanical Systems and Signal Processing* (Elsevier).

Dr. Suryanarayana K. is currently a Professor in the Department of Electrical and Electronics Engineering, NMAMIT, Nitte, Karnataka, India. He obtained his B.E. in EEE from Mangalore University, M.Tech. in System Analysis, and Computer Applications from Karnataka Regional Engineering College, Surathkal, and Ph.D. in the area of Power Electronics from VTU, Belagavi. His major areas of research interest include—Power Electronics, Control Systems, and Signal Processing. He has published papers in International Journals, presented papers at international and national conferences. He served as a consultant for PiOCtave Solutions, Bengaluru, Pii-Tech Solutions, Bengaluru, and HEXMOTO Controls Pvt. Ltd., Mysuru. He has worked at various industries like Central ResearchLaboratory (Bharat Electronics Limited), Sling Media Pvt. Ltd., Mistral Solutions Pvt. Ltd. and Tandberg Technology India Pvt. Ltd. He is a member of IEEE and ISSI.

20W Multi-output Isolated Power Supply Using Secondary Regulated Flyback Topology

Dhanush Acharya and Suryanarayana K.

1 Introduction

In linear power supplies, semiconductor devices operate in the linear region. To maintain the constant voltage across the load, the load current is varied by changing the base or gate drive of the semiconductor devices. This results in increased power dissipation across the device, thus increasing the conduction losses resulting into poor system efficiency. Thus, the linear power supplies are suitable only for low-power applications [1].

In switched-mode power supplies (SMPS), the power semiconductor device operates in cut-off or saturation region. Thus, the major disadvantage of linear power supplies regarding lower efficiency is overcome by SMPS [2]. So, the flyback topology is preferred in low and medium power offline applications which provides higher efficiency compared to linear power supplies [3]. A secondary regulated current-controlled closed loop 20 W, 200–400 V, flyback-based SMPS suitable to drive an H-bridge is designed and developed.

The closed loop configuration consists of a flyback transformer, output filter, PWM IC UCC28C44DR and opto-isolator LIA130S. In the proposed system, opto-isolator monitors the secondary winding voltage, and if it deviates from the reference voltage due to changes in the supply voltage, output current variation due to change in load, temperature variation is considered as error. This error is fed back to PWM IC, and duty cycle is varied to maintain the constant output voltage.

The paper organized in VII sections. Section 1 gives introduction, and Sect. 2 deals with basics of flyback converter. System overview and design considerations

D. Acharya (✉) · Suryanarayana K.
Department of Electrical and Electronics Engineering, NMAM Institute of Technology, Nitte 574110, Karkala, India

Suryanarayana K.
e-mail: suryanarayana@nitte.edu.in

Sanjeevikumar et al. (eds.), *Advances in Renewable Energy and Electric Vehicles*, Lecture Notes in Electrical Engineering 767,
https://doi.org/10.1007/978-981-16-1642-6_1

are discussed in Sect. 3. Section 4 deals with protection circuits and designing a closed loop converter. Section 5 provides the simulation and hardware results. Section 6 will conclude the design.

2 Flyback Converter

The basic block diagram of flyback converter is as in Fig. 1. The DC input voltage is obtained by rectifying sinusoidal alternating voltage. Flyback converter works on the principle of coupled inductor mechanism due to which the induced voltage depends on inductance and rate of change of current. Since in the DC supply rate of change of current is zero, voltage will not be induced. Thus, a switch is included in the primary of the flyback converter and is triggered such that induced voltage is nonzero. For the generation of pulsating DC supply in the converter, the primary switch is switched on and off at certain switching frequency using gate pulses. The coupled inductor configuration helps to adopt the necessary polarity by changing the dot position and diode orientation. This also provides the galvanic isolation between the mains and connected loads. By adjusting the turns ratio of the flyback transformer, the output voltage could be stepped up or down with respect to the input voltage.

The flyback operation can be classified into two modes of operation, i.e. continuous conduction mode (CCM) and discontinuous conduction mode (DCM).

In CCM, transformer does not dissipate all the stored energy to secondary. So, the current in the primary side of inductor never falls to zero. This mode is used in high-current, low-voltage applications.

In DCM transformer dissipates all the stored energy to secondary. So, the current in the primary side of inductor falls to zero during each cycle. This mode is used in low-current, high-voltage applications.

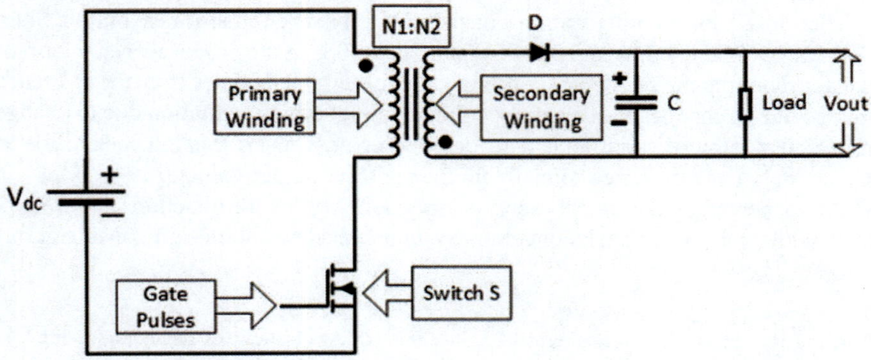

Fig. 1 Basic diagram of flyback converter

In proposed design, flyback is implemented using DCM operation. This DCM operation depends on the on and off condition of primary switch as well as secondary diode.

3 System Overview

The block diagram of the proposed system is as in Fig. 2: it is specifically designed to drive an H-bridge switches used to drive a stepper motor. Since the H-bridge switches require the gate turn-on voltage between 15 and 20 V, flyback auxiliary supply is designed for 15 V.

The purpose of each flyback winding is listed in Table 1. The PWM IC UCC28C44DR can withstand a maximum supply voltage of 18 V, and an auxiliary

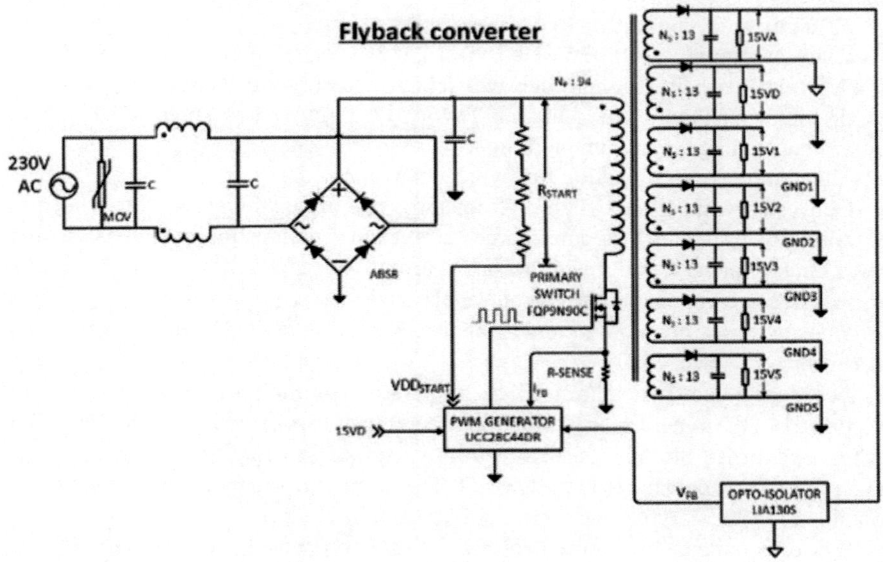

Fig. 2 Block diagram

Table 1 Purpose of flyback windings

S. No	Winding	Power	Purpose
1	15VD	1 W	Supply for PWM IC
2	15VA	1 W	Feedback to opto-isolator
3	15V1,15V2,15V3,15V4	1 W	For H-bridge TOP switches
4	15V5	4 W	For H-bridge BOTTOM switches

supply of 15VD is generated to power the PWM IC. However, in order to protect the PWM IC from over voltage, a Zener diode of 18 V is used. The regulation winding, 15VA, is fed back to opto-isolator LIA130S.

The proposed system has the DC input voltage range from 200 and 400 V. The system is operated with a switching frequency of 50 kHz and is capable of delivering 20 W.

The universal AC input is given to the rectifier through EMI filter to minimize the high-frequency noise present in power lines. It also protects the circuit from noise in the signal transmission and ensures the intended performance of electrical equipment. The diode bridge rectifier converts AC supply to DC. Metal Oxide Varistor (MOV) is used to protect the system from transient voltages.

The proposed flyback converter has two control loops. The first being the outer voltage loop and later is inner current loop. The instantaneous current of the circuit is sensed using R_{SENSE} resistor. R_{START} is the star-tup circuit that provides the initial supply to PWM IC till it gets sufficient voltage from auxiliary winding of the converter.

To regulate the output to 15 V irrespective of input voltage or load variations, feedback is provided to a PWM IC via LIA130S opto-isolator as in Fig. 2. The PWM controller has error amplifier which provides the control signal to the current mode controller. For the stable and satisfactory dynamic response, precision feedback components must be used in the design.

The parameter specification for flyback converter is as in Table 2.

Converter with higher switching frequency, size of magnetics and filters become smaller. However, core loss, gate charge current and switching losses increase. Lower switching frequency results in higher peak currents and conduction losses. A compromise between component size, current levels and losses must be made by the designer in choosing the switching frequency. In this design, a switching frequency of 50 kHz is chosen.

Leakage inductance is caused by imperfect coupling between primary and secondary of the transformer. Leakage inductance appears in series with the MOSFET when it is turned off. This results in turn-off spikes and ringing across the device and might cause damage to the MOSFET. The design procedure followed and values obtained for 20 W flyback converter are listed in Table 3.

Based on inductance requirement, EE 30/15/7 core with $A_c A_w$ of 7700 mm^4 and A_e of 57.3 mm^2 is chosen. Based on A_e value, primary and secondary turns calculated for each winding as listed in Table 4.

Table 2 Parameter specification for flyback converter

Parameter	Specification
Input voltage	200–400 V DC
Maximum output power (P_o)	20 W
Switching frequency (f_s)	50 kHz
Output voltage	15 V

Table 3 Design calculation

S. No.	Component	Formula	Value
1	Primary side inductance	$L_m = \frac{(V_{dc(min)} D_{max})^2}{2 P_{in} F_s K_{rf}}$	1.92 mH
2	Maximum peak current of MOSFET	$I_{d(peak)} = I_{EDC} + \frac{\Delta I}{2}$	0.834 A
3	Core window product area	$A_c A_w = \frac{V_1 I_1}{2 K_w B_m J F_s}$	2213 mm^4
4	Primary turns	$N_{p(min)} = \frac{L_m I_{over}}{B_s A_e}$	94
5	Turns ratio	$n = \frac{N_p}{N_{s1}} = \frac{V_{RO}}{V_{01} + V_{F1}}$	7.45
6	Secondary turns 15 V	$N_{S1} = \frac{N_P}{n}$	13
7	Cross-sectional area of winding	$A = \frac{I}{J}$	0.0166 m^2 (Primary) 0.022 m^2(1 W) 0.089 m^2(4 W)
8	Strands for primary winding	$P_{strand} = \frac{A}{0.0591}$	1 (Primary) 1 (1 W) 2 (4 W)

Table 4 Winding parameters with EE core of 30/15/7

S. No.	Winding	Voltage	No. of turns	Wire gauge
1	Primary	400	94	SWG32X1
2	Secondary	15 (1 W)	13	SWG32X1
3	Secondary	15 (4 W)	13	SWG32X2

Lower forward drop of Schottky diodes improves the converter efficiency due to reduced conduction losses in comparison with silicon diodes. Schottky is selected based on peak reverse voltage rating, peak repetitive forward current and average forward current rating of the device.

The peak reverse voltage that the device will be subjected to is as in (1):

$$V_{rpk} = n V_{in(max)} - V_{drop(FET)} \tag{1}$$

Winding technique is another important and critical parameter in flyback converter. Poorly designed flyback transformer will reduce the converter efficiency. The winding pattern followed in the design is as in Fig. 3.

The major consideration in the EE core bobbin is to ensure the proper creepage distance between primary and secondary pins. The procedure followed for the winding is as follows:

1. Primary winding is the first winding on the bobbin. Since this winding is connected to the drain of the MOSFET, this ensures the point with maximum voltage swing will be shielded by other winding so that EMI is reduced. Primary

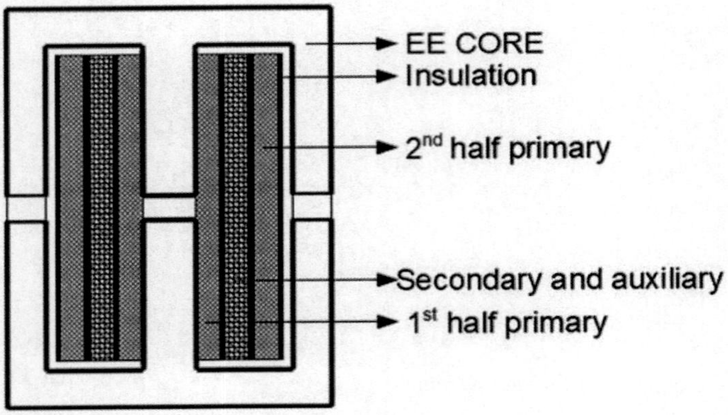

Fig. 3 Flyback transformer layout

winding is split into two separate windings so that leakage inductance is reduced by increasing coupling between primary and all bias windings.

2. Three layers of insulation are placed between primary and secondary windings in order to reduce the possibility of breakdown between the layers and also to reduce the interwinding capacitance between primary.
3. All the bias windings are placed on top of the primary insulation.
4. Three layers of basic insulation are placed on top of the secondary windings in order to pass the safety standards for electrical strength requirements.
5. Remaining primary winding is wound, and three layers of basic insulation are placed.

By following the above procedure, leakage inductance measured is less than 2% of primary inductance.

4 Protection Circuits and Design of Closed Loop

During the primary switch turn off process, ringing in the voltage across MOSFET will be observed due to primary inductance, leakage inductance and output capacitance. The ringing frequency could be written as in (2):

$$f_{\text{ring}} = \frac{1}{2 * \pi * \sqrt{((L_{\text{leak}} + L_m) * C_{\text{out(MOSFET)}})}} \tag{2}$$

Since the output capacitance and primary inductance are fixed, leakage inductance will contribute more to the amplitude and ringing frequency.

MOSFET snubber circuit is necessary in order to clamp the ringing voltage caused by leakage inductance. The snubber should be designed such that the voltage across

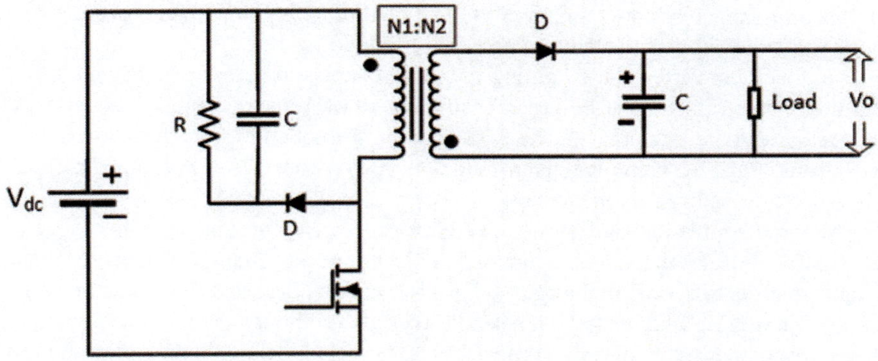

Fig. 4 RCD snubber clamping

the switch is less than the breakdown voltage. Two widely used snubber circuits are RCD clamping and Zener or transil clamping.

The RCD snubber consists of high voltage diode D, high voltage capacitor C and resistor R as in Fig. 4. RCD network absorbs the current in the leakage inductance by turning on the snubber diode after particular voltage, and thus, it will charge the capacitor C. Thus, the ringing may be damped to a level lower than the circuit without snubber. The snubber capacitor C is large enough that its voltage does not change significantly during one switching cycle. RCD clamp is a cheaper solution; however, it dissipates power even under no-load conditions. There is at least the reflected voltage VR across the clamp resistor at all times.

Another easy solution is Zener clamping in which TVS diode is used to clamp the ringing voltage as in Fig. 5. The TVS diode is connected in series with a reverse polarity diode. The TVS must clamp the voltage to sum of reflected voltage and input voltage. It will dissipate all the power as clamping resistor. However, the cost

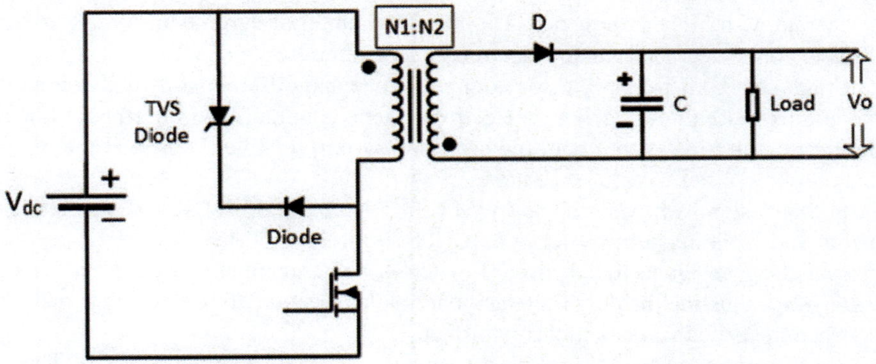

Fig. 5 TVS diode clamping

of the components is more compared to RCD clamping. It gives a better and defined clamping level but dissipates more power at full load.

In flyback converter, the clamping circuit is connected across primary winding, and the device rating must be larger than the sum of reflected voltage and voltage caused by leakage inductance. The RCD clamping frequently fails during transient conditions, and TVS clamping gives good dynamic response. In the proposed circuit, TVS diode clamping is employed.

The transformer leakage inductance also causes ringing and overshoot in the secondary diode. Leakage inductance and diode output capacitance form the resonant circuit. In order to damp out the oscillations, RC snubber is used across the secondary diode. The ringing frequency on the secondary side (without snubber) is determined by the resonant tank frequency created by the reflected leakage inductance and the junction capacitance of the output diode. The ringing frequency is written as in (3):

$$f_{\text{ring}} = \frac{1}{2 * \pi * \sqrt{\left(L_{\text{leak}} * C_{\text{junction(diode)}}\right)}} = 3.82\,\text{MHz} \tag{3}$$

where $C_{\text{junction(diode)}} = 12$ pF and $L_{\text{leak}} = 2\%$ of L_m.

For making Q factor near to 1, snubber capacitance must be at least three times the output capacitance of the diode as in (4), and to provide critical damping, the snubber resistor could be computed as in (5):

$$C_{\text{snub}} = 3 * C_{\text{out(diode)}} = 135\,\text{pF} \tag{4}$$

$$R_{\text{snub}} = \sqrt{\frac{L_{\text{leak}}}{N^2}\frac{1}{C_{\text{out(diode)}}}} = 19\,\Omega \tag{5}$$

Switching frequency of the converter decides the filter requirement of any converter. To achieve higher power density and reduced system size, converter needs to operate with higher frequency. The rapid switching of currents in the converter leads to conducted and radiated electromagnetic interference (EMI).

Conducted EMI is classified as common mode and differential mode. Common mode noise essentially occurs by electromagnetic coupling, and differential noise occurs by pulsating currents. In the proposed system, $\prod$ filter is used at the input and output in order to reduce the noise.

In this design, two ICs such as PWM controller UCC28C44DR of Texas Instruments and Littlefuse opto-isolator LIA130S are used. PWM controller is used to provide the gate signals to the primary switch as well as maintain the constant output voltage by adjusting the duty ratio, and opto-isolator is used to provide the isolation between measured signal and the controller.

Start-up circuit is used to provide the supply to the controller until MOSFET gets the required pulses from PWM IC. At power on, the current is delivered to the IC by start-up resistors, and once the capacitor voltage reaches under voltage lockout level, the IC starts to give gate pulses. Then, the controller takes the supply from auxiliary winding. PWM IC UCC28C44DR has RT/CT pin to select the necessary switching frequency, and system has been designed to operate at a switching frequency of 50 kHz.

The current in the primary winding is converted to voltage signal by connecting R_{SENSE} resistor and given to CS pin of PWM IC. The IC will stop generating the gate pulses if set overcurrent limit is reached. The R_{SENSE} resistor is calculated as in (6):

$$I_{SENSE} = \frac{V}{R_{SENSE}} \tag{6}$$

Slope compensation is required in closed loop current-controlled systems when the duty cycle exceeds 50% else inductor current will oscillate at half the switching frequency.

Slope compensation fixes these problems by adding ramp to the current signal and helps in damping out the oscillations.

Regulated output of winding is calculated using potential divider circuit as in Fig. 6.

To select potential divider values, (7) and (8) are used

$$R_{upper} > \frac{V - V_{ref}}{80\,\mu A} \tag{7}$$

where $V_{ref} = 1.24\,V$ and 80 μA is input offset current for LIA130S.

$$\frac{\text{Rupper}}{\text{Rlower}} = \frac{V}{V_{ref}} - 1 \tag{8}$$

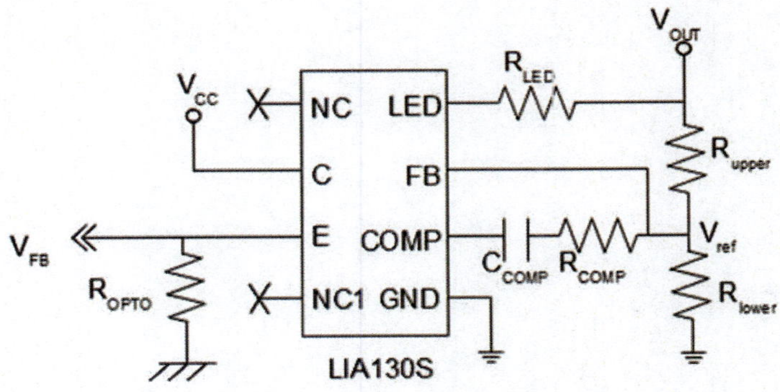

Fig. 6 Opto-isolator circuit diagram

5 Simulation and Hardware Results

To validate the proposed design considerations, simulation as well as experimental tests is carried out. The proposed system works well at input DC voltage of 400 V. Feedback control is necessary in order to remove the error in the output. The closed loop simulation is carried out as in Fig. 7.

The output voltage of 15 V is as in Fig. 8. It is clearly seen that the curve is smooth and oscillations are removed. Error in the output voltage is removed using closed loop configuration.

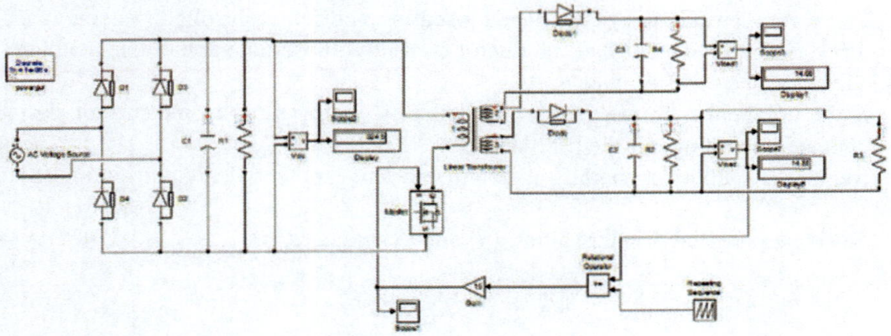

Fig. 7 Simulink model of closed loop flyback converter

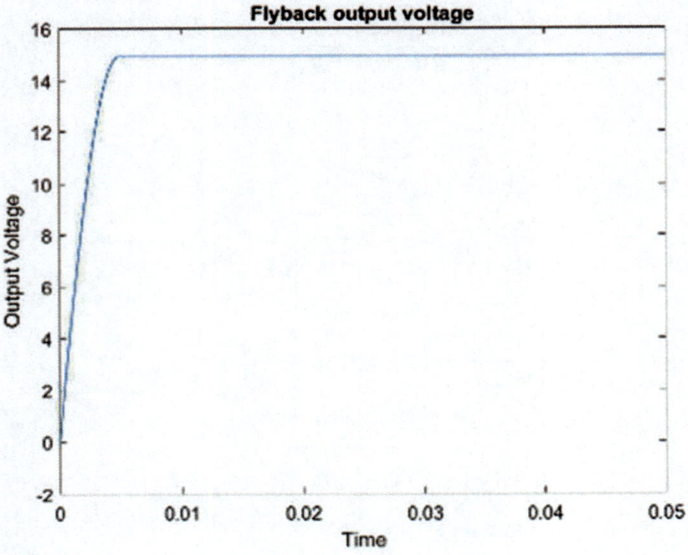

Fig. 8 15 V output voltage

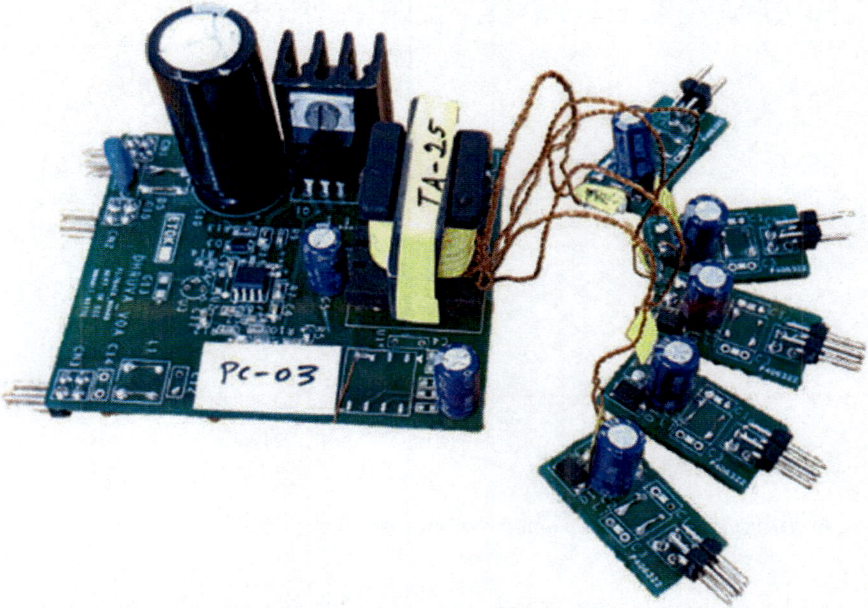

Fig. 9 Hardware set-up of flyback

The hardware approach in this design includes schematic entry using Orcad Capture and layout using PCB Editor tool. The hardware prototype of proposed system is as in Fig. 9. It is a four-layer PCB that consists of top, Vcc, ground and bottom layers. Daughter boards have two layers consisting of Vcc and ground. The size of main board is 6.5 cm × 5.5 cm and daughter board is 2.5 cm × 1.35 cm.

The flyback converter implements UCC28C44DR IC to accomplish constant output voltage by varying the duty cycle. The drain to source voltage of primary switch with TVS diode protection is shown in Fig. 10. It can be observed that maximum V_{DS} is 625 V with TVS diode protection.

The 18 V regulated winding voltage and unregulated winding voltage waveform is as in Fig. 11. The regulated winding is provided to opto-isolator, and full load is applied to unregulated winding. It can be seen that the proposed converter system is able to regulate the output when the load jumps from no load to full load. This flyback output voltages are used to drive the H-bridge switches.

Without EMI filter, the noise in the output voltage is 150mv peak to peak as in Fig. 12. By putting the common mode and differential mode filter, the noise is reduced to 30 mV as in Fig. 13.

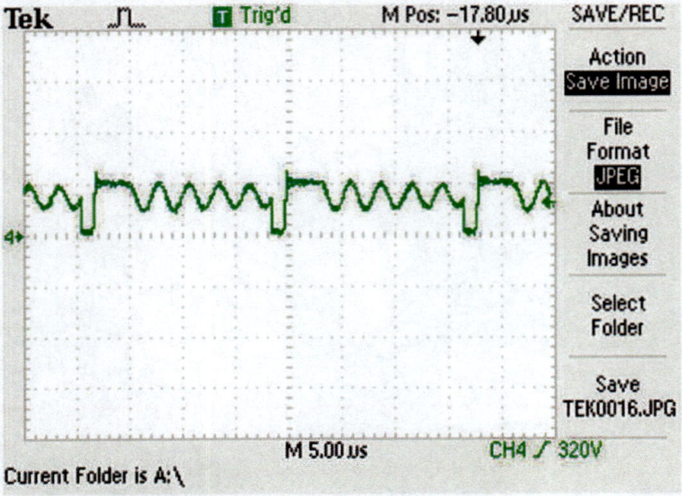

Fig. 10 Drain to source voltage of primary switch waveform

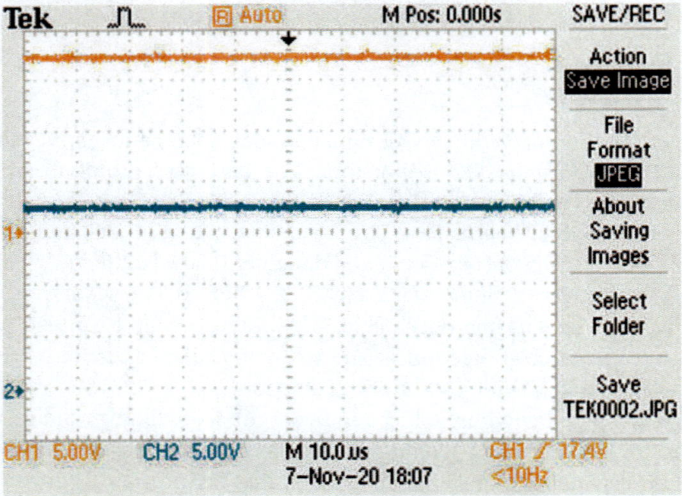

Fig. 11 Regulated winding voltage (blue colour) and unregulated winding voltage (orange colour) waveform

6 Conclusion

This paper deals with design and development of secondary regulated flyback converter. The proposed system was simulated using MATLAB Simulink tool to

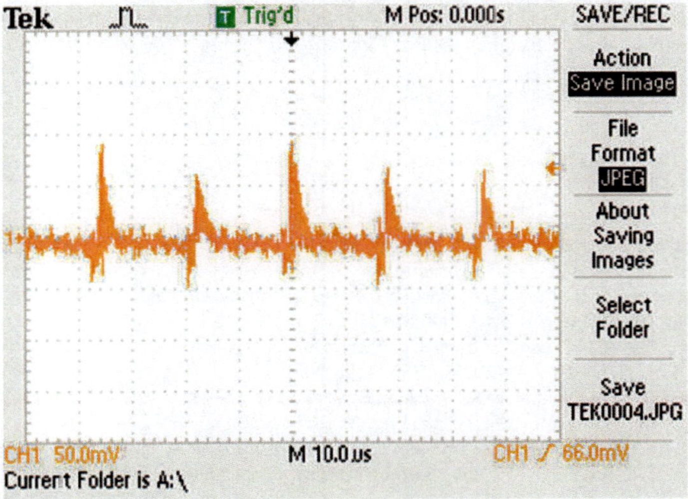

Fig. 12 Output waveform without EMI filter

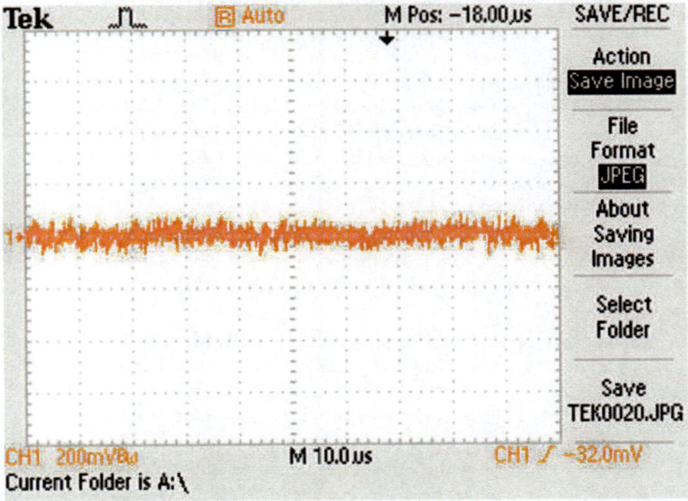

Fig. 13 Output waveform with EMI filter

obtain the theoretical results. Hardware implementation is done using OrCAD Capture tool, and obtained practical output is similar to theoretical results obtained.

Acknowledgements The author would like to express the gratitude to Nitte Education Trust and beloved Principal Dr. Niranjan N Chiplunkar for giving encouragement to all activities that held in the college. The author would express sincere gratitude to Mr. Ravikiran Rao, Mrs. Swathi Hatwar,

Mrs. Raksha Adappa for guidance and constructive feedback after reviewing this paper. The author would like to thank Research and Innovation Center, Nitte, for providing opportunity to this work.

References

1. R. Perez, A comparative assessment between linear and switching power supplies in portable electronic devices, in *IEEE International Symposium on Electromagnetic Compatibility. Symposium Record (Cat. No.00CH37016)*, vol. 2, Washington, DC, (2000), pp. 839–843. https://doi.org/10.1109/ISEMC.2000.874731.
2. J. Ahmad, S. Sonar, Design of flyback converter for DC-DC application. Int. J. Innov. Technol. Explor. Eng. (IJITEE) **9**(3) (2020). ISSN 2278-3075
3. G. Medha, S. Vijayshankar, N.K. Diwakar, B. Venkatesh, 18W non-isolated closed loop flyback converter. Int. J. Ind. Electron. Electr. Eng. (IEEE) **3** (2015). ISSN 2347-6982
4. A.A. Mohammed, S.M. Nafie, Flyback converter design for low power application, in *2015 International Conference on Computing, Control, Networking, Electronics and Embedded Systems Engineering (ICCNEEE)*, Khartoum, (2015), pp. 447–450. https://doi.org/10.1109/ICCNEEE.2015.7381410
5. M. Salimi, M. Hamedi, Adaptive nonlinear control of the flyback switch mode power supplies, in *2017 International Conference on Mechanical, System and Control Engineering (ICMSC)*, St. Petersburg, Russia, (2017). https://doi.org/10.1109/ICMSC.2017.7959508
6. Application note for "flyback transformer design" International Rectifiers
7. M. Ravikiran Rao, K. Suryanarayana, H. Swathi Hatwar, R. Adappa, Design and implementation of 400W flyback converter using SiC MOSFET, in *First International Conference on Advances in Electrical and Computer Technologies 2019 (ICAECT 2019)*
8. W. Kleebchampee, C. Bunlaksananusorn, Modeling and control design of a current-mode controlled flyback converter with opto-isolator feedback, in *2005 International Conference on Power Electronics and Drives Systems*, Kuala Lumpur, (2005), pp. 787–792. https://doi.org/10.1109/PEDS.2005.1619792
9. Miloudi, S. Nemmich, H. Slimani, Analysis and reduction of common-mode and differential-mode EMI noise in a Flyback switch-mode power supply (SMPS), in *2012 20th Telecommunications Forum (TELFOR)*, Belgrade, (2012), pp. 1080–1083. https://doi.org/10.1109/TELFOR.2012.6419398
10. A. Connaughton, A.P. Talei, K.K. Leong, K. Krischan, A. Muetze, Variable on-time control scheme for the secondary-side controlled flyback converter. IEEE Trans. Power Electron. **34**(3), 2416–2426 (2019). https://doi.org/10.1109/TPEL.2018.2844021
11. M. Ferdowsi, A. Emadi, Pulse regulation control technique for flyback converter. IEEE Trans. Power Electron. **20**(4) (2005)
12. Application note for "Fifth generation fixed frequency design guide" Infineon
13. Application note for "Bi-CMOS Low Power Current Mode PWM Controller" Texas Instruments

A Review on Social Group Optimization Technique for Power Capability Enhancement with Combined TCSC-UPFC

A. V. Sunil Kumar, R. Prakash, R. S. Shivakumara Aradhya, and Mahesh Lamsal

1 Introduction

With the demand for electrical power increasing every year, the electric supply industry has undergone an extreme and overwhelming transformation worldwide. Due to the ever-changing load pattern, the conventional generating stations like hydro, thermal, nuclear as well as renewables like wind, solar, geothermal, tidal, etc., are installed to meet the consumer's demand. The power generated by these stations needs transmission lines to connect the generating stations to load centers [1, 2]. While transmitting power, some transmission systems may be overloaded, loaded to the full capacity, or may not be loaded to the full capacity. Consequently, the system voltage profile deteriorates, and in some extreme cases, the system may collapse affecting the security of the system.

Power system operators use FACTS-based solutions to maintain the stability of the system. These devices operate to control different electrical parameters in the transmission networks. There are many types of power electronic controllers used in FACTS [3]. These controllers operate very fast and are powerful in maintaining the stability limit in the transmission system. UPFC and TCSC are considered as the most efficient controllers among the different FACTS controllers. UPFC controls the power flow in the system by compensating the line impedance, voltage magnitude,

A. V. Sunil Kumar (✉) · R. Prakash · R. S. Shivakumara Aradhya · M. Lamsal
Department of Electrical and Electronics, Acharya Institute of Technology, Bengaluru, India
e-mail: sunilkumarav@acharya.ac.in

R. Prakash
e-mail: prakashrp1960@gmail.com

R. S. Shivakumara Aradhya
e-mail: rssaradhya@gmail.com

M. Lamsal
e-mail: lamsalmahesh99@gmail.com

Sanjeevikumar et al. (eds.), *Advances in Renewable Energy and Electric Vehicles*,
Lecture Notes in Electrical Engineering 767,
https://doi.org/10.1007/978-981-16-1642-6_2

and phase angle. TCSC also provides efficient power flow control in addition to fault current limitation at some point. It is economical in terms of cost for solving various stability problem.

Optimal allocation of FACTS is particularly important in power system for its economic effectiveness and system performance. To get the maximum benefits, FACTS devices of suitable types should be placed at the most appropriate location. Various optimization techniques like evolutionary programming (EP) [4], optimal power flow (OPF) [5], genetic algorithm (GA) [6], teaching learning-based optimization (TLBO) [7], particle swarm optimization (PSO), biogeography-based optimization (BBO), and weight-improved particle swarm optimization (WIPSO) [8] have been utilized to solve the allocation problem. Social group optimization (SGO) is another method which can be used to solve for locating the FACTS devices [1].

2 Articulation of the Problem

2.1 *Intent of the Optimization*

Real power losses P_L and net voltage deviation (NVD) affect the operation of transmission line. The reduction of both helps in enhancing stability during transmission. Hence, the objective function (OF) is written as

$$\mathrm{OF} = W_1 * P_L + W_2 * \mathrm{NVD} \tag{1}$$

where W_1 and W_2 are weight factors of loss and, Knowing conductance G of the line joining buses i and j **whose voltages are** V_i and V_j, δ_{ij} and NLB being phase difference between the buses voltages and total number of buses, respectively, the real power loss is expressed as

$$P_L = \sum_{j=1}^{\mathrm{NLB}} G(V_i^2 + V_j^2 - 2V_i V_j \cos \delta_{ij} \tag{2}$$

If VD_i is the potential deviation at bus I, the net voltage deviation is

$$\mathrm{NVD} = \sum_{j=1}^{\mathrm{NBL}} \mathrm{VD}_i \tag{3}$$

where

$$VD_i = \begin{cases} 0 \text{ if } 0.95 < V_i < 1.05 \\ (1 - V_i)^2 \text{if } 0.9 < V_i < 0.95 \text{ or } 1.05 < V_i < 1.10 \\ 5 * (1 - V_i)^2 \text{if } V_i < 0.9 \text{ or } V_i > 1.1 \end{cases}$$

2.2 Equality Constraints

The equality constraints in the transmission line are as follows:

$$P_{Gi} - P_{Di} - V_i \sum_{j=1}^{\text{NLB}} V_j (G \cos \delta_{ij} + B \sin \delta_{ij}) = 0 \quad (4)$$

$$Q_{Gi} - Q_{Di} - V_i \sum_{j=1}^{\text{NLB}} V_j (G \cos \delta_{ij} - B \sin \delta_{ij}) = 0 \quad (5)$$

In Eqs. (4) and (5), P_{Gi} and P_{Di} are the active power generation and requirement, respectively, V_i and V_j are the absolute voltage values at buses i and j. G and B are the properties of transmission lines (what properties? Be specific), and δ_{ij}= is the phase difference between the bus voltages [2].

2.3 FACTS Devices Considered

Two FACTS devices have been considered in the present study, namely TCSC and UPSC. Brief description of them is presented below.

3 Social Group Optimization (SGO)

3.1 Overview of Social Group Optimization

Satapathy and Naik have developed social group optimization technique by impersonating the behavior and knowledge in human groups, for example, behavioral traits on human life like honesty, fear, tolerance, etc. In SGO, each person in the population has some knowledge on solving the complex problem, and the remaining persons in the group acquire knowledge through that person in the group. The person having knowledge is known as best person or best solution. The best person propagates knowledge among the entire persons involved in the group improving their knowledge level. This technique has improving phase and acquiring phase. The improving phase synchronizes the positions of people, and the acquiring phase allows the person in that group to discover the best solution for the population under concern. The optimization technique is formulated mathematically as [9]:

Let X_i be considered as the initial knowledge of persons in the group and $i = 1, 2, 3, \ldots, N$ with N being the total number of persons in the group. The best person who is identified as g_{best} in the population tries to pass on the knowledge to all persons which will eventually improve the knowledge in the group.

Hence,

$$g_{\text{best}} = \min\{f_i, i = 1, 2, \ldots, N\} \quad (7)$$

The updated position or knowledge of every population in the group is expressed by the following relation:

$$X_{\text{new}_{i\,j}} = c * X_{\text{old}_{i\,j}} + r * \left(g_{\text{best}}(j) - X_{\text{old}_{i\,j}}\right) \quad (8)$$

where X_{new} is the new knowledge, X_{old} is the old knowledge, G_{best} is the best knowledge, r is the random number in the range [0, 1], and c is the self-introspection parameter which can be set from [0,1].

In the acquiring phase, the person will find the global solutions based on knowledge updating in the improving phase. Here, select one person from the group (X_r) based on i not equal to r. Once the fitness value becomes $f(X_i) < f(X_r)$, then following procedure is followed:

$$X_{\text{new}_{i,j}} = X_{\text{old}_{i,j}} + r_1 * \left(X_{i,j} - X_{r,j}\right) + r_2 * \left(g_{\text{best}_j} - X_{i,j}\right) \quad (9)$$

Here, r_1 and r_2 are random number in the range [0, 1]. Using trial-and-error approach, r_1 and r_2 is set as 0.4 and 0.2. $X_{r,j}$ is the knowledge value of the chosen individual.

The SGO algorithm is illustrated in Fig. 1, where the person shown inside the circle is the one with best knowledge or g_{best}.

3.2 OPF with FACTS Devices by Social Group Optimization

SGO is implemented to determine the location of FACTS devices. The proposed method is listed below and is also shown in Fig. 2.

Step 1: Initialization of population and design variables. Population N is assigned as $N = 6$ for 6 bus system, $N = 14$ for 14 bus system, and $N = 57$ for 57 bus system. Similarly, design variable indicates the total number of FACTS devices to be included in the particular bus system.

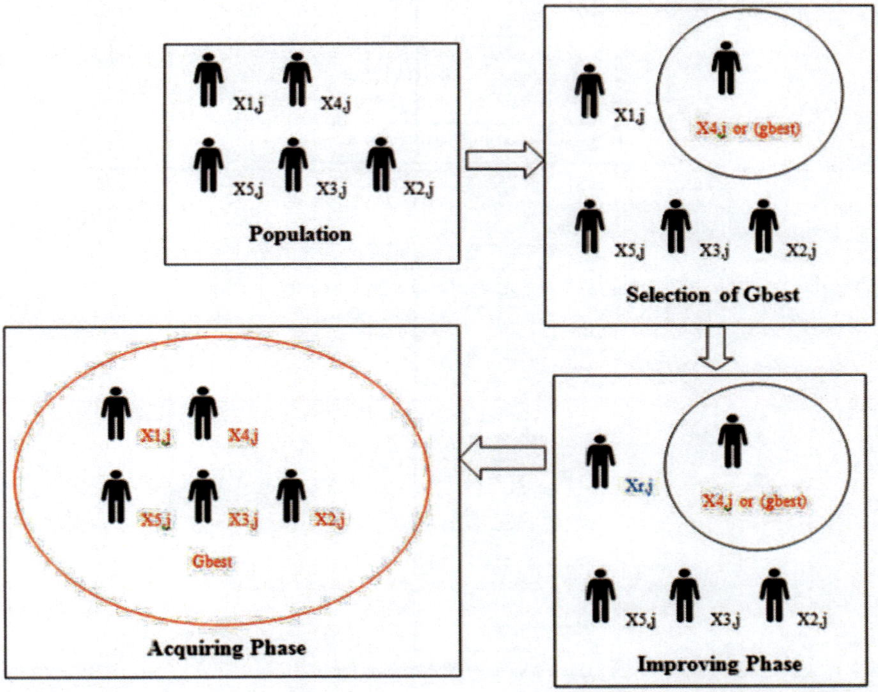

Fig. 1 Basic understanding of SGO

Step 2: Performing the base case load flow. This step involves to solve base case power flow by Newton–Raphson power flow method.
Step 3: Identifying g_{best} and calculating the fitness function. The g_{best} is considered best value calculated from the step 2, and fitness function is calculated using Eq. (3) in terms of net voltage deviation.
Step 4: Improving and acquiring phase. The best value is compared with other population using the Eqs. (8) and (8). The g_{best} is updated accordingly.
Step 5: Solving power flow with FACTS device. This step solves the power flow once again after the inclusion of multi-type FACTS devices—TCSC and UPFC.

4 Result and Discussion

The solution to find optimal allocation of FACTS controllers to minimize the losses and voltage deviation for the IEEE-6, IEEE-14, and IEEE-57 bus systems is obtained and discussed. The simulation was carried out by MATLAB software.

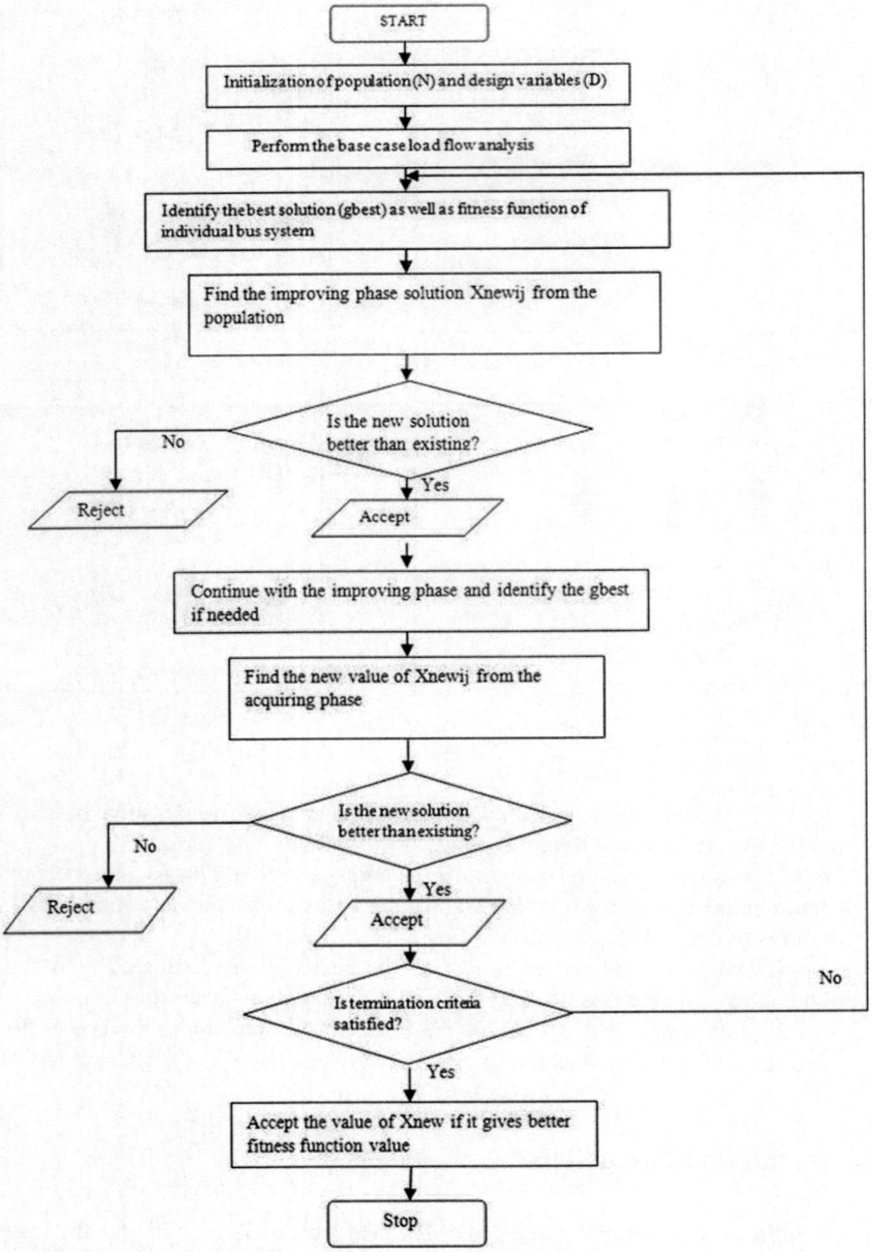

Fig. 2 Implementation of social group optimization

Table 1 Location of UPFC and voltage profile with/without FACTS (IEEE-6)

Bus number	Device connected	Voltage (p.u) without facts	Voltage (p.u) with facts after SGO
1		1	1
2		1	1
3	**UPFC**	**1.0819**	**1.04378**
4		1.0239	1.00064
5		0.9543	0.96256
6		0.95232	0.95484

Table 2 Location of TCSC and line flows before and after FACTS devices (IEEE-6)

From bus	To bus	Device connected	P flow without facts	Q flow without facts	P flow with facts	Q flow with facts	P loss without facts	P loss with facts
1	2	–	0.0071	−0.00227	−0.00258	0.00015	0	0
2	**3**	**TCSC**	**−0.20246**	**−0.11628**	**−0.95886**	**−0.46862**	**0.00544**	**0**
3	4	–	0.0921	0.05761	0.28445	1.62019	0.00202	0
5	**4**	**TCSC**	**−0.27992**	**−0.11906**	**−0.90740**	**−0.43997**	**0.01016**	**0**
6	5	–	0.01348	−0.01137	0.00512	−0.02094	7e−05	0.0010
2	5	–	0.10956	0.04969	0.10452	0.03520	0.00289	0.00243
1	6	–	0.32518	0.10618	0.31601	0.09541	0.0117	0.01090

4.1 IEEE-6 Bus System

The allocation of the FACTS controllers is identified from SGO algorithm. Three numbers of FACTS controllers; one UPFC and two TCSC are used for 6 bus system. In order to do so, the self-retrospective coefficient c in Eq. (7) is chosen as 0.2. Similarly, the value of r_1 and r_2 in equation is taken as 0.2 and 0.4 based on trial-and-error approach. After performing the algorithm successfully, we were able to identify the location of UPFC at bus number 3 and two TCSC at lines 2–3 and 5–4. Table 1 shows the impact of UPFC in improving the voltage profile of not only bus number 3 but also the overall system. Table 2 shows the addition of TCSC on the line 2–3 and 5–4 which minimizes the active power loss to a very low number.

4.2 IEEE-14 Bus System

The 14 bus and line data are standard IEEE data. Four FACTS devices—two UPFC and two TCSC—are used for this system. After implementing the algorithm, we were able to identify the location of UPFC at bus numbers 2 and 14. Similarly, TCSC was placed at line 2–3 and line 5–4.

Table 3 shows the impact of UPFC. The voltage level at bus number 2 after inserting UPFC is almost similar to the voltage level without UPFC. However, the voltage level at bus number 14 is drastically improved after the insertion of UPFC. Similarly, Table 4 shows the impact of TCSC in minimizing the active power losses at lines 2–3 and 5–4. Figure 3 shows the voltage profile of the IEEE-14 bus system with and without FACTS controllers. Similarly, Fig. 4 compares the active power loss before the insertion of FACTS device and after the FACTS device. In all the case, it is observed that these FACTS devices improve the voltage profile and line

Table 3 Location of UPFC and voltage profile with/without FACTS (IEEE-14)

Bus number	Device connected	Voltage (p.u) without facts	Voltage (p.u) with facts after SGO
1	–	1.06	1.06
2	**UPFC**	**1.045**	**1.04**
3	**–**	**1.0**	**1.0**
4	–	1.0027	1.0108
5	–	1.0066	1.007
6	–	1.0	1.0
7	–	1.0043	1.005
8	–	1.0	1.0
9	–	0.99885	1.00017
10	–	0.99117	0.99228
11	–	0.99187	0.99246
12	–	0.98509	0.98661
13	–	0.98082	0.98387
14	**UPFC**	**0.97181**	**0.98485**

Table 4 Location of TCSC and line flows with/without FACTS (IEEE-14)

From bus	To bus	Device connected	P without facts	Q without facts	P with facts	Q with facts	P loss without facts	P loss with facts
1	2	–	1.5823	−0.19236	1.56608	−0.18857	0.04473	0.04282
1	5	–	0.72748	0.27835	0.69224	0.27065	0.00298	0.00269
2	**3**	**TCSC**	**0.73676**	**0.1004**	**1.83575**	**0.79844**	**0.02388**	**0**
2	5	–	−0.40845	−0.08489	−0.37238	−0.09521	0.0097	0.0082
5	**4**	**TCSC**	**0.6208**	**−0.09785**	**0.71228**	**−0.10989**	**0.0052**	**0**
5	6	–	0.43618	0.34666	0.42192	0.34927	0.0031	0.00257
12	13	–	0.01522	0.00422	0.0124	−0.00017	6e−05	0.0003

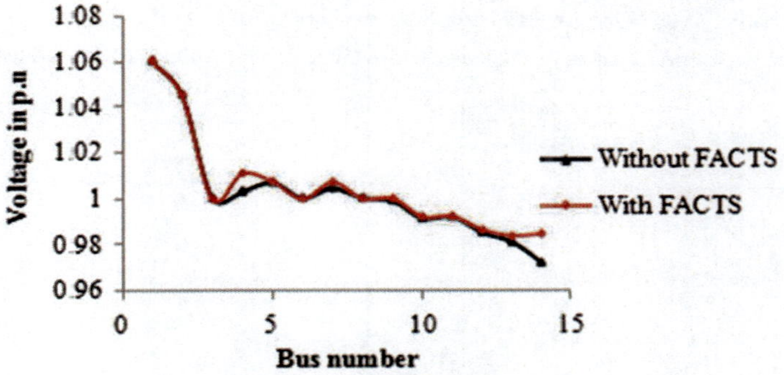

Fig. 3 Voltage profile of IEEE-14 with and without FACTS

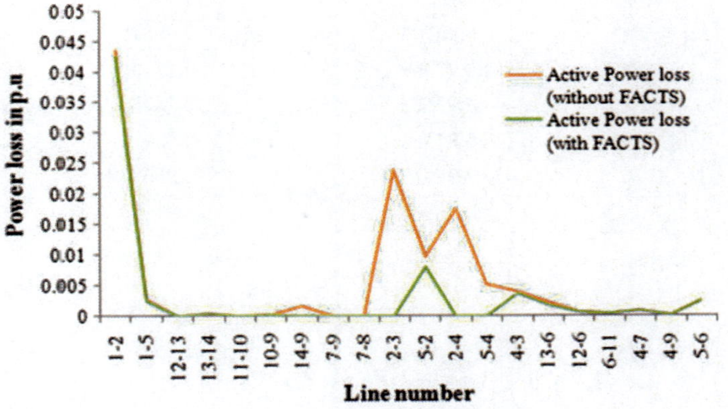

Fig. 4 Comparison of power loss before and after FACTS devices

flows of the system enhancing system security and minimizing voltage deviations and line loadings.

4.3 IEEE-57 Bus System

In this scenario, eight FACTS devices—five UPFC and three TCSC—are used. After performing the algorithm, we were able to locate UPFC at bus numbers 21, 25, 26, 27, and 31, while TCSC is located at line numbers 12–13, 14–15, and 22–23. Table 5 shows the impact of UPFC in maintaining the voltage level at bus numbers 21, 25, 26, 27, and 31. The worst-case scenario is observed at bus number 31 as its voltage limit before inserting UPFC is 0.87408 and is under limit. Not only the voltage level at bus 31 improved but also the voltage profile of the overall system is improved. This can be seen at Fig. 5. Similarly, we can observe from Table 6 that the use of TCSC

Table 5 Location of UPFC and voltage profile with/without FACTS (IEEE-57)

Bus number	Device connected	Voltage (p.u) without facts	Voltage (p.u) with facts after SGO
1	–	1.04	1.04
7	–	0.98166	0.9818
9	–	0.98	0.98
10	–	0.97547	0.97611
11	–	0.97798	0.9794
15	–	0.98382	0.9835
18	–	0.97051	0.97158
19	–	0.94012	0.94473
21	**UPFC**	**0.97439**	**0.9824**
24	–	0.95273	0.97056
25	**UPFC**	**0.90743**	**0.93618**
26	**UPFC**	**0.95389**	**0.97289**
27	**UPFC**	**0.97653**	**0.98256**
31	**UPFC**	**0.87408**	**0.92411**
52	–	0.96913	0.9669
57	–	0.96668	0.97189

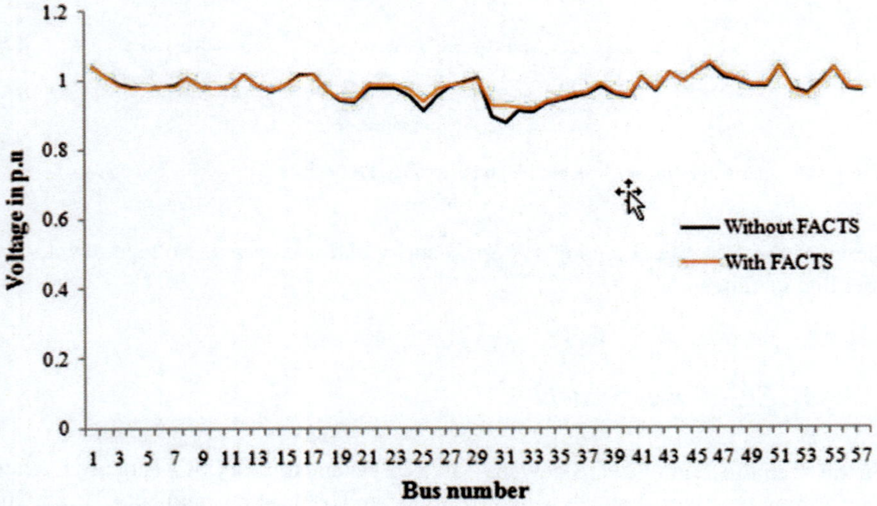

Fig. 5 Voltage profile of IEEE-57 with and without FACTS

Table 6 Location of TCSC and line flows with/without FACTS (IEEE-57)

From bus	To bus	Device connected	P without facts	Q without facts	P with facts	Q with facts	P loss without facts	P loss with facts
1	2	–	0.88029	0.85856	0.9754	0.83223	0.0116	0.01262
3	4	–	0.61759	0.10879	0.56054	0.05841	0.00454	0.00367
9	12	–	0.03824	−0.12404	0.02327	−0.12101	0.00114	0.00102
11	13	–	−0.06285	−0.06179	−0.00947	−0.05541	0.00018	7e−05
12	**13**	**TCSC**	**−0.14121**	**0.64539**	**−0.18069**	**0.8588**	**0.00754**	**0**
13	15	–	−0.5036	−0.0072	−0.43218	0.15922	0.00709	0.0059
14	**15**	**TCSC**	**−0.63894**	**−0.23802**	**−0.94229**	**−0.28349**	**0.00843**	**0**
21	22	–	0.00598	0.0567	−0.0179	−0.00142	0.00035	2e−05
22	**23**	**TCSC**	**−0.00518**	**−0.07547**	**0.0865**	**0.05387**	**9e−05**	**0**
23	24	–	−0.06826	−0.0977	0.03662	0.05915	0.00356	0.00084

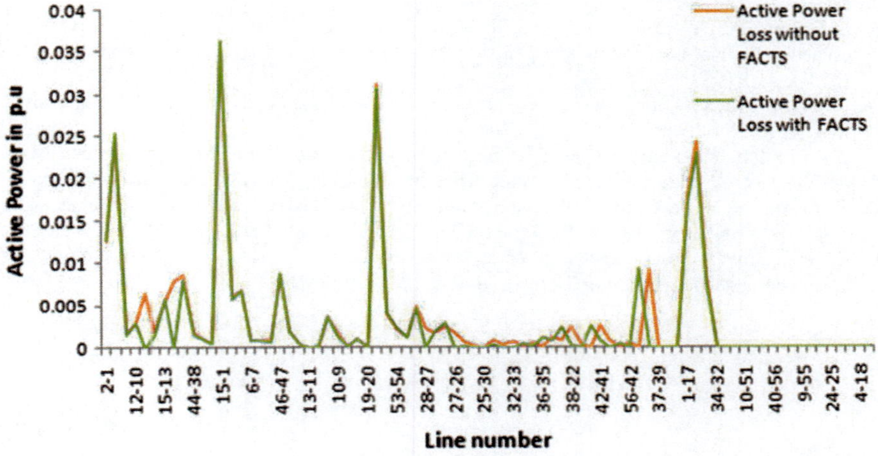

Fig. 6 Comparison of power loss before and after FACTS devices (IEEE-57)

has clearly improved the line flows and minimized the losses. This can be observed at Fig. 6. Overall, the complete system has enhanced their security by minimizing losses and maintaining voltage limit.

5 Conclusion

In this paper, the locations of FACTS devices are found by using the SGO technique to attain the maximum benefit. Simulations were performed on the IEEE-6, IEEE-14, and IEEE-57 buses, and the results found were recommended. This work observed the new technique for optimal location of FACTS device and the effectiveness of these devices in terms of minimizing the active power losses and load voltage deviation. We can conclude from this work that the installation of FACTS devices using SGO is beneficial for power system restructuring.

References

1. R. Mohan Mathur, R. K. Verma, *Thyristor-Based FACTS Controllers for Electrical Transmission Systems*. (Wiley, 2002)
2. N.G. Hingorani, L. Gyugyi, *Understanding FACTS: Concepts and Technology of Flexible AC Transmission Systems* (Wiley, New York, 1999)
3. K.R. Padiyar, *FACTS Controllers in Transmission and Distribution, India* (New Age Int, New Delhi, 2007)
4. W. Ongasukel, P. Jirapong, Optimal allocation of FACTS devices to enhance total transfer capability using evolutionary programming. Proc. IEEE Int. Symp. Circ. Syst. **5**, 4175–4178 (2005)
5. L. Ippolito, A.L. Cortiglia, M. Petrocelli, Optimal allocation of FACTS devices by using multi-objective optimal power flow and genetic algorithms. Int. J. Emerg. Electr. Power Syst. **7**(2) (2006). Article 1
6. L. Cai, I. Erlich, Optimal choice and allocation of FACTS devices using genetic algorithms. ISAP Intell. Syst. Appl. Power Syst. (2003)
7. A.R. Jordehi, Optimal setting of TCSCs in power systems using teaching-learning-based optimization algorithm. Neural Comput. Appl. https://doi.org/10.1007/s00521-014-1791-x
8. K. Kavitha, R. Neela, Optimal allocation of multi-type FACTS devices and its effect in enhancing system security using BBO, WIPSO and PSO. J. Electr. Syst. Inform. Technol. **5**, 777–793 (2018)
9. S. Satapathy, A. Naik, Social group optimization (SGO): a new population evolutionary optimization technique. Comples Intell. Sys. **2**, 173–203 (2016)

Analysis and Evaluation of the Impacts of FACTS Devices on the Transmission Line Protection

B. R. Rajeev

1 Introduction

We know from the fundamentals of power system that, transmission lines carries bulk power from far generating stations to the local distribution substation. Hence transmission line protection plays an important role in reliability and security of the power system. Over-current protection is very appealing and attractive because of its inherent simplicity; however reach of the over current relay depends on type of fault as well source impedance which are variable may lead to mal-operate in long transmission lines [1]. As EHV lines are the part of interconnecting grid, any mal- operation is not tolerable and any mal-operation may jeopardize the stability and security of the grid [1]. This led to the search of a relaying principle whose reach is not dependent on the type of fault current and its magnitude but depends on the ratio of voltage and current measured at relay location [1]. This is nothing but impedance based protection system called distance protection which can work effectively and accurately for transmission line protection. This protection technique works effectively in the conventional power system but in modern power system, performance parameters of the transmission lines can be controlled by reactive power compensating devices [2]. Transmission lines with compensating devices may leads to mal-operate the distance relay and may affect the security of the grid [3].

With the application of power electronic devices and controllers, it's made possible to control the power flow dynamically by controlling the levels of compensation. This led to the Flexible AC Transmission technology (FACTS) [4, 5] and commissioning of these FACTS device to improve the power transfer capability, stability and controllability is being carried in most of the grid transmission lines around the globe as well in India [3]. Due to the presence of these FACTS devices in transmission lines,

B. R. Rajeev (✉)
EEED SIT Tumakuru, Tumakuru, Karnataka, India
e-mail: rajeevbr@sit.ac.in

Sanjeevikumar et al. (eds.), *Advances in Renewable Energy and Electric Vehicles*, Lecture Notes in Electrical Engineering 767,
https://doi.org/10.1007/978-981-16-1642-6_3

may affect the proper functioning of distance relay which may lead to under reach and over reach problems [6]. During some worst cases, mal- operation of relay may leads to series of tripping and finally the system black-outs [7, 8]. From the view of security of the grid system, it's very much needed to investigate the new attributes of compensated transmission lines. Analytical calculation of apparent impedances seen by the relay in presence of midpoint TCSC and midpoint STATCOM during both ground faults and phase faults are performed by resolving sequence impedance networks, effect of shunt current in case of shunt compensation and series voltage in case of series compensation are analyzed by obtained equations. Using GPS synchronized measurements [9], measured current/voltage signal at the FACTS location, is transmitted with negligible communication delay by using fiber optic communication link offers a fast and quick intelligent relaying to react adaptively to the effects of FACTS compensation during faults [10, 11]. Based on the obtained analytical equations and GPS synchronized measurement a novel algorithm can be derived for adaptive relaying [12]. To perform the analysis, a modified IEEE-9 bus system is built and a conventional relaying is implemented using PSCAD/EMTDC [13]. STATCOM and TCSC is modeled and placed in mid-point to investigate its impacts on protection [14, 15].

2 Apparent Impedance Calculation

2.1 In Presence of STATCOM at the Midpoint

Let STATCOM be located at the midpoint of line whose impedance $Z_L = 1$ p.u. and a fault occurs at 'n' distance from the relay as shown in the Fig. 1

$$(nI_{s0}Z_{L0} + nI_{sh0}Z_{L0} - 0.5I_{sh0}Z_{L0} + R_f I_{f0});$$

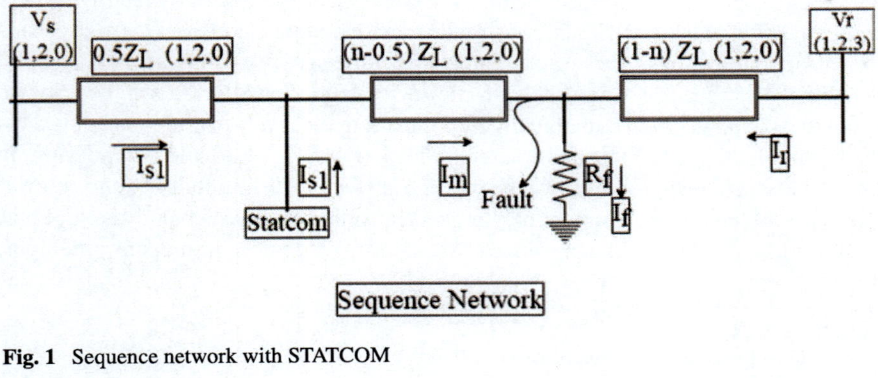

Fig. 1 Sequence network with STATCOM

Referring to Fig. 1, we have the following

$Z_{ss} = 0.5\, Z_L$ [Impedance from sending end to STATCOM];
$Z_{\text{SF}} = n - 0.5Z_L$ [Impedance from statcom to till fault];
$Z_{\text{FR}} = (1 - n)Z_L$ [Impedance from fault to receiving end].

During Ground Fault: An L-G fault occur just after STATCOM (n > 0.5 p.u)

Positive negative and zero sequence components of voltages are expressed as,

$$V_{s1} = 0.5\, I_{s1} Z_{L1} + I_{m1}(n - 0.5)Z_{L1} + R_f I_f$$
$$V_{s2} = 0.5\, I_{s2} Z_{L1} + I_{m2}(n - 0.5)Z_{L1} + R_f I_{f2}$$
$$V_{s0} = 0.5\, I_{s0} Z_{L0} + I_{m0}(n - 0.5)Z_{L0} + R_f I_{f0}$$

In transmission line $Z_{L1} = Z_{L2}$.
Also,

$$I_{m1} = I_{s1} + I_{sh1};\ I_{m2} = I_{s2} + I_{sh2};\ \text{and}$$
$$I_{m0} = I_{s0} + I_{sh0};$$

Sending end voltage/relay measuring voltage is given by,

$$V_s = V_{s0} + V_{s1} + V_{s2};$$
$$V_s = \left(nI_{s1}Z_{L1} + nI_{sh1}Z_{L1} - 0.5I_{sh1}Z_{L1} + R_f I_{f1}\right)$$
$$+ \left(nI_{s2}Z_{L1} + nI_{sh2}Z_{L1} - 0.5I_{sh2}Z_{L1} + R_f I_{f2}\right)$$
$$+ \left(nI_{s0}Z_{L0} + nI_{sh0}Z_{L0} - 0.5I_{sh0}Z_{L0} + R_f I_{f0}\right);$$

Add and subtract $(nI_{S0}Z_{L1})$, $(nI_{Sh0}Z_{L1})$, and $(0.5I_{S0}Z_{L1})$, we get,

$$V_s = nI_s Z_{L1} + (n - 0.5)(Z_{L0} - Z_{L1})(I_{sh} + I_{Sh0}) + nI_{s0}(Z_{L0} - Z_{L1}) + \left(R_f I_{fa}\right)$$

Ignoring I_{sh0}, and making $R_f = 0$ as there will be no neutral to ground current in delta connected side of the transformer connected in the STATCOM assembly.
We have,

$$V_s = nI_s Z_{L1} + (n - 0.5)(Z_{L0} - Z_{L1})(I_{sh}) + nI_{s0}(Z_{L0} - Z_{L1})$$

$$Z_{\text{app}} = \frac{V_{\text{relay}}}{I_{\text{relay}}};\ V_{\text{relay}} = V_S;\ I_{\text{relay}} = I_S + KI_{S0};\ Z_{\text{app}} = \frac{V_s}{I_s + k\, I_{s0}};$$

Finally we have

$$Z_{\text{app}} = nZ_{L1} + \frac{(n - 0.5)I_{sh}Z_{L1}}{I_s + k\, I_{s0}}$$

where $K = \frac{Z_{L0}-Z_{L1}}{Z_{L1}}$; '$K$' is the residual current compensation factor. By observing above equation we can infer that, the error introduced in the apparent impedance due to shunt compensation is given by

$$Z_{\text{error}} = \frac{(n-0.5)I_{sh}Z_{L1}}{I_s + KI_{S0}};$$

Actually, nZ_{L1} represents line impedance up to fault point.

During Phase Fault: An L-L fault occur just after STATCOM (n > 0.5 p.u)

Through similar approach as described above, apparent impedance seen by the relay is given by

$$Z_{\text{app}} = nZ_{L1} + \frac{(n-0.5)I_{sh}Z_{L1}}{I_{S1}}$$

During L-L fault there will be no zero sequence component and hence it is neglected.

2.2 *In Presence of TCSC at the Midpoint*

Let $Z_L = 1$ p.u and a fault is imposed at 'n' distance from relay (Fig. 2).

With a similar approach, after calculation using sequence network reduction for an LG/LL fault, we get in general

$$Z_{\text{app}} = nZ_{L1} + \frac{V_T}{I_{\text{relay}}};$$

where $I_{\text{relay}} = I_s + KI_{S0}$; in case of LL fault, we should ignore we should ignore I_{S0}.

As in shunt compensation, series compensation also affects relay operation and likely to mal-operate.

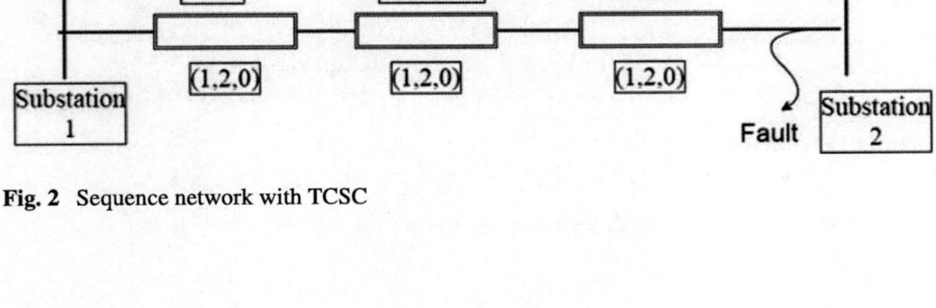

Fig. 2 Sequence network with TCSC

3 Under-Reach and Over-Reach Effects

Case 1: During Shunt compensation with STATCOM:

- When injected current is +Ve, i.e. Ish is inductive (+Ve), during which STATCOM is injecting capacitive reactive power into the system which increases the Zapp, hence relay suffers under-reach problem. i.e. apparent impedance trajectory falls outside the zone of protection during fault.
- When injected current is −Ve, i.e. Ish is −Ve, during which statcom draws capacitive reactive power from the system which decreases the Zapp, hence relay suffers over-reach problem. Apparent impedance trajectory falls within the zone of protection even though fault occurs outside the zone.

Case 2: During series compensation with TCSC:

Due to the series voltage injected, in presence of series compensation, relay suffers over reach problem.

Thus, analytically by sequence network reduction method, we have proved in presence of FACTS device in transmission line, causes relay to suffer overreach and under reach problems. So an adaptive novel algorithm is required to mitigate the problems.

4 Modeling of Test System

4.1 Modeling of Study System

IEEE-9 bus system is modeled in PSCAD and power flows are checked. FACTS device is placed in mid-point of the line which connects two weak buses.

4.2 Modelling of STATCOM and TCSC

A twelve pulse STATCOM using PI controller which regulates the weak bus voltage is modeled in PSCAD/EMTDC. Using the power flows obtained for the test system, the KVAR ratings of the STATCOM is decided so that the voltages at the weak buses are maintained at 1 p.u. The constants of the PI controller are obtained from nyquist plot plotted for the control system in PSIM platform where the system is found stable.

For different firing angles, PSCAD model of TCSC is simulated with 20% fixed compensation inserted at the mid of the line connecting the weak buses.

5 Simulation Results

The PSCAD/EMTDC model of WSCC 9 bus system is rigged up in PSCAD/EMTDC and after looking into power flows, Bus-8 found to have less voltage of 0.79 p.u and at the midpoint of the line connecting Bus-7 and Bus-8, FACTS device is inserted and Simulated with insertion of various faults at different distances to evaluate the performance of the relay for zone-1 operation. Results are summarized as follows.

- From Fig. 3a, it is cleared that, relay is operating absolutely fine and the apparent impedance trajectory is falling inside the zone-1 when fault is injected within zone-1 only.
- From Fig. 3b, when a fault occurs within zone-1, but relay suffers under reach problem, apparent impedance trajectory falls beyond zone-1.
- From Fig. 3c, when a fault is inserted at zone-3, still the apparent impedance trajectory falls very near to zone-1 suffering overreach effect.
- Similarly, from Fig. 3d, apparent impedance trajectory falling in zone-2 when a fault is imposed at zone-3 suffering overreach problem in presence of TCSC.
- In Fig. 3e. when Phase-a to Ground fault is applied (A-G) fault, a trip signal is initiated and fault current(Ia) peak can be noted during fault and after trip signal, current become zero.

Thus distance relay will prone to have errors in presence of shunt and series compensation which is verified both analytically and practically by simulation.

6 Adaptive Relay Setting Flow Chart

- With changing system condition, a relay has to adapt to the new conditions without incurring any errors and for the same we have to design a new adaptive relay setting.
- After understanding the impacts of the compensation on distance protection, the other objective is to obtain an adaptive relay algorithm flow chart by taking the feedbacks of shunt current I_{sh} by STATCOM and series injected Voltage (Vt) by TCSC.

  Using PMU's at both relaying bus as well at the FACTS bus, the phasor estimations are done at both the buses for voltage and current samples and communicated to the relay without any delay.
- A relay algorithm flow chart is presented in the Fig. 4 through phasor estimation by PMU's at both relay bus and FACTS bus, a new setting zone is determined without any communication delay and accordingly new mho circle for the new zone is plotted. A new trip law is given to the relay to mitigate the errors. This mean the synchronized measurements(sequence components using Fourier Transforms) from phasor measurement units (PMU) i.e. samples of current and voltage magnitudes and phasors at relay bus as well at the FACTS bus without any

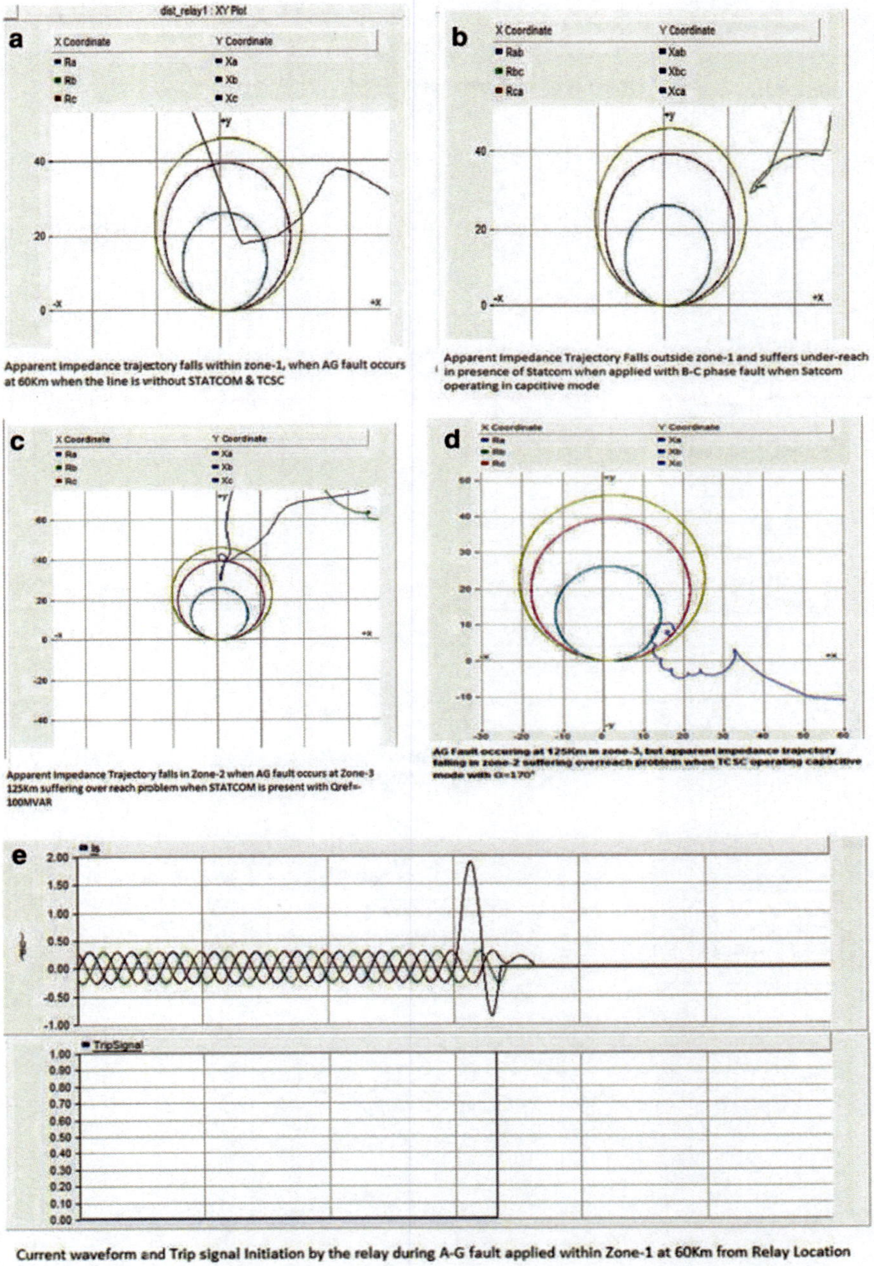

Fig. 3 **a** Impedance trajectory without STATCOM and TCSC (base case), **b** Under-reach effect in presence of STATCOM, **c** Over-reach effect in presence of STATCOM, **d** Over-reach effect in presence of TCSC, **e** Current waveform and trip signal initiation by the relay during A-G fault applied within Zone-1

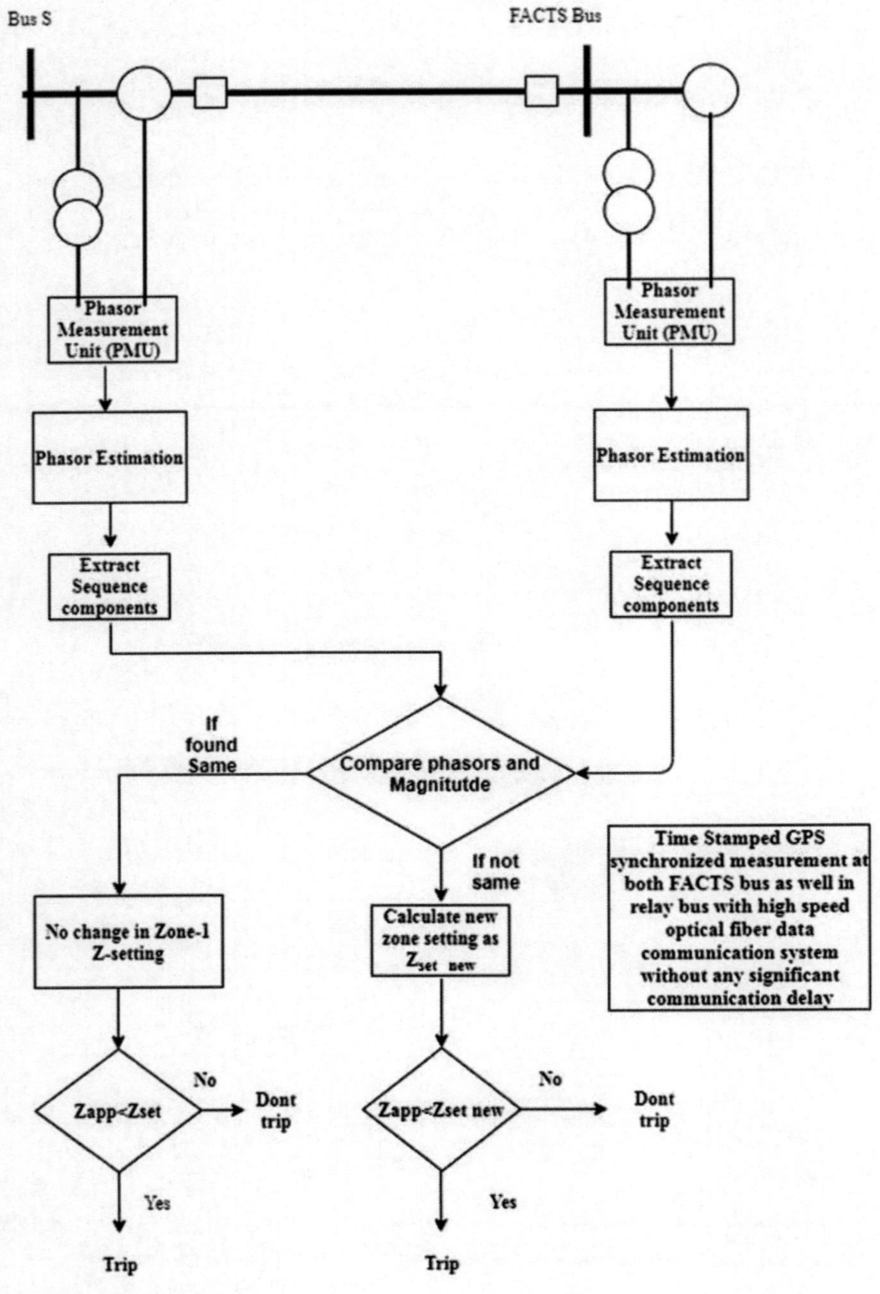

Fig. 4 Relay algorithm flow chart

communication delay, are extracted and apparent impedances are calculated and compared, if it is as per zone-1 setting, no changes in relay function but FACTS device is contributing reactive currents (may be during fault), then after comparison it will not be same, hence relay adapt by itself by changing to a new zone setting for zone-1 and hence avoiding the under reach and overreach effects. It is clearly depicted in Fig. 4.

7 Conclusion and Future Scope

A shunt/series compensated transmission line will prone to have under reach and overreach effects. It's been verified analytically step by step using sequence network methods during unsymmetrical faults (both ground and phase faults). This analysis depicted the under reach and overreach effects during presence of series and shunt compensations separately. The same analytical results are verified using PSCAD modelling and simulations. A WSCC 9 bus system is rigged up to see the power flows and bus-8 found critical with a voltage of 0.79 p.u. Hence a line connecting bus-7 and bus-8 is taken for the experimental verification, FACTS devices TCSC (for series compensation) and STATCOM placed at the midpoint of the weak line. In both cases without FACTS and with FACTS a mho relay characteristics are obtained and observed that it shown under reach and overreach effects. An algorithm flow chart to determine a new zone setting to compensate the errors due to the presence of FACTS device is proposed. It uses synchronized measurement principle using PMU's and communication without any delay using fiber optic communication link to calculate new zone setting so that relay act adaptively to the prevailing changes in the system due to interactions of the FACTS devices. But practical implementation and verification of the adaptive relay algorithm flow chart as a proof for mitigating the errors is the future work identified.

Acknowledgements This work is done during the Summer Faculty Fellowship-2019 at IIT Delhi. I would like to profusely thank Prof. B K Panigrahi Department of Electrical Engineering IIT Delhi for the opportunity and support.

References

1. Y.G. Paithankar, S.R. Bhide, *Fundamentals of Power System Protection*, 2 ed. (Prentice Hall India Learning Private Limited, 2010)
2. T.J.E. Miller, *Reactive Power Control in Electrical Systems*. (Wiley, India)
3. S. Mukhopadhyay, A.K. Tripathy, Application of FACTS in Indian power system, in *IEEE Conference 2002*
4. N.G. Hingorani, L. Gyugyi, *Understanding FACTS: Concepts and Technology of Flexible AC Transmission Systems*. (Wiley India, 2011)
5. K.R. Padiyar, *FACTS Controllers in Power Transmission and Distribution*. (New age International Publishers). ISBN (13) 978-81-224-2541-3

6. A.R. Singh, N.R. Patne, V.S. Kale, Adaptive distance protection setting in presence of mid-point STATCOM using synchronized measurement. Int. J. Electr. Power Energ. Syst. **67**, 252–260 (2015)
7. S. Biswas, P.K. Nayak, State-of-the-art on the protection of FACTS compensated high-voltage transmission lines: a review. Int. J. High Volt. IET (2018)
8. A. Ghorbani, M. Khederzadeh, B. Mozafari, Impact of SVC on the protection of transmission lines. Int. J. Electr. Power Energ. Syst. **42**(1), 702–9 (2012)
9. S. Karn, S.R. Samantaray, A. Sharma, Supervising zone-3 operation of the distance relay using synchronised phasor measurements. IET Int. J. Gener. Transm. Distrib. (2018)
10. G. Debomita, T. Ghose, D.K. Mohanta, Communication feasibility analysis for smart grid with phasor measurement units. IEEE Trans Ind. Inform. **9**(3), 1486–96 (2013)
11. IEEE Standard for Synchrophasors for Power Systems, IEEE Std C37.118-2005 (Revision of IEEE Std 1344- 1995), vol. 57. (2006), pp. 0–1
12. B. Kumar, A. Yadav, A.Y. Abdelaziz, Synchrophasors assisted protection scheme for the shunt-compensated transmission line. Int. J. Gener. Transm. Distrib. IET (2017)
13. Manitoba HVDC Research Centre, PSCAD Users Guide V4.6
14. M. Khederzadeh, T.S. Sidhu, Impact of TCSC on the protection of transmission lines. IEEE Trans. Power Deliv. **21**(1) (2006)
15. F.A. Albasri, T.S. Sidhu, R.K. Varma, Performance comparison of distance protection schemes for shunt-FACTS compensated transmission lines. IEEE Trans. Power Deliv. **22**(4) (2007)

Carbon-Based Textile Dry and Flexible Electrodes for ECG Measurement

Newton Rai, Habibuddin Shaik, N. Veerapandi, Veda Sandeep Nagaraj, and S. Veena

1 Introduction

The rapid growth of cardiovascular diseases (CVDs), including diabetes mellitus, hypertension, myocardial infarction, can lead to the death of human life. According to the WHO, CVDs are the number one cause of death globally and annually more people die from CVDs than any other disease. In 2016, around 17.9 million people died from CVDs which represent 31% of all global deaths. Out of these, heart attack is the major cause for the deaths. It is observed that these deaths take place mostly in developing and underdeveloping countries [1]. Thus, a consistent monitoring electrocardiography of patients is required. ECG is a device which measures the electrical activity of the human's heart over a period of time. The originator of the electric signal in the human's heart is the sinus node, and the signal generated by the sinus node is responsible for the contraction of cardiac muscles. These electrical pulses are detected by electrode either by an invasive manner or non-invasive manner, and the graph recorded in the ECG is represented as a series of electric waves, with specific

N. Rai (✉) · S. Veena
Department of Electrical and Electronics Engineering, Nitte Meenakshi Institute of Technology, Bangalore, India

S. Veena
e-mail: veena.s@nmit.ac.in

H. Shaik
Department of Physics, Nitte Meenakshi Institute of Technology, Bangalore, India

N. Veerapandi
Centre for Nano Science and Engineering(CENSE), Indian Institute of Science (IISc), Bangalore, India

V. S. Nagaraj
Department of Electronics and Communication Engineering, Nitte Meenakshi Institute of Technology, Bangalore, India

Sanjeevikumar et al. (eds.), *Advances in Renewable Energy and Electric Vehicles*, Lecture Notes in Electrical Engineering 767,
https://doi.org/10.1007/978-981-16-1642-6_4

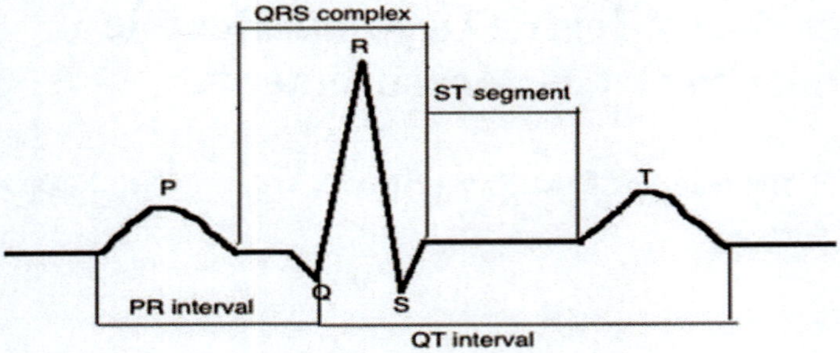

Fig. 1 Standard electrocardiogram (ECG) wave

structures and period that repeat in each cardiac cycle such as the P, T and the QRS complex wave as shown in Fig. 1. These waves represent to the various successions of cardiovascular muscle exercises, which make it conceivable to assess the cardiac state and can be utilized to measure the rate and rhythm of heartbeats, the size and position of the heart chambers, the presence of any damage to the heart's muscle cells, the effect of heart drugs and the capacity of artificial pacemakers [2].

The conventional ECG sensor is classified into two types—implantable-type ECG sensor and patch-type ECG sensor. In the case of the implantable-type ECG sensor, medical procedure is required to place the electrode in the heart for ECG measurements. Consequently, it is pricey and intrusive while in patch-type ECG sensor at least three electrodes should be fixed to a human body to record the ECG signal. In this strategy, two electrodes are utilized for detecting the voltage difference on the body surface and the third electrode is utilized as a ground electrode [3–5]. Usually ECG signals are commonly obtained through wet Ag/AgCl electrodes, which are accessible at low cost and can give high-calibre and stable signals. The coupling with the skin is upheld by an ionic component for instance solid or fluid hydrogel in order to improve the quality of the contact, thereby diminishing the skin impedance. This sort of electrodes cannot be used for long-term monitoring as they can cause skin irritations and also spreading the conductive gel may provoke a short circuit when electrodes are put near to each other. Therefore, alternative dry flexible ECG electrodes are required for ECG recording [6, 7].

Dry electrodes are the option in contrast to the ordinary Ag/AgCl anode [8, 9]. Various types of dry electrodes are available, viz. capacitive electrodes, small spike electrodes, conductive polymer electrodes. The capacitive electrodes are susceptible to motion artefacts [10]. Furthermore, these kinds of electrodes are complex and its volume is bigger than that of other types of electrodes. The spike electrodes with spikes array could penetrate through the external skin layer. However, this type of electrodes may cause uneasiness and skin injury to users. While dry carbon electrodes are flexible and comfortable as well as harmless to the skin [11–16], this paper presents the novel carbon-based textile and dry flexible electrodes for measuring the ECG signal. The design and fabrication of dry electrodes are discussed in Section two followed by its characterization. The working principle and ECG signal processing are discussed in Sections four and five, respectively. The circuit simulation and circuit hardware are discussed in Sections six and seven, respectively. Similarly, the results and conclusion are discussed in consecutive sections.

2 Design and Fabrication

2.1 Design

This section describes methodology for the design of carbon-coated ECG electrodes. The material selection, design of ECG electrodes and also the steps for fabricating the electrodes are discussed.

2.1.1 Materials

Kapton is a polyimide film introduced by DuPont in the late 1960. It remains stable across a wide variety of temperature from -269 to + 400 °C. [17, 18]. Kapton tape of thickness 120 μm is used to prepare the mask file. Carbon is used as the conducting material and is coated on a cotton fabric. Although graphene acts as a good conducting material than carbon, in this work carbon is used as the sensing material as it is cheap, compatible and easily available. The cotton fabric has good adhesive properties than the Kapton substrate. It gives protection against short circuit caused from the body sweats and electrodes. Therefore, the cotton fabric has been chosen for carbon coating [19]. Figure 2 shows the coating of carbon material on both cotton and Kapton substrate.

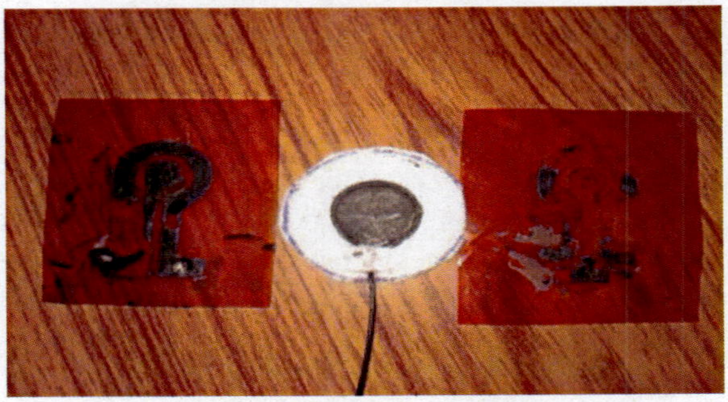

Fig. 2 Carbon pasted on both Kapton and cotton fabric substrate

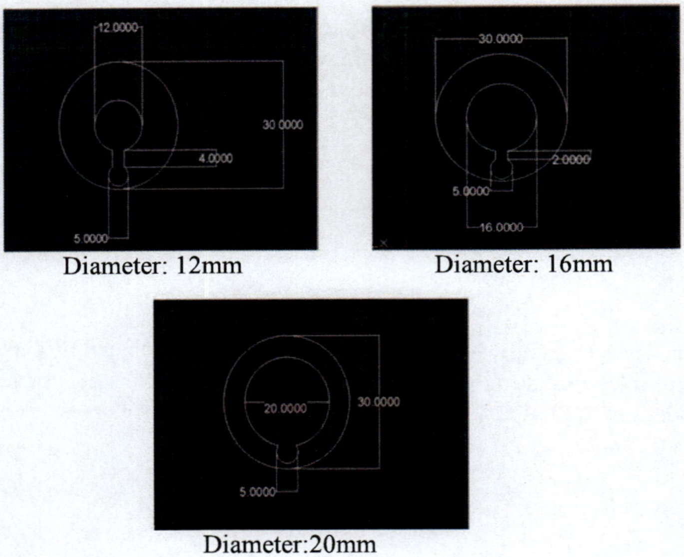

Fig. 3 AutoCAD design

2.1.2 AutoCAD-2D Design

The 2D structure for the mask file is designed using AutoCAD [20]. Mask files of various diameters (E1-20, E2-16, E3-12) mm are designed as shown in Fig. 3 [21].

2.2 *Fabrication*

2.2.1 Preparation of Mask File

The Kapton tape which is used as the mask file is cut into different sizes of diameters using a laser cutter [22]. After cutting a Kapton tape, the mask file is washed using an Isopropanol solution to remove the dust particles attached. The mask is kept in the oven for 1 h at a temperature of 150 °C. The final mask of different diameters is as shown in Fig. 4. The cotton fabric substrates are also cut in round shape with different diameters of (20, 16, 12) mm for coating carbon on it.

2.2.2 Coating of Conducting Materials

Various methodologies are available for the screen printing. Commercially ECG electrodes can be produced by injecting ink of carbon on a substrate using a screen

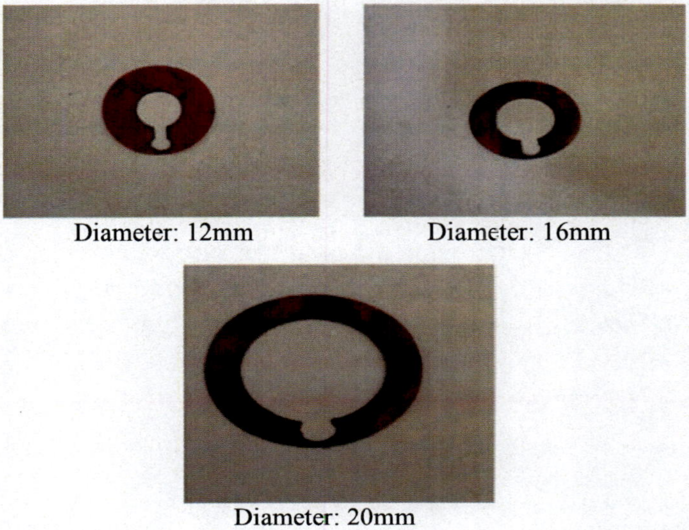

Fig. 4 Mask file design

printing machine but in this work for the prototype model, the carbon is coated manually on the substrate with the help of a mask file. The thickness of carbon ink coated on a substrate is 120 μm. This type of coating carbon layer on the substrate is adopted from a concept of pattern casting. In casting, a pattern is a replica of the object to be cast, which is utilised to set up the hole into which molten material will be poured during the casting process. Similarly, the mask here is similar to the pattern required and the thick paste of carbon is applied on a cavity created by the mask [23].

3 Characterization of Skin–Electrode Impedance

The characterization of the fabricated dry ECG electrodes and commercial wet Ag/AgCl electrodes was performed to compare their performance. The skin–electrode impedance should be as low as possible to minimize the common-mode rejection noise for the better quality of the ECG signal. The skin–electrode impedance can be modelled as a parallel combination of the RC circuit where R1 and C1 represent the impedance and the capacitance formed due to the skin and electrode coupling. R2 represents the resistance of the conducting gel. E represents the potential applied between the two electrodes [24]. The equivalent circuit diagram of the skin–electrode impedance for dry and wet electrodes is shown in Fig. 5.

The total skin impedance is given as

For dry ECG electrode,

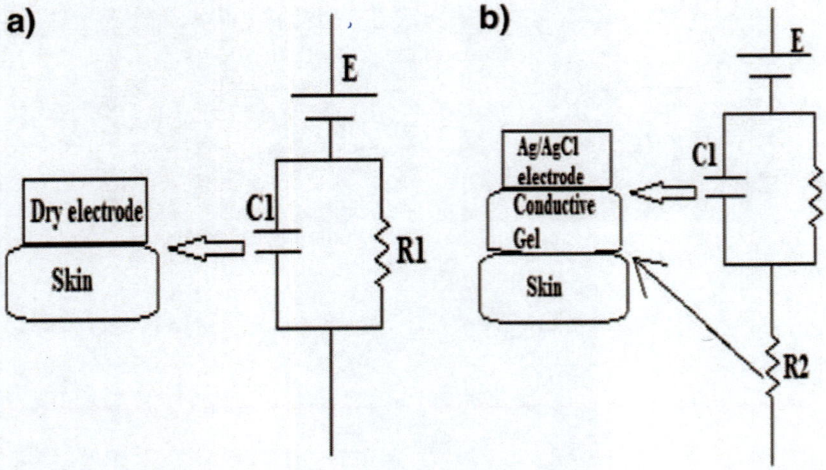

Fig. 5 **a** Equivalent skin–electrode circuit for dry electrode. **b** Equivalent skin–electrode circuit for Ag/AgCl electrode

$$Z(w) = \frac{R_1 X_c}{R_1 + X_C} \quad (1)$$

For wet Ag/AgCl electrode,

$$Z(w) = R_2 + \frac{R_1 X_c}{R_1 + X_C} = \frac{R_1 + R_2 + jwC_1R_1R_2}{1 + jwC_1R_1} \quad (2)$$

where

$$X_c = \frac{1}{jwC}, \quad w = 2\pi f.$$

From Eq. (1), the total impedance is the function of frequency. Thus, the measurement of impedance was performed in the frequency range from 1 to 700 Hz. An electrode is placed on a human's wrist with the help of a rubber band at a nearby distance (<2 cm) from the other electrode as shown in Fig. 6. A potential of 4v was applied to the skin through two electrodes in the frequency range from 1 to 700 Hz. The skin impedance frequency response curves for the dry ECG electrodes (E_1, E_2, E_3) and the Ag/AgCl electrodes are shown in Fig. 7.

From Fig. 7, it is seen that the skin–electrode impedance is dependent on the size of electrodes and it can be concluded that the skin–electrode impedance is inversely proportional to the diameter of the electrodes [20]. For instance, the skin impedance of dry electrodes having diameters of 12 mm, 16 mm, 20 mm is 585 kΩ, 571 kΩ and 565 kΩ, respectively, at a common frequency of 10 Hz. The results also concluded that skin impedance of carbon-coated dry electrodes (E_1, E_2, E_3) were higher than

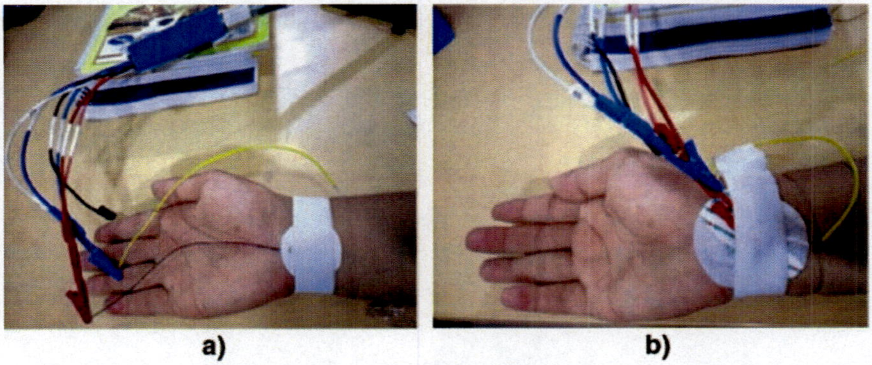

Fig. 6 **a** Experiment setup to measure skin–electrode impedance for dry ECG electrode. **b** Experiment setup to measure skin–electrode impedance for Ag/AgCl electrode

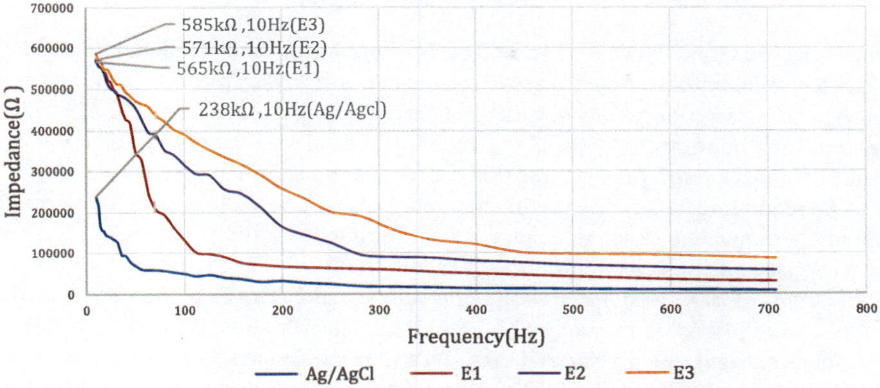

Fig. 7 Skin–electrode impedance with respect to frequency

that of wet Ag/AgCl electrodes at a frequency of 10 Hz. This is due to the fact that wet Ag/AgCl electrodes contain electrolyte gel. Though the fabricated dry electrodes have high skin impedance than wet electrodes, they are in well range of impedance value to measure ECG signal.

4 Working Principle

ECG electrodes record the electrical activity generated by the heart's muscle contraction and relaxation which propagate in/as pulsating electrical waves through the skin. The contraction and relaxation of cardiac muscles are due to electrical signals received from the sinus node. Electrodes are suitably placed on a body, thereby the

Fig. 8 Experimental set-up for recording ECG

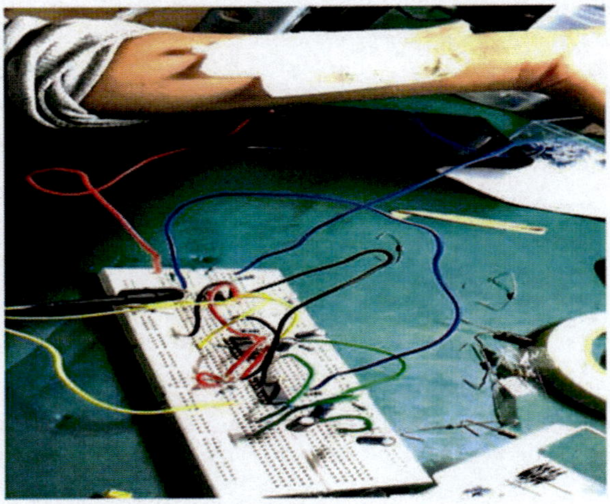

small electric signal which is generated by the polarization, and depolarization of the heart's muscle can be detected. The voltage change is sensed by measuring the current change across two electrodes positive and negative [25]. A poor choice of electrode material and improper location of electrodes could lead to unnecessary polarization. The potential caused by the polarization could cause noise which weakens the signal.

The requirements for the experimental setup are as follows. It consists of a power supply, a digital oscilloscope(display screen), printed dry ECG electrodes, and an ECG data acquisition electronic circuit as shown in Fig. 8. The experiment set-up for estimations of ECG were led at room temperature. Three dry electrodes, with same radius, were placed on the body of a healthy volunteer at three different positions: left forearm, right forearm and right leg, without removing the hair and with no skin preparation as shown in Fig. 9 [26]. The experiment was carried out for different diameters of electrodes, and their performances were compared. The output from

Fig. 9 Placement of electrodes

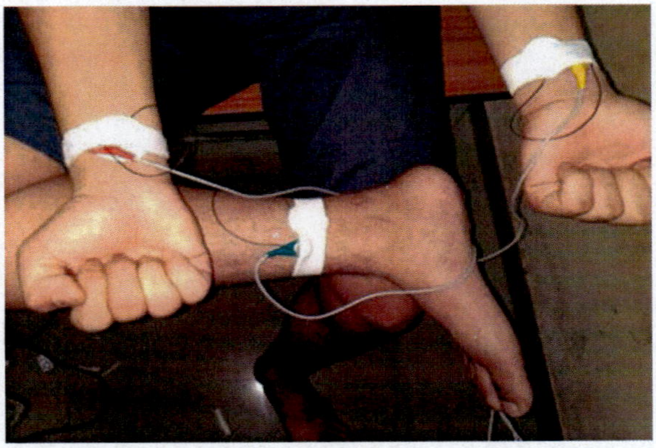

the ECG acquisition circuit was connected with an Arduino board for imaging and recording the ECG signal on the laptop screen.

5 ECG Signal Processing

The ECG signal detected by the electrodes was very weak electrical signals, and its amplitude was varying from 0.03 to 4 mv. These signals are usually affected by noise, such as low-frequency noise, muscle noise and electromagnetic noise. The signal sensed by the electrodes is within the range of 0.05 to 200 Hz signals [20, 27]. Due to the high impedance of the human body, the impedance of the ECG signal can reach several Ω to KΩ. So, the main challenge in signal processing of the ECG signal is to eliminate the noises and amplify the small magnitude ECG signal to properly stabilize ECG signals [25, 28]. The complete block diagram for the signal processing of the ECG signal is shown in Fig. 10.

For the acquisition of the ECG signal, the electrical signal obtained from the human body passes through various filtering and amplifying circuits. Proper care must be taken while designing the data acquisition circuit to suppress interference. The different circuits involved in the acquisition of ECG data and their design are shown below.

5.1 Design of Instrumentation Amplifier

The instrumentation amplifier is the main part of the ECG signal processing circuit. It has very high common-mode rejection ratios (CMRR) of the order of 100 dB and

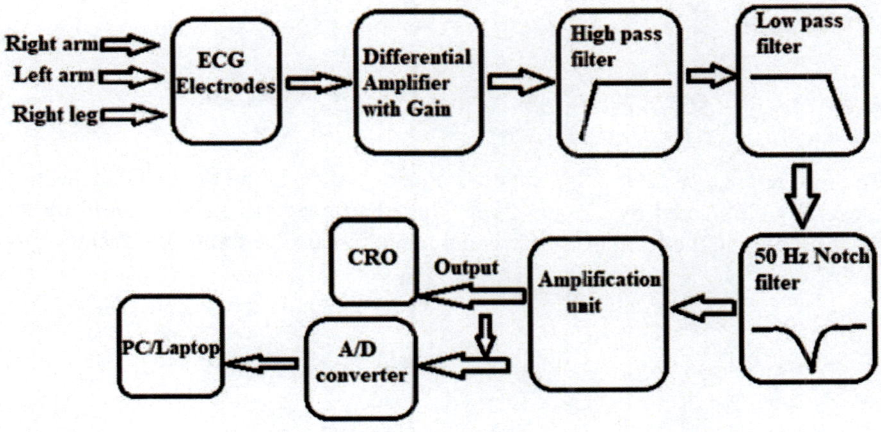

Fig. 10 Block diagram of ECG signal processing

high input impedance which is required for the acquisition of electrical signals from the human body [29]. The AD620AN instrumentation amplifier which is used in this project is low price, low power consumption and easily available amplifier. While designing this amplifier, the gain can be adjusted by varying the external resistor. The gain of this amplifier is [30]

$$\text{Gain} = 1 + \frac{50\,\text{k}\Omega}{R} \tag{3}$$

5.2 Design of High-Pass and Low-Pass filter

In order to eliminate the DC offset present in the output signal obtained from the instrumentation amplifier, the signal is passed through the high-pass filter. The operation amplifier LM324AN is used to design the second-order high-pass filter for this task which eliminates the low-frequency interference. It is designed with a cut-off frequency of 2 Hz. The second-order low-pass filter is also designed using LM324AN to filter high-frequency noise. In this project, the low-pass filter is designed to filter the frequency above 200 Hz. The cut-off frequency for second-order low-pass and high-pass filter is given as [31]:

$$f_{\text{cutoff}} = \frac{1}{2\pi\sqrt{R_1 R_2 C_1 C_2}} \text{(for low pass filter)} \tag{4}$$

if $R_1 = R_2 = R$ and $C_1 = C_2 = C$, then

$$f_{\text{cutoff}} = \frac{1}{2\pi RC} \text{(high pass filter)} \tag{5}$$

5.3 Design of Notch Filter

The interference caused by the supply frequency of 50 Hz in the ECG acquisition process is eliminated by a notch filter. This interference is mainly caused by the power supply in the form of a differential mode signal. The frequency designed for this filter is

$$f = \frac{1}{2\pi RC}$$

6 Circuit Simulation

The circuit is simulated on a Multisim software. The input and output waveforms are analysed. The circuit simulation is shown in Fig. 11a, and their corresponding input and output waveforms is shown in Fig. 11b.

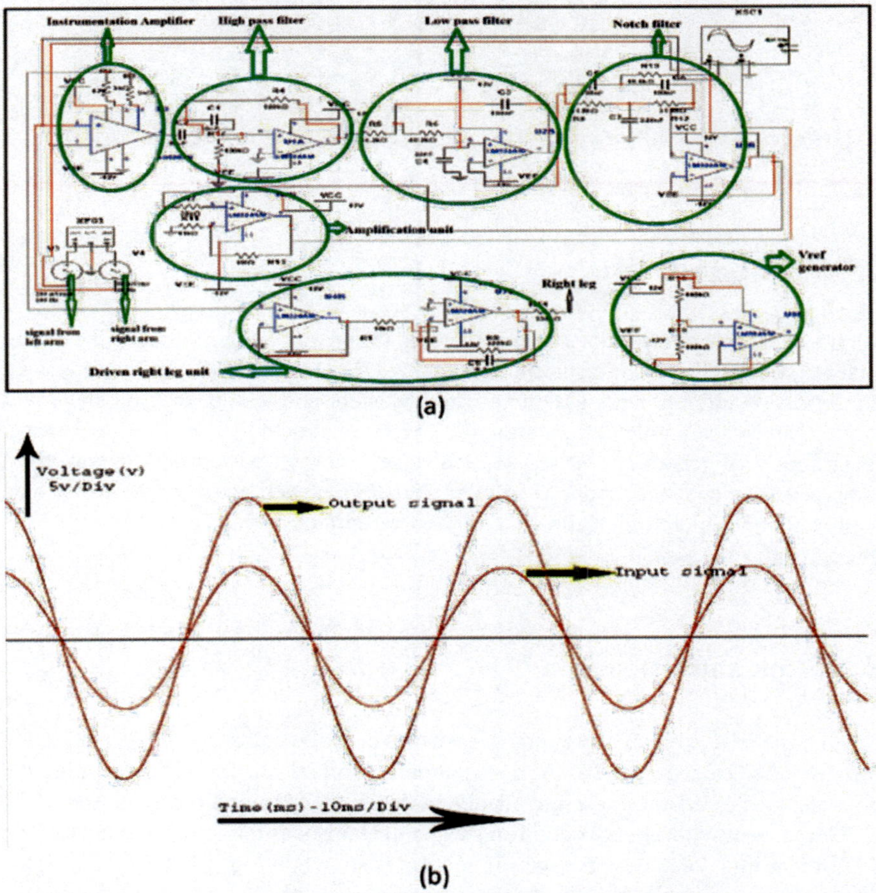

Fig. 11 **a** System circuit simulation on Multisim software. **b** Input and output waveforms simulated on Multisim software

Fig. 12 ECG acquisition circuit assembly

7 Circuit Hardware

All the electronic components were assembled on a zero PCB board as per the circuit design. The ECG acquisition circuit assembly is shown in Fig. 12. All the filters and amplification circuits were assembled step by step, and testing of each stage was carried out by testing the output signal for each constructed stage. The input signal from the signal generator was fed to each stage, and their corresponding waveform is observed on an oscilloscope. The two 9 V batteries were used for the ±9 V dual supply for the operation while its common terminal provides the ground for the circuit [32].

8 Result and Discussion

Finally, the ECG signals were measured using both the commercial electrodes and proposed ECG electrodes. For the commercially available electrodes, the conducting gel was used to decrease the impedance between the skin and the electrodes. The ECG was measured for both conditions when the body was in motion and in rest. The ECG recording when the body was in motion is shown in Fig. 13. This ECG signal contains lots of interference due to the motion of arms and is difficult to identify the P, Q, R, S, T waves.

The working of wet electrodes and dry electrodes was compared since the ECG signal obtained from both types of electrodes contains visible P, Q, R, S, T waves. Therefore, they were easily compared and analysed by calculating their respective peak and amplitude with respect to time [33]. Figure 14a shows the ECG obtained from the wet Ag/AgCl electrode, and Fig. 14b shows ECG obtained from a carbon-coated dry electrode when the body is at rest.

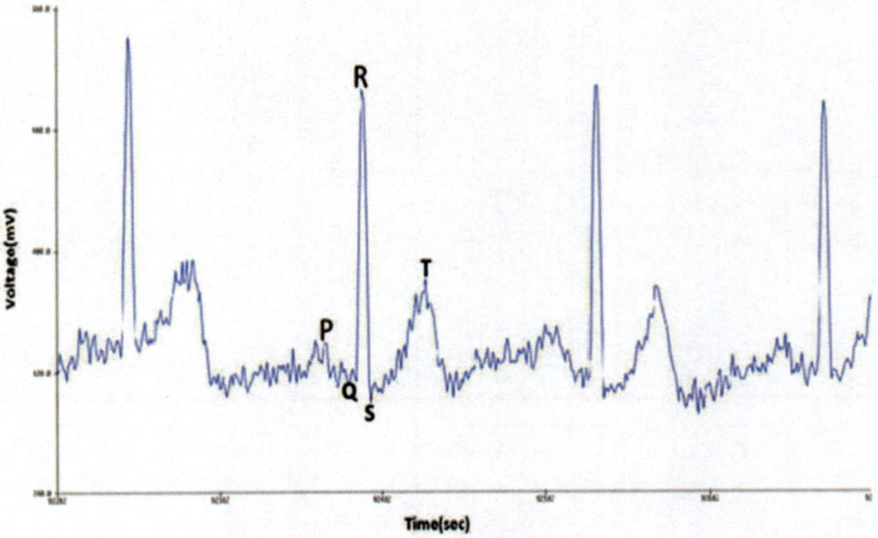

Fig. 13 ECG recorded when the body in motion

MATLAB is useful and handy on ECG signal processing that everyone can easily monitor his/her heart condition. Normally, the ECG signal procurement equipment can eliminate the power line interference. However, the baseline wandering and other wideband noises cannot be easy to suppress by hardware, while on the other hand, the MATLAB software is more powerful for suppressing all type of band noises and feasible for offline ECG signal processing. After recording the ECG signals, the recorded signal was processed through MATLAB processing software. The raw ECG signal extracted from the human body is subjected to various filtering stages like baseline wander removal, wavelet decomposition, noise cancellation (filtering), bandpass filtering, removal of a linear trend to remove unwanted signals. The MATLAB signal processing is based on software coding rather than using the hardware components. All processing units are designed mathematically, and working of each stage was analysed. Figure 15 shows the entire signal processing stages for recording ECG signal of the patient in MATLAB.

9 Conclusion

In this work, a detailed process for developing carbon-based textile dry and flexible electrodes for recording ECG signals is presented in a simple way. The fabricated dry ECG electrodes are wearable, adaptable, biocompatible, reusable and gel-free. The proposed ECG measuring device is eco-friendly and easily handled by common

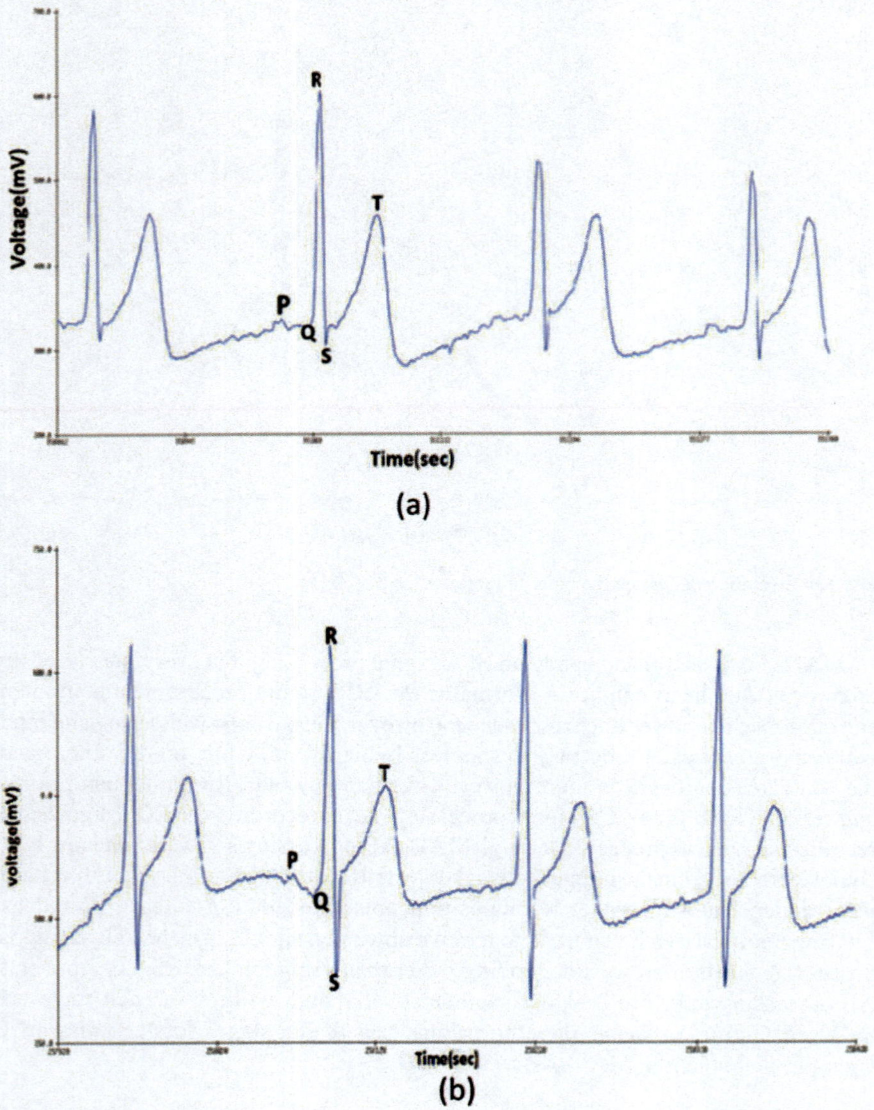

Fig. 14 **a** ECG obtained from wet Ag/AgCl electrodes. **b** ECG obtained from carbon-coated dry electrodes

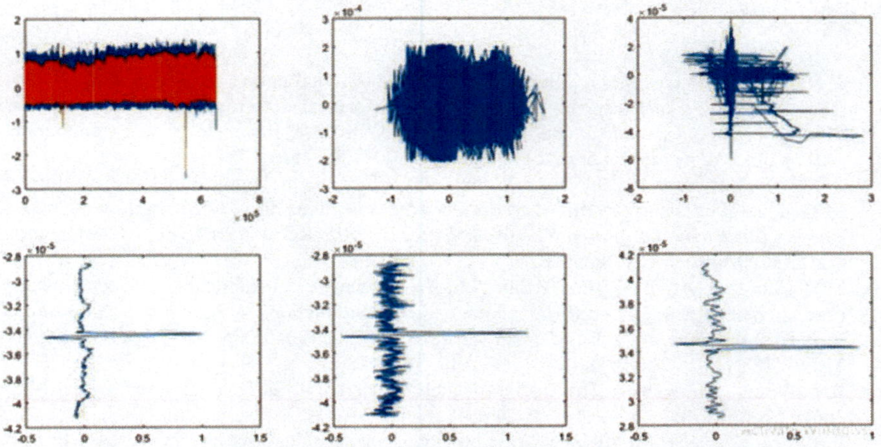

Fig. 15 MATLAB simulation of ECG acquisition circuit

consumers without any medical knowledge. The electrical characteristics of fabricated dry electrodes (E_1, E_2, E_3) were compared with wet Ag/AgCl electrodes by measuring skin–electrode impedance and found that the skin impedance of dry electrodes having a diameter of 12 mm, 16 mm, 20 mm are 585 kΩ, 571 kΩ and 565 kΩ, respectively, at a common frequency of 10 Hz. This experiment strongly proves that the electrodes with the largest diameter exhibit better performance in terms of signal intensity. This can be attributed to the fact that the skin–electrode impedance is inversely proportional to the diameter of electrodes. Also from the results, it was observed that skin impedance measured by dry electrodes (E_1, E_2, E_3) was higher than that of wet Ag/AgCl electrodes at a common frequency of 10 Hz. However, the measured values are still within the range of low impedance values to measure ECG signal. The reason for the low skin impedance of the Ag/AgCl electrodes than dry electrodes is due to the fact that the Ag/AgCl electrodes contain conducting gel. Experimentally, the ECG was recorded at two different conditions, motion and relaxed (rest), and found that the ECG recorded during motion displayed lot of disturbances than in a relaxed position. The recorded ECG graph also confirmed that the signals acquired from the dry electrodes have high quality in terms of peak intensity therefore easy to analyse than the signals acquired from conventional Ag/AgCl electrodes. Therefore, the proposed dry ECG electrodes are economic, convenient and have high potential to replace conventional wet electrodes.

Acknowledgements The authors gratefully acknowledge the facilities, support and guidance extended by the faculty and staff of Packaging Lab, CeNSE, IISc, Bangalore. Also, the authors would like to thank the Center for Nanomaterials and MEMS, NMIT, Bangalore, for extending the facilities to carry out this work and for the continuous support and encouragement.

References

1. S. Balambigai, P. Jeevitha, A survey on investigation of vital human parameters from photoplethsmography signal to predict the risk of cardiovascular diseases, in *2017 International Conference on Advanced Computing and Communication Systems (ICACCS)*, Coimbatore, (2017), pp. 1–4. https://doi.org/10.1109/ICACCS.2017.8014561
2. S. Mejhoudi, R. Latif, A. Toumanari, W. Jenkal, A. Elouardi, Implementation and evaluation of ECG signal processing algorithms on embedded architectures, in *2017 International Conference on Electrical and Information Technologies (ICEIT)*, Rabat, (2017), pp.1–6. https://doi.org/10.1109/EITech.2017.8255287
3. M.H. Chowdhury, Q.D. Hossain, P. Saha, M.M. Rahaman, Design, fabrication and performance evaluation of a three electrode ECG recorder, in *2016 International Conference on Innovations in Science, Engineering and Technology (ICISET)*, Dhaka, (2016), pp. 1–4. https://doi.org/10.1109/ICISET.2016.7856500
4. B.B. Winter, J.G. Webster, Driven-right-leg circuit design. IEEE Trans. Biomed. Eng. **BME-30**(1), 62–66 (1983)
5. E.M. Spinelli, N.H. Martinez, M.A. Mayosky, A transconductance driven-right-leg circuit. IEEE Trans. Biomed. Eng. **46**(12), 1466–1469 (1999)
6. H. Kim, Y. Kim, Y. Kim, Miniature electrocardiography sensor using a flexible printed circuit and MEMS technology, in *2008 IEEE International Conference on Multisensor Fusion and Integration for Intelligent Systems*, Seoul, (2008), pp. 545–550. https://doi.org/10.1109/MFI.2008.4648052.
7. A. Searle, L. Kirkup, A direct comparison of wet, dry and insulating bioelectric recording electrodes. Physiol. Meas. **21**(2), 271 (2000)
8. A. Gruetzmann, S. Hansen, J. Müller, Novel dry electrodes for ECG monitoring. Physiol. Meas. **28**(11), 1375 (2007)
9. Y. Zhou, X. Ding, J. Zhang, Y. Duan, J. Hu, X. Yang, Fabrication of conductive fabric as textile electrode for ECG monitoring. Fiber Polym. **15**(11), 260–2264 (2014)
10. H. Jung, J. Moon, D. Baek, J. Lee, Y. Choi, J. Hong, S. Lee, CNT/PDMS composite flexible dry electrodes for long-term ECG monitoring. IEEE Trans. Biomed. Eng. **59**(5), 1472–1479 (2012)
11. A. Karilainen, J. Hansen, S. Muller, Dry and capacitive electrodes for long-term ECG-monitoring, in *Proceedings of the 8th Annual Workshop on Semiconductor Advances for Future Electronics*, Veldhoven, (2005), pp. 155–161
12. P. Griss, H.K. Tolvanen-Laasko, P. Meriläinen, G. Stemme, Characterization of micromachined spiked biopotential electrodes. IEEE Trans. Biomed. Eng. **49**(6), 597–604 (2002)
13. K.P. Hoffmann, R. Ruff, Flexible dry surface-electrodes for ECG long-term monitoring, in *Engineering in Medicine and Biology Society, Proceedings of the 29th Annual International Conference of the IEEE, EMBS*, Lyon, (2007), pp. 5739–5742
14. C. Fonseca, J.P. Silva Cunha, R.E. Martins, V.M. Ferreira, J.P. Marques de Sá, M.A. Barbosa, A. Martins da Silva, A novel dry active electrode for EEG recording. IEEE Trans. Biomed. Eng. **54**(1), 162–165 (2007)
15. J.Y. Baek, J.H. An, J.M. Chio, K.S. Park, S.H. Lee, Flexible polymeric dry electrodes for the long-term monitoring of ECG. Sens. Actuators A **143**, 423–429 (2008)
16. W.C. Ng, H.L. Seet, K.S. Lee, N. Ning, W.X. Tai, M. Sutedja, J.Y.H. Fuh, X.P. Li, Microspike EEG electrode and the vacuum-casting technology for mass production. J. Mater. Process. Technol. 209, 4434–4438 (2009)
17. M.Q. Lokman, F. Ahmad, H. Yahaya, O. Mikami, F.K.C. Harun, S. W. Harun, Printed silver nanoparticles on kapton tape as passive saturable absorber, in *TENCON 2017—2017 IEEE Region 10 Conference*, Penang, (2017), pp. 2383–2387. https://doi.org/10.1109/TENCON.2017.8228260
18. K. Yang et al., Comparison between KAPTON polyimide insulation layers and nylon plates in superconducting tapes winding, in *2015 IEEE International Conference on Applied Super*

conductivity and Electromagnetic DevicesASEMD), Shanghai, (2015), pp. 294–295. https://doi.org/10.1109/ASEMD.2015.7453580
19. C.L. Lam, N.N.Z.M. Rajdi, D.H.B. Wicaksono, MWCNT/cotton-based flexible electrode for electrocardiography, in *Sensors, 2013 IEEE*, Baltimore, MD, (2013), pp. 1–4. https://doi.org/10.1109/ICSENS.2013.6688179
20. J. Yoo, H.J. Yoo, Fabric circuit board-based dry electrode and its characteristics for long-term physiological signal recording. Ann. Int. Conf. Eng. Med. Bio. Soc. 2497–2500 (2011)
21. V. Perumal, Mask design and simulation: computer aided design for lab-on-chip application. Adv. Mater. Res. **832**, 84 (2013). https://doi.org/10.4028/www.scientific.net/AMR.832.84
22. M.T. Shahid, M. Ammar Khan, M.Z. Khan, Design and development of a computer numeric controlled 3D Printer, laser cutter and 2D plotter all in one machine, in *2019 16th International Bhurban Conference on Applied Sciences and Technology (IBCAST)*, Islamabad, Pakistan, (2019), pp. 569575. https://doi.org/10.1109/IBCAST.2018667138
23. M. Abu-Saude, B.I. Morshed, Polypyrrole (PPy) conductive polymer coating of dry patterned vertical CNT (pvCNT) electrode to improve mechanical stability, in *2016 IEEE Topical Conference on Biomedical Wireless Technologies, Networks, and Sensing Systems (BioWireleSS)*, Austin, TX, (2016), pp. 84–87. https://doi.org/10.1109/BIOWIRELESS.2016.7445569
24. P. Das, Md.F. Hossain, J.-Y. Park, Chemically reduced graphene oxide-based dry electrodes as touch sensor for electrocardiograph measurement. Microelectron. Eng. **180** (2017). https://doi.org/10.1016/j.mee.2017.05.048
25. M. G. Srinivasa, P.S. Pandian, Dry electrodes for bio-potential measurement in wearable systems, in *2017 2nd IEEE International Conference on Recent Trends in Electronics, Information & Communication Technology (RTEICT)*, Bangalore, (2017), pp. 270–276. https://doi.org/10.1109/RTEICT.2017.8256600
26. O.J. Escalona et al., Wrist and arm body surface bipolar ECG leads signal and sensor study for long-term rhythm monitoring, in *2017 Computing in Cardiology (CinC)*, Rennes, (2017), pp. 1–4. https://doi.org/10.22489/CinC.2017.071-458
27. J. Datta, S. Saha, S. Chowdhury, A. Acharya, Development of an ECG signal acquisition module, in *2017 Devices for Integrated Circuit (DevIC)*, Kalyani, (2017), pp. 419-421. https://doi.org/10.1109/DEVIC.2017.8073983
28. W. Baochun, L. Min, Y. Yaning, Z. Weiwei, ECG acquisition circuit design based on C8051F330, in *Proceedings of 2012 IEEE-EMBS International Conference on Biomedical and Health Informatics*, Hong Kong, (2012), pp. 93–96. https://doi.org/10.1109/BHI.2012.6211515
29. E. Spinelli, R. Pallàs-Areny, M. Mayosky, AC-coupled front-end for biopotential measurements. IEEE Trans. Biomed. Eng. **50**(3), 391–395 (2003). [FILTER] B. Winter, J. Webster, Driven-right-leg circuit design. IEEE Trans. Biomed. Eng. **30**(1), 62–66 (1983)
30. W. Wang, Z. Wu, H. Huang, C. Luo, Low-Power Instrumental Amplifier for Portable ECG, in *2009 IEEE Circuits and Systems International Conference on Testing and Diagnosis*, Chengdu, (2009), pp. 1–4. https://doi.org/10.1109/CAS-ICTD.2009.4960836
31. W.Y. Du, W. Jose, Design of an ECG sensor circuitry for cardiovascular disease diagnosis. Int. J. Biosen. Bioelectron. **2**(4), 120–125 (2017). https://doi.org/10.15406/ijbsbe.2017.02.00032
32. M. Fernandes, K. Lee, R. Ram, J. Correia, P. Mendes, Flexible PDMS-based dry electrodes for electro-optic acquisition of ECG signals in wearable devices, in *Proceedings of IEEE Engineering* in *Medicine* and *Biology Society* (*EMBC*), (2010), pp. 3503–3506
33. A. Arcelus, M. Sardar, Mihailidis, Design of a capacitive ECG sensor for unobtrusive heart rate measurements, in *IEEE International Instrumentation and Measurement Technology Conference (I2MTC)*, (2013), pp. 407–410
34. H. Jeong, J. Seo, Electrode design on plastic substrates using laser patterned double-sided tape and gold leaf, in *2019 41st Annual International Conference of the IEEE Engineering in Medicine and Biology Society (EMBC)*, Berlin, Germany, (2019), pp. 1286–1289. https://doi.org/10.1109/EMBC.2019.8857635

Comparative Analysis of MPPT Techniques in Grid-Connected and Stand-alone PV System

B. B. Tara and H. L. Suresh

1 Introduction

Increase in energy demand to fulfill the requirements of the mankind made to look into alternative source of energy which is replaceable to depleting source of energy on which one could not able to depend for long time. Solar energy is the key source of energy in the electrical sector especially for distant applications. Considerable breakout in the semiconductor industry field in the year 1970s slowly declined the price of solar cell in the twenty-first century [1].

In order to extract the maximum power solar PV array at all the conditions, the maximum power point tracking is necessary. Solar PV system are used in various arrangements such as grid-connected network and stand-alone PV systems. Ample amount of sunrays required to fall on the PV panels to get maximum efficiency out of it. PV panel output is not only dependent on the irradiance level but also the temperature of the panel. Load characteristic varies according to the variation in the solar irradiance and temperature of the panel, so there is a requirement of maximum power point tracking (MPPT) control.

Various conventional techniques discussed in this paper are incremental conductance method, perturb and observe techniques, open circuit and short-circuit voltage method. Some of these methods have certain drawback in extracting maximum power from sunrays on the earth. Modern techniques such as artificial neural network, fuzzy logic control, ANFIS, and PSO can overcome from major shortcomings compared to conventional techniques. Few techniques are sophisticated with grid-connected network, and some other strategies are flexible with stand-alone PV system.

B. B. Tara (✉) · H. L. Suresh
Department of Electricals and Electronics Engineering, Sir. M VIT, Bengaluru, Karnataka, India
e-mail: tara@canaraengineering.in

Sanjeevikumar et al. (eds.), *Advances in Renewable Energy and Electric Vehicles*,
Lecture Notes in Electrical Engineering 767,
https://doi.org/10.1007/978-981-16-1642-6_5

2 Significance of MPPT

As sunrays hits the solar PV module, the solar cell converts sun light into required electrical energy. Efficiency of the solar cell around 35%, the main objective is to improve the efficiency of the PV cell or module. There may be several factors that affect the solar irradiance on the panel viz latitude, solar axis orientation, in addition to that there are other affects, namely clouds, building shadows, tree shadows nearby panel, etc. Solar panel output is dependent on the irradiance and temperature of the PV panel, and it is necessary to recognize a certain I–V characteristic point called maximum power point (MPP). This MPP shifts according to the environment condition, so there is a necessity of tracking of MPP in I–V characteristic. The MPPT technique not only adjusts the duty cycle of the converter by tracking the maximum power; furthermore, it improves the efficiency of the solar energy conversion system [2].

Figures 1 and 2 show the block diagram of stand-alone PV system and block diagram of grid-connected PV arrangement using converter circuit, respectively.

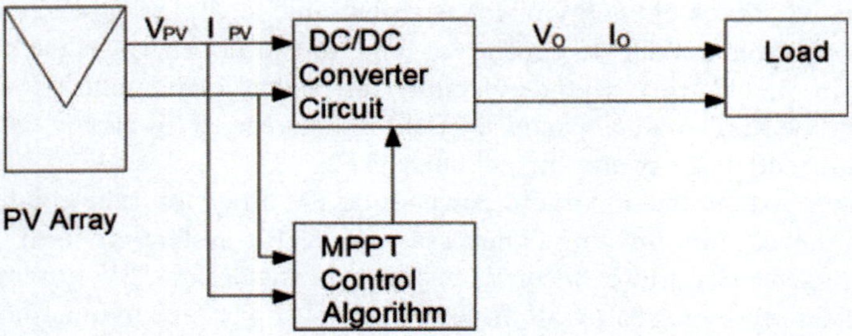

Fig. 1 Block diagram of stand-alone PV System

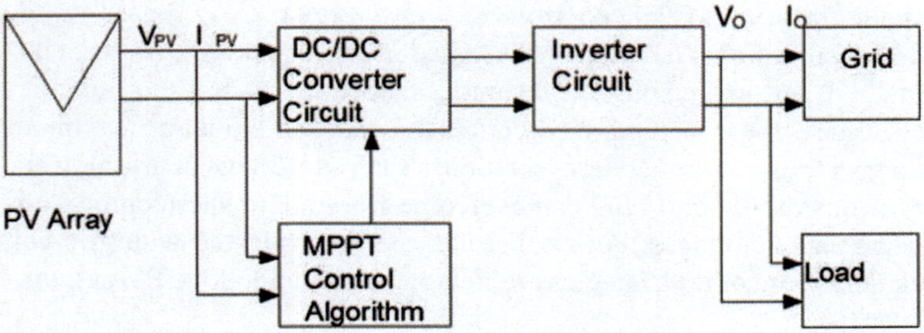

Fig. 2 Block diagram of grid-connected PV System

3 MPPT Techniques Overview

The MPPT techniques are used to extract the maximum power available from sun. The following techniques are widely used in the grid-connected and stand-alone PV systems.

3.1 *Incremental Conductance*

The incremental voltage and current changes in the PV array output, i.e., $\mathrm{d}I$ and $\mathrm{d}V$ are sensed in INC method. This has much faster tracking than P&O algorithm. The maximum power point occurs when $\frac{\mathrm{d}P}{\mathrm{d}V} = 0$, the (1) becomes $\frac{\mathrm{d}I}{\mathrm{d}V} = \frac{-I}{V}$. Further computation of INC method involves the comparison between $\frac{\mathrm{d}I}{\mathrm{d}V}$ and $\frac{I}{V}$ (array conductance) as in (2), (3), (4).

$$\frac{\mathrm{d}P}{\mathrm{d}V} = \frac{\mathrm{d}(IV)}{\mathrm{d}V} = I + V\frac{\mathrm{d}I}{\mathrm{d}V} = I + V\frac{\Delta I}{\Delta V} \tag{1}$$

$$\frac{\Delta I}{\Delta V} = \frac{(-I)}{V} \text{ at MPP} \tag{2}$$

$$\frac{\Delta I}{\Delta V} < \frac{(-I)}{V} \text{ at right MPP} \tag{3}$$

$$\frac{\Delta I}{\Delta V} > \frac{(-I)}{V} \text{ at left MPP} \tag{4}$$

The output voltage is at maximum power point when $\frac{\mathrm{d}I}{\mathrm{d}V}$ and $\frac{I}{V}$ are same, voltage is maintained by the controller until there is a change of irradiance level, and process repeats for the change in irradiance level. This is the optimized method under rapidly changing atmospheric condition [3]. In grid-connected system, power delivered to the grid is very fast as it follows the insolation change [4].

3.2 *Perturbation and Observation (P&O)/Hill-Climbing Technique*

This method is easy to implement; therefore, this is the commonly used MPPT strategy. The block diagram of P&O method is as shown in Fig. 3. The hill-climbing technique/P&O controller measures power and adjusts the small amount array voltage until increased power is equal to zero. This method results in power output oscillations, but it is suitable for rapidly changing atmospheric condition [5].

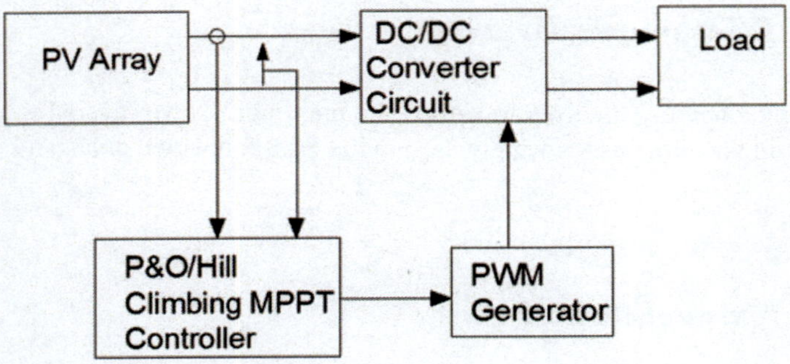

Fig. 3 Block diagram of P&O based MPPT technique for stand-alone PV system

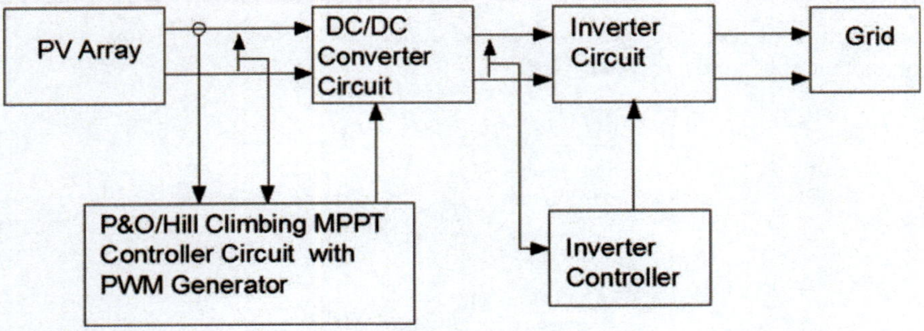

Fig. 4 Block diagram of P&O based MPPT technique for grid-connected system

The converter circuit in Fig. 4 having inverter control system has both current loop and voltage loop. The DC bus voltage is maintained constant by the PI controller present in the voltage loop. The reference current is generated by the PI controller with the help of voltage error, and it is used to synchronize with grid voltage.

This MPPT technique is also appropriate for the grid-connected converter system to get the fast dynamic performance and PV output voltage [6, 7].

3.3 Fuzzy Logic-Based MPPT

In the grid-connected PV, the nonlinear problems are common; this issue is avoided by fuzzy logic-based MPPT controller [8]. There are two processes involved a. Fuzzification and b. Defuzzifiaction. $e(k)$, $C_e(k)$ are error and change of error, respectively, these are the two input variables, and increased value of the duty cycle (ΔD) is the output as shown in (5), (6), (7).

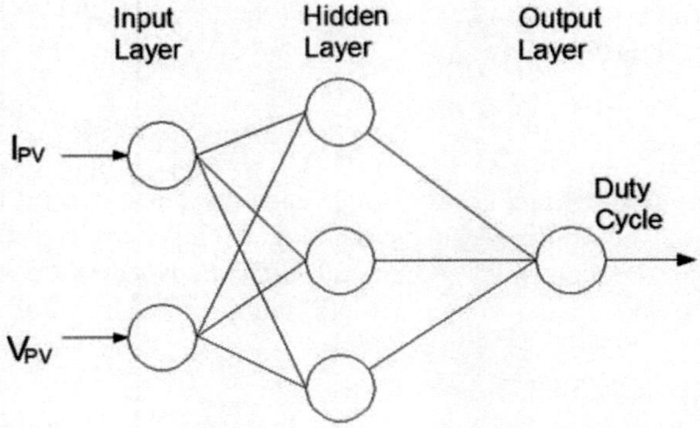

Fig. 5 Configuration of ANN

$$e(k) = \frac{dP}{dV}(k) - \frac{dP}{dV}(k-1) \tag{5}$$

$$C_e(k) = e(k) - e(k-1) \tag{6}$$

$$D(k) = D(k) - \Delta D(k-1) \tag{7}$$

$e(k)$ outlines location of the maximum power point in the PV module, it is given by the P–V curve slope, and direction of MPP is recognized by the $C_e(k)$ and ΔD can be obtained by the controller [9].

3.4 Artificial Neural Network-Based MPPT

The NN based algorithm automatically calculates the MPP and which is capable of modeling both nonlinear and linear problem [10], and the ANN network is as shown in Fig. 5. The power extraction efficiency of the stand-alone PV is more accurate using ANN based MPPT than other types. The average power obtained using this method is considerably good compared to other techniques. Tracking efficiency is good under fast changing atmosphere. Stable operation hybrid power generation is possible using neural network-based MPPT control [11]. For the grid-connected PV system, ANN based MPPT controller has very good efficiency with good amount of power delivered to the grid [12].

3.5 Fractional Open Circuit MPPT

Fractional voltage MPPT strategy is based on the point that maximum power point voltage V_{mpp} linearly relating with open circuit voltage V_{OC} as in Eq. (8). K_v is the voltage factor, and the value varies between 0.7 and 0.8 based upon the characteristic of the PV array[13].

$$V_{mpp} = K_v V_{OC}. \tag{8}$$

This method is very simple and considered as most effective method in reducing cost, but it is not efficient method in solar power conversion technology [14]. In the grid-connected PV system, a DC-DC boost converter with PI controller is added to circuit in order to increase and stabilize the output DC voltage [15].

3.6 Beta Method

The beta method uses the variable β which is given by (9), I and V are the output variables measure, and c is the diode constant. The $\boldsymbol{\beta}_a$ in (10) must be calculated repeatedly and should be between $\boldsymbol{\beta}_{\min}$ and $\boldsymbol{\beta}_{\max}$, then the system is under steady state, otherwise the system is under transient state. If it is in latter state $\Delta \boldsymbol{D}$ need to be calculated.$\boldsymbol{\beta}_g$, $\boldsymbol{\beta}_{\min}$ and $\boldsymbol{\beta}_{\max}$ values are based on the solar irradiation and the temperature of the PV panel [16]. The method is moderately complex and works better under fast changing atmosphere condition.

$$\beta = \ln \frac{I}{V} - c \times V \tag{9}$$

$$\Delta D = N \times \left(\beta_a - \beta_g\right) \tag{10}$$

In grid-connected system, three modes of operations can be considered, namely (a) continuous conduction mode, (b) discontinuous conduction mode, and (c) just discontinuous conduction mode. The continuous conduction mode is efficient than discontinuous modes because of reduced switching and conduction losses, reduced peak current stress [17].

3.7 Fixed Duty Cycle

In this method, the system impedance is adjusted only once to track the maximum power, so which is not suitable under change in weather conditions. Advantages of this method are easy to implement and does not need any feedback system.

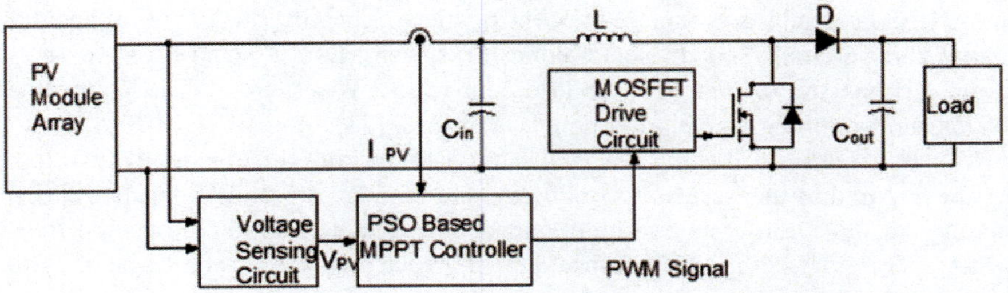

Fig. 6 Generalized block diagram of PSO based MPPT technique

3.8 PSO

Using particle swarm optimized technique, PSO has several advantages viz., time taken to reach MPP is faster under partial shading condition, time to reach MPP is faster under partial shading condition, reduction in power oscillations is under steady-state condition, and comparatively slight increase in power extraction at MPP and tracking efficiency is almost 100% with and without partial shading condition [18, 19]. The generalized block diagram of PSO based method is as shown in Fig. 6.

3.9 ANFIS Technique

The technique is the combination of artificial neural network and fuzzy logic control, called adaptive neuro-fuzzy inference system (ANFIS) shown in Fig. 7. The error signal obtained by taking the current and voltage feedback from solar array is sent to

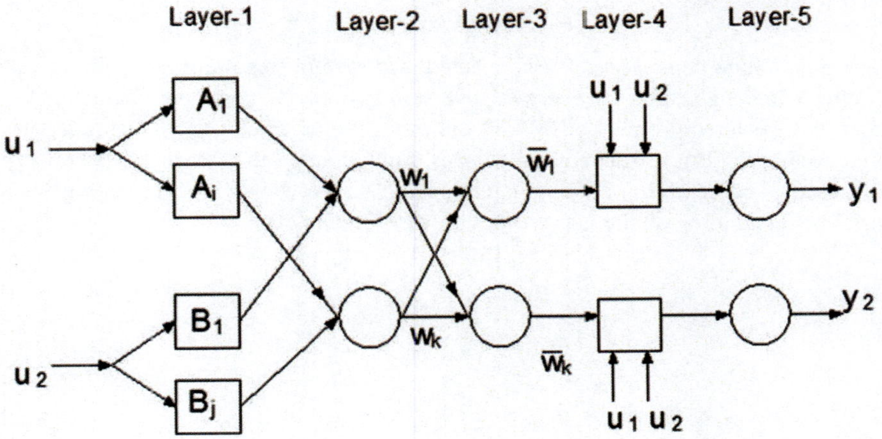

Fig. 7 ANFIS structure

the ANFIS controller to generate the control signal. Thereby, the DC/DC converter output can be controlled. In stand-alone PV system, DC/DC converter circuit is directly connected to load [20]. In grid-connected PV system, ANFIS controlled DC/DC converter varies the grid output via inverter [21].

This method has following advantages (i) Fast tracking of MPP, (ii) no external sensor required to measure solar irradiance and temperature on PV panel, and (iii) smooth steady-state output can be obtained. The data acquisition allows the forecasting of optimal sizing of PV, future signal for each compounds of the system. The predicted signal can be used in performance analysis of PV system.

In grid-connected system, ANFIS technique has MPP tracking efficiency of 100% in with less tacking time [22].

4 Comparison of Various Energy Management Strategies

The different MPPT techniques analyzed based on the following parameters viz. (i) according to number of sensors, (ii) according to tracking speed, (iii) according to circuit type, (iv) according to the cost, etc.

Two different parameters must be sensed every duty of MPPT viz. voltage and current/temperature/irradiance. Current sensor is costly, and voltage sensor is cheaper in cost. In modern applications, MPP tracking speed is significant to a great extent. There are slow responding and fast responding MPPT techniques which are available. Some application is comfortable with analog circuit and few are flexible with digital circuit, some needs both. Ultimate aim is to reduce the cost of the solar energy conversion system. Different MPPTs are analyzed as given in Table 1.

5 Conclusion

There is a huge demand for PV connected grid system/stand-alone system in near future. Almost all the solar energy conversion system has nonlinear characteristic; thus, there is a requirement of MPPT control technique. Each control technique has unique feature. But as technology evolves, the reduction in cost, increased energy efficiency, and flexibility are the major concern. In that view, the discussion gives a brief understanding on the major types of MPPT techniques.

Table 1 Comparison table of various MPPT techniques

MPPT technique	Sensor	Application grid-connected/stand-alone	Complexity level	Circuit type-A/D	Cost	Tracking speed
Incremental conductance	V and I	Better suited for stand-alone	Complex	D	High	Medium
Perturbation and observation (P&O)	V and I	Better suited for stand-alone	Complex	Both	Moderate	Medium
Fuzzy logic based	V / I	Both	Medium	D	High	Fast
Neural network	V /I	Both	Medium	D	High	Fast
Fractional open circuit	V / I	Better suited for stand-alone	Simple	D	Low	Slow
Beta method	V and I	Both	Medium	D	High	Fast
PSO	V and I	Both	Complex	D	High	Fast
ANFIS	V and I	Both	complex	D	High	Fast

References

1. L.S. Zabo, The history of using solar energy, in *7th International Conference on Modern Power Systems (MPS 2017)*
2. F. Liu, S. Duan, F. Liu, B. Liu, Y. Kang, A variable step size INC MPPT method for PV systems. IEEE Trans. Industrial Electronics **55**(7), 2622–2628 (2018)
3. B. Liu, S. Duan, F. Liu, P. Xu, *Analysis and Improvement of Maximum Power Point Tracking Algorithm Based on Incremental Conductance Method for Photovoltaic Array*, 7th International
4. S.V. Rajani, V.J. Pandya, Simulation and comparison of perturb and observe and incremental conductance MPPT algorithms for solar energy system connected to grid. Sadhana **40**, 139–153 (2015)
5. L. Egiziano, N. Femia, D. Granozio, G. Petrone, G. Spagnuolo, M. Vitelli, Photovoltaic inverters with Perturb&Observe MPPT technique and one-cycle control, in *IEEE International Symposium on Circuits and Systems*, Island of Kos, p. 4, p. 3721 (2006)
6. N. Femia, G. Petrone, G. Spagnuolo, M. Vitelli., A technique for improving P&O MPPT performances of double-stage grid-connected photovoltaic systems. IEEE Trans. Industrial Electronics **56**(11), 4473–4482, Nov 2009
7. F. Liu, Y. Kang, Y. Zhang, S. Duan, Comparison of P&O and hill climbing MPPT methods for grid-connected PV converter, in *2008 3rd IEEE Conference on Industrial Electronics and Applications*, Singapore, pp. 804–807 (2008)
8. C. Robles Algarín, J. Taborda Giraldo, O. Rodríguez Álvarez, Fuzzy logic based MPPT controller for a PV system. Energies **10**(12), 2036 (2017)
9. Q. Zeng, L. Chang, R. Shao, Fuzzy-logic-based maximum power point tracking strategy for Pmsg variable-speed wind turbine generation systems, in *Canadian Conference on Electrical and Computer Engineering*, Niagara Falls, ON, pp. 000405–000410 (2008)

10. Y.H. Liu, C.L. Liu, J.W. Huang, J.H. Chen, Neural-network-based maximum power point tracking methods for photovoltaic systems operating under fast changing environments. Sol. Energy **89**, 42–53 (2013)
11. W. Lin, C. Hong, C. Chen.,"Neural-network-based MPPT control of a stand-alone hybrid power generation system. IEEE Trans. Power Electronics **26**(12), 3571–3581, Dec 2011
12. M. Sunar, C. Nithya, J.P. Roselyn, Study of intelligent MPPT controllers for a grid connected PV system, in *IEEE International Conference on Intelligent Techniques in Control, Optimization and Signal Processing (INCOS)*, Srivilliputhur, pp. 1–6 (2017)
13. P. Das, Maximum power tracking based open circuit voltage method for PV system. Energy Proc. **90**, 2-13M (2016)
14. J. Ahmad, A fractional open circuit voltage based maximum power point tracker for photovoltaic arrays, in *2nd International Conference on Software Technology and Engineering*, San Juan, PR, pp. V1-247–V1-250 (2010)
15. H. Trabelsi, M. Elloumi, H. Abid, M. Kharrat, MPPT controllers for PV array panel connected to Grid, in *18th International Conference on Sciences and Techniques of Automatic Control and Computer Engineering (STA)*, Monastir, pp. 505–510 (2017)
16. X. Li; H. Wen, L. Jiang, E.G. Lim., Photovoltaic modified β-parameter-based MPPT method with fast tracking. J. Power Electronics, Jan. 2016
17. S. Jain, V. Agarwal, A new algorithm for rapid tracking of approximate maximum power point in photovoltaic systems. IEEE Power Electronics Lett. **2**(1), 16–19, March 2004
18. F.M. de Oliveira, S.A. Oliveira da Silva, F.R. Durand, L.P. Sampaio, V.D. Bacon, L.B.G. Campanhol, Grid-tied photovoltaic system based on PSO MPPT technique with active power line conditioning. IET Power Electronics **9**(6), 1180–1191 (2016)
19. K. Ishaque, Z. Salam, M. Amjad, S. Mekhilef, An improved Particle Swarm Optimization (PSO)–based MPPT for PV with reduced steady-state oscillation. IEEE Trans. Power Electronics **27**(8), 3627–3638, Aug. 2012
20. N. Khaehintung, P. Sirisuk, W. Kurutach, A novel ANFIS controller for maximum power point tracking in photovoltaic systems, in *5th International Conference on Power Electronics and Drive Systems*, PEDS 2003, Vol. 2, Singapore, pp. 833–836 (2003)
21. A. Mellit, Artificial intelligence based-modeling for sizing of a stand-alone photovoltaic power system: proposition for a new model using neuro-fuzzy system (ANFIS), in *3rd International IEEE Conference Intelligent Systems*, London, pp. 606–611 (2006)
22. A. Arora, P. Gaur, Comparison of ANN and ANFIS based MPPT Controller for grid connected PV systems, in *2015 Annual IEEE India Conference (INDICON)*, New Delhi, 2015, pp. 1–6. https://doi.org/10.1109/INDICON.2015.7443568

Comparative Study of Linear Induction Motor Guns and Coil-Guns for Naval and Ground-Based Artillery

Shreyas Maitreya, Bhushan Raghuwanshi, and Priyanka Paliwal

U_s	Stator supply voltage
R_s	Stator resistance
L_s	Stator inductance
R_r	Rotor resistance referred to stator side (assumed equal to stator resistance)
L_r	Rotor inductance
Q	edge effect correction factor
I_s	Stator current
I_r	Rotor current
L_m	magnetizing inductance
Φ	flux
p	no. of poles
τ	pole pitch
v	speed of runner
a, d	length and diameter of solenoid
$x(t)$	position of runner inside the solenoid from one end
$i_s(t)$, $i_r(t)$	stator and runner current
μ_0	magnetic permeability of free space
l	length of the side of the runner that is perpendicular to the magnetic lines of force in the solenoid
B	magnetic field
n	number of turns per unit length of the solenoid

S. Maitreya (✉) · B. Raghuwanshi · P. Paliwal
Department of Electrical Engineering, Maulana Azad National Institute of Technology, Bhopal, India

Sanjeevikumar et al. (eds.), *Advances in Renewable Energy and Electric Vehicles*,
Lecture Notes in Electrical Engineering 767,
https://doi.org/10.1007/978-981-16-1642-6_6

Table 1 Conventional medium Caliber Cannon versus electromagnetic launcher

Projectile parameters	Mauser	Electromagnetic
V_{max} (m/s)	1405	2300
Max. Acceleration (1000*g)	84	149
Mass (kg)	0.235	0.090
Energy at Target (MJ)	0.232	0.238
Launcher Parameters	*Mauser*	*Electromagnetic*
Bore diameter (mm)	30	14.5 × 32.7
Barrel length (m)	3.41	3.35

1 Introduction

In the case of artillery, conventional weapons launch projectiles carrying explosive warheads at high velocities to destroy and eliminate enemy targets. Such systems are expensive, dangerous, pose the risk of premature detonation and have a limited range and energy available on the target [1]. To mitigate the above-mentioned problems, the use of electromagnetic launchers capable of launching projectiles at hyper-velocities is being studied as a viable alternative to conventional chemical propellant-driven artillery [2]. Table 1 presents technical superiority of electromagnetic launchers over conventional chemical propellant-driven launchers used in ground combat. The comparison is carried out between a Mauser 30 mm MK-32 Cannon and a traditional electromagnetic launcher.

The higher projectile velocity, higher acceleration and higher energy to mass ratio of the electromagnetic launcher in comparison to the Mauser cannon [23] make it a superior alternative to conventional artillery. Electromagnetic launchers especially railguns and coil-guns have been a subject of research since the 1970s, but current research on electromagnetic launcher mainly focussed on the electromechanical parameters of the coil-guns [4, 5]. Along with presenting both computer simulations and hardware implementation of the coil-gun, this paper also speaks on their on-field deployability[5] In the case of electromagnetic launchers, three types of hyper-velocity electromagnetic launchers are being extensively studied viz. railguns, linear induction motor guns and coil-guns [6].

2 Basic Design of Coil Gun

Railguns are beyond the scope of this study, and the main area of focus shall be exclusively on linear induction motor guns and coil-guns. The main operating principle of both these weapons is indicated, i.e. Lenz's law and the key differences between these weapons are mainly in their mechanical performance which is discussed in the subsequent sections. As shown in Fig. 1, coil-gun is an electromagnetic launcher that consists of a solenoid(s) that acts as the stator and a runner. The principle of operation

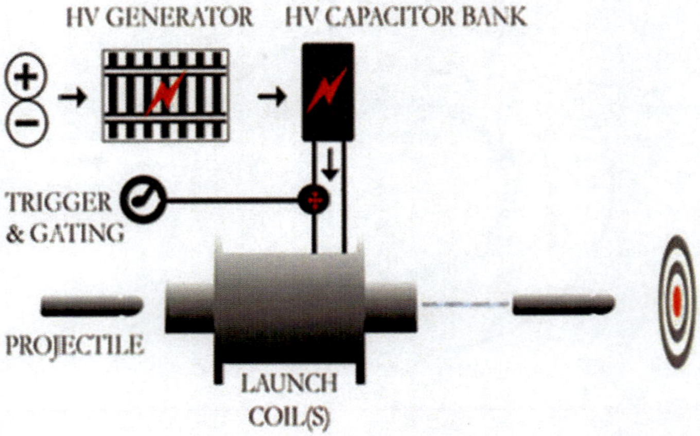

Fig. 1 Schematic of coil gun [2]

of a coil-gun is identical to that of a conventional linear induction motor except for the fact that a linear induction motor works on AC whereas coil-gun works on DC.

The use of DC in place of AC reduces the effects of the oscillatory nature of the force that is produced because of the alternating nature of the current that is supplied to the stator in the case of a linear induction motor gun. This aspect will be covered in greater detail in the subsequent sections.

3 Modelling of Linear Induction Motor Gun

This section presents a discussion on the modelling of a linear induction motor gun. Figure 2 presents a schematic of a rotary induction motor [1]. The development of a rotary induction motor into a linear induction motor is presented in Fig. 3.

As shown in Fig. 3, a linear induction motor is essentially a conventional rotary induction motor (Fig. 2) that has been cut along an axial plane and then laid on a flat surface [6]. Such a motor consists of a stator which is the stationary part and a runner which is the travelling part.

The stator in a linear induction motor produces a travelling magnetic field instead of a rotating one which is in the case of a rotary induction machine. The runner chases the travelling magnetic field and tries to catch it as per Lenz's law [6]. The stator and the runner are often referred to as the primary and secondary, respectively, and these terms will be used interchangeably hereafter throughout the paper. The runner shall also be referred to as the projectile. A linear induction motor's performance can be studied by analysing its equivalent circuit both in the a–b–c frame and stationary frame [6]. The equivalent circuit of linear induction motor in the a–b–c frame is presented in Fig. 4.

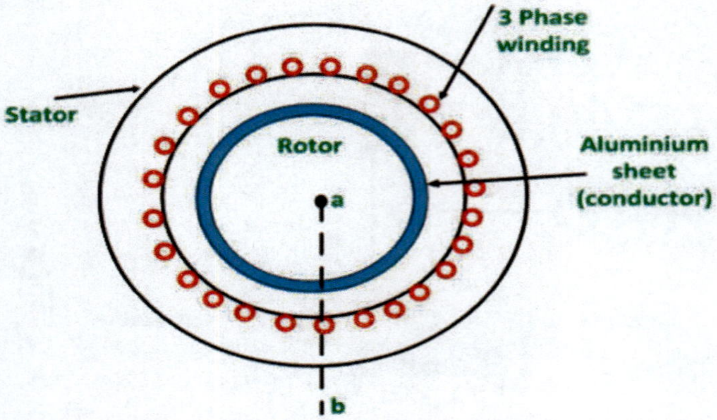

Fig. 2 Schematic of a rotary induction motor [1]

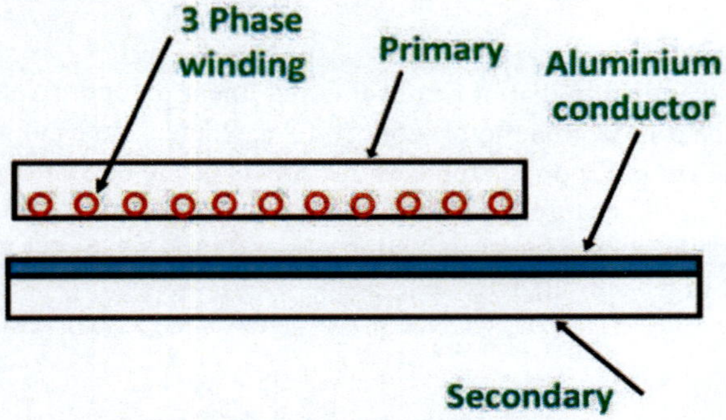

Fig. 3 Rotary induction motor developed into linear induction motor [1]

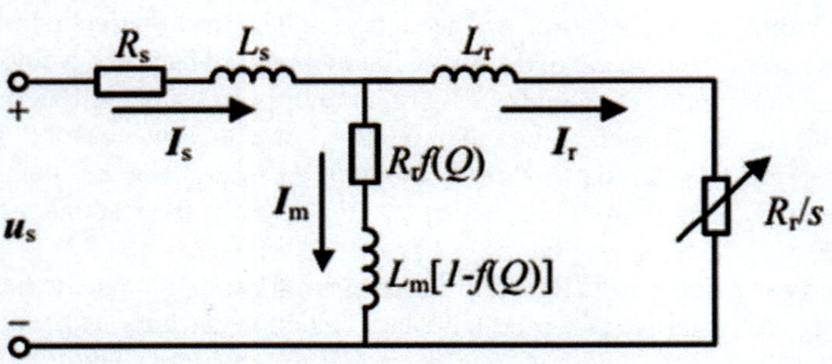

Fig. 4 Per phase equivalent circuit of linear induction motor in a–b–c reference frame

The edge effect comes into play in a linear induction motor because the distribution of flux in a linear induction motor is asymmetrical unlike that in a rotary induction motor because of a uniform air gap between the stator and rotor of a rotary induction motor which is absent in a linear induction motor. The end effect factor of the linear induction motor is determined from the expression as follows:

$$Q = \frac{\mathrm{DR_r}}{(L_\mathrm{m} + L_\mathrm{r})v} \tag{1}$$

The function that governs the effect of the edge effect factor on the various properties of the linear induction motor is given by:

$$f(Q) = \frac{1 - e^{-Q}}{Q} \tag{2}$$

Now, the correction values of rotor resistance and magnetising induction can be determined with the help of Eq. (2

$$R\prime_\mathrm{r} = R_\mathrm{r} f(Q) \tag{3}$$

$$L\prime_\mathrm{m} = L_\mathrm{m}(1 - f(Q)) \tag{4}$$

Now, writing the voltage equation of the linear induction motor in the stationary reference frame, the voltage matrix can be constructed as follow:

$$\begin{bmatrix} u_{s\alpha} \\ u_{s\beta} \\ 0 \\ 0 \end{bmatrix} = \begin{bmatrix} i_{s\alpha} & i_{r\alpha} & \Phi_{s\alpha} & 0 \\ i_{s\beta} & i_{r\beta} & \Phi_{s\beta} & 0 \\ i_{r\alpha} & i_{s\alpha} & \Phi_{r\alpha} & \Phi_{r\beta} \\ i_{r\beta} & i_{s\beta} & \Phi_{r\beta} & -\Phi_{r\alpha} \end{bmatrix} \begin{bmatrix} R_\mathrm{s} + R\prime_\mathrm{r} \\ R\prime_\mathrm{r} \\ \frac{\mathrm{d}}{\mathrm{d}t} \\ \omega_r \end{bmatrix} \tag{5}$$

From the above matrix, the equations for flux are determined as follows:

$$\Phi_{s\alpha} = L_s i_{s\alpha} + L'_\mathrm{m}(i_{s\alpha} + i_{r\alpha}) \tag{6}$$

$$\Phi_{s\beta} = L_s i_{s\beta} + L'_\mathrm{m}\left(i_{s\beta} + i_{r\beta}\right) \tag{7}$$

$$\Phi_{r\alpha} = L_r i_{r\alpha} + L\prime_\mathrm{m}(i_{s\alpha} + i_{r\alpha}) \tag{8}$$

$$\Phi_{r\beta} = L_r i_{r\beta} + L\prime_\mathrm{m}\left(i_{s\beta} + i_{r3}\right) \tag{9}$$

From Eq. (6), the electromagnetic thrust acting on the runner is given by,

Table 2 Parameters of linear induction motor gun

Parameter	Value
Inverter Bus Voltage	400 V
Mass of runner(kg)	2
Pole pairs	8
Pole pitch	0.15
T_s (sSeconds)	1e–5

$$F_e = \frac{3\pi p}{8\tau}\left(\Phi_{s\alpha} i_{s\beta} - \Phi_{s\beta} i_{s\alpha}\right) \tag{10}$$

The net mechanical force acting on the block is given by

$$F_{net} = m\frac{dv}{dt} = F_e - F_{losses} \tag{11}$$

$$m\frac{d^2 x}{dt^2} = \frac{3\pi p}{8\tau}\left(\Phi_{s\alpha} i_{s\beta} - \Phi_{s\beta} i_{s\alpha}\right) - F_{losses} \tag{12}$$

From the above equation, it can be inferred that the force on the runner is oscillatory since the currents in both α and β axes are sinusoidal.

Since, a dedicated block for a linear induction motor is not available on SIMULINK, the above equations were used, and a MATLAB code was written in the MATLAB workspace to study the behaviour of a linear induction motor. The parameters of linear induction motor gun are presented in Table 2.

Figure 5 presents the thrust-time graph of a linear induction motor. From Fig. 5, it is clear that the thrust, i.e. the force acting on the runner is highly oscillatory, thereby grossly limiting the destructive force of the gun by limiting the peak velocity due to the fluctuating nature of the direction of the acceleration vector of the projectile.

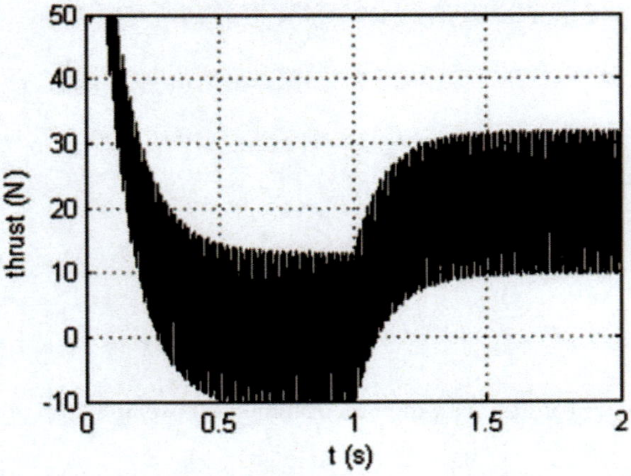

Fig. 5 Thrust-time graph of linear induction motor

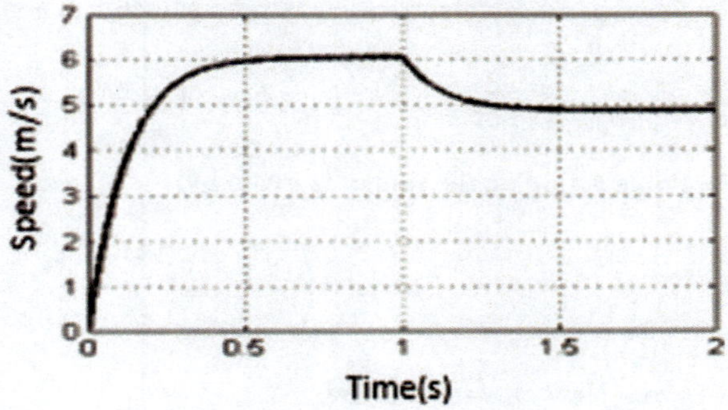

Fig. 6 Speed-time graph of the runner in a linear induction motor

Figure 6 presents the speed-time graph of the runner in a linear induction motor. Figure 6 shows another key limitation of a linear induction motor gun which is the low muzzle velocity of the projectile.

4 Performance Analysis of Coil Gun

Figure 7 presents the magnetic field inside a solenoid at any point on the axis of the solenoid. As shown in Fig. 7, each of the coils of a coil-gun can be regarded as a solenoid of finite length with the runner at point P, and its electromagnetic parameter can be determined from Eq. (13).

From Biot-Savart law, the expression for the magnetic field inside the solenoid comes to be:

$$B = \frac{\mu_0 i(t)_s}{2}\left(\frac{a}{\left(x(t)^2 + a^2\right)^{0.5}}\right) \tag{13}$$

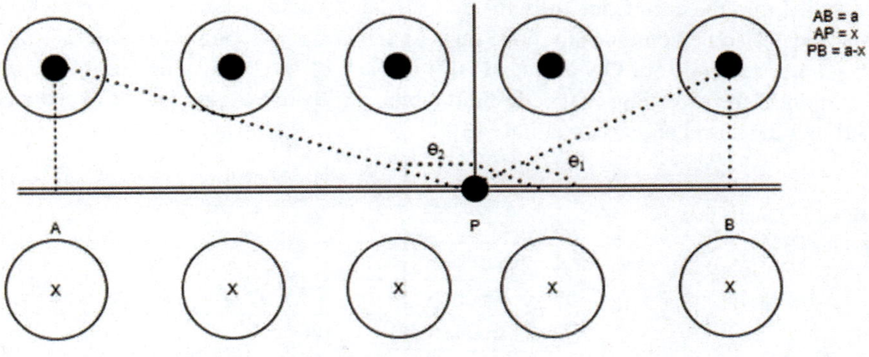

Fig. 7 Magnetic field inside a solenoid

Force acting on a current-carrying conductor in a magnetic field is given by

$$F = i\left(\vec{l} \times \vec{B}\right) \tag{14}$$

Force acting on the runner is given by,

$$F = i_r(t)\left(\frac{l\mu_0 i_s(t)}{2}\left(\frac{a}{\left(x(t)^2 + a^2\right)^{0.5}}\right)\right) \tag{15}$$

From Newton's second law,

$$m\frac{d^2x}{d^2t} = i_r(t)\left(\frac{l\mu_0 i_s(t)}{2}\left(\frac{a}{\left(x(t)^2 + a^2\right)^{0.5}}\right)\right) \tag{16}$$

Solving the above differential equation gives us an estimate of the position of the runner. It must be noted that the above equations have been derived by assuming that the size of the runner is negligible in comparison to that of the stator [11].

5 Simulation Results and Discussion

A coil-gun can be treated as a series R-L load. To meet the extremely high current demand of a coil-gun, it is generally powered by a capacitor bank of high voltage and high capacitance [7, 8] (Table 3).

A coil-gun is generally set up as shown in Fig. 8.

The capacitors need to be charged up to 400 V DC, and once the capacitors are charged, they are disconnected from the primary power supply [9, 10] and connected across the coils [2, 11]. Which are triggered as per the following sequence:

COIL 1 is turned ON initially when the projectile is initially at rest. The projectile is pulled into the coil. Once fully inside COIL-1, switching circuit 1 is turned OFF and the projectile continues to move through the coil on account of its inertia [12].

The sequential turn ON and turn OFF of each of the coils is necessary for the creation of the travelling magnetic field which will cause the projectile to advance through the barrel and acceleration [13].

Table 3 Simulation parameters

Parameter	Value
Number of stages	3
Power supply	400 V, 3900 *mF* capacitor bank
Total barrel length	100 mm

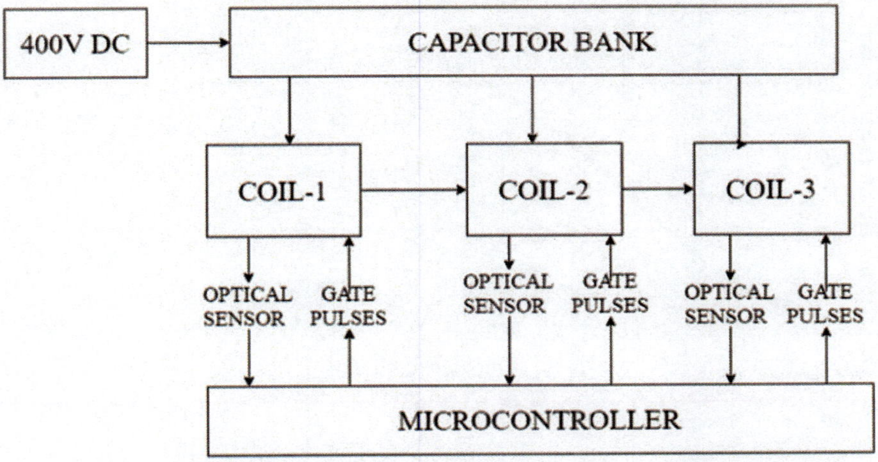

Fig. 8 Generic block diagram of 3 stage coil-gun

Figures 9 and 10 represent the force–time and speed-time graph, respectively, for coil gun. From Fig. 9, it can be observed that the force acting on the projectile although variable in magnitude maintained a positive direction up to 2.71428 ms after reversing its direction for a brief period before becoming zero. This allows the projectile to have a greater peak velocity, i.e. 35 m/s as seen in Fig. 10 for the same supply voltage of 400 V because of the positive direction of the thrust vector for most of its operating period. Most importantly, the projectile fired by a coil-gun reaches its peak velocity within 3 ms, whereas one fired by a linear induction motor gun takes almost 1 s apart from being several orders of magnitude slower. A lower firing time allows for quick reloading of the gun, in this case resulting in a theoretical firing capacity of more around 368.42 shots per second instead of a mere 1 shot per

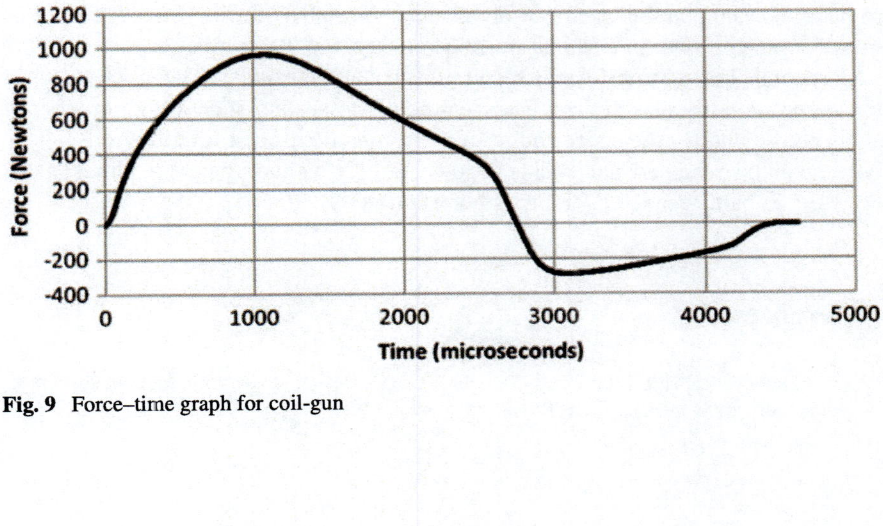

Fig. 9 Force–time graph for coil-gun

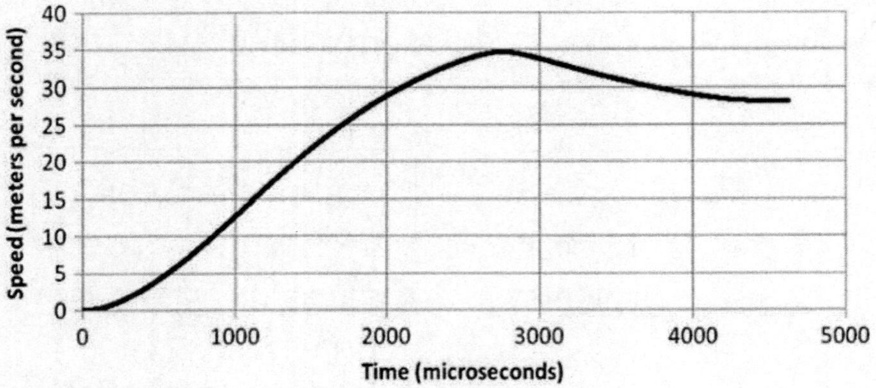

Fig. 10 Speed-time graph for coil-gun

second which could be achieved by the linear induction motor gun. This ensures that a greater number of targets can be engaged with further bolstering the destructive capabilities of the coil-gun w.r.t a linear induction motor gun.

6 Conclusion

The increased destructive power, reduced wear and tear and improved efficiency of a coil-gun combined with its reduced operating cost per shot make it a viable alternative to conventional chemical-driven weapons as seen in the introductory section(s) of the paper. Further, in comparison to other electromagnetic weapons, it can be concluded from the above simulations, it can be concluded that coil-guns are superior to linear induction motor gun as the projectile fired by the coil-gun is 35 times faster and can fire 368 shots per second such a high number of shots per second allows a coil-gun to have the range and capacity of an artillery gun but with the firing speed of an infantry weapon both in terms of muzzle velocity and the number of shots fired in each second. The Achilles' heel for coil-guns is not their range or destructive power but the availability of sufficient electrical power to sustain such a high number of shots per second for the entire course of a given battle if not a war. For coil-guns to become mainstream, advancements in power-supply technology are required so that they can be battle-ready under any circumstances.

References

1. A. Balikci, Z. Zabar, L. Birenbaum, D. Czarowski, On design of electromagnetic launchers for super velocity launchers. IEEE Trans. Magn. **43**(1), 107–110 (2007)

2. J.T. Tzeng, E.M. Schmidt, Comparison of electromagnetic and conventional launchers based on Mauser 30-mm MK 30–2 Barrels. IEEE Trans. Plasma Sci. **39**(1), 149–152 (2011)
3. N.S. Brahmbhatt, *Design and Optimization of an Electromagnetic Railgun*, Open Access Master's report, 2008, Houghton
4. Y.A. Dreizin, *Electromagnetic Launcher with Advanced Rail and Barrel Design*, United States of America, US5483863A, Patent 16 January 1996
5. S.J. Tatake, K.J. Daniel, K.R. Rao, A.A. Ghosh, I.I. Khan, Railgun. Defence Sci. J. **47**(3), 257–262
6. F. Korkmaz, I. Topaloglu, R. Gurbuz, Simulink model of vector controlled linear induction motor with end effect for electromagnetic launching system. Elektronika Ir Elektrotechnika **20**(1), 29–32 (2014)
7. Z. Zabar, Y. Naot, L. Birenbaum, E. Levi, P.N. Joshi, Design and power conditioning for the Coil-gun. IEEE Trans. Magnetics (1989). ieeexplore.ieee.org
8. J.L. He, Z. Zabar, E. Levi, B. Birenbaum, Transient performance of linear induction launchers fed by generators and capacitor banks. IEEE Trans. Magn. **27**(1), 585–590 (1991)
9. K. Chan, G. Clark, D. Helmlinger, D. Ng, S.A. McAfee, *Electromagnetic Coil-gun Final Report*, 2013, Reutgers School of Engineering, New Jersey
10. A. Balikci, High-velocity linear induction launchers. Int. J. Impact Eng., 1405–1409 (2008)
11. J.H. Gully, Power supply technology for electric guns. IEEE Trans. Magn. **27**(1), 329–334 (1991)
12. J.L. He, E. Levi, Z. Zabar, L. Birenbaum, Y. Naot, Analysis of induction-type coil-gun performance based on cylindrical current sheet model. IEEE Trans. Magn. **27**(1), 579–583 (1991)
13. H.M. Mohamed, M.A. Abdalla, A.A. Mitkees, W. Sabry, *Investigation on Electromagnetic Launching for Single Stage Coil*-gun, 2014, Military Technical College, Cairo

Comparative Study on Flyback Converter with PID Controller and Neural Network Controller

Nutana Shetty and Pradeep Kumar

1 Introduction

Flyback converter is most widely used switched mode power supply (SMPS) topology. Flyback converter because of its simplicity, less number of components and providing the isolation between the source and load it is used in offline converters, computers, printers. Flyback converter can provide the output voltage which depends on the duty ratio of switch and turns ratio of flyback transformer.

To provide a stable DC output, the regulation of output voltage of converter is necessary. The most widely used control methods, to control the voltage of the system are voltage mode control and current mode control. In voltage mode control, the duty ratio is directly controlled, in which the output voltage is sensed, and pwm signal for the switch is adjusted accordingly. In a current mode control in addition to voltage sensing loop, it has inner current sensing loop, which enables the inherent current protection to the converter.

To analyze the system behavior, mathematical modeling of the converter is carried out using state space method. Frequency response of converter is obtained using bode plot through MATLAB simulation. Compensator for the system is designed to obtain stable operation of converter.

N. Shetty (✉) · P. Kumar
Electrical and Electronics Engineering, NMAMIT Nitte, Karkala, India
e-mail: nutanakshetty@nitte.edu.in

P. Kumar
e-mail: pradeep@nitte.edu.in

Sanjeevikumar et al. (eds.), *Advances in Renewable Energy and Electric Vehicles*,
Lecture Notes in Electrical Engineering 767,
https://doi.org/10.1007/978-981-16-1642-6_7

2 Flyback Converter Operating Principle

The isolated type converters provide a galvanic isolation between the source and load. Because of the isolation, the load is protected from high voltage/current operation of the source. The flyback converter provides the isolation by using a high-frequency transformer A4 paper size Fig. 1 shows the basic flyback converter circuit diagram with transformer equivalent model [1]. Depending on the position of switch flyback converter, operation is divided in two different states. When switch is in ON state, the DC source is connected across the primary magnetizing inductor. The energy is stored in the primary magnetizing inductor. The energy will not transfer into the secondary side with the switch closed position. Because of dot polarity, secondary diode is reverse biased; load is supplied by the capacitor.

When switch is in OFF state, the energy which is stored in the magnetizing inductor is delivered to the transformer secondary winding. With OFF state of switch secondary side diode is forward biased, energy is supplied to the load, and capacitor is charged. The relation between converter input and output voltage is given Eq. 1 [2]

$$V_g = V\left(\frac{D}{1-D}\right)\left(\frac{N_2}{N_1}\right) \tag{1}$$

V_g—Converter DC input voltage
V—Converter DC output voltage
D—Duty ratio of the switch
N_2/N_1—Turns ratio of flyback transformer.

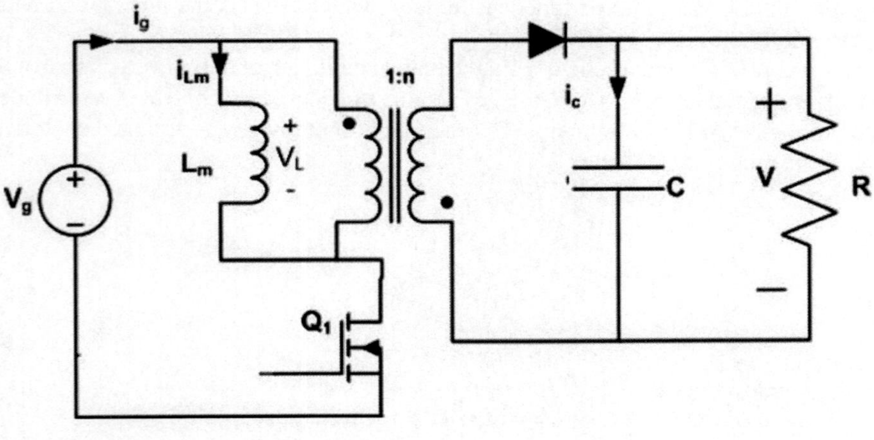

Fig. 1 Flyback converter circuit diagram

3 Peak Current Mode Control

The current mode control is a commonly used method to control DC output of the switched mode power supplies [2]. The peak current through the inductor or switch is controlled in a peak current mode control. The current mode control involves the two feedback loops. Inner loop is a current loop, and outer loop is voltage loop [3]. Figure 2 shows the circuit of flyback converter with peak current mode control.

The fixed frequency clock signal is given to the switch to close it initially, with switch closed the inductor current starts to increase. The error signal is generated by the outer voltage loop by comparing DC output voltage of converter and the required reference voltage. The compensator circuit gives the control signal v_C, and it is compared with the sensed inductor current. When peak current through the inductor reaches the control signal level, at that instant, the turning OFF of switch takes place and inductor current starts to decrease. Figure 3 shows the wave form of current through the inductor $i_{L(t)}$ with slope m_1 during switch closed position, m_2 during switch open position, control signal v_C, and gating signal generated for switch.

Current mode control method has a problem of subharmonic oscillation for the duty cycle exceeding the 50%. To overcome this problem, additional compensating

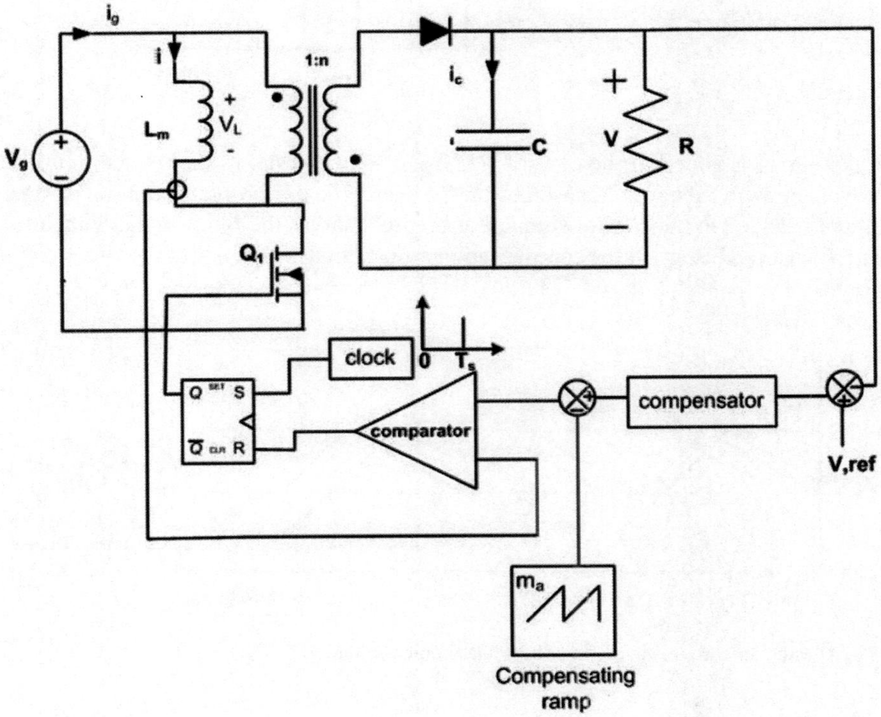

Fig. 2 Flyback converter with peak current mode control

Fig. 3 **a** Inductor current and control voltage, **b** Initial clock set signal to turn on the switch, **c** reset signal to rest the latch, **d** PWM signal given to the switch

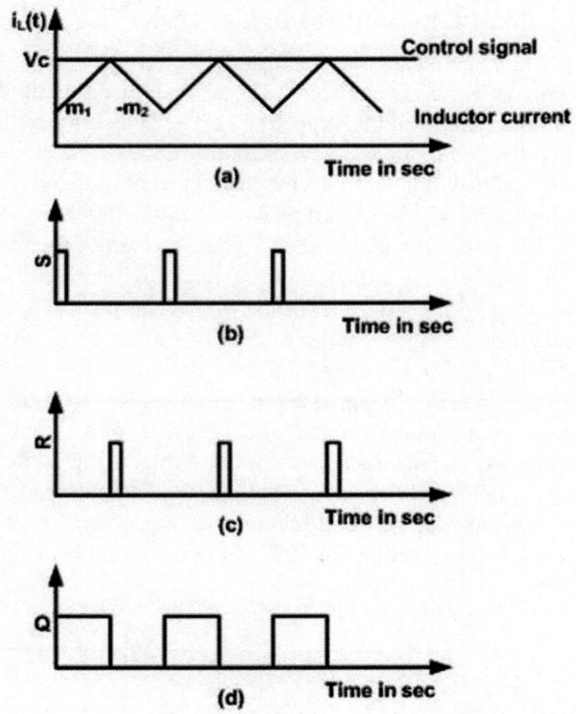

ramp signal is added to the control signal v_C. Figure 4 shows the inductor current waveform with subharmonic oscillation. To overcome this problem, additional ramp signal is added to the control signal V_C. Figure 5 shows the ramp signal with slope m_a. The slope chosen for the compensating ramp signal is ½ of m_2, to achieve a stable control.

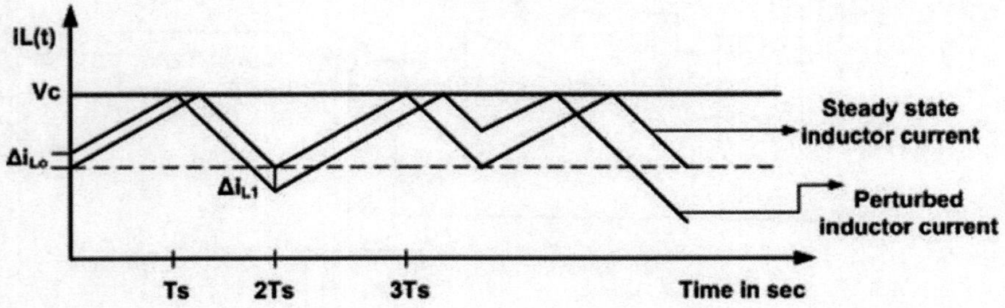

Fig. 4 Inductor current waveform with subharmonic oscillation

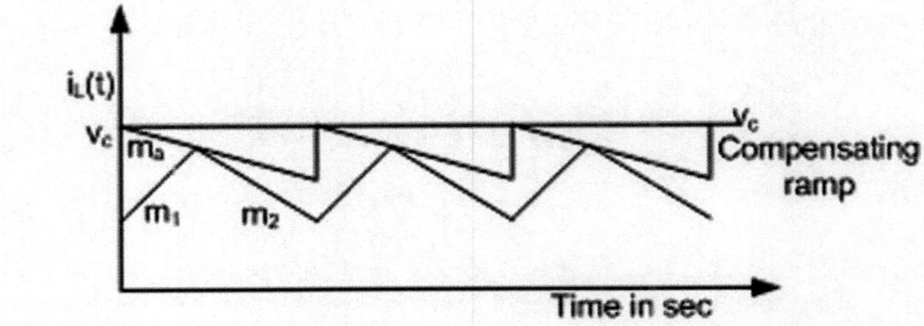

Fig. 5 Compensating ramp signal to suppress the subharmonic oscillation

4 State Space Model of Flyback Converter

The flyback converter equivalent circuit with non-ideal condition, during the switch closed is shown in Fig. 6. For the state space model, the current through the inductor $i_L(t)$ and voltage across the capacitor $v_C(t)$ are considered as state variables. For the interval when switch is closed, the differential equations are given by,

$$L\frac{\mathrm{d}i_L(t)}{\mathrm{d}t} = v_g(t) \tag{2}$$

$$C\frac{\mathrm{d}v_C(t)}{\mathrm{d}t} = \frac{-v_C(t)}{(R + R_C)} \tag{3}$$

$$v(t) = \frac{R}{(R + R_C)}v_C(t) \tag{4}$$

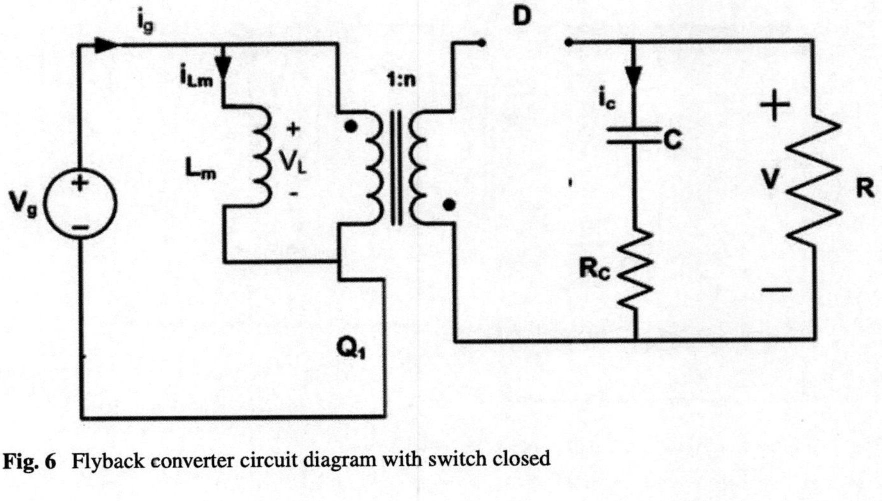

Fig. 6 Flyback converter circuit diagram with switch closed

The representation of the converter in state space model is given by

$$\begin{bmatrix} \frac{di_L(t)}{dt} \\ \frac{dv_C(t)}{dt} \end{bmatrix} = \begin{bmatrix} 0 & 0 \\ 0 & \frac{-1}{(R+R_C)C} \end{bmatrix} \begin{bmatrix} i_L(t) \\ v_C(t) \end{bmatrix} + \begin{bmatrix} \frac{1}{L} \\ 0 \end{bmatrix} v_g \tag{5}$$

$$v(t) = \begin{bmatrix} 0 & \frac{R}{(R+R_C)} \end{bmatrix} \begin{bmatrix} i_L(t) \\ v_C(t) \end{bmatrix} \tag{6}$$

Figure 7 shows the flyback converter circuit with switch open. The equations of current though the inductor and voltage across the capacitor are given by

$$L\frac{di_L(t)}{dt} = -\frac{1}{n}\left[\frac{R}{(R+R_C)}v_C(t) + \frac{RR_C}{n(R+R_C)}i_L(t)\right] \tag{7}$$

$$C\frac{dv_C}{dt} = -\frac{1}{(R+R_C)n}v_C(t) + \frac{R}{n(R+R_C)}i_L(t) \tag{8}$$

$$v(t) = \frac{R}{(R+R_C)}v_C(t) + \frac{RR_C}{(R+R_C)n}i_L(t) \tag{9}$$

The state space representation of Eq. (7) to (9) is given by

$$\begin{bmatrix} \frac{di_L(t)}{dt} \\ \frac{dv_C(t)}{dt} \end{bmatrix} = \begin{bmatrix} \frac{-RR_C}{n^2(R+R_C)L} & \frac{-R}{n(R+R_C)L} \\ \frac{R}{n(R+R_C)C} & \frac{-1}{(R+R_C)C} \end{bmatrix} \begin{bmatrix} i_L(t) \\ v_C(t) \end{bmatrix} + \begin{bmatrix} \frac{1}{L} \\ 0 \end{bmatrix} v_g \tag{10}$$

$$v(t) = \begin{bmatrix} \frac{RR_C}{n(R+R_C)} & \frac{R}{(R+R_C)} \end{bmatrix} \begin{bmatrix} i_L(t) \\ v_C(t) \end{bmatrix} \tag{11}$$

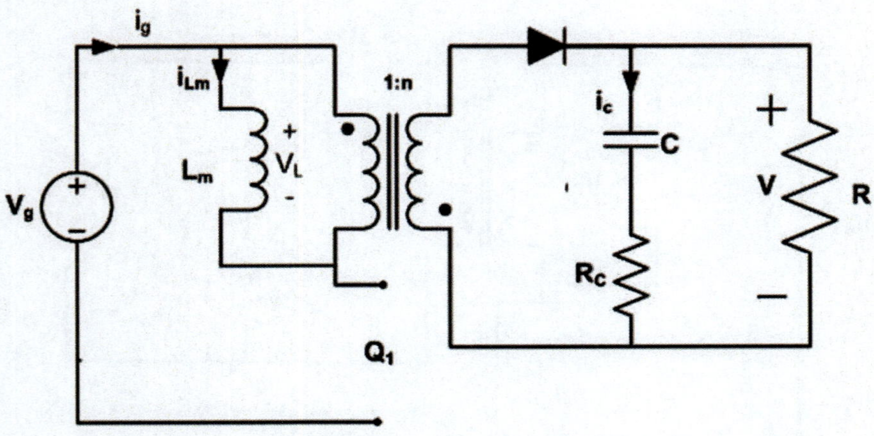

Fig. 7 Flyback converter circuit diagram with switch open

The average current through inductor and voltage across the capacitor is obtained by averaging the subintervals for a switching period.

$$L\frac{\mathrm{d}i_L(t)}{\mathrm{d}t} = v_g(t)\mathrm{d}(t) - \frac{1}{n}\left[\frac{R}{(R+R_C)}v_C(t) + \frac{RR_C}{n(R+R_C)}i_L(t)\right]\mathrm{d}'(t) \quad (12)$$

$$C\frac{\mathrm{d}v_C(t)}{\mathrm{d}t} = \frac{-v_C(t)}{(R+R_C)} + \frac{R}{n(R+R_C)}i_L(t)\mathrm{d}'(t) \quad (13)$$

In order to analyze the dynamic behavior of a converter and to achieve a small signal model, the perturbation is introduced in the input $v_g(t)$ and duty cycle d(t),

$$v_g(t) = V_g + \hat{v}_g(t)$$

$$\mathrm{d}(t) = D + \hat{\mathrm{d}}(t) \quad (14)$$

In response to input perturbation, the current through the inductor, voltage across the capacitor, and output voltage are also perturbed.

Transfer function of power stage is obtained as

$$H_{\text{open}}(s) = G_o \times f_P \times f_h \quad (15)$$

$$f_P = \frac{\left(1 + \frac{s}{w_{\text{ESRz}}}\right)\left(1 - \frac{s}{w_{\text{RHPz}}}\right)}{\left(1 + \frac{s}{w_{P1}}\right)} \quad (16)$$

$$f_h = \frac{1}{\left(1 + \frac{s}{w_{p2}Q_p} + \frac{s^2}{w_{p2}^2}\right)} \quad (17)$$

G_o is the open-loop gain of the power stage. The capacitor ESR zero w_{ESRz}, converter right half plane zero w_{RHPz} is given in Eqs. (18) and (19)

$$w_{\text{ESRz}} = \frac{1}{R_{\text{ESR}} \times C} \quad (18)$$

$$w_{\text{RHPz}} = \frac{R \times (1-D)^2 \times n^2}{L \times D} \quad (19)$$

The damping factor Q_P is given by

$$Q_P = \frac{1}{\Pi \times [m_C \times (1-D) - 0.5]]} \quad (20)$$

$$m_C = 1 + \frac{m_a}{m_1} \quad (21)$$

m_C is the slope compensation factor, m_a is the slope of compensating ramp.

The frequency response of the power stage which has gain of 16 dB is shown in Fig. 8. Converter has a zero due to capacitor ESR at $f_{ESRz} = 2.17$ kHz a right half plane zero at $f_{RHPz} = 23.38$ kHz.

The closed-loop bode plot is shown in Fig. 9. In the compensation loop, a pole is placed at a frequency of 2.17 kHz to cancel the ESR zero of converter. For good phase margin, a compensator zero is placed at 584.5 Hz.

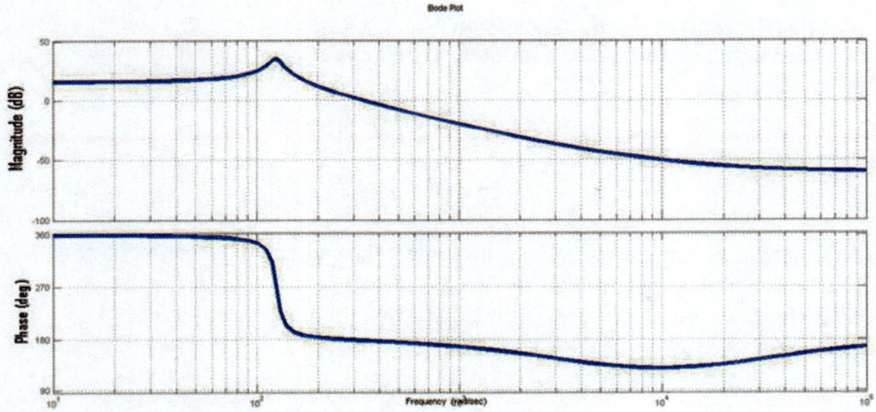

Fig. 8 Open-loop bode plot of converter

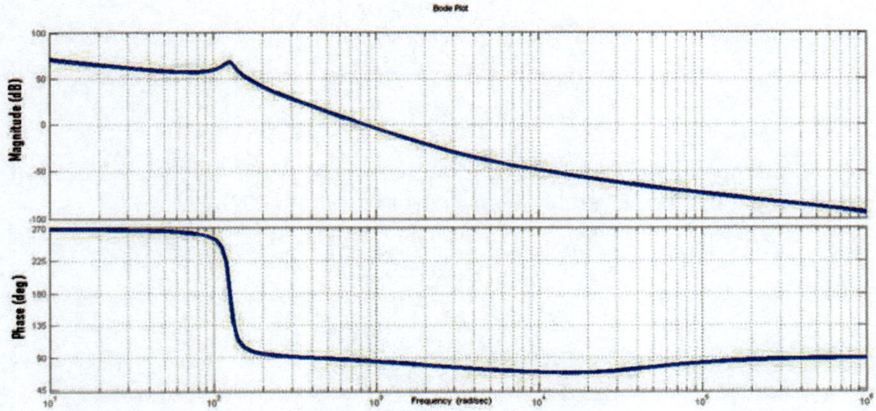

Fig. 9 Closed-loop bode plot of converter

5 Performance Comparision of PID Controller and Artificail Neural Netwok

Artificial neural network has a capability to meet the requirement of various environments which are not linear always.

A neural network block which works on the backpropagation model is used. A general backpropagation neural network model has three layers that is input, output, and middle layer as shown in Fig. 10.

The multi-layer feed forward network has to be trained with different set of examples or data, to obtain a required network behavior. The training process involves tuning the value of biases of the network and weights of the network. Mean square error(mse) is the default network function.

5.1 Neural Network(NN) Predictive Controller

NN predictive controller predicts the future plant performance by using the neural network model of nonlinear plant. The control input is then calculated by the controller to optimize the plant performance. Generally, a neural network model uses previous plant output and input to predict the future output [4].

Table 1 gives the parameter details of the neural network.

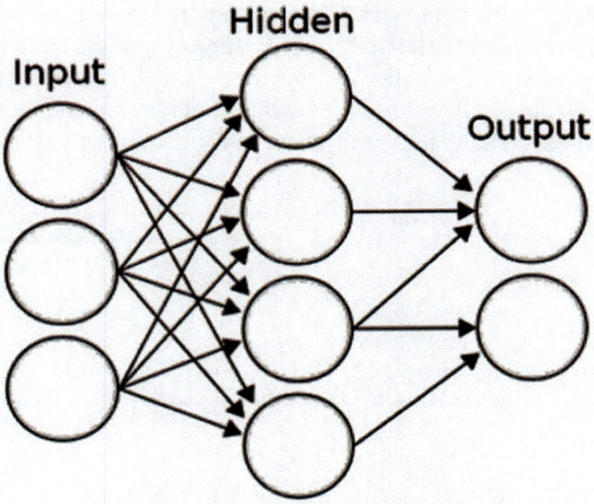

Fig. 10 Feed forward neural network layer

Table1 NN predictive controller parameter

Parameters	Values	Description	Value
Size of hidden layers	4	Cost horizon (N_2)	7
Training epochs	200	Control horizon (N_u)	2
Training function	'trainlm'	Control weighting factor (ρ)	0.005
Training samples	1000	Search parameter (α)	0.001

Table 2 Design specification

Parameter	Value
Switching frequency	40 kHz
Input voltage	325 V
Output voltage	30 V
Output Current	8A

5.2 *Simulation Results*

A flyback converter is simulated in MATLAB/simulink to analyze the performance of PID controller and the NN predictive controller. The converter design parameters are given in Table 2.

Figure 11 shows the output voltage of flyback converter which is simulated in with a PID controller. It is observed that the settling time of the output is about 4 ms.

The circuit is simulated by replacing the traditional PID controller with a NN predictive controller the network trained for the different input and output values. Figure 12 shows the output voltage of the converter with the NN predictive controller it observed that the settling time for the output is about 3 ms.

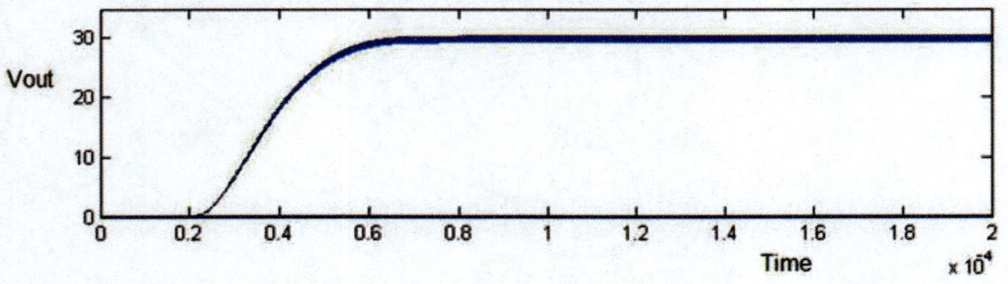

Fig. 11 Flyback converter output with a PI controller

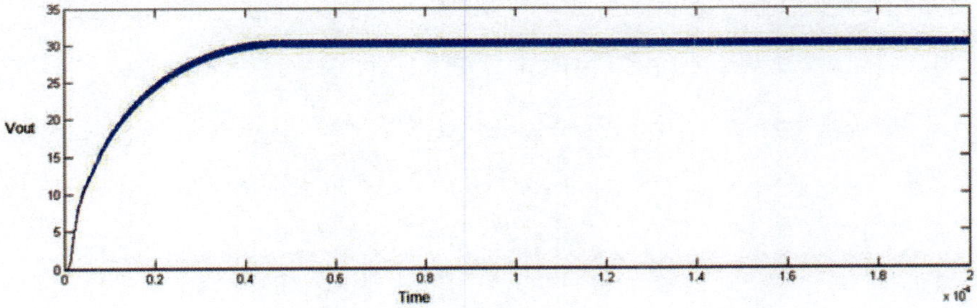

Fig. 12 Flyback converter output with a NN predictive controller

6 Conclusion

The flyback converter is described with peak current mode control, and modeling of flyback converter is explained by the state space model. Closed-loop simulation of converter is done in MATLAB/Simulink with PID controller and NN predictive controller. The result obtained through NN predictive controller is well accepted than the PID controller.

References

1. G. Abbas, U. Farooq, M.U. Asad, Appilcation of neural network based model predictive controller to power switching converters, in *International Conference and Workshop on Current Trends in Information Technology (CTIT)* (2011)
2. R.W. Erickson, *Fundamentals of Power Electronics*, 2nd edn (Kluwer Academic Publishers, 2004). ISBN 0-306-480-48-4
3. D.W. Hart, *Power Electronics,* Valparaiso, University Valparaiso, Indiana
4. K. Cheon, J. Kim, M. Hamadache, On replacing PID controller with deep learning controller for DC motor system. J. Autom. Control Eng., January 2015
5. I.J. Prasuna, M.S. Kavya, K. Suryanarayana, B.R. Shrinivasa Rao, Digital peak current mode control of boost converter, in *International Conference on Magnetics,Machines & Drives* (2014) K. Elissa
6. K. Suryanarayana, L.V. Prabhu, S. Anantha, K. Vishwas, Analysis and modelling of digital peak current mode control, in*IEEE International Conference on Power Electronics, Drives and Energy Systems*
7. U. Suprabha Padiyar, Vedavyasa Kamath,Design and implementation of a universal input flyback converter, in *International Conference on Electrical, Electronics and Optimization Techniques*

Comparison of Five Fuel Cell Electric Vehicles

Mayank Gautam and K. V. S. Rao

1 Introduction

Transport sector accounts for a major component of environmental pollution. Conservation of environment and resources through out the globe are grabbing the attention of the people. According to the International Energy Agency (IEA), 17% of CO_2 emissions are due to transportation, which also includes automobiles [1]. Conventional vehicles run on hydrocarbon fuels causing green house gas (GHG) emissions and environmental pollution. This leads to discover alternate fuel options for automobile and transportation sector [2]. For the last two centuries, water is used as a fuel, which is a combination of hydrogen and oxygen and can be easily separated by using water electrolysis technology [3].

Hydrogen can be a clean and environmental friendly fuel. It can be used in internal combustion engines directly or can be used by mixing with other fuels. Some amount of water is also generated, when H_2 is used in internal combustion engines (ICE) along with other emissions [2]. In the past few years, many researchers have shown interest on fuel cell electric vehicles (FCEVs). This is due to their inherent advantages of zero pollution, CO_2 emission free, and low noise. Conventional vehicles vent out huge amount of CO_2 in the atmosphere and are harmful to environment and human health. FCEV is zero emission vehicles and emits only water vapor as an exhaust. So, the use of FCEVs can be eco-friendly approach for automotive and transport sector. FCEV provides a solution for the issues related to oil dependence, GHG emission, and air pollution by using chemical energy of H_2 and FC technology in vehicles. FCEVs are better than the present conventional vehicles in terms of energy conversion efficiency, driving range, and carbon emission. Also FCEVs have the

M. Gautam (✉)
Department of Renewable Energy, Rajasthan Technical University, Kota, India

K. V. S. Rao
Gayatri Vidya Parishad College of Engineering for Women, Visakhapatnam, India

Sanjeevikumar et al. (eds.), *Advances in Renewable Energy and Electric Vehicles*,
Lecture Notes in Electrical Engineering 767,
https://doi.org/10.1007/978-981-16-1642-6_8

advantage in terms of refueling time compared to electric vehicles making them first choice for the vehicular applications [4].

This paper deals with different operating parameters of vehicles, which are beneficial to choose to get full advantage. To get better advantage, electrolyzer as well as FC vehicles should also be selected carefully in case of vehicular applications. Selection of electrolyzer is necessary to get maximum amount of hydrogen for the same amount of water compared to others, to cover more driving range. On the other hand, reduction in mileage or fuel economy indirectly affects the amount of CO_2 emission, life cycle fuel cost. In this paper, a comparative analysis of different FCEVs and gasoline vehicles is made, and necessary results are described briefly. It is also concluded that in terms of CO_2 emissions, FCEVs are always advantageous. However, every FCEV is not superior over gasoline vehicles in terms of cost. Hence, careful selection is needed.

2 Types of Commercially Available Vehicles for Automobile Sector

2.1 Conventional Vehicles

They have internal combustion engines (ICE) using gasoline as a fuel and are responsible for the CO_2 emissions.

2.2 Plug-in Hybrid Electric Vehicles

They also have an ICE that uses gasoline as a fuel and additionally have a battery to extend the range upto 20 miles, which have the advantage of 25% CO_2 reduction [5].

2.3 Battery-Operated Electric Vehicle

They have a large battery as a single power source, which vent out zero emissions and high efficiency. Short driving range and long charging times are the major disadvantages of battery-operated electric vehicle [5].

2.4 *Fuel Cell Electric Vehicle (FCEV)*

FCEVs are the vehicles that convert the chemical energy of H_2 into electrical energy to power the motor of the vehicle [6].

3 Hydrogen Fuel Cell Electric Vehicle

H_2 is used as a fuel, and fuel cell is used as an engine in FCEV. They have zero emission, large driving range, and less fueling time compared to the conventional vehicles [5]. FCEV comprises of mainly six components, namely fuel cell system, H_2 storage tank, air intake system, electric motor, power control unit, and power control unit. Stored H_2 from the storage tank and oxygen taken from the air by air intake system produces current. This current is controlled by the power control unit as per need. According to the operating conditions, control unit can either supply power to the motor or charge the battery or both. The main function of battery is just storage of excess electricity which can be used whenever needed. The generated current rotates the motor for the mechanical work in terms of the rotation of wheels [4]. Table 1 shows the specifications of five FCEVs. Fuel cell works as an energy converter, while the conventional batteries are an energy source. Therefore, the FC always requires an external feeding source [2]. Figure 1 shows the basic diagram of the system.

Proton electrolyte membrane (PEM) fuel cell using H_2 as a fuel have low start up time, low temperature, small size, high power density, and require negligible maintenance work making them prime for transport and automotive applications [2]. Presently, so many FCEV models are available in the market as Hyundai Tucson fuel cell, Mercedes B-Class fuel cell, Honda FCX Clarity, and Toyota Mirai. These vehicles have almost same features of power levels and hydrogen storage system (storage tank of 700 bars with fiber-wrapped composite) with battery [12].

In case of battery electric vehicles, the weight of energy storage system (ESS) increases with the driving range of the vehicle thus limiting the driving range. Fuel saving of a vehicle with 100 kg weight reduction is almost 0.3–0.5 L/100 km which is almost 6–10% of the total fuel for a fuel economy of 5 L/100 km. In other words, increase in 100 kg mass accounts to 0.3–0.5 L/100 km (7.5–12.5 g CO_2/km) increased fuel consumption for a passenger car of 1500 kg [13].

Battery electric vehicles are cost effective compared to FCEV up to 100 km travel where as beyond 100 km, FCEV are the cheapest [12]. The major components of these vehicles are the fuel cell, convertors, H_2 storage tank, electric motor, batteries, or super capacitors.

Table 1 Specifications of five fuel cell passenger vehicles [7–11]

S. no	Vehicle model	Fuel cell type	Fuel type	Fuel cell stack power (kW)	Motor type & power	H_2 storage capacity (kg)	Pressure (MPa)	Maximum driving range per tank (km)	CO_2 emission (g/km)
1	Tucson fuel cell	PEM	Compressed gaseous hydrogen	100	Induction	5.6	70	424	0
2	Honda Clarity	Proton electrolyte fuel cell	Compressed gaseous hydrogen	103	Synchronous electric motor	5.46	70	585.6	0
3	Toyota Mirai	Solid polymer electrolyte fuel cell	Compressed gaseous hydrogen	114	AC synchronous electric generator	5.0	70	499.2	0
4	Mercedes Benz	PEM	Compressed gaseous hydrogen	100	NA	3.7	70	385	0
5	Hyundai ix35 fuel cell	NA	Compressed gaseous hydrogen	100	Induction motor	5.63	70	594	0

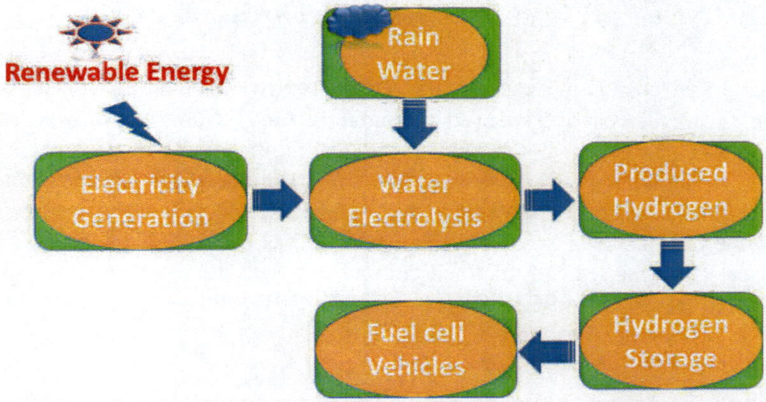

Fig. 1 Basic diagram of the system

4 On Board Storage of H_2 Fuel in Vehicles

The average fuel consumption of new cars is in the range of 8.5−12.75 km/L (20 to 30 miles per gallon). Current conventional gasoline vehicles have storage capacity of 30–45 L (10–16 gallons) of space. Since hydrogen has twice the efficiency of gasoline vehicles, they would store between 5 and 8 kg of hydrogen, which is equivalent to between 200 and 400 L, which is a sizable reduction in the space needed for fuel. Liquid H_2 storage tanks are light in weight; they can also be used for onboard storage, but storage at extremely low temperature is a difficult task [14].

5 CO_2 Emission Savings by Using H_2 as a Fuel in FCEV

Transport sector plays an important role in the CO_2 emission and global warming. So, it is necessary to move toward the green transportation to reduce green house gas emission due to automobiles. Hydrogen is a clean energy carrier which is abundantly available in the atmosphere. H_2 when used as a fuel in the FC generates zero CO_2 emissions and only water vapor is generated as exhaust. The quality of H_2 as a clean fuel which makes it a better option for the current transportation and automotive field. According to [6, 13], the European CO_2 emission target upto 2015 is 130 g/km in gasoline vehicles. By considering CO_2 emission per km, one can find the total CO_2 emission from the vehicles, and it can be compared with H_2 fuel cell vehicles which have zero emission during operating range. One can see a significant amount of CO_2 emission can be reduced by using H_2 as a fuel.

6 Life Cycle Fuel Cost of Gasoline Vehicles and FCEV

Life cycle fuel cost is the cost of fuel, consumed by a vehicle during its whole lifetime. Levelized cost of dispensed hydrogen (LCODH) from solar or wind energy is 2.34–4.68 US$/kg$H_2$ or 2–4 €/kgH_2 [15]. The lifetime of both FCEV and gasoline vehicles is assumed to be 15 years on the basis of 40 km/day driving distance, overall driving distance will be 219,000 km or 137,188 miles in 15 years. The average price of gasoline in Canada in April 2015 is considered as US$1.08/L or €0.92/L [16]. In Europe, Govt. has already defined compulsory fuel efficiency standards of 5.6 L/100 km for petrol and 4.9 L/100 km for diesel in 2015, and 4.1 L/100 km for petrol and 3.6 L/100 km for diesel by 2021 [17].

7 CO_2 Emissions and Fuel Economy Targets for Automobiles

CO_2 emission is a major concern worldwide. Policies for the reduction of CO_2 are made by different countries by imposing some specific targets of CO_2 emission per km. In Europe, one fifth of total emission is by road transport of which cars contribute to 75% [13].

In Europe, 130 g CO_2/km for year 2015 and 95 g CO_2/km for year 2020 is targeted by European Commission in 2009 [6, 13, 18, 19]. But the difference between certified and actual CO_2 emission in 2016 for passenger cars is 25–35 g CO_2/km which is almost 25% higher than the established target [6]. For light commercial vehicles, the target is 175 g/km for the year 2017 and 147 g/km by 2020 [13]. Table 2 shows CO_2 emission targets set by different countries.

According to [20], greenhouse gas emissions from a conventional gasoline vehicle can be calculated by multiplying total fuel consumption with CO_2 emission factor

Table 2 CO_2 emission standards of different countries [13]

S. no	Country	CO_2 target (g/km)	Year
1	European Union (Passenger cars)	95	2021
2	European Union (Light commercial vehicles)	147	2020
3	United States & Canada	97	2025
4	Japan	122	2020
5	China	117	2020
6	India	113	2021
7	South Korea	97	2020
8	Brazil	138	2017
9	Mexico	145	2016

(8887 g CO_2/gallon). United States Environmental Protection Agency (USEPA, 2011) and Intergovernmental Panel on Climate Change (IPCC, 2006) also used the same process. USEPA in 2011 also used another method for calculation of CO_2 emission per mile by applying the fuel economy/mileage data or miles per gallon (mpg), using (1), (2) and, (3).

$$CO_2 \text{ emissions per mile} = \frac{CO_2 \text{ per gallon}}{\text{MPG}} \quad (1)$$

$$CO_2 \text{ emissions per vehicle} = \frac{CO_2 \text{ per gallon}}{\text{MPG}} \times \text{Miles per vehicle} \quad (2)$$

$$\text{Total emissions} = \text{Number of vehicles} \times \text{Distance travelled} \times \text{Emission per vehicle distance travelled} \quad (3)$$

Europe already defined compulsory fuel efficiency standards for new cars, fuel economy of 5.6 L/100 km for petrol or 4.9 L/100 km for diesel in 2015, and 4.1 L/100 km for petrol or 3.6 L/100 km for diesel by 2021. US also proposed for passenger vehicles fuel economy of 54.5 mpg, or 5.2 L/100 km by 2025, which can correspond to 50% improvement in fuel economy [17]. According to [13], failure of a manufacturer to follow the preset standards will be penalized in the range of US$5.85–US$111.15 or €5–€95 on per gram extra CO_2 emission per vehicle sold. Incorporation of some alternative green fuel vehicles can also be an approach for the reduction of preset CO_2 emission targets.

8 Methodology

For estimation of CO_2 emission savings by means of FCEV, amount of hydrogen produced from any source is required. Water electrolysis technology is considered in the present study for the production of hydrogen. Specifications of some typical electrolyzers are given in Table 3 [21–25]. Ten million liters (10,000 m^3) of water is assumed for the production of H_2 through electrolysis. The produced H_2 can be used onsite or transported to different distances. The whole calculation has been done by considering driving distances in 'km' and fuel economy/driving range/mileage is in 'km/L'. So, the fuel economy and driving distances are converted from miles per gallon (MPG) to kilometer per liters (km/L) and 'miles' to 'km' for the whole calculation. The whole methodology is divided in the following steps [26]:

Step1: Electrolysis efficiency and amount of H_2 produced is calculated using [26–29]. Isothermal compression efficiency of hydrogen is considered to be 95% [26–31].

Step2: By using Table 1, driving range/fuel economy/mileage of vehicles is calculated, and results are given in Table 5.

Table 3 Specifications of electrolyzers [21–25]

S. no	Model	H_2 pro. rate (Nm^3/h)	Water cons (L/Nm^3 H_2)	Energy cons (kWh/Nm^3 H_2)	H_2 purity (%)	Pressure (Bar)
1	E1	300	0.9	4.4	99.9	0.02
2	E2	250	1	4.5	99.5	16
3	E3	170.6	0.84	5.3	99.5	*5/12/20
4	E4	45	2	5.2	99.9	10
5	E5	150	1	5.9	> 99%	10

a* cons. denotes consumption, b *pro. denotes production

Step3: For an average distance of 40 km/day, amount of hydrogen required for each vehicle is calculated (considering their driving range/fuel economy/mileage). Then, by utilizing the quantity of H_2 produced (per day) from different electrolyzers as a fuel in each type of FCEVs, number of FCEV that can be operated (per day) is calculated.

Step4: For the purpose of comparison with calculated number of FCEVs, by considering the permissible CO_2 emission target of 130 g/km in the equivalent gasoline vehicles (upto 2015) [6, 13], the total per day CO_2 emission for an average distance of 40 km/day is calculated.

Step5: Finally, total annual CO_2 emission saving is estimated using (1), (2), (3). Further the calculations of CO_2 emissions involve following steps:

(1) For the calculation of CO_2 emission, first we use (1) and find the value of CO_2 emission per mile or km. For this, driving range/fuel economy/MPG (km/L) is calculated by considering the hydrogen tank storage capacity and driving range per tank from Table 1, and CO_2 emission factor 8887 g CO_2/gallon of gasoline is considered [20].

(2) CO_2 emission per vehicle for a driving distance of 40 km/day is calculated using (2) by multiplying per day driving distance to CO_2 emission per mile or km.

(3) Now finally, the total emissions can be found using (3). For this, number of vehicles are considered from Table 6, driving distance is 40 km/day, and the CO_2 emission (130 g CO_2/km) occurs during the driving distance is considered.

So from (1), if the fuel economy is given, then amount of CO_2 emission can be found. In the same way, if amount of CO_2 emission is given, then also it is possible to find out fuel economy of a vehicle. Figure 2 explained the strategy of calculation in the form of flowchart.

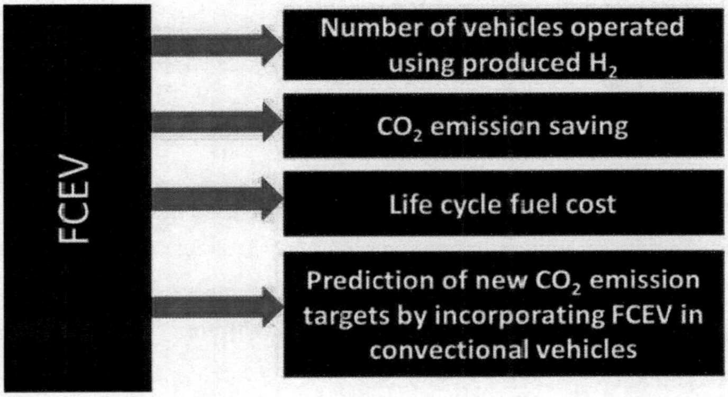

Fig. 2 Flowchart for calculations

Table 4 Different parameters of electrolyzers

S. no	Model	Amount of water required for per kg H_2 (L)	Efficiency of electrolysis (%)	Amount of H_2 produced at source (kg/day)
1	**E1**	10.01	68.1	9,98,667
2	**E2**	11.13	66.6	8,98,800
3	**E3**	9.346	56.54	10,70,000
4	**E4**	22.25	57.62	4,49,400
5	**E5**	11.13	50.79	8,98,800

9 Results and Discussions

9.1 Electrolysis of Water

In electrolysis, DC current is passed through the electrolyzers for the production of hydrogen and oxygen. For the electrolysis of water, five commercial electrolyzers are considered, and their hydrogen production (onsite) and efficiency is calculated as per the given specifications in Table 3, and the results are given in Table 4. It was found that the electrolyzer E1 has maximum electrolysis efficiency or lowest power consumption which is 68.1%. But amount of H_2 produced is maximum by E3 which is 10,70,000 kg/day because lowest water required for per kg of H_2 production.

9.2 Number of Vehicles Operated Using Produced Hydrogen

Hydrogen produced from water electrolysis is in its purest form and can be utilized in FC vehicles. By using Table 1, driving range/fuel economy (km/kg) of the vehicles

Table 5 Driving range calculated for different FCEVs

S. no	Vehicle model	Maximum driving range per tank (km)	Driving range (km/kg)
1	Tucson fuel cell	424	75.71
2	Honda Clarity	585.6	107.25
3	Toyota Mirai	499.2	99.84
4	Mercedes Benz	385	104.05
5	Hyundai ix35 fuel cell	594	105.51

Table 6 Number of vehicles operated using onsite produced hydrogen

S. no	Model	Amount of hydrogen produced at source (kg/day)	Tucson Fuel Cell	Honda Clarity	Toyota Mirai	Mercedes Benz	Hyundai ix35 Fuel cell
1	E1	9,98,667	18,90,226	26,77,675	24,92,673	25,97,782	26,34,233
2	E2	8,98,800	17,01,204	24,09,908	25,13,045	23,38,004	23,70,810
3	E3	10,70,000	20,25,243	28,68,938	26,70,720	27,83,338	28,22,393
4	E4	4,49,400	8,50,602	12,04,954	11,21,702	11,69,002	11,85,405
5	E5	8,98,800	17,01,204	24,09,908	25,13,045	23,38,004	23,70,810

is calculated and results are given in Table 5. For an average distance of 40 km/day, amount of hydrogen required is calculated. Then by using produced H_2, number of vehicles that can be run are calculated. The results obtained for number of vehicles operated by using onsite produced H_2 are given in Table 6.

9.3 Estimation of CO_2 Emission Savings

By considering number of FCEVs that can be run using produced hydrogen, one can estimate CO_2 emission savings. FCEVs are zero emission vehicles, so by the comparison, CO_2 emitted by the same number of gasoline vehicles is equal to CO_2 saving by FCEVs. For this, amount of CO_2 emitted by a gasoline vehicle for the distance of 40 km (25 miles) by considering 130 g CO_2/km [6, 13, 18, 19] is calculated. Then, per day and annual CO_2 emission reduction by the total number of vehicles is calculated, and results are given in Table 7.

Table 7 Overall CO_2 emission savings per day and annual CO_2 emission savings for different vehicles for onsite hydrogen production

S. no	Model	Tucson fuel cell		Honda Clarity		Toyota Mirai		Mercedes Benz		Hyundai ix35 fuel cell	
		CO_2 in ton/day	Annual CO_2 reduction (Lac ton)	CO_2 in ton/day	Annual CO_2 reduction (Lac ton)	CO_2 in ton/day	Annual CO_2 reduction (Lac ton)	CO_2 in ton/day	Annual CO_2 reduction (Lac ton)	CO_2 in ton/day	Annual CO_2 reduction (Lac ton)
1	E1	9,829	35.90	13,924	50.82	12,962	47.31	13,509	49.31	13,698	50.0
2	E2	8,846	32.30	12,532	45.74	11,666	42.58	12,158	44.38	12,328	45.0
3	E3	10,531	38.44	14,919	54.45	13,888	50.69	14,473	52.83	14,676	53.57
4	E4	4,423	16.14	6,266	22.87	5,833	21.29	6,079	22.19	6,164	22.5
5	E5	8,846	32.30	12,532	45.74	11,666	42.58	12,158	44.38	12,328	45.0

Table 8 Fifteen years life cycle fuel cost comparison of FCEV and gasoline vehicles

S. no	Vehicle	Fuel cost (US$)	Per day driving distance (km)	Fuel economy (km/kg) or (km/L)	Lifetime driving distance (km)	Amount of fuel L or Kg	Lifetime cost (US$)
1	Diesel	1.08/ L	40	20.41	2,19,000	10,730	11,588
2	Petrol	1.08/L	40	17.85	2,19,000	12,269	13,250
3	Tucson	4.68/ kg	40	75.71	2,19,000	2,893	13,537
4	Honda	4.68/kg	40	107.25	2,19,000	2,042	9,556
5	Toyota Mirai	4.68/kg	40	99.84	2,19,000	2,194	10,268
6	Mercedes	4.68/kg	40	104.05	2,19,000	2,105	9,850
7	Hyundai	4.68/kg	40	105.51	2,19,000	2,076	9,714

9.4 Life Cycle Fuel Cost of Gasoline Vehicles and FCEV

H_2 dispensed cost US$4.68/kg$H_2$ is considered for the calculation. The average price of gasoline in Canada in April 2015 is considered as US$1.08/L [20]. The fuel economy (km/L) for petrol and diesel vehicles is calculated considering 5.6 L/100 km for petrol and 4.9 L/100 km for diesel [15], and fuel economy of hydrogen vehicles (km/kg) is taken from Table 5. Considering a lifetime period of fifteen years, the fuel cost of gasoline and FCEVs is calculated. Table 8 shows fifteen years life cycle fuel cost comparison of FCEV and gasoline vehicles. It can be seen that FCEV TUCSON has the highest life cycle fuel cost among the FCEV and gasoline vehicles considered here.

9.5 Estimation of New CO_2 Emission Targets

Different countries fixed their future CO_2 emission targets from vehicles. To achieve such targets, either one should improve the fuel economy or incorporate some percentage of alternative vehicles in number of gasoline vehicles.

So, for the same number of vehicles, the new reduced CO_2 emission target can be set. This study has concern about both the issues. In this study, FCEVs are considered for incorporation. Fuel economy targets of European cars are 17.85 km/L of petrol or 20.41 km/L of diesel in 2015 and 24.39 km/L of petrol and 27.77 km/L of diesel for the year 2021 [18]. The percentage reductions in the consumption of petrol and diesel per km from 2015 to 2021 for petrol and diesel vehicles are 26.82% and 26.50%, respectively. In terms of CO_2 emission, the target is 131.71 g/km for petrol vehicles or 115.09 g/km for diesel vehicles in 2015, and by 2021, the corresponding values are 96.39 g/km, 84.66 g/km for petrol and diesel vehicles respectively.

Table 9 CO_2 emission target reduction when 25%, 50%, or 75% FCEVs are incorporated in conventional vehicles

S. no	Country	CO_2 Target (g/km)			
		Preset targets	New target with 25% FCEV	New target with 50% FCEV	New target with 75% FCEV
1	European (Passenger Cars)	95	71.25	47.50	23.75
2	European (Light commercial vehicles)	147	110.25	73.20	36.75
3	U.S. & Canada	97	72.75	48.5	24.25
4	Japan	122	91.5	61	30.5
5	China	117	87.75	58.5	29.25
6	India	113	84.75	56.5	28.25
7	South Korea	97	72.75	48.5	24.25
8	Brazil	138	103.5	69	34.5
9	Mexico	145	108.75	72.5	36.25

Another method of CO_2 emissions reduction is using some percentage of H_2 in conventional vehicles. In this study, incorporation of 25%, 50%, and 75% H_2 in conventional vehicles is considered. New CO_2 emission from these vehicles are predicted, and the results are given in Table 9. Also the fuel economy for these vehicles is also calculated. On the basis of given CO_2 emission targets of different countries, fuel economy to achieve such targets is calculated and is given in Table 10.

10 Conclusions

To reduce CO_2 emissions, hydrogen can be a better option compared to the conventional fuels in automobiles and for transportation sector. In this study, it is considered that CO_2 emission from each vehicle are 130 g/km. Among the five electrolyzers used for hydrogen production, E1 has the highest electrolysis efficiency or lowest power consumption. But amount of H_2 produced is maximum by E3 because minimum water is required for per kg of H_2 production. Hence, more vehicles can be run by using H_2 produced by E3, and thus, large amount of CO_2 emission can be reduced. It can be concluded that choice of electrolyzer is also essential, and the electrolyzer which uses minimum amount of water for the production of hydrogen be the first choice. For the maximum reduction of CO_2 emission, the electrolyzer with the lowest water consumption for per kg of hydrogen generation shall be considered.

The life cycle total fuel cost of the vehicles is in the order of tucson > petrol > diesel > toyota > mercedes > hyundai > honda. The cost for Tucson FC vehicle is

Table 10 Fuel economy of conventional vehicles required to achieve same CO_2 emission targets

S. no	Country	Preset fuel economy targets (km/L)	Fuel economy required for gasoline vehicles to achieve (km/L)		
			Targets same as 25% of vehicles are FCEV	Targets same as 50% of vehicles are FCEV	Targets same as 75% of vehicles are FCEV
1	European (Passenger cars)	24.75	33	49.5	98.99
2	European (Light commercial vehicles)	15.99	21.32	32.12	63.97
3	U.S. & Canada	24.24	32.32	48.48	96.95
4	Japan	19.27	25.69	38.54	77.08
5	China	20.09	26.79	40.19	80.38
6	India	20.81	27.74	41.61	83.22
7	South Korea	24.24	32.32	48.48	96.95
8	Brazil	17.04	22.72	34.07	68.15
9	Mexico	16.21	21.62	32.43	64.86

US$13,597, and for Honda FC vehicle, it is US$9556 for fifteen years of life time with a driving range of 40 km per day. If the fuel economy/driving range/mileage of the vehicle is low, then it can also happen that the life cycle fuel cost of FCEV would become higher than gasoline vehicle due to higher price of hydrogen compared to petrol or diesel. Obviously, it will reduce the CO_2 emission, but it will not be cost effective compared to gasoline vehicles. We can also see from Table 8 that for some FCEV the overall life cycle fuel cost of FCEV will be higher than diesel/petrol vehicles.

Two methods of CO_2 emission reduction are there. First one is by incorporating some percentage of alternative green fuel vehicles in conventional vehicles, and second one is improving the fuel economy of gasoline vehicles. By adding 25%, 50%, or 75% FCEV in conventional vehicles, the CO_2 emission targets of different countries can be reduced by 25%, 50%, and 75%, respectively. If the FCEV are not added, then to achieve such targets improvement in fuel economy by 133%, 200%, and 400% is essential.

References

1. S.H. Ha, W. Liu, H. Cho, S.H. Kim, Technological advances in the fuel cell vehicle: patent portfolio management, in *Technological Forecasting & Social Change*, pp. 1–13, Jul 2015
2. M. Gurz, E. Baltacioglu, Y. Hames, K. Kaya, The meeting of hydrogen and automotive: a review. Int. J. Hydrogen Energy, 1–13, Feb 2017

3. B. Tanc¸, H. T. Arat, E. Baltacıoglu, K. Aydın, Overview of the next quarter century vision of hydrogen fuel cell electric vehicles. Int. J. Hydrogen Energy **44**(20), 10120–10128, Apr 2019
4. X. Liu, K. Reddi, A. Elgowainy, H.L. Busch, M. Wang, N. Rustagi, Comparison of well-to-wheels energy use and emissions of a hydrogen fuel cell electric vehicle relative to a conventional gasoline-powered internal combustion engine vehicle. Int. J. Hydrogen Energy **45**(10), 972–983 (2020)
5. B. Lane, B. Shaffer, G.S. Samuelsen, Plug-in fuel cell electric vehicles: a California case study. Int. J. Hydrogen Energy, 1–7, Mar 2017
6. S. Tsiakmakis, A. Marotta, J. Pavlovic, K. Anagnostopoulos, The difference between reported and real-world CO_2 emissions: the difference between reported and real-world CO_2 emissions: How much improvement can be expected by WLTP introduction?, in *World Conference on Transport Research – WCTR*, Shanghai, China, pp. 3937–3947, Jul 2016
7. BMW. (2017, Sep. 11) [Online]. Available: http://www.hyundai.com
8. Mercedes. (2020, Apr. 15) [Online]. Available: http://www.daimler.com/
9. Toyota. (2020, Apr. 15) [Online]. Available: https://ssl.toyota.com/mirai/faq.html
10. Hyundai USA. (2020, Apr. 15) [Online]. Available: lhttps://www.hyundaiusa.com/tucsonfuelcell/index.aspx
11. Honda. (2020, Apr. 15) [Online]. Available: lhttps://www.honda.com
12. B. Bendjedia, N. Rizoug, M. Boukhnifer, F. Bouchafaa, M. Benbouzid, Influence of secondary source technologies and energy management strategies on energy storage system sizing for fuel cell electric vehicles. Int. J. Hydrogen Energy, 1–15, Mar 2017
13. G. Fontaras, N.G. Zacharof, B. Ciuffo, Fuel consumption and CO_2 emissions from passenger cars in Europe À Laboratory versus real-world emissions I. Prog. Energy Combust. Sci. **60**, 97–131 (2016)
14. B. Abderezzak, K. Busawon, R. Binns, Flows consumption assessment study for fuel cell vehicles: towards a popularization of FCVs technology. Int. J. Hydrogen Energy, 1–7, Dec 2016
15. V. Oldenbroek, L.A. Verhoef, A.J.M.V. Wijk, Fuel cell electric vehicle as a power plant: fully renewable integrated transport and energy system design and analysis for smart city areas. Int. J. Hydrogen Energy, 1–31, Jan 2017
16. P. Ahmadi, E. Kjeang, Comparative life cycle assessment of hydrogen fuel cell passenger vehicles in different Canadian provinces. Int. J. Hydrogen Energy **40**, 12905–12917 (2015)
17. N. Kholod, M. Evans, Reducing black carbon emissions from diesel vehicles in Russia: an assessment and policy recommendations. Environ. Sci. Policy **56**, 1–8 (2015)
18. A. Roberts, R. Brooks, P. Shipway, Internal combustion engine cold-start efficiency: a review of the problem, causes and potential solutions. Energy Convers. Manage. **82**, 327–350 (2014)
19. S. Tsiakmakis, G. Fontaras, B. Ciuffo, Z. Samaras, A simulation-based methodology for quantifying European passenger car fleet CO_2. Appl. Energy **199**, 447–465 (2017)
20. M. Alsabbagh, Y.L. Siu, J. Barrett, I.A. Gelil, CO_2 emissions and fuel consumption of passenger vehicles in Bahrain: current status and future scenarios. Sustain. Res. Inst. Univ. Leeds, Leeds, UK **53**, 1–30 (2013)
21. NEL Hydrogen. (2020, Apr. 15) [Online]. Available: http://nelhydrogen.com/
22. Suzhou Jiangli Hydrogen Production Equipment Co. Ltd. (2020, Apr. 15) [Online]. Available: http://www.jingli-hydrogenplant.com/hydrogengenerator/
23. Mercury Hydrogen Generators. (2020, Apr. 15) [Online]. Available: http://www.erreduegas.com/en/hydrogen-generators.html
24. Hydrogenics. (2017, Apr. 15) [Online]. Available: http://www.hydrogenics.com/
25. Next Hydrogen. (2017, Apr. 15) [Online]. Available: https://www.nexthydrogen.com/
26. M. Gautam, Hydrogen energy and it's applications for power generation, water production, energy storage, and as a fuel in fuel cell vehicles for CO_2 reduction, M.Tech. dissertation, Dept. of Renewable Energy, Rajasthan Tech. Univ., Kota, India, 2018
27. M. Gautam, K.V.S. Rao, B.K. Saxena, Relevance of hydrogen as an alternative of electricity for energy transmission and transportation of water in India, Presented at the Int. Conf. on Circuits Power and Computing Technologies, Kollam, Kerala, Apr 2017. https://doi.org/10.1109/ICCPCT.2017.8074235

28. M. Gautam, K.V.S. Rao, B.K. Saxena, Selection of electrolyzer-fuel cell combination for supply of water and electricity in remote areas, Presented at the Int. Conf. on Smart Technologies for Smart Nations, Bengaluru, Karnataka, Aug 2017. https://doi.org/10.1109/SmartTechCon.2017.8358380
29. Gupta Oxygen Pvt. Ltd. (2020, Apr. 15) Gas conversion chart [Online]. Available: http://www.guptagases.com/guptagases/download/conversion.pdf
30. P. Mazza, R. Hammerschlag, Carrying the energy future: comparing hydogen and electricity for transmission, storage, and transportation, Institute for Lifecycle Environmental Assessment, Seattle, Washington, Jun 2004
31. R. Hammerschlag, P. Mazza, Questioning Hydrogen. Energy Policy **33**, 2039–2043 (2005)

Design and Fabrication of Hybrid System for Highway Power Generation

N. Kaustubhasai and T. C. Balachandra

1 Introduction

With the developing concerns in global warming and usage of fossil fuels, the solutions are being looked upon to save the earth for future sustainable power sources for example, wind, solar, etc. and increasing the usage of these renewable energy systems is the solution being considered for the years to come. Inexhaustible sources such as wind and solar are contributing to a greater extent throughout the world. The Hybrid Renewable energy systems have become famous as stand-alone power systems for giving power in remote territories because of advances in these renewable energy technologies.

A Hybrid energy system is the one which for the most part comprises of at least two sustainable power sources utilized together to give expanded system effectiveness just as greater balance in energy balance [13]. This Paper presents this hybrid energy system i.e. Vertical axis wind turbine (VAWT) and PV panel for highway power generation. A VAWT is a type of turbine where the principle rotor shaft is designed to be transversal to the wind while the fundamental parts are stationed at the base. The challenges faced with wind energy is the amount of clearance area required for placing the turbines. VAWT blades will be designed to harness the wind energy along the highways. Generally, there are many blade designs such as straight bladed, H-type, J-type, C-type, etc [14].

In this model, the blades are designed as similar to that of a shape of a cup and are made up of plastic such that it rotates even at lower speeds and generates power.

N. Kaustubhasai (✉) · T. C. Balachandra
Department of Electrical and Electronics Engineering, Nitte Meenakshi Institute of Technology, Bangalore, Karnataka, India

T. C. Balachandra
e-mail: balachandra.tc@nmit.ac.in

Sanjeevikumar et al. (eds.), *Advances in Renewable Energy and Electric Vehicles*, Lecture Notes in Electrical Engineering 767,
https://doi.org/10.1007/978-981-16-1642-6_9

Solar energy also the largest contributor in terms of power produced per KWH per person per day.

Highway power generation inherently depends on SPV and wind for obvious reasons and an optimal combination of the two gives raise to challenging design and development problems. Ensuring steady consistent power output has been one of the issues.

To address these issues, efficient design of mechanical and electrical systems becomes necessary. Solar tracking and the use of LDRs (Light Dependent Resistors) help in ensuring steady output from SPV systems. The model is designed such that the main components such as DC dynamo, battery and other electromechanical systems are kept in the base for easier maintenance. The bearing is designed to withstand low to high wind speeds.

The gearing system plays a major role as it enhances the speed of the turbine and gear ratio used is 1:7. Whole assembly from the base mounting to the Solar panel mounting is made up of mild steel. All the parts are designed and assembled using Catia V5 and analysis of the individual parts and whole assembly is created using Ansys. The analysis is done at different speeds from low to high wind speeds.

Fabrication generally refers to manufacturing, specifically the crafting of individual parts as part of an individual or larger combined product. In the hybrid system, base mounting is fabricated initially in a cone like structure. The next phase in fabrication process is the fabrication of shaft, dc dynamo mounting, gear mounting and rods for fixing the blades.

The last phase is the solar panel mounting. The various processes utilised for developing this prototype are cutting, electric arc welding for assembling and surface finishing [15].

2 Literature Review

The process of designing the system described above using MATLAB simulation combined with manual calculations wherever appropriate is presented in this work [1].

VAWT blades were designed in semi-circular shape and was connected to the alternator through a shaft and pulley mechanism and proposed that the efficiency of VAWT could be increased by changing the shape and size of the blade in [2]. VAWT was designed such that it was coupled with two generators at the top and bottom to produce electricity in both directions in [3].

An insight into the hybrid highway power generation is given in [4]. The hybrid model comprised of Cygnus type VAWT and blades were designed in J-shape such that wind could also be captured at the back of smooth aerofoil and two polycrystalline Solar panels mounted on the top of the hybrid model were connected to the battery through the PWM charge controller. The Cygnus type VAWT is the blend of gyro mill type and savonius type VAWT. The hybrid model was fabricated for generation of large amount of power for highways and domestic applications.

The collaboration of hybrid green energy driven highway lighting system with Internet of Things (IoT) was observed by Rahman et al. [5]. The hybrid system was designed such that the Solar panel mounted on the top and the mounting structure had two lights attached too at the sides.

In this hybrid system, the Polycrystalline type Solar panel was considered as a core source of energy and the wind turbine was the secondary source of energy. Two sustainable power sources to be specific solar, wind was combined to supplement one another under a supervision of IoT driven controller.

While designing a Vertical axis wind turbine, an aspect ratio of the wind turbine plays an important role as the power coefficient value depends on it and is defined as the ratio between height of the blade and rotor radius (AR = h/R). Aspect ratio also impacts the Reynolds number and the wind turbines with lower aspect ratio have many advantages such as higher power coefficients, thicker blades, greater strength from the enhanced inertia movement of the rotor [6].

Pandey and Sharma [7] proposed generation of power using the road and transport in two different ways and their storage. The VAWT was designed in Creo software and comprised of blades, collars, shafts, gears and generators and the system also comprised of solar panel mounted at the top in [8].

Maximum power coefficient can be obtained by increasing the number of blades at lower wind velocity [9]. Li et al. [10] proposed that VAWT is capable as a small-scale energy system that occupies less space in the populated urban area.

CFD analysis study proposed that the turbine blades is capable of withstanding the forces without deforming under extensive pressures. The designed blades are strong and simple in structure and can capture wind from any direction and can withstand any weather condition [11, 12].

3 Methodology

The various stages of the development and testing methodology is shown in the following flow diagram: (Fig. 1)

4 Design Parameters

4.1 Design of Blades

$$\begin{aligned} &\text{Height of the blades, } h = 457.2\,\text{mm} = 0.457\,\text{m} \\ &\text{Radius of the rotor. } R = 422.4\,\text{mm} = 0.422\,\text{m} \end{aligned} \tag{1}$$

Swept area,

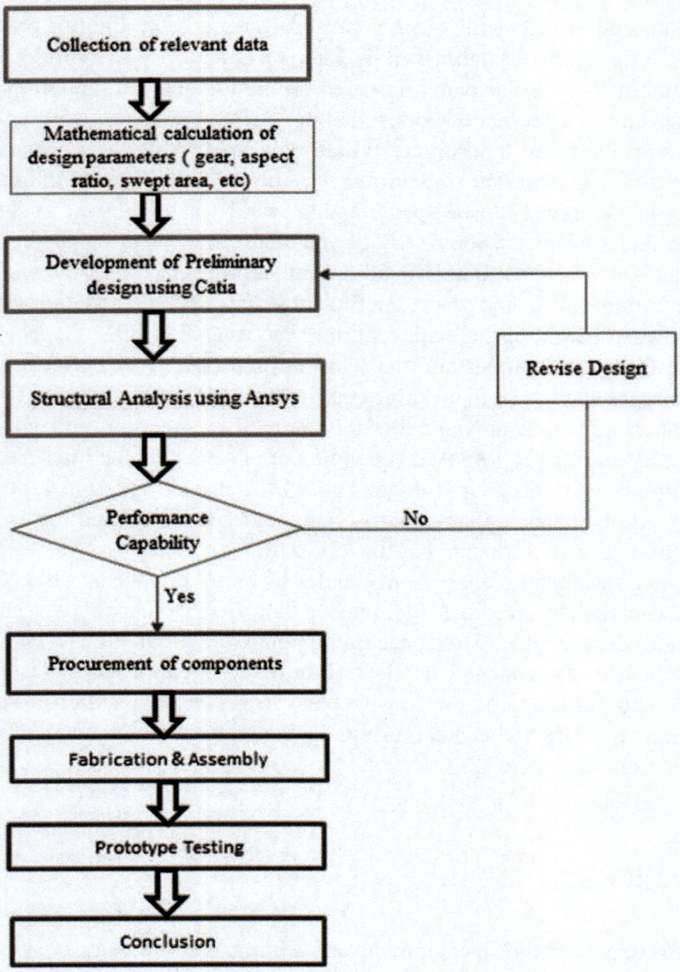

Fig. 1 Product development life cycle diagram

$$S = 2^{*} R^{*} 1$$
$$S = 2^{*}0.422^{*}0.457$$
$$S = 0.38\,\mathrm{m}^2 \tag{1}$$

where "*S*" denotes the Swept area (m^2), "*R*" denotes the rotor radius (m) and "*l*" the length/height of the blade (m) [18].

4.2 Design of Gear

$$\text{Outer radius of driver gear, } r_o = 33.81\text{ mm}$$
$$\text{Inner radius of driver gear, } r_i = 31.23\text{ mm.}$$
$$\text{Outer radius of driven gear, } R_o = 67.49\text{ mm.}$$
$$\text{Inner radius of driven gear, } R_i = 69.94\text{ mm.}$$
$$D = 25/4.716$$
$$D = 5.30\text{ mm}$$
$$\text{Circular pitch,} \quad P = 4.716\text{ mm} \tag{2}$$

where "$n1$" denotes number of teeth of driver gear and "P" denotes Circular pitch (mm) [17].

Addendu,

$$a = 0.3183{*}P$$
$$a = 0.3183{*}4.716$$
$$a = 1.501 \tag{3}$$

Dedendum,

$$b = 0.3979{*}P$$
$$b = 0.3979{*}4.716$$
$$b = 1.876 \tag{4}$$

where "P" is the Circular pitch (mm).

Root diameter,

$$D_R = D - 2b$$
$$D_R = 5.30 - 2{*}1.876$$
$$D_R = 1.548\text{ mm} \tag{5}$$

where "D" denotes the Pitch diameter and "b" denotes the dedendum.

Gear ratio,

$$G_r = n_1/n_2$$
$$G_r = 1:7 \tag{6}$$

where "$n2$" denotes the number of teeth of driven gear [17].

4.3 *Design of Shaft*

Generally designing of shaft involves the determination of inner and outer diameter in case of hollow shaft and diameter in case of solid shaft [16].

$$\text{Outer diameter } d_o = 50.8\,\text{mm}$$
$$\text{Inner diameter } d_i = 25.4\,\text{mm}$$
$$\text{Length} \quad L = 1500\,\text{m}$$

4.4 *Aspect Ratio*

Aspect ratio,

$$\begin{aligned} \text{AR} &= h/R \\ \text{AR} &= 457.2/422.4 \\ \text{AR} &= 1.08 \end{aligned} \tag{7}$$

where "AR" denotes the aspect ratio, "*h*" denotes the height of the blade and "*R*" denotes the radius of the rotor. The 2D & 3D model were designed by considering the design parameters.

5 2D and 3D CAD Model

Generally, the designing process starts with sketching, 2D drawing, 3D modelling and assembly. Catia consists of many workbenches to perform this design operations. The design procedure & 2D drawing of the assembled model in isometric view and front view are shown in the: (Figs.2, 3 and 4)

The next step in the designing process is the 3D modelling. 3D modelling is also an important step as the 3D model will be further analysed for stress, strain, deformation, etc. The following figure shows the 3-dimensional model of the assembly.

6 System Components

The hybrid model comprises of many components such as solar panel, battery, DC dynamo, gear system, VAWT, light dependent resistors, etc, which contribute to power generation. The details of the main components are specified as follows:

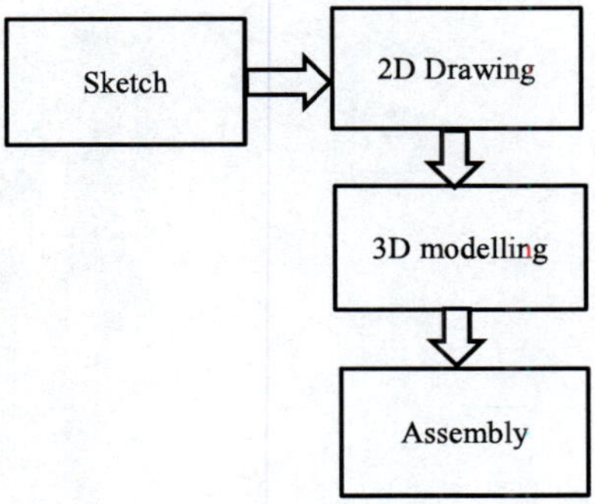

Fig. 2 Design procedure

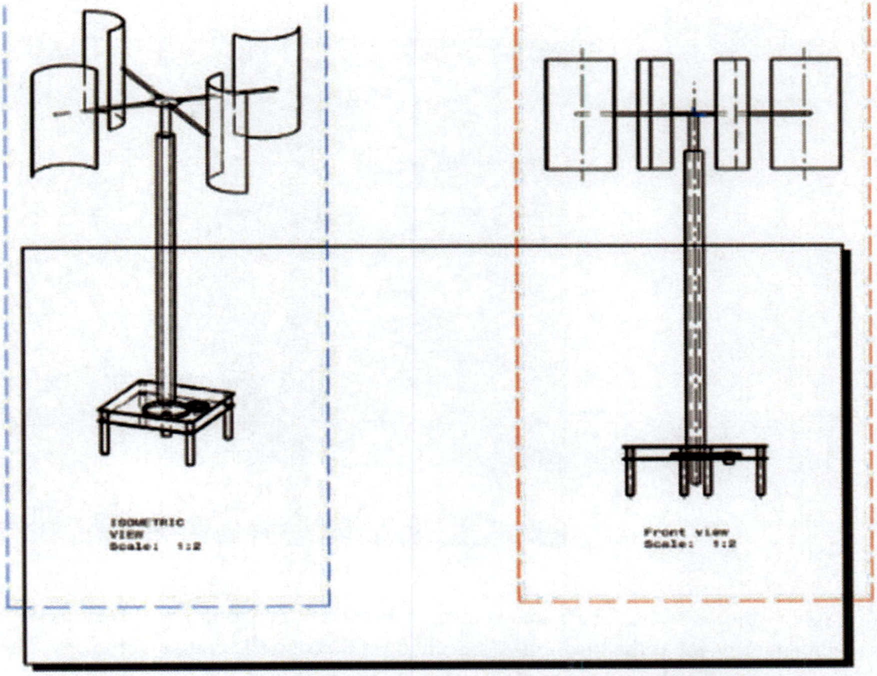

Fig. 3 2D model of assembly

6.1 Solar Pv Panel

A polycrystalline type solar panel is utilised in this model to reduce the overall cost and with a specification of 12V, 10W. The solar panel is mounted on the top and is connected to light detector resistors which helps in maximum absorption of sunlight

Fig. 4 3-Dimensional model

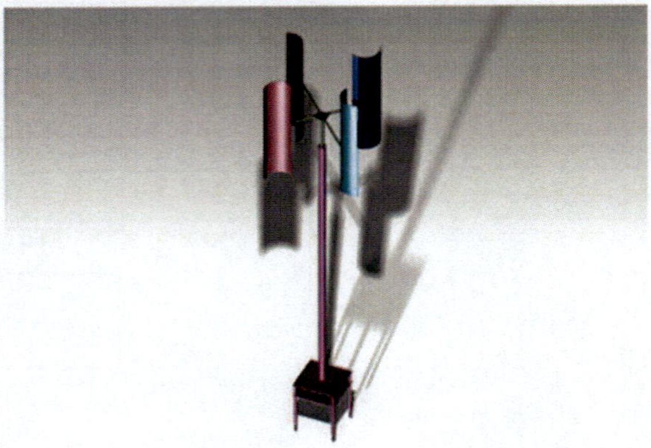

Fig. 5 Solar panel

throughout the day. The solar panel is also connected with a motor which helps with movement of panel as the location of sun changes through the course of the day. The solar panel used in the hybrid model is shown in the (Fig. 5).

6.2 Vertical Axis Wind Turbine (VAWT)

The VAWT's have their shaft axis perpendicular to the ground and hence no yaw mechanism is required. VAWT's have a simplified geometry and are also suitable for small scale applications as it can work even when operated in unstable wind

Fig. 6 Spur gear

conditions. In this hybrid model, the vertical hub wind turbine is connected to a DC dynamo through a gearing system and due to the wind flow from the moving vehicles the turbine blades rotate to generate useful power.

6.3 Gears

Spur gear terminology is utilised which gives certain specifications of the gear such as pitch diameter, circular pitch, contact ratio, addendum, dedendum, etc. which is used for designing a spur gear. In the hybrid model, gear and pinion are designed to provide a driven to driver ratio of 1:7. Out of the two gears, driver is connected to a DC dynamo and driven is connected to shaft which in turn is connected to ball bearing because of its simple design, low cost, carries radial and axial loads and creates less friction torque. It enhances the speed of rotation of shaft and also turbine blades. The following figure shows the driven gear and the driver gear with an attachment for connecting the DC dynamo (Fig. 6).

6.4 Battery

As part of the design of our model, 12 Volt Lead acid battery is used as they are cheap and also easily available. In the hybrid model, battery can be mounted at the base for easier maintenance and utilised for storing energy and it is also connected with the relays through LDR's for the movement of solar panel.

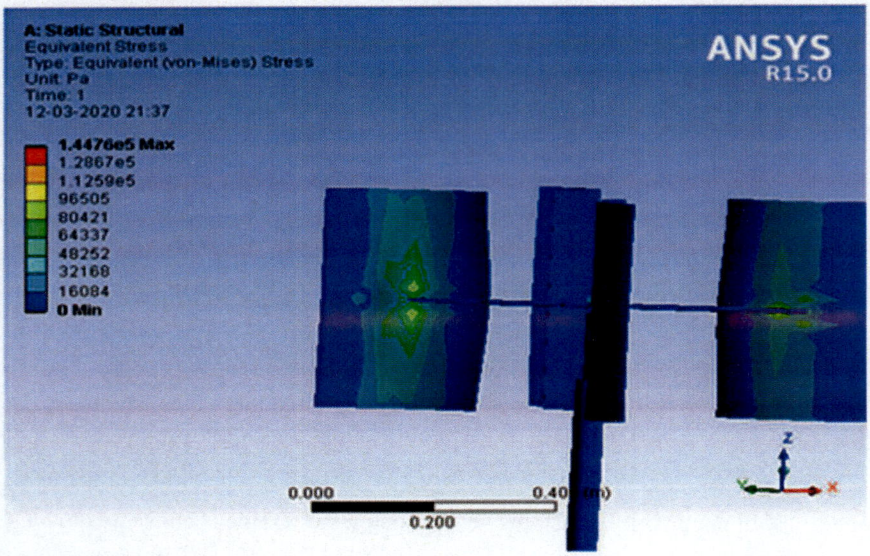

Fig. 7 Stress at 3 m/s

7 Performance Analysis

The Analysis of the 3-dimensional model is conducted in Ansys to study the performance of the model under different wind velocities, forces and pressure conditions. Different mechanical parameters such as stress, strain and deformation acting on the model at different velocities are measured by considering different boundary conditions (x, y, z axis) for effective final prototype. The stress–strain distribution on the 3-dimensional model at different wind speeds is observed in the (Figs. 7, 8, 9 and 10).

Stress–Strain graph at different wind speeds is observed in the figure: (Fig. 11)

8 Result

8.1 Power Generation and Calculation

A proving ground model is developed utilizing VAWT, solar PV panel, gear system, DC dynamo, battery, ESP 32, etc. This section illustrates our hybrid model power generation process which is achieved through both sustainable power sources i.e. solar and wind. The solar panel absorbs the solar radiation through the day and helps in generation of power. It is connected with light detector resistors which senses the light and also connected to the dc motor which helps in the movement

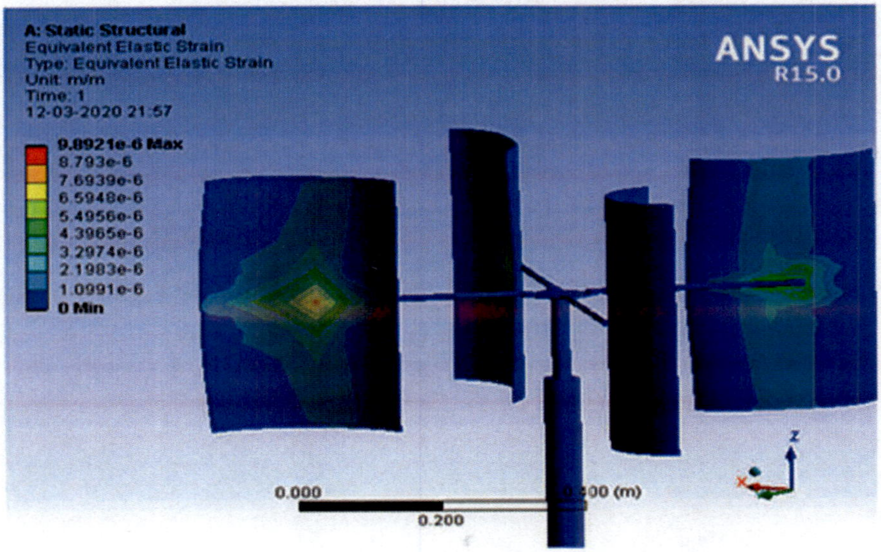

Fig. 8 Strain at 3 m/s

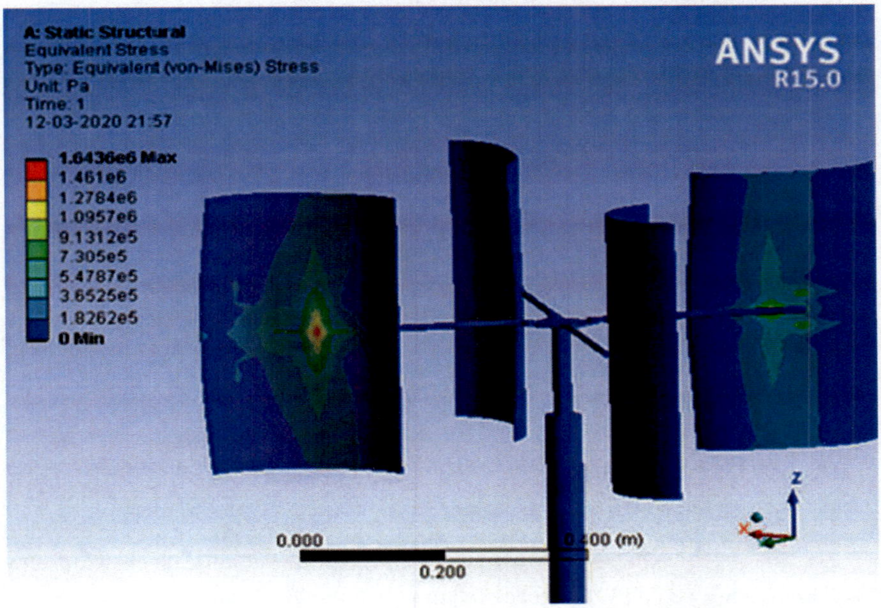

Fig. 9 Stress at 13 m/s

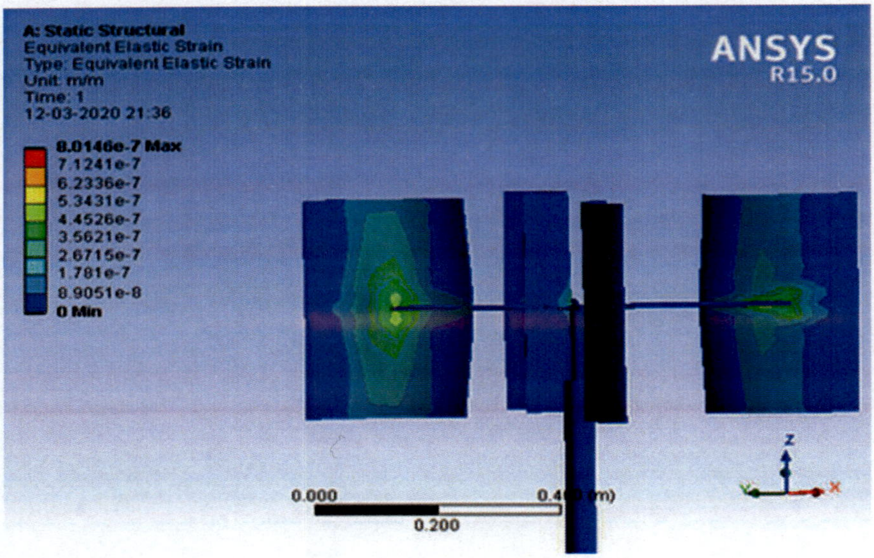

Fig. 10 Strain at 13 m/s

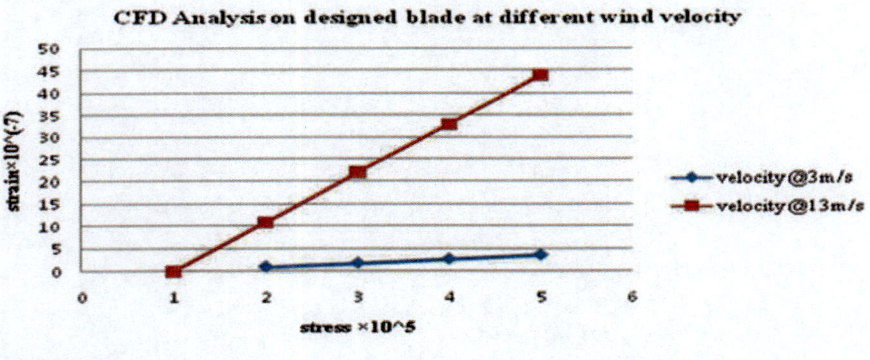

Fig. 11 Stress–strain curve

of solar panel according to the location of the sun, this tracking mechanism helps in maximum absorption of sunlight and adheres to the power generation from the PV panel which is stored in the battery. The VAWT is connected to the dynamo through the gear system. As the blades rotate due to the unobstructed wind flow from the moving vehicles, the gear system enhances the speed of rotation and the power from VAWT is stored in the battery connected to the dynamo. The NodeMCU (ESP32) wi-fi microchip is connected with the battery, solar panel and vertical axis wind turbine. When the internet and mobile hotspot is turned on, the ESP connects to the hotspot and is programmed such that it uploads the battery, solar and wind

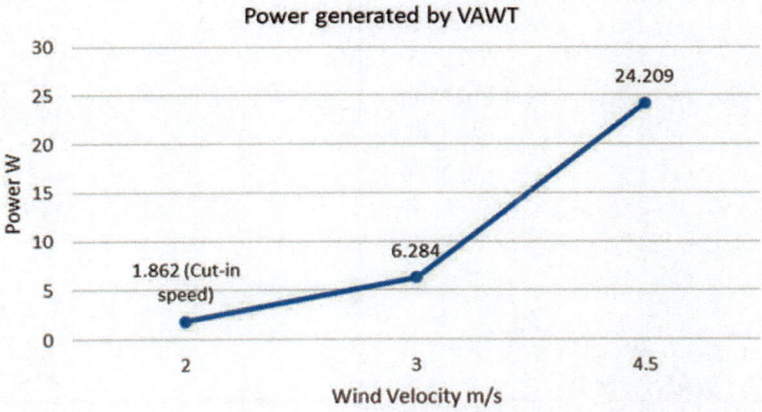

Fig. 12 Power versus wind velocity

voltages respectively to the cloud. The VAWT is designed to start-up at 1.35m/s and start producing power at 2m/s (cut-in). The theoretical power from the VAWT can be arrived at, using the formula below:

$$Pw = 0.5^{*} \rho^{*} S^{*} V^{3} \tag{8}$$

where "*Pw*" denotes the power produced by VAWT, "*V*" denotes the wind speed (m/s), "ρ" air density (kg/m^3) and "*S*" swept area (m^2).

The graph of power v/s wind velocity is observed in the figure: (Fig. 12).

Torque produced by turbine blades,

$$\begin{aligned} \text{Torque } (\tau) &= 0.5^{*} \rho^{*} S^{*} R^{*} V^{2} \\ \tau &= 0.5^{*}1.225^{*}0.38^{*}0.422^{*}2^{2} \\ \tau &= 0.392\, N - m \end{aligned} \tag{9}$$

where "*S*" denotes swept area (m^2), "*V*" denotes wind speed (m/s), "*R*" rotor radius (m) and "ρ" air density (kg/m^3) (Table 1)

Table 1 Turbine parameters

Wind turbine parameters	Wind speed (m/s)		
	2	3	4.5
Torque (N-m)	0.392	0.883	1.988
Power (Watt)	1.862	6.284	24.209
Angular velocity (rad/sec)	4.75	7.109	10.688

$$\begin{aligned}
&\text{Power produced}(P_w) = 0.5^* \rho^* S^* V^3 \\
&P_w = 0.5^* 1.225^* 0.38^* 2^3 \\
&P_w = 1.862\,\text{Watt} \\
&\text{Angular Velocity}(\omega) = P_w/\tau \\
&\omega = 1.862/0.392 \\
&\omega = 4.75\,\text{rad/sec}
\end{aligned} \tag{10}$$

.

8.2 *Hybrid System*

The figure shows completely fabricated hybrid system: (Fig. 13).

9 Conclusion

This paper proposes designing, analysis and fabrication of the hybrid solar and wind turbine for highway power generation in order to contribute to green energy solutions and to reduce the overdependency on stand-alone VAWT and/or solar panel-based solutions. The focus of this work is to combine two sustainable power sources such as wind and solar. The designing of VAWT is done by considering design parameters and analysis is performed for the design to know the mechanical parameters acting on the VAWT and different parts and prototype of the hybrid system was developed. The blades were designed to capture maximum wind from the moving vehicles and were connected to the gear system and generator. Sun tracking mechanism was implemented for solar panel to increase the absorption of solar radiation by the panel eventually helps in more generation of power. This type of small-scale hybrid system can be used for provincial expressway applications as well as for urban highways and is more efficient as compared to the contemporary design.

Fig. 13 Hybrid system

References

1. Subhash, Ragavendhan, Vinoth, S. Krishna, J. Nijnathan, Design and fabrication of highway wind power generation. Int. J. Emerg. Technol. Comput. Sci. Electron. (IJETCSE) **25**(5), (2018) ISSN:0976–1353
2. S.A. Kulkarni, M.R. Birajdar, Vertical axis wind turbine for highway application. Imperial J. Interdisciplinary Res. (IJIR) **2**(10) (2016)
3. K. Chavda, T. Thakar, J. Parekh, J. Satwara, H. Panchal, Design and fabrication of highway wind turbine. Int. J. Sci. Res. Develop. (IJSRD) **1**(10) (2013)
4. A. Bavchakar, K.N. Chougale, S.S. Belanekar, S.P. Rane, N.B. Sawant, A hybrid model of vertical axis wind turbine—solar power generation for highway and domestic application. in *International Conference on Computation of Power Energy Information and Communication (ICCPEIC)*, IEEE (2018)

5. M.A. Rahman, M.Y. Mukta, A. Yousuf, A.T. Asyhari, M.Z.A. Bhuiyan, C.Y. Yaakub, IoT based hybrid green energy system driven highway lighting system. in *International Conference on Dependable Autonomic and Secure Computing, International Conference on Pervasive Intelligence and Computing*, IEEE (2019)
6. S. Brusca, R. Lanzafame, M. Messina, Design of a vertical axis wind turbine: how the aspect ratio affects the turbines performance. Int. J. Energy Environ Eng **5**, 333–340 (2014)
7. A. Pandey, J.P. Sharma, Power generation and automation system designing for highway. IEEE, (2015)
8. P. Desai, M. Mahida, K. Patel, A. Chauhan, Hybrid energy generation on national highway 6, India. Int. J. Adv. Eng. Res. Develop. **4**(4), (2017)
9. A. FauziSaat, N. Rosly, Aerodynamic analysis of vertical axis wind turbine. J. Aviation Aerospace Technol. **1**(1), 1–6 (2019)
10. Z. Li, R. Han, P. Gao, C. Wang, Analysis and implementation of a drag-type vertical-axis wind turbine for small distributed wind energy systems, (2019)
11. V.M.N. Punugu, T.S. Kabra, Design and analysis of adaptable flexed-cup vertical axis wind turbine. IJEDR. **6**, (2018)
12. P. Malge, T. Ganesha, Study and analysis of savonious vertical axis wind turbine with neodymium permanent magnet rotor. Int. Res. J. Eng. Technol. (IRJET) **04**, (2017)
13. https://en.wikipedia.org/wiki/Hybrid_renewable_energy_system
14. J. Earnest, in *Wind Power Generation*, 2nd edn. (PHI learning private ltd, 2015)
15. https://en.wikipedia.org/wiki/Metal_fabrication
16. J.B.K. Das, P.L. Srinivas Murthy, in *Design of machine elements-1* (2014)
17. J.B.K. Das, P.L. Srinivas Murthy, in *Design of machine elements-2* (2014)
18. J. Castillo, Small-scale vertical axis wind turbine design (2011)

Design and Modelling of 1 kW, 200–400 V, Multiphase Boost Converter

A. Soubhagya, Ravikiran Rao M, and Suryanarayana K.

1 Introduction

Many high-power applications require boost converters as an integral interface between the available low voltage sources and the output loads, which operate at higher voltages. Based on the type of application, either single-phase or multiphase boost converters are used. In single-phase boost converter for higher power applications, voltage and current stress on the switch increases, and also, voltage and current ripples in the output and input side increase leading to huge filter requirement and increased EMI issues. As a result, losses in the system increase causing poor efficiency and higher cost. This issue can be solved by increasing the number of phases (multiphase topology), which divides the current according to the number of phases and reduces the stress on switches and ripples in the input and output parameters.

Silicon switches suffer from drawbacks such as low bandgap energy, switching frequency limitations and low thermal conductivity. Wide bandgap semiconductors, such as silicon carbide (SiC) and gallium nitride (GaN), provide larger bandgaps, higher breakdown electric field and higher thermal conductivity. SiC switches provide higher blocking voltage, higher junction temperature and higher switching frequencies while compared to Si switches [1]. Operation of the system at higher switching frequency reduces the size of passive components required in the converter, and size of heat sink required will be less because of higher thermal conductivity [2].

A. Soubhagya (✉) · Ravikiran Rao M · Suryanarayana K.
Department of Electrical and Electronics Engineering, NMAM Institute of Technology, Nitte, Karkala, India

Ravikiran Rao M
e-mail: ravikiranraom@nitte.edu.in

Suryanarayana K.
e-mail: suryanarayana@nitte.edu.in

Sanjeevikumar et al. (eds.), *Advances in Renewable Energy and Electric Vehicles*, Lecture Notes in Electrical Engineering 767,
https://doi.org/10.1007/978-981-16-1642-6_10

In this paper, a 1 kW, 200–400 V, multiphase boost converter which is quite suitable to use in electric vehicles, battery charging, motor drivers, PV systems, etc., is designed and modelled, and the improved efficiency and reliability of the system due to SiC switch are also observed. The main focus is on reducing the ripples, losses and cost of the system. The PWM signals and control algorithms required for the converter are generated using the microcontroller MC56F84789. It has a full set of programmable peripherals such as PWM, ADC, DAC and timers [3].

Section 2 presents the design parameters and conditions of the system and also the reasons to choose multiphase topology. Sections 3 and 4 present state space modelling and small signal analysis, respectively, and also arrive at the transfer function of the converter. Controller design conditions and calculation of parameters are shown in Sect. 5. The simulation and hardware implementation of the system are presented in Sects. 6 and 7, respectively. Section 8 presents the conclusion about the converter and results.

2 System Overview

The system under consideration is a 200–400 V, 1 kW boost converter. To achieve the desired goal of highly efficient and reliable system, a boost converter will be developed using SiC devices which will help in improving the system efficiency. Desired control signals will be generated using in-house developed control card that houses NXP make MC56F84789 controller. The signals from controller are fed to a driver board which houses driver IXDN609SI IC. The input power and necessary isolated power supply are generated from 100 to 300 V flyback SMPS designed in-house. The block diagram of the proposed converter is shown in Fig. 1. To achieve rugged and stable operation of the system, voltage and current at input and output are continuously monitored using signal conditioning units and Hall effect sensors. These feedback signals help to achieve complete control of the system and also to regulate the output to desired value by generating appropriate PWM signals.

Employing multiphase provides advantages that include

1. Thermal performance related to conduction losses of the power supply is proportional to square of current. If multiple phases are used, these losses can be reduced.
2. The system will be more compact in size.
3. Ripple current cancellation
4. Improved transient response
5. Optimized efficiency over the load range
6. Lower output ripple voltage [4].

The design requirements are as in Table 1.

The calculated values of the components /parameters of the system are shown in Table 2 [5, 6].

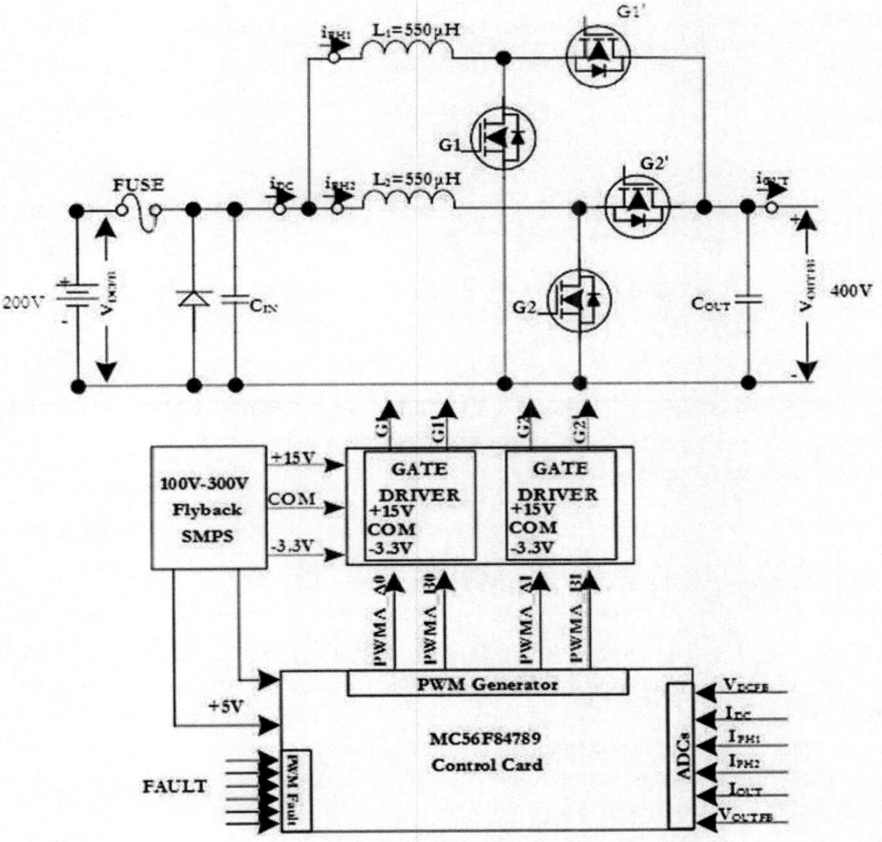

Fig. 1 Block diagram of the system

Table 1 Design parameters of the converter

Sl.No	Parameter	Description	Value
1	V_{in}	Input voltage	200 V
2	V_{out}	Output voltage	400 V
3	P_o	Output power	1 kW
4	f_s	Switching frequency	100 kHz
5	η	Efficiency (assumed)	95%
6	ΔI	Ripple current (assumed)	30%

3 State Space Model

An efficient mathematical modelling of the system plays a crucial role in optimal and accurate design of the controller.

Table 2 Values of components/parameters of the converter

Sl. No.	Parameters	Equation	Value
1	Duty cycle	$D = 1 - \frac{\eta * V_{in}}{V_o}$	0.5725
2	Input current	$I_{in} = \frac{P_o}{\eta * V_o}$	5.84 A
3	Average current per phase	$I_{ph} = \frac{I_{in}}{2}$	2.92 A
4	Inductor per phase	$L = \frac{V_i * D}{\Delta I * f_s}$	555 μH
5	Capacitor	$C = \frac{V_o^* D}{\Delta V^* R_L^* 2^* f_s}$	10 μF
6	Resistor	$R_L = \frac{V_o^2}{P_o}$	160 Ω

It is clear that there is a strong need for a fast and accurate model that can tackle the many difficult design trade-offs and also ensure accuracy at the limits. One such method is state space modelling.

3.1 Single-Phase Boost Converter

Consider the boost converter circuit shown in Fig. 2 with source voltage v_g, source current i_g, capacitor current i_c, capacitor voltage v_c, output voltage v_o and output current i_o.

Let the state variables be inductor current i_L and capacitor voltage v_c and output variables be v_o and i_c [7–10].

When the switch S is in ON condition,

$$L\frac{\mathrm{d}i_L(t)}{\mathrm{d}t} = v_g \tag{1}$$

$$C\frac{\mathrm{d}v_c(t)}{\mathrm{d}t} + \frac{v_c}{R} + i_0 = 0 \tag{2}$$

Rearranging Eqs. 1 and 2,

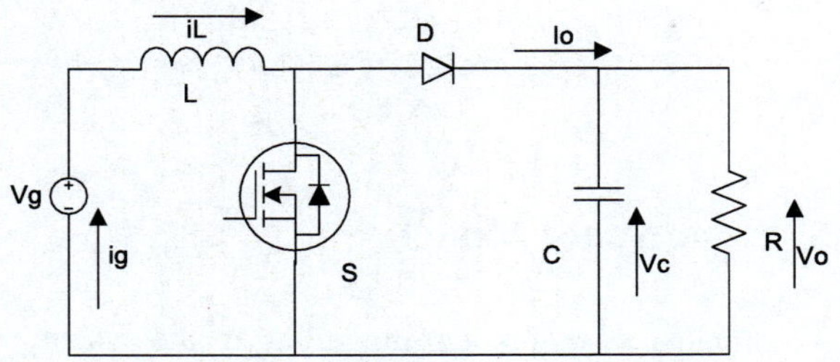

Fig. 2 Single-phase boost converter circuit

$$\frac{\mathrm{d}i_L(t)}{\mathrm{d}t} = \frac{v_g}{L} \tag{3}$$

$$\frac{\mathrm{d}v_c(t)}{\mathrm{d}t} = -\frac{v_c}{RC} - \frac{i_0}{C} \tag{4}$$

Arranging the Eqs. 3 and 4,

$$\begin{bmatrix} \frac{\mathrm{d}i_L(t)}{\mathrm{d}t} \\ \frac{\mathrm{d}v_c(t)}{\mathrm{d}t} \end{bmatrix} = \begin{bmatrix} 0 & 0 \\ 0 & -1/RC \end{bmatrix} \begin{bmatrix} i_L \\ v_c \end{bmatrix} + \begin{bmatrix} 1/L \\ 0 \end{bmatrix} v_g + \begin{bmatrix} 0 \\ -1/C \end{bmatrix} \tag{5}$$

When the switch is in OFF condition,

$$L\frac{\mathrm{d}i_L(t)}{\mathrm{d}t} = v_g - v_c \tag{6}$$

$$C\frac{\mathrm{d}v_c(t)}{\mathrm{d}t} + \frac{v_c}{R} + i_0 - i_L = 0 \tag{7}$$

Rearranging Eqs. 6 and 7,

$$\begin{bmatrix} \frac{\mathrm{d}i_L(t)}{\mathrm{d}t} \\ \frac{\mathrm{d}v_c(t)}{\mathrm{d}t} \end{bmatrix} = \begin{bmatrix} 0 & -1/L \\ -1/C & -1/RC \end{bmatrix} \begin{bmatrix} i_L \\ v_c \end{bmatrix} + \begin{bmatrix} 1/L \\ 0 \end{bmatrix} v_g + \begin{bmatrix} 0 \\ -1/C \end{bmatrix} i_0 \tag{8}$$

The output equations during ON and OFF conditions can be, respectively, written as

$$\begin{matrix} v_o \\ i_g \end{matrix} = \begin{pmatrix} 0 & 1 \\ 1 & 0 \end{pmatrix} \begin{matrix} i_L \\ v_c \end{matrix} \tag{9}$$

Equations 5 and 8 are in the form

$$\frac{\mathrm{d}x(t)}{\mathrm{d}t} = Ax(t) + Bu(t)$$

And Eq. 9 in the form

$$y(t) = Cx(t) + Eu(t)$$

Therefore, to obtain the final form,

$$\begin{aligned} A &= A_1 d + A_2(1-d) \\ B &= B_1 d + B_2(1-d) \\ C &= C_1 d + C_2(1-d) \\ E &= E_1 d + E_2(1-d) \end{aligned}$$

where d is the duty ratio of the converter.

By substituting the values, the state space equation can be written as

$$\begin{bmatrix} \frac{\mathrm{d}i_L(t)}{\mathrm{d}t} \\ \frac{\mathrm{d}v_c(t)}{\mathrm{d}t} \end{bmatrix} = \begin{bmatrix} 0 & \frac{-(1-d)}{C} \\ \frac{(1-d)}{C} & -1/RC \end{bmatrix} \begin{matrix} i_L \\ v_c \end{matrix} + \begin{bmatrix} \frac{1}{L} \\ 0 \end{bmatrix} v_g + \begin{bmatrix} 0 \\ \frac{1}{-C} \end{bmatrix} i_O \tag{10}$$

This equation is also known as average large signal model.

The output equation remains the same as shown in Eq. (9), since $E = 0$.

3.2 *Multiphase Boost Converter*

Consider the multiphase boost converter circuit shown in Fig. 3 with source voltage v_g, source current i_g, capacitor current i_c, capacitor voltage v_c, output voltage v_o and output current i_o. Let the state variables be capacitor voltage v_c and inductor currents of phase 1 and 2, i.e. $i_{L1,}$ $i_{L2,}$ respectively, and output variables be v_o and i_c [7–10].

The operation of two-phase boost converter can be divided into four modes. Consider the state space variables similar to used in single-phase boost converter.

(1) *Mode 1*

When switch $S1$ is in ON condition and Mode 12 is in OFF condition,

$$L_1 \frac{\mathrm{d}i_{L1}(t)}{\mathrm{d}t} = v_g \tag{11}$$

Since Mode 12 is in OFF position,

$$L_2 \frac{\mathrm{d}i_{L2}(t)}{\mathrm{d}t} = v_c \tag{12}$$

$$i_{L2} = C \frac{\mathrm{d}v_c(t)}{\mathrm{d}t} + \frac{v_c}{R} \tag{13}$$

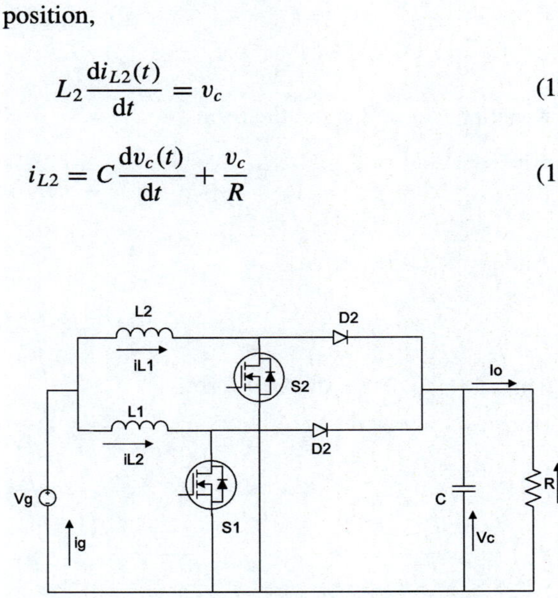

Fig. 3 Multiphase boost converter circuit

$$\frac{dv_c(t)}{dt} = \frac{i_{L2}}{C} - \frac{v_c}{RC} \tag{14}$$

(2) *Mode 2*

When both the switches $S1$ and $S2$ are in OFF condition,

$$L_1 \frac{di_{L1}(t)}{dt} = v_c \tag{15}$$

$$L_2 \frac{di_{L2}(t)}{dt} = v_c \tag{16}$$

$$i_{L1} + i_{L2} = C\frac{dv_c(t)}{dt} + \frac{v_c}{R} \tag{17}$$

(3) *Mode 3*

When switch $S1$ is in ON condition and $S2$ is in OFF condition,

$$L_2 \frac{di_{L2}(t)}{dt} = v_g \tag{18}$$

Since $S1$ is in OFF condition,

$$L_1 \frac{di_{L1}(t)}{dt} = v_c \tag{19}$$

$$i_{L1} = C\frac{dv_c(t)}{dt} + \frac{v_c}{R} \tag{20}$$

$$\frac{dv_c(t)}{dt} = \frac{i_{L1}}{C} - \frac{v_c}{R} \tag{21}$$

(4) *Mode 4*

It follows the same procedure as of mode 2 resulting in same set of equations. Mode 1 can be rearranged and written as

$$\begin{bmatrix} \frac{di_{L1}(t)}{dt} \\ \frac{di_{L2}(t)}{dt} \\ \frac{dv_c(t)}{dt} \end{bmatrix} = \begin{bmatrix} 0 & 0 & 0 \\ 0 & 0 & 1/L2 \\ 0 & 1/C & -1/RC \end{bmatrix} \begin{bmatrix} i_{L1} \\ i_{L2} \\ v_c \end{bmatrix} + \begin{bmatrix} 1/L1 \\ 0 \\ 0 \end{bmatrix} v_g + \begin{bmatrix} 0 \\ 0 \\ 1/C \end{bmatrix} i_0 \tag{22}$$

Mode 2 and Mode 4 can be written as

$$\begin{bmatrix} \frac{di_{L1}(t)}{dt} \\ \frac{di_{L2}(t)}{Et} \\ \frac{dv_c(t)}{dt} \end{bmatrix} = \begin{bmatrix} 0 & 0 & 1/L1 \\ 0 & 0 & 1/L2 \\ 0 & 0 & -1/RC \end{bmatrix} \begin{bmatrix} i_{L1} \\ i_{L2} \\ v_c \end{bmatrix} + \begin{bmatrix} 0 \\ 0 \\ 0 \end{bmatrix} v_g + \begin{bmatrix} 0 \\ 0 \\ 1/C \end{bmatrix} i_0 \tag{23}$$

Mode 3 can be written as

$$\begin{bmatrix} \frac{di_{L1}(t)}{dt} \\ \frac{di_{L2}(t)}{dt} \\ \frac{dv_c(t)}{dt} \end{bmatrix} = \begin{bmatrix} 0 & 0 & 1/L1 \\ 0 & 0 & 0 \\ 1/C & 0 & -1/RC \end{bmatrix} \begin{bmatrix} i_{L1} \\ i_{L2} \\ v_c \end{bmatrix} + \begin{bmatrix} 0 \\ 1/L2 \\ 0 \end{bmatrix} v_g + \begin{bmatrix} 0 \\ 0 \\ 1/L1 \end{bmatrix} i_0 \quad (24)$$

To obtain the complete state space matrix, by referring to Eqs. (22), (23), (24).

$$\text{We have } A1 = \begin{bmatrix} 0 & 0 & 0 \\ 0 & 0 & 1/L2 \\ 0 & 1/C & -1/RC \end{bmatrix}$$

$$A2 = A4 = \begin{bmatrix} 0\,0 & 1/L1 \\ 0\,0 & 1/L2 \\ 0\,0 & -1/RC \end{bmatrix}, \; A3 = \begin{bmatrix} 0\,0 & 1/L1 \\ 0\,0 & 1/L2 \\ 0\,0 & -1/RC \end{bmatrix}$$

$$B1 = \begin{bmatrix} 1/L1 \\ 0 \\ 0 \end{bmatrix} B2 = B4 = \begin{bmatrix} 0 \\ 0 \\ 0 \end{bmatrix}, \; B3 = \begin{bmatrix} 0 \\ 1/L2 \\ 0 \end{bmatrix}$$

Therefore,

$$A = A_1 d_1 + A_2 d_2 + A_3 d_3 + A_4 d_4$$
$$B = B_1 d_1 + B_2 d_2 + B_3 d_3 + B_4 d_4$$

d_1, d_2, d_3, d_4 are duty ratio in mode 1, mode 2, mode 3, mode 4, respectively. By substituting respective values

$$\begin{bmatrix} \frac{di_{L1}(t)}{dt} \\ \frac{di_{L2}(t)}{dt} \\ \frac{dv_c(t)}{dt} \end{bmatrix} = \begin{bmatrix} 0 & 0 & \frac{(1-d_1)}{L1} \\ 0 & 0 & \frac{(1-d_3)}{L2} \\ \frac{(1-d_1)}{C} & \frac{(1-d_3)}{C} & -1/RC \end{bmatrix} \begin{bmatrix} i_{L1} \\ i_{L2} \\ v_c \end{bmatrix} + \begin{bmatrix} \frac{d_1}{L1} \\ \frac{d_3}{L2} \\ 0 \end{bmatrix} v_g + \begin{bmatrix} 0 \\ 0 \\ 1/C \end{bmatrix} i_0 \quad (25)$$

4 Small Signal Model

For single-phase boost converter, at the equilibrium condition, $\frac{dx(t)}{dt} = 0, d = D, v_g = V_g, i_g = I_g$, accordingly [10].
Therefore, Eq. (10) becomes

$$\begin{matrix} 0 \\ 0 \end{matrix} = \begin{pmatrix} 0 & \frac{-(1-D)}{C} \\ \frac{(1-D)}{C} & -1/RC \end{pmatrix} \begin{matrix} I_L \\ V_c \end{matrix} + \begin{matrix} \frac{1}{L} \\ 0 \end{matrix} V_g + \begin{matrix} 0 \\ \frac{1}{-C} \end{matrix} I_0 \quad (26)$$

This is in the form of

$$AX + BU = 0$$

Small signal model is obtained by substituting in average large signal model, for every variable a steady state part and a small signal variation about the equilibrium point,

i.e.

$$\begin{aligned} d &= D + d^{\wedge} \\ v_o &= V_o + v_o^{\wedge} \\ v_g &= V_g + v_g^{\wedge} \\ i_o &= I_o + i_o^{\wedge} \\ i_L &= I_L + i_L^{\wedge} \end{aligned}$$

Since $AX + BU = 0$ at equilibrium, Eq. 10 can be written as

$$\begin{matrix} \frac{di_L^{\wedge}(t)}{dt} \\ \frac{dv_c^{\wedge}(t)}{dt} \end{matrix} = \begin{pmatrix} 0 & \frac{-(1-D-d^{\wedge})}{C} \\ \frac{(1-D-d^{\wedge})}{C} & -1/RC \end{pmatrix} \begin{matrix} I_L + i_L^{\wedge} \\ V_c + v_c^{\wedge} \end{matrix} + \begin{matrix} \frac{1}{L} \\ 0 \end{matrix} (V_g + v_g^{\wedge}) + \begin{matrix} 0 \\ \frac{1}{-C} \end{matrix} (I_o + i_o^{\wedge}) \tag{27}$$

By simplification,

$$\begin{matrix} \frac{di_L^{\wedge}(t)}{dt} \\ \frac{dv_c^{\wedge}(t)}{dt} \end{matrix} = \begin{pmatrix} 0 & \frac{-(1-D)}{C} \\ \frac{(1-D)}{C} & -1/RC \end{pmatrix} \begin{matrix} i_L^{\wedge} \\ v_c^{\wedge} \end{matrix} + \begin{matrix} V_c/L \\ (-I_L)/C \end{matrix} \mathrm{d}^{\wedge} + \begin{matrix} \frac{1}{L} \\ 0 \end{matrix} v_g^{\wedge} + \begin{matrix} 0 \\ \frac{1}{-C} \end{matrix} i_o^{\wedge} \tag{28}$$

Therefore, state equation of small signal model is

$$\begin{matrix} \frac{di_L^{\wedge}(t)}{dt} \\ \frac{dv_c^{\wedge}(t)}{dt} \end{matrix} = \begin{pmatrix} 0 & \frac{-(1-D)}{C} \\ \frac{(1-D)}{C} & -1/RC \end{pmatrix} \begin{matrix} i_L^{\wedge} \\ v_c^{\wedge} \end{matrix} + \begin{matrix} \frac{1}{L} & 0 & V_c/L \\ 0 & \frac{1}{-C} & \frac{(-I_L)}{C} \end{matrix} \begin{matrix} v_g^{\wedge} \\ i_o^{\wedge} \\ \mathrm{d}^{\wedge} \end{matrix} \tag{29}$$

Similarly, for the two-phase boost converter,

$$\begin{bmatrix} \frac{di_{L1}^{\wedge}(t)}{dt} \\ \frac{di_{L2}^{\wedge}(t)}{dt} \\ \frac{dv_c^{\wedge}(t)}{dt} \end{bmatrix} = \begin{bmatrix} 0 & 0 & \frac{(1-D_1)}{L1} \\ 0 & 0 & \frac{(1-D_3)}{L2} \\ \frac{(1-D_1)}{C} & \frac{(1-D_3)}{C} & -1/RC \end{bmatrix} \begin{bmatrix} i_{L1}^{\wedge} \\ i_{L2}^{\wedge} \\ v_c^{\wedge} \end{bmatrix} + \begin{bmatrix} \frac{d_1}{L1} & 0 & V_c/L1 \\ \frac{d_3}{L2} & 0 & \frac{(-I_{L2})}{C} \\ 0 & 1/C & 0 \end{bmatrix} \begin{bmatrix} v_g^{\wedge} \\ i_o^{\wedge} \\ d^{\wedge} \end{bmatrix} \tag{30}$$

We have state space model given by

$$\dot{x} = Ax + Bu \tag{31}$$

$$y = Cx \tag{32}$$

Taking Laplace transform of the above equations,

$$sX(s) = AX(s) + BU(s) \tag{33}$$

$$Y(s) = CX(s) \tag{34}$$

By further simplification,

$$Y(s) = C(sI - A)^{-1}BU(s) \tag{35}$$

Therefore, by substituting corresponding values from Eqs. (26) and (30), the transfer function of output voltage to duty ratio for both single and multiphase can be obtained as

- Single-phase converter

$$\frac{\hat{v}_o(s)}{\hat{d}(s)} = \frac{D'v_c - sI_L L}{L\left[s^2 + \frac{s}{RC} + D'^2\right]} \tag{36}$$

where $D' = 1{-}D$.

- Multiphase converter

$$\frac{\hat{v}_o(s)}{\hat{d}(s)} = \frac{RL_1L_2\left[\frac{D_1' * v_c}{L_1} - \frac{D_3' * I_{L2}}{C}\right]s}{s^3L_1L_2 + \left(L_1L_2 - D'^2_1 RL_2\right)s + RL_1D'^2_3} \tag{37}$$

where $D_1' = 1{-}D_1$ and $D_3' = 1{-}D_3$.

5 Controller Design

To maintain the constant and stable output of the system, closed loop controllers are employed. Controllers are designed in such a way to regulate the output parameters based on the changes in input and output conditions (Fig. 4).

V_g, V_o, d are the input voltage, output voltage and duty cycle of the system, respectively. The feedback of the parameter to be controlled is taken and compared with the reference value, and this gives error output. The ideal control occurs when there is zero error, but in practical scenario, it is quite impossible [10].

If K is controller gain,v_c is control voltage, then

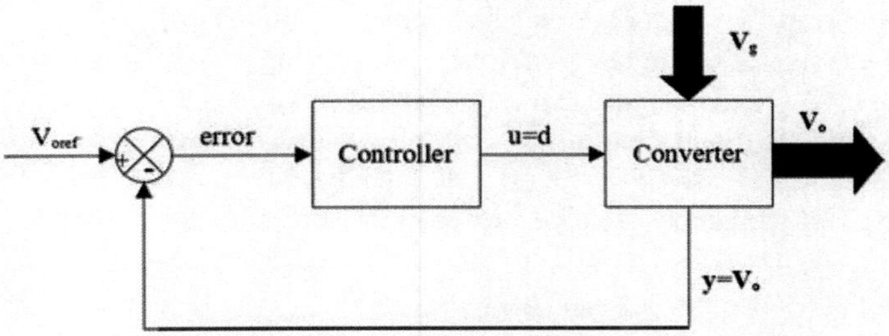

Fig. 4 Block diagram of generic controller

$$e = \frac{v_c}{K}$$

For e = 0, either $v_c = 0$ or $K =$ infinity. But $v_c = 0$ will result in broken loop, i.e. there will be no enclosed negative feedback resulting in no use of controller. So, to have lesser error value, K should be very high. Therefore, based on K and error value, different types of controllers such as P, I, PI, PD, PID are designed. For the multiphase boost converter under consideration (Sect. 2), a PI controller is designed.

The controller can be designed using different methods.

- Tuning the controller directly on the system (hardware or simulation), also known as Ziegler–Nicholas technique
- By formal approach, i.e. methods such as root locus technique are used to deduce the controller parameters and then included in hardware or simulation. This method is mainly used for non-minimum phase system such as boost converter. Initially, the state space equations and small signal model of the plant are determined. The root locus of the system is found, and the system is tested for satisfactory step response. Later, the values from these calculations are used to determine the controller parameters.

Using a PI controller provides better transient response and nearly zero steady state error. It also has the advantages such as removal of high frequency noise, improved phase margin and gain margin. Thus, for the proposed multiphase boost converter, a PI controller is designed using Ziegler–Nicholas technique.

For PI controller,

$$v_c = K_p + K_i \int e \, \mathrm{d}t$$

where K_p is proportionality scaling factor, and K_i is integrator scaling factor. For tuning the controller,

- Set up the open loop system.
- Introduce limiters across the integrator and v_c. This is done so that v_c swing can fall in sync with that of PWM.

- Keeping the system in open loop, check phase relationship.
  Let $K_i = 0$ and $K_p =$ finite value. Phase relationship is stable if y increases with increase in $V_{oref}$and decreases with decrease in $V_{oref.}$
- Now, enable the integration initially with low value by keeping $K_p = 0$. Slowly increase the value, till the system reaches steady state.
- After the system reaches steady state, calculate the value of K_p and K_i.

$$K_p = 0.45 * K_u$$

$$K_i = \frac{K_p}{T_u}$$

where K_u is ultimate gain, and T_u is oscillation period.

6 Simulation

6.1 *Simulation of Open Loop System*

The simulation is conducted in MATLAB/Simulink, and the circuit is as shown in Fig. 5. The system is designed to have an input voltage of 200 V, with duty ratio of 57.25% and switching frequency of 100 kHz.

The input and output DC currents and the phase currents through inductors shown in Fig. 6, depicts that with two-phase topology shows the ripple at input side is minimum.

Gating sequence of the system is designed such that Q2 pulses are delayed by an angle of 180° with respect to Q1 which results in reduction in the input current ripple (Fig. 7).

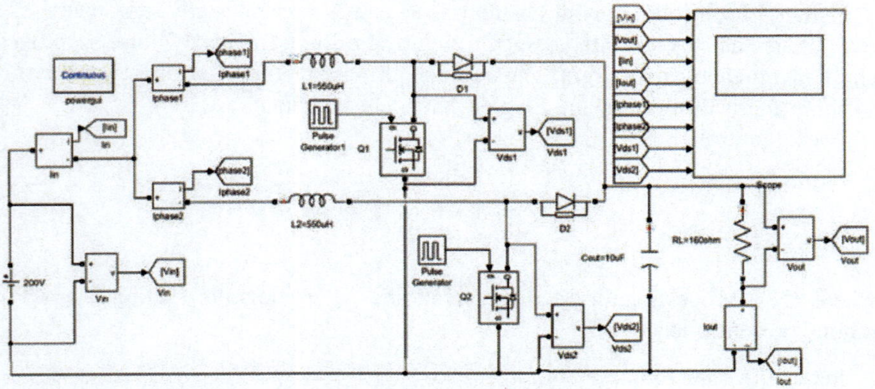

Fig. 5 Simulated circuit of multiphase boost converter (open loop)

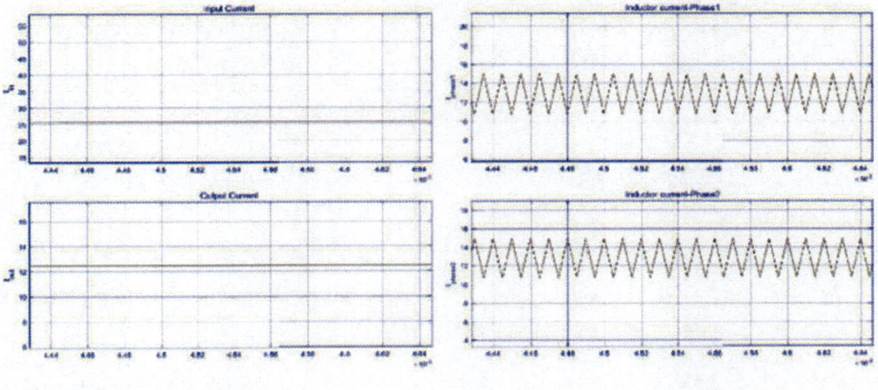

Fig. 6 Current waveforms

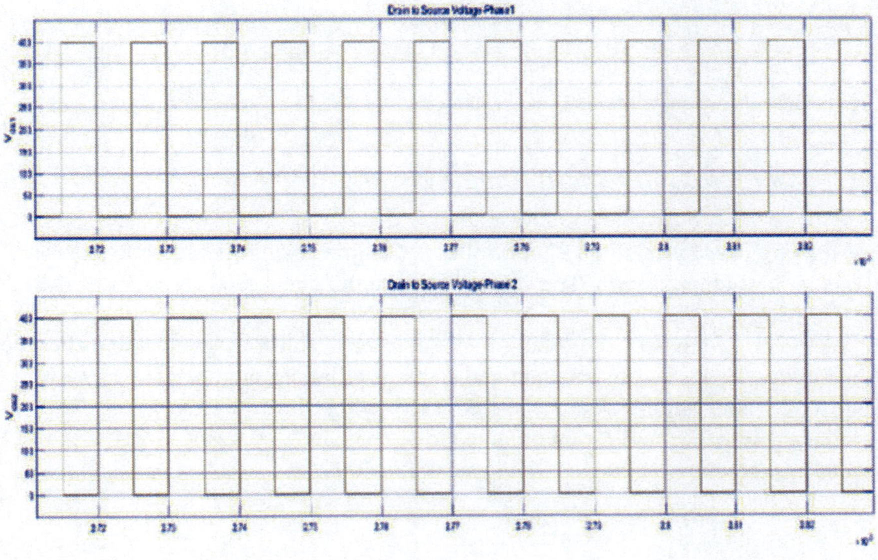

Fig. 7 Drain to source voltages of phase 1 and phase 2

From the results of output voltage, it can also be concluded that output ripples are comparatively less. The output voltage ripple is found to be 1.61% which is in the limits of 1% (Fig. 8).

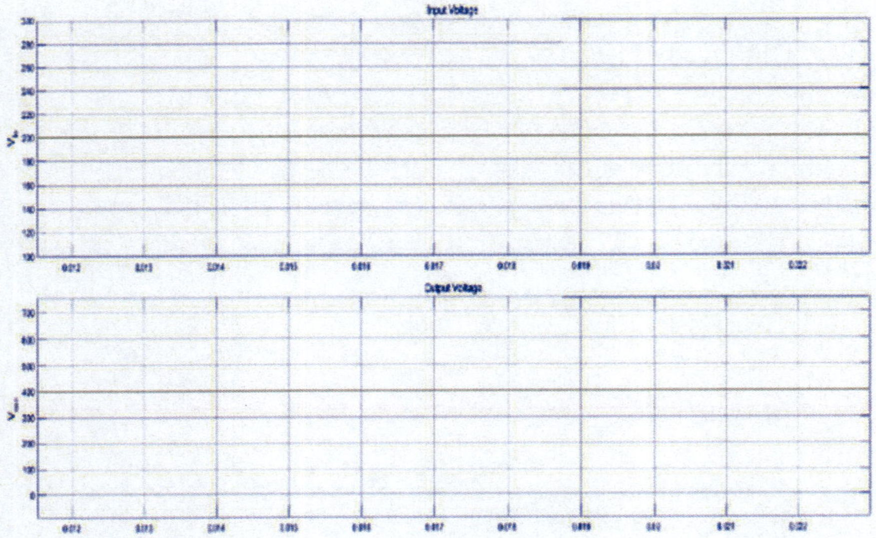

Fig. 8 Input and output voltages

6.2 Simulation of Closed Loop System

For the proposed system, the PI controller is designed using Ziegler–Nichols technique in Simulink platform. The values of the components used are as mentioned in Sect. 2 (Fig. 9).

As mentioned in Sect. 5, initially K_i is set to zero, and K_p is set to 1 and checked for stability. At $K_p = 1.3$, constant and stable oscillations are obtained as shown in Fig. 10. For this value of K_p, $K_u = 1.3$ and $T_u = 0.001$ s.

The values of K_p and K_i are obtained to be 0.58 and 700, respectively. Figure 11 represents the output voltage of the system, i.e. 400 V.

7 Hardware Implementation

The system is implemented on PCB for higher switching frequency operation by doing schematics followed by layout using OrCad tool. This system is used in an in-house developed electric vehicle for battery charging. The required PWM signals for the switches are provided by the microcontroller MC56F84789. Inherent features such as short circuit protection, protection from inrush current and soft start are implemented. Input to the system can be a DC supply or solar panel. The implemented multiphase boost converter is shown in Fig. 12.

The existing system has been implemented with hysteresis control. The required control signals are generated using MC56F84789, and low pass filter of 40 kHz

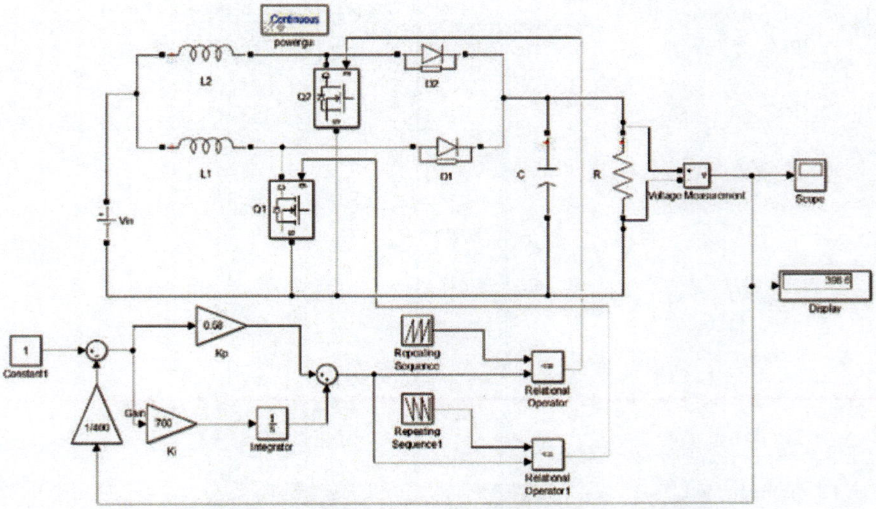

Fig. 9 Simulated model of the system

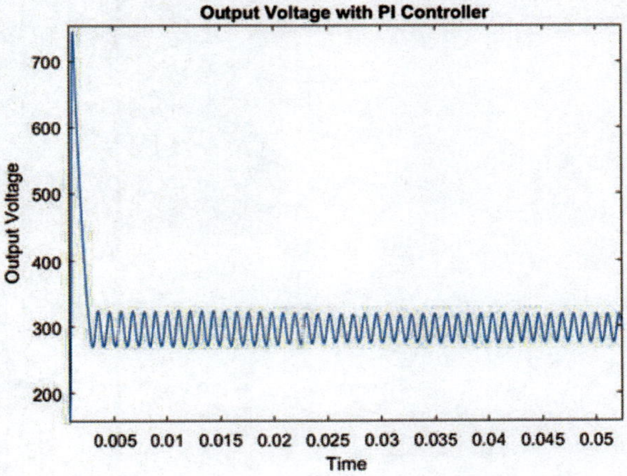

Fig. 10 Output voltage with $K_p = 1.3, K_i = 0$

frequency is used for signal conditioning. The output voltage is thus stabilized. Figures 13 and 15 represent the outputs obtained under hysteresis control.

Multiphase topology of the converter is mainly chosen for the reason of reducing ripples in input and output side. It is proven that delaying the sequence of gating pulses by 180° in two-phase converter results in reduction in the input current ripple. This fact is realized both in simulation and practical mode. Figure 14 shows gating pulse, inductor currents in two phases and input current.

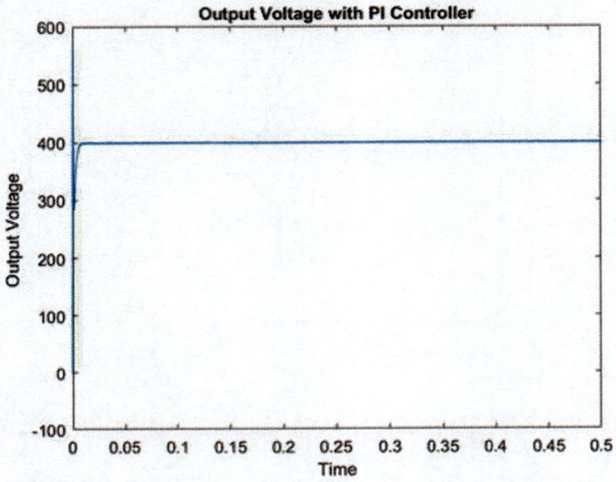

Fig. 11 Output voltage with $K_p = 0.58$, $K_i = 700$

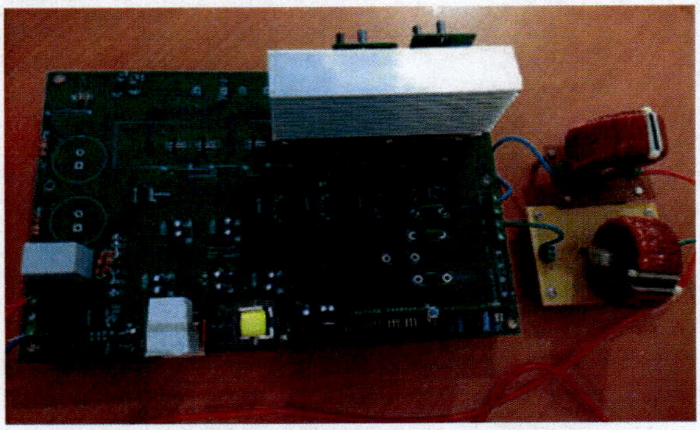

Fig. 12 Implemented multiphase boost converter

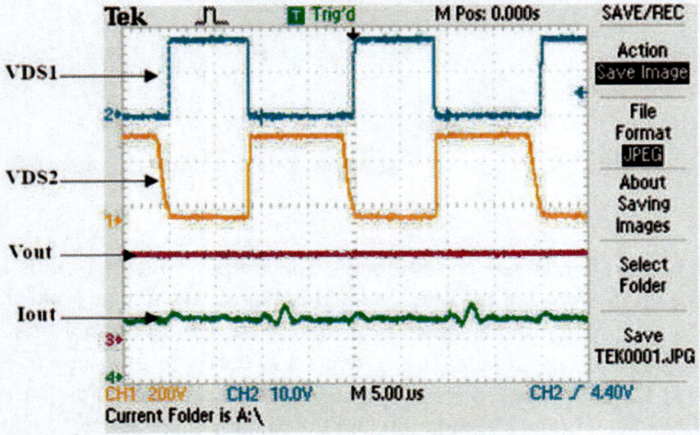

Fig. 13 Converter loaded up to 1 kW with $V_o = 400$ V and $I_o = 2.5$A

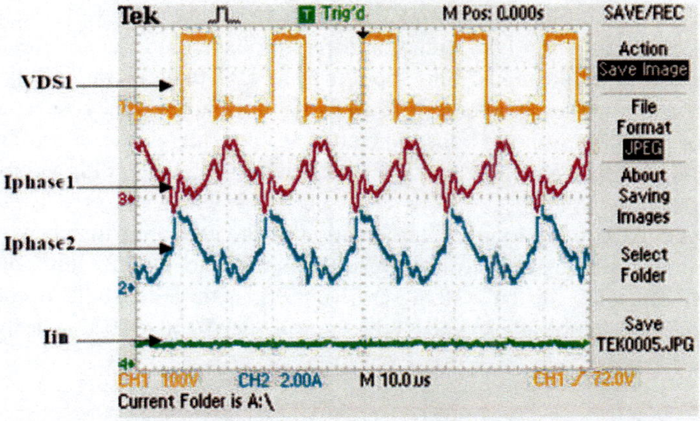

Fig. 14 Boost converter with reduced ripple in input current

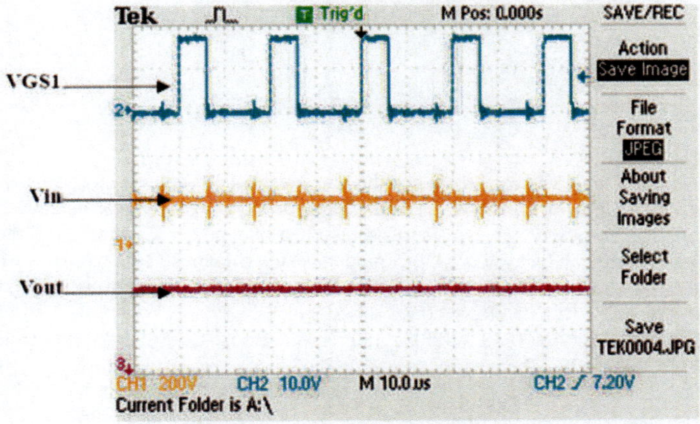

Fig. 15 Light load condition with $V_{in} = 200$ V; $V_{out} = 400$ V

The efficiency of the converter is high during full load condition, and it is low during light load condition. The response of the system during these extreme conditions is observed. Figures 13and15 show system response under full load and light load, respectively.

8 Conclusion

Multiphase boost converter is highly promising for higher current applications with the advantage of reduced ripple current and reduced hot spots on the printed circuit board. The use of multiphase topology for the proposed boost converter has resulted in the total current getting distributed in different phases and reducing current and thermal stress on the switches. This topology has also reduced the requirement of output filter due to reduced current ripple and higher ripple frequency of the converter.

Usage of SiC switch in the boost converter has considerably reduced the size of the PCB and also usage of heat sinks. The efficiency obtained is high compared to that of a Si switch.

The results from open loop implementation of the converter match with that of the simulation. Hysteresis control has resulted in a stable and efficient output. The designed PI controller for the system has given satisfactory and stable output. Thus, a reliable and stable multiphase boost converter of 1 kW, 200–400 V is designed and implemented meeting all the desired conditions.

Acknowledgements The author would like to thank Dr. Niranjan Chiplunkar, Principal NMAMIT, for always encouraging and supporting the research works. The author would also like to thank 'Centre for Design of Power Electronic Systems' (CDPES), Research and Innovation Centre, NMAMIT.

References

1. A. Elasser, T.P. Chow, Silicon carbide benefits and advantages for power electronics circuits and systems. Proc. IEEE **90**(6), 969–986 (2002). https://doi.org/10.1109/JPROC.2002.1021562
2. G. Vacca, Benefits and advantages of silicon carbide power devices over their silicon counterparts. Semiconductor TODAY Compounds Adv Silicon 12 Issue 3, April/May (2017)
3. Freescale Semiconductor, Inc., "MC56F847xx Reference Manual with Addendum", Document Number: MC56F847XXRM Rev. 2.0, 03/2016
4. Texas Instruments, "AN-1820 LM5032 Interleaved Boost Converter", Application Report SNVA335A–May 2008–Revised May (2013)
5. D.W. Hart, *Power Electronics* (McGraw-Hill, United States, 2001)
6. U. Ned Mohan, in *Power Electronics*, (2nd ed., New York, Wiley inc) (1989)
7. Texas Instruments, Voltage Mode Boost Converter Small Signal Control Loop Analysis Using the TPS61030. SLVA274A–May 2007–Revised January (2009)
8. B. Surya Prabha, S. Ramprasanth, Mathematical modelling and performance analysis of quadratic boost converter. Int. J. Sci. Eng. Res. **9**(3) (2018)
9. P. Tyagi, V.C. Kotak, B. Mathew, V.P. Sunder Singh, State space modelling of high gain DC-DC boost converter with coupling inductor. Int. J. Eng. Res. Technol. (IJERT) **03**(01) (2014)
10. R.W. Erickson, *Fundamentals of Power Electronics* (Kluwer Academic/Plenum Publishers, New York, 2001)

Digital Twinning of the Battery Systems—A Review

H. C. Gururaj and Vasudha Hegde

1 Introduction

Battery is an electrochemical device used to store electrical energy. Battery undergoes numerous charging and discharging cycles during the lifetime affecting its parameters [1]. Among various types of batteries, lithium ion batteries (LIB) have carved out a niche for themselves, thanks to their high energy density, allowing them last longer in between charges while providing high output current. Having a low self-discharge rate allows them to retain the charge when not in use. They demand very little maintenance. For the same size, LIBs are much lighter than their counterparts. The most valuable attribute of a LIB is the versatility with shapes and sizes offering users a plethora of options to suit their needs [2]. Since the battery is a major chunk in the entire product cost [3], it is imperative to accurately estimate the battery parameters employing battery management system (BMS) to cut down the design costs and to extract maximum energy.

NASA pioneered the concept of pairing technology, the predecessor of the present-day digital twin technology. Digital twin technology is so vital to the business today that in the year 2017 it was declared as one of the Gartner's Top 10 Strategic Technology Trends [4]. Digital twin is much more than just a simulation. Simulation is static, whereas a digital twin is active as it receives real-time data from its physical counterpart. Simulation focuses just on the product, whereas a digital twin focuses on the entire business, helps improve the process and make better decisions [5].

H. C. Gururaj (✉)
Department of Electrical and Electronics Engineering, DRR Government Polytechnic, Davangere, India

V. Hegde
Department of Electrical and Electronics Engineering, Nitte Meenakshi Institute of Technology, Bengaluru, India
e-mail: vasudha.hegde@nmit.ac.in

Sanjeevikumar et al. (eds.), *Advances in Renewable Energy and Electric Vehicles*,
Lecture Notes in Electrical Engineering 767,
https://doi.org/10.1007/978-981-16-1642-6_11

Endeavour of this paper is to emphasize the importance of digital twin for the battery systems. Organization of this paper is as follows: Section 2 introduces contemporary publications on digital twins for batteries, and Sect. 3 has the summary.

2 Digital Twin

The digital twin platform from some of the most recent publications is discussed in this section.

2.1 Digital Twin for Assessment of Battery Pack Degradation

In [6], an economical platform for the digital twin for lithium ion battery pack employed in the spacecraft was developed to assess its degradation. Figure 1 shows the architecture for the digital twin platform comprising the assessment unit and the visual software unit. Remote sensing link transmits real-time data to ground station, which is then processed by the assessment unit to provide the degradation result. As shown in Fig. 2, visual software unit displays both the real-time data and the results from the assessment unit. Temperatures, voltage, current, etc., of the battery are the state parameters for real-time data transmission that are stored first and later used for the analysis.

The output of the unit for assessment can be the prediction of state of charge (SoC) [7], remaining useful life (RUL)/state of health (SoH) [8]. SoC associates with cells of the battery, while RUL is pertaining to entire battery pack.

Testing and verification of accurate mirroring of the battery state in real-time is the basic function of digital twin. Visual software will present effects after assessment of real-time data which can be correlated to test data for verifying the functioning of

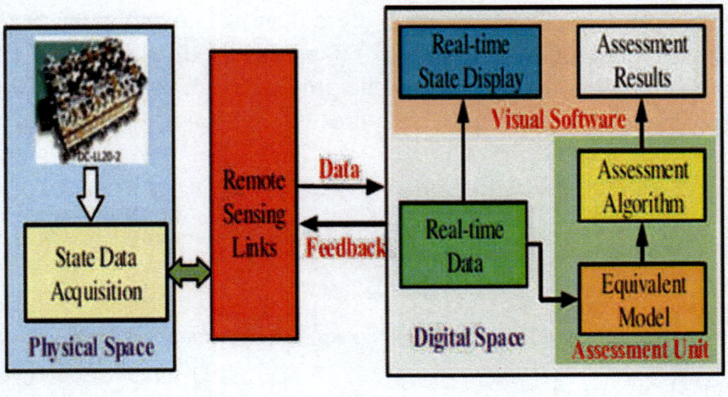

Fig. 1 Architecture of the digital twin platform [6]

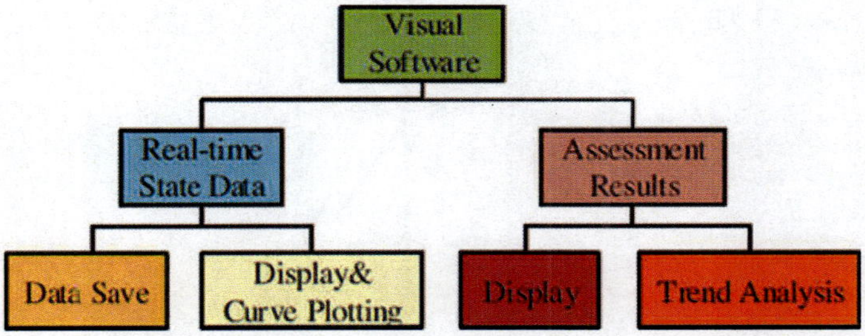

Fig. 2 Function framework of visual software [6]

digital twin. The assessment of degradation of the battery pack (SoC prediction and RUL estimation) is the core function of digital twin platform.

Status of the LIB pack is assessed by the assessment unit based on algorithms and models. SoC of the cells is estimated using the algorithm of Kalman filter–least squares support vector machine (KF-LSSVM). Electric current, terminal voltage (real-time data), and historical test data (training data) are the inputs to the algorithm. Output of the algorithm and current value are fed as the input to the Kalman filter, whose yield is SoC estimation.

SoH is evaluated using the algorithm of auto regression model-particle filter (AR-PF). Historical test data (training data), real-time data are the input for auto regression (AR) model. Yield of the model is the Health Index (HI). HI along with real-time data are inputs to particle filter, whose yield is SoH/RUL prediction.

Figure 3 shows the plots for true value obtained from test data and SoC estimation using KF-LSSVM [9], both being consistent with one other proving accurate prediction of SoC can be carried using the specified model.

Figure 4 shows the plots for true value obtained from test data and RUL estimation using AR-PF algorithm. As seen from the plots estimation of RUL, test data's true values are in close agreement.

Fig. 3 Results of KF-ALSSM algorithm [6]

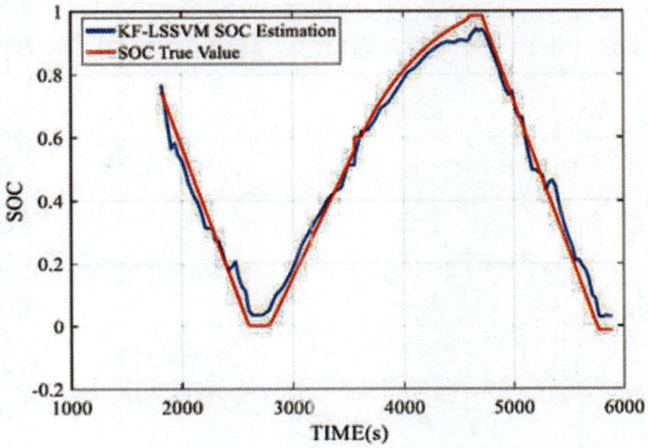

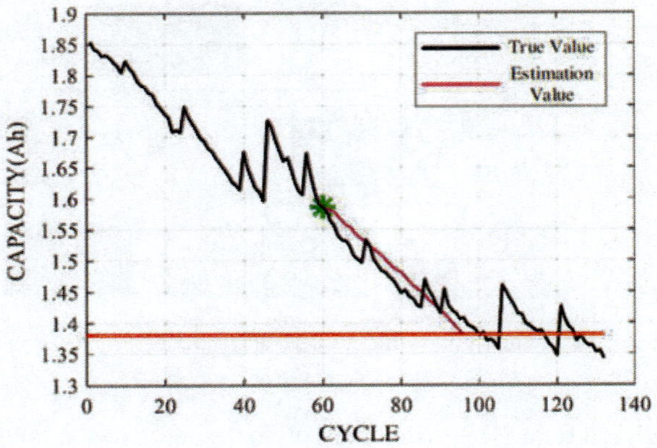

Fig. 4 Results of AR-PF algorithm [6]

The proposed platform needs improvement as it is incompatible with different algorithms and types of batteries.

2.2 *Digital Twin For Digital Services of Battery System*

In [10], a digital twin model was presented to cover the complete life cycle of the battery. Figure 5 shows the proposed architecture comprising three levels: service level, twin level, and hardware + connectivity level. Hardware + connectivity level serves the purpose of data accumulation. In a modern industry, the relevant data may be obtained through manufacturing execution system (MES)/enterprise resource planning (ERP) [11]. Supplementary data loggers and sensors are used to supervise the ambient conditions during production and logistics. Data loggers amass measured data which can be automatically/manually read out or transmitted live data through Internet of Things (IoT) networks [12].

The twin level is placed in the environment of cloud computing [13]. Twin level offers micro-services [14], each performing one task and corresponding through numerous interfaces. Figure 6 shows outline of the architecture's data types. Master

Fig. 5 Reference architecture of a digital twin [10]

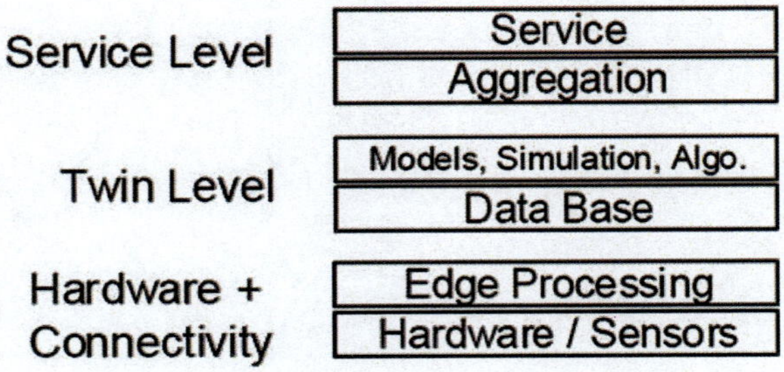

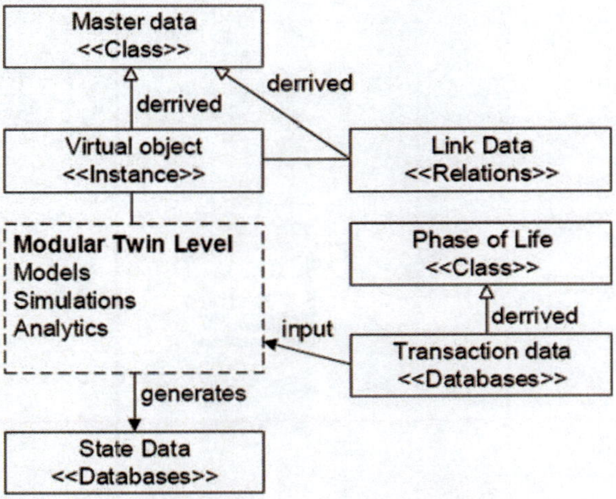

Fig. 6 Outline of datatypes pertaining to architecture of digital twin [10]

data is the technical description of the physical objects, represented in the virtual world. Transaction data is used to acquire the continued experience concerning the physical component. Simulations or models are fed with transaction data to obtain insight regarding state of real system. Results are composed as state data. Parent–child relationship among the components is monitored using the link data. Knowledge created in twin level can be utilized to cater to particular user's needs in service level. Service level contains aggregation layer. The collected information is first conditioned and later formatted during aggregation. The output layer acts as a link to the service layer. Output layer can be an application programming interface (API) or graphical interface.

The proposed metamodel has subset high voltage battery system (HVBS), logistics subset, production subset, monitoring, and testing. Subset HVBS is prevalent in every life phase. The life cycle begins with bringing together of relevant components of the system. During the onset of the manufacturing process, every single component is the subject of twin architecture, and once assembled, the entire battery system will be the physical subject. In order to mimic the process of manufacturing, a virtual assembly is implemented. Vehicle life cycle temperature, voltage, current, etc., are the transaction data. During the second life, criterion for transaction data remains the same, but only the ambient system changes.

The logistics subset has logistic entity, transport, and storage as the sub-modules. Shock, vibrations, temperature, and humidity during the transport form the transaction data. The production subset is comprised of various states pertaining to production. The focus is on system assembly. Transaction data is the measurement of the effect of vibration, shock on the battery components. Data in every subset is logged in a uniform way in the monitoring subset. Standardized classes are used for logging different parameters. Testing subset provides analysis on the documented transaction data and generates the state data specific to each test carried out. Further research is suggested with regard to use-cases, stakeholders, and implementation.

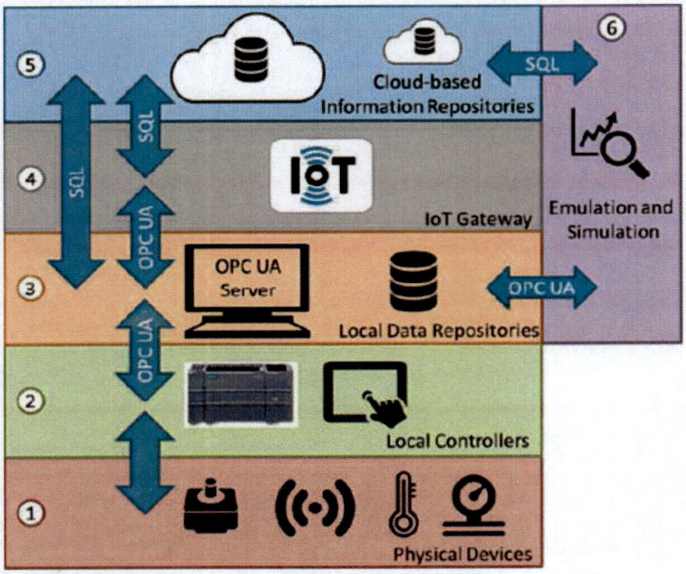

Fig. 7 Connection architecture for a digital twin [15]

2.3 6 Layer Digital Twin Architecture

In [15], a digital twin model with six layers for a manufacturing cell was presented. As shown in Fig. 7, layer 01 consists of numerous physical devices like sensors, actuators, etc., which feed to and receive signals from the next layer. Layer 02 comprises local controller such as a programmable logical controller (PLC) [16] in order to lend some additional capabilities to the digital twin. Layer 03 is the local data repository such as local database and Open Platform Communications United Architecture (OPCUA) [17]. This layer provides characteristics like global connectivity, reliability, real-time performance, and security. Layer 04 is the IoT gateway in the midst of connected world and the physical twin. This layer connects information from layer 05 to data in layer 03[18]. The layer 04 provides multi-dimensional data correlation/data intelligence [19], which modifies the data pertaining to different twin architectures into generic information, prevents bandwidth bottlenecks, avoids enormous database, makes sure that physical twin is in relevant state before passing any commands from layer 06 to layer 03, and resolves any arising conflict of data from repositories in layer 06. The layer 05 is the information repository for digital as well as physical twin. It has information regarding latest state of physical twin.

Layer 06 adds the intelligence to the digital twin. The layer can realize the following roles: Remote monitoring, predictive analytics, simulation of future behaviour, optimization along with validation, documentation along with communication, connection among disparate systems, and digital twin control. The effectiveness of the proposed architecture is tested through a case study where physical equipment's change of state is communicated from layer 01 to layer 06 and a command for data change was transmitted from layer 06 to layer 01 proving the operational capability of the architecture. However, this research raised concerns over incompatible versions and the complex installation procedure for drivers and the connection being slow to online cloud server.

3 Summary

A digital twin can not only perform the basic function of mirroring the present state of a physical object using the real-time data but also perform the core function like predicting the future state. LIBs are extensively used as power source in satellites, electric vehicles, and consumer electronics. The reliability and safety of the LIBs are directly linked to its degradation. Hence, the assessment of battery's degradation, estimation of SoH, SoC of the battery is critical. The SoH, SoC estimation can be achieved through suitable algorithm. A digital twin can provide digital services to different stakeholders like component and logistic suppliers, original equipment manufacturer, vehicle owner, fleet operator, and second life users covering the entire life cycle of the battery. In the reference architecture of a digital twin, hardware layer acquires the data, twin layer transforms data to knowledge, and service layer provides the access to this knowledge. Digital twin is defined by a metamodel comprising: Subset HVBS having various classes alongside attributes exhibiting master data of respective components in the physical world. The logistic subset provides the information on the logistic entity during storage and transportation. The production subset comprises all the states of the production pertaining to HVBS. Data in every subset is logged in a uniform way in the monitoring subset. Testing subset provides a digital platform for comprehensive testing under various conditions before the series production of the HVBS. Digital twin can be implemented using a multi-layer architecture comprising the physical twin, local controllers, local data repository, IoT gateway, data repository (cloud-based), and emulation and simulation layer. Data is collected at the physical twin level, processed by the local controllers and later stored in the local data repository employing an OPC-UA server. IoT provides the gateway between local data repository and the cloud storage. The brain of the digital twin is the sixth layer modelling both current and future behaviours. Going forward each product manufactured may possess its own digital twin generating the data for analysis, helping create real-time predictions relating to predictive maintenance, product life cycle, etc., and hence, revolutionizing product development and testing in almost every field.

References

1. *A Guide to Understanding Battery Specifications* Dec 2008. Accessed on Nov 24 (2020). [online]. Available : http://web.mit.edu/evt/summary_battery_specifications.pdf
2. What is a lithium-ion battery and how does it work?Accessed on Nov 24 (2020) [online]. Available :https://www.cei.washington.edu/education/science-of-solar/battery-technology/
3. Lithium-Ion Batteries—Price Trend and Cost Structure Nov 26 (2019). Accessed on Nov 24,2020. [online]. Available :https://www.beroeinc.com/article/lithium-ion-batteries-price-trend-cost-structure
4. What Is Digital Twin Technology—And Why Is It So Important? Mar 6 (2017). Accessed on Nov 30 2020 [online]. Available :https://www.forbes.com/sites/bernardmarr/2017/03/06/what-is-digital-twin-technology-and-why-is-it-so-important/#75e8ab7a2e2a
5. The difference between a simulation and a digital twin Oct 23 (2019). Accessed on Nov 30 2020.[online].Available : https://blogs.sw.siemens.com/mindsphere/the-difference-between-a-simulation-and-a-digital-twin/
6. Y. Peng, X. Zhang, Y. Song, D. Liu, A low cost flexible digital twin platform for spacecraft lithium-ion battery pack degradation assessment. in *2019 IEEE International Instrumentation and Measurement Technology Conference (I2MTC)*, Auckland, New Zealand, pp. 1–6. (2019). https://doi.org/10.1109/I2MTC.2019.8827160
7. L. Zheng, J. Zhu, G. Wang, D. D. Lu, T. He, Differential voltage analysis based state of charge estimation methods for lithium-ion batteries using extended Kalman filter and particle filter. Energy 1028–1037 (2018)
8. Y. Song, D. Liu, Y. Hou, J. Yu, Y. Peng, Satellite lithium-ion battery remaining useful life estimation with an iterative updated RVM fused with the KF algorithm. Chin. J. Aeronaut. **31**(1), 31–40 (2018)
9. J. Meng, G. Luo, F. Gao, Lithium polymer battery state-of-charge estimation based on adaptive unscented kalman filter and support vector machine. IEEE Trans. Power Electron. **31**(3), 2226–2238 (2016). https://doi.org/10.1109/TPEL.2015.2439578
10. L. Merkle, A.S. Segura, J. Torben Grummel, M. Lienkamp, Architecture of a digital twin for enabling digital services for battery systems. in *2019 IEEE International Conference on Industrial Cyber Physical Systems* (ICPS), Taipei, Taiwan, pp. 155–160. (2019) https://doi.org/10.1109/ICPHYS.2019.8780347
11. S. Dumbrava, I.M. Valova, The design of an enterprise resource planning software application for manufacturing control. in *EUROCON 2005—The International Conference on "Computer as a Tool*, Belgrade, pp. 587–590. (2005). https://doi.org/10.1109/EURCON.2005.1629997
12. S. Gupta, N. Mudgal , R. Mehta, Analytical study of IoT as emerging need of the modern era. in *2016 3rd International Conference on Computing for Sustainable Global Development (INDIACom)*, New Delhi, pp. 233–235. (2016)
13. T. Kushida, G.S. Pingali, Industry Cloud–effective adoption of cloud computing for industry solutions. in *2014 IEEE 7th International Conference on Cloud Computing*, Anchorage, AK, pp. 753–760. (2014). https://doi.org/10.1109/CLOUD.2014.105
14. N. Cui, Y. Hu, D. Yu, F. Han, Research and implementation of intelligent workshop IoT cloud platform based on micro-services. in *2019 IEEE International Conference on Signal Processing, Communications and Computing (ICSPCC)*, Dalian, China, pp. 1–5. (2019). https://doi.org/10.1109/ICSPCC46631.2019.8960804
15. A. Redelinghuys, A. Basson, K. Kruger, in *A Six-Layer Digital Twin Architecture for a Manufacturing Cell: Proceedings of SOHOMA* (2018). https://doi.org/10.1007/978-3-030-03003-2_32
16. M. Chattal, V. Bhan, H. Madiha, S.A. Shaikh, Industrial automation and control trough plc and labview. in *2019 2nd International Conference on Computing, Mathematics and Engineering Technologies (iCoMET)*, Sukkur, Pakistan, (pp. 1–5). (2019). https://doi.org/10.1109/ICOMET.2019.8673448

17. *IoT Basics: What is OPC UA*? 07 Feb 2019. Accessed on Nov 24 2020. [online]. Available : https://www.spotlightmetal.com/iot-basics-what-is-opc-ua-a-42878/#:~:text=OPC%20UA%20and%20Industry%204.0,services
18. J. Lee, B. Bagheri, H.-A. Kao, A cyber-physical systems architecture for industry 4.0-based manufacturing systems. Soc. Manuf. Eng. **3**, 18–23 (2014)
19. *Data intelligence* Accessed on Nov 24 (2020). [Online].Available:https://www.techopedia.com/definition/28799/data-intelligence

Droop Control Strategies for Microgrid: A Review

Neha Bhatt, Ritika Sondhi, and Sudha Arora

1 Introduction

Our electricity grid has seen revolutionary transformation in its conventional structure. Microgrids are making their place in the conventional grid structure and playing important role in improving system efficiency and reliability and generating clean energy [1–3]. These microgrids consist distributed energy resources (DERs), storage devices, and loads and can operate in both grid connected as well as islanded mode [4]. Need of distributed generation is also being felt because of the reason that conventional energy sources are depleting and exhausting. Hence, alternatives are needed to satisfy continuously rising demand of electricity. Other advantages include higher efficiency, better load regulation, capability to expand if need arises and ease of maintenance [5].

Although distributed generation may look like a convenient solution for convention grid structure, but in practice, operation of these parallel connected distributed units faces many control challenges like inverse power flow, harmonic current circulation, voltage and frequency fluctuations, etc. Therefore, need of efficient control strategies cannot be ignored. Microgrid control can be classified as centralized and decentralized. In centralized control, all units are connected through communication channel (e.g.,-master slave control) and communicate continuously to control network parameters, active power, reactive power, voltage, and frequency. This control is capable to perform under different working conditions but has complex structure and less scope for system expansion. Another, efficacy of this control relies heavily on the efficacy of communication network, failing of which may result in adversely affecting the network reliability. On the contrary, decentralized control is based on locally received information. This control is easy in implementation and compatible to system expansion. Droop control is one such control strategy that is

N. Bhatt (✉) · R. Sondhi · S. Arora
College of Technology, GBPUAT, Pantnagar, India

Sanjeevikumar et al. (eds.), *Advances in Renewable Energy and Electric Vehicles*,
Lecture Notes in Electrical Engineering 767,
https://doi.org/10.1007/978-981-16-1642-6_12

based on the drooping characteristic of traditional synchronous generators. These characteristics follow linear relation between active power and frequency and reactive power and voltage. But these conventional droop characteristics suffer from various drawbacks.

- Voltage and frequency deviations
- Slow transient response
- Unable to give good performance with nonlinear and unbalanced load.
- Affected by line impedance variation
- Does not consider harmonic current content
- Improper current sharing between units.

Many droop control strategies have been suggested in literature to overcome the shortcomings of conventional droop method. A method is proposed in [6], which improves current sharing by increasing droop gains as load current increases. A comparative analysis [7] between frequency droop and angle droop shows lesser deviation in frequency with later technique. A control technique based on virtual power source and virtual impedance has been proposed in [8]. The negative virtual resistance compensates the line resistance and decouples active power, and virtual inductance improves reactive power sharing accuracy. Similarly, [9, 10] propose control based on virtual resistance. A piecewise linear droop strategy is proposed in [11]. It follows different droop characteristics for different load conditions to get better current sharing and voltage regulation. As linear droop techniques fail to incorporate the effect of noise and disturbances, therefore, [12] proposed a robust droop control method to improve both current sharing as well as voltage stability under these conditions. Another paper [13] proposes adaptive droop control strategy for linear loads which compensates the effect of feeder impedance mismatch by adaptive droop coefficients.

2 Droop Control

Linear drooping relation exists between active power–frequency and reactive power–voltage in synchronous generator. Droop control for microgrids is based on the similar approach. Operating point moves on the characteristic depending on load condition. For a change in active power and reactive power demand, there will be a corresponding change in frequency and voltage, respectively.

Conventional droop control is a simple and reliable control method for highly inductive network, but as microgrid is resistive in nature, hence performance of conventional droop control suffers.

2.1 Limitation with Conventional Droop Control

When converter modules are operating in parallel, current sharing is a major concern among these parallel connected modules. In its simplest form, droop control introduces internal or external resistance for proper sharing of load current.

Consider a simple microgrid structure consisting of two parallel connected distributed generators (DG) along with load as shown in Fig. 2, where V_1 and V_2 represent output voltages of converters, R_1, R_2 are cable resistances, and R_{D1}, R_{D2} are droop resistances [11].

For the given circuit, the bus voltage V_{bus} is given as

$$V_{\text{bus}} = \frac{R_L R_2 V_1 + R_L R_1 V_2}{R_L R_1 + R_L R_2 + R_1 R_2} \tag{1}$$

And the difference in currents shared by converters is given by,

$$\begin{aligned} \Delta I &= I_1 - I_2 \\ &= \frac{R_L(V_1 - V_2) + V_1 R_2}{R_L R_1 + R_L R_2 + R_1 R_2} - \frac{R_L(V_2 - V_1) + V_2 R_1}{R_L R_1 + R_L R_2 + R_1 R_2} \end{aligned} \tag{2}$$

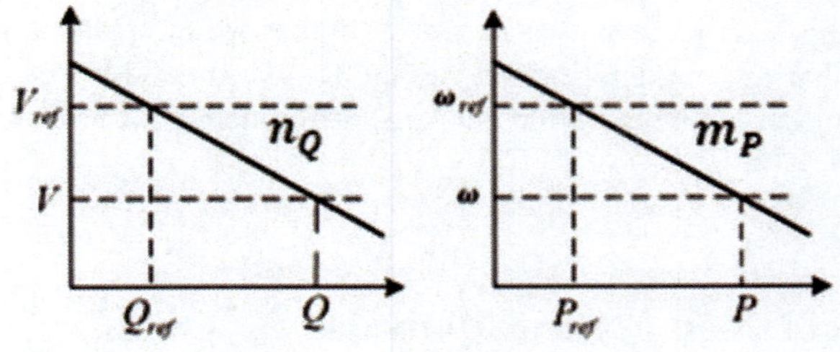

Fig. 1 Voltage and frequency droop characteristic [14]

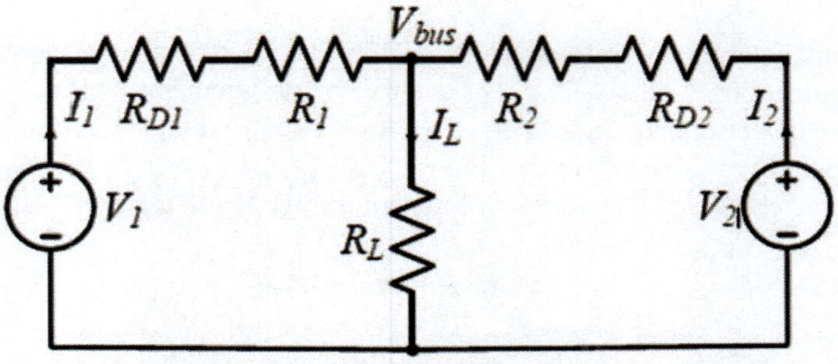

Fig. 2 Equivalent model of parallel connected DGs [11]

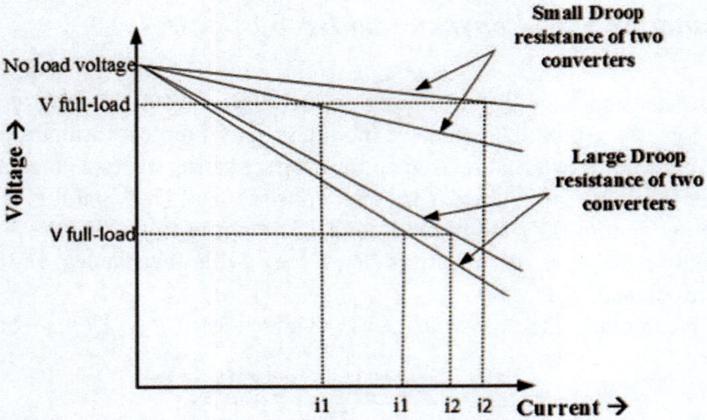

Fig. 3 Droop characteristics of parallel connected converters [16]

Ideally, all units should share the load uniformly, and from (2), it is clear that it is possible only when voltages V_1, V_2 and resistances R_1, R_2 are equal as ΔI becomes zero in that case. But conventional droop control is only a compromise between voltage regulation and current sharing as there is always some variation in cable resistances or some other parameter which results in voltage mismatch or current sharing differences [15]. As shown in Fig. 3, higher value of droop gain results in better current sharing, but deteriorated voltage regulation and lower value of droop gain ensure better voltage regulation but large difference in current sharing between converters.

3 Review of Droop Control Methods

3.1 $P-\omega/V-Q$ Droop

Conventional droop control methods include $P-\omega/V-Q$ control strategies for parallel operation of DERs. In $P-\omega$ control, output frequency reduces with the increase in power output, and for $V-Q$ control, output voltage magnitude reduces with increase in reactive power demand.

Relation between $P-\omega$ and $V-Q$ is shown in Fig. 4, and relation is given by,

$$\omega = \omega_0 - m_P(P - P_0) \tag{3}$$

$$V = V_0 - n_Q(Q - Q_0) \tag{4}$$

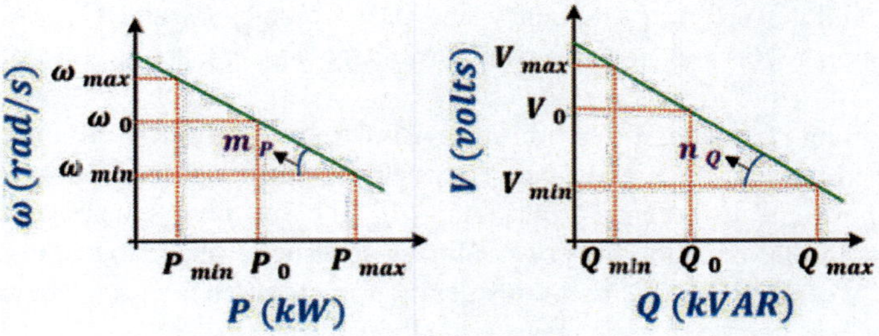

Fig. 4 $P-\omega$ and $V-Q$ droop characteristics [17]

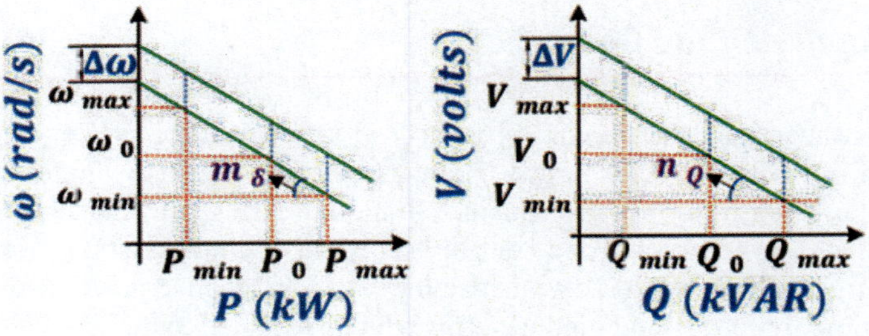

Fig. 5 Adaptive droop characteristic [17]

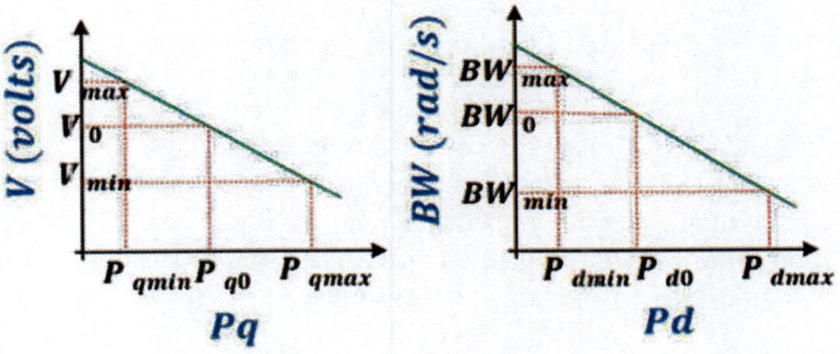

Fig. 6 Droop characteristic with signal injection [17]

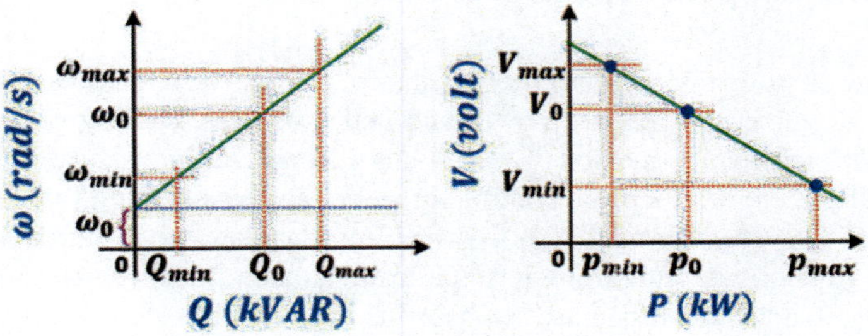

Fig. 7 $P–V/Q–\omega$ droop [17]

where ω_0 and V_0 are base frequency and base voltage, P_0 and Q_0 are nominal operating points for real and reactive power, and m_P, n_Q are droop coefficients [17–19].

As drooping characteristic is not present inherently in DERs, drooping feature is introduced through buck boost converter, series resistors, introduction of voltage droop proportional to output current, etc. [20, 21] But these schemes show poor current regulation, unsuitability for nonlinear and single phase loads, poor power sharing between DER units, non-consideration of harmonics current content, and high power loss in case of series resister methods [22].

3.2 Adaptive Droop Control

Although conventional droop control is easy to implement, but suffers from poor power sharing between DER units. These techniques are also limited to linear loads, get affected by impedance mismatch and have sluggish transient response. On the other side, adaptive droop control is capable of compensating these issues. These control techniques identify mismatch in real and reactive power and generate corresponding frequency and voltage terms which are added to the droop equations.

$$\omega = \omega_0 - m_\delta(P - P_0) + \Delta\omega \tag{5}$$

$$V = V_0 - n_Q(Q - Q_0) + \Delta V \tag{6}$$

In [23–26], adaptive droop control is introduced through virtual impedance concept. Literature [12, 27] consider resistive impedance droop control, whereas [28] considers virtual inductive impedance control. Literature [23] proposes self-adaptive droop control strategy which utilizes energy storage systems to track power mismatch and adjust droop coefficient accordingly.

3.3 Virtual Impedance Method

Unlike power grid, microgrids line impedance is resistive which leads to power coupling of active and reactive power and hence reduces stability of the micro-grid. Virtual impedance methods [29–32] consider virtual impedance in feedback path of voltage control loop and control output voltage by adjusting value of virtual impedance. However, introduction of virtual impedance in the system may lead to increased voltage drop and harmonic amplification, therefore [33] suggests introduction of negative voltage resistance. This results in better performance of the system with smaller value of virtual impedance.

3.4 Signal Injection Method

Virtual impedance methods are based on calculation of exact value of line impedances to decide droop coefficient, which leads to increased computation burden and increased implementation difficulty. To avoid this problem, a small AC current is injected into the parallel connected inverters. This results in better power sharing under unbalance condition [29, 32, 34].

$$V = V_0 - D_V P_q \tag{7}$$

$$BW = BW_0 - D_{bw} P_d \tag{8}$$

where D_v, D_{bw} are boost and droop coefficients, BW_0 is nominal bandwidth of voltage loop, and P_q, P_d are injected signals.

3.5 Virtual Frame Transformation Method

Conventional droop methods assume lines to be mainly inductive; but considering various types of power electronics-based distributed system, actual system impedance may vary in a range from inductive to resistive. This leads to power coupling problem. Therefore [19, 35], use orthogonal frame transformation that transforms actual real and reactive power into virtual real and reactive power. This frame transformation can be realized by

$$\begin{bmatrix} P' \\ Q' \end{bmatrix} = T_{PQ} \begin{bmatrix} P \\ Q \end{bmatrix} = \begin{bmatrix} \sin\theta & -\cos\theta \\ \cos\theta & \sin\theta \end{bmatrix} \begin{bmatrix} P \\ Q \end{bmatrix} \tag{9}$$

But power sharing is not efficient in this method, therefore a virtual reference frame method is proposed in [36–39]. This technique transforms actual voltage and frequency into virtual voltage and frequency using an orthogonal transformation matrix.

$$\begin{bmatrix} \omega' \\ E' \end{bmatrix} = T_{\omega E} \begin{bmatrix} \omega \\ E \end{bmatrix} = \begin{bmatrix} \cos\theta & \sin\theta \\ -\sin\theta & \cos\theta \end{bmatrix} \begin{bmatrix} \omega \\ E \end{bmatrix} \tag{10}$$

3.6 Angle Droop Control

Angle droop control methods use power angle (δ) as control variable instead of frequency [40–42]. This technique is capable of providing efficient power sharing without significant change in the frequency [7] but gets affected by line impedances, therefore [43] suggests a topology-independent inter-node-communication-based technique. Angle droop control equation is given as

$$\delta = \delta_0 - m_i(P - P_0) \tag{11}$$

3.7 P–V/Q–ω Droop Control

Conventional $P\!-\!f/V\!-\!Q$ droop equations work well in case of highly inductive line impedance, and with high voltage microgrids, low voltage microgrids are generally resistive; hence, same equation does not give same performance. Although active power is still dependent on frequency, but dependency of active power on voltage and similarly dependency of reactive power on frequency is more as compared to voltage [44–46]. Therefore, droop equations are modified accordingly.

$$\omega = \omega_0 + m_q(Q_1 - Q_0) \tag{12}$$

$$V = V_0 - n_q(P_1 - P_0) \tag{13}$$

3.8 H-infinity Control

Performance of control techniques gets affected by the selection of system parameters, and these parameters may vary within a certain range in a distributed system [47–49]. H-infinity method is based on minimization of the effect of these variations, which are expressed as an optimization problem. A stabilizing feedback controller along with properly selected weighting functions allow shaping of open-loop frequency response of feedback control system to match desired loop shape as closely as possible [50]. A current controller design using H-infinity repetitive control is suggested for reduction of total harmonic distortion (THD) for grid connected inverters [51].

3.9 *ANN and Fuzzy-Based Control*

Variation in line parameter and microgrid structure can affect the performance of control techniques. Artificial neural network (ANN) and fuzzy logic-based control techniques offer the benefits of independency from above-mentioned parameters. Self-learning feature of ANN algorithms allows feasibility of control under varying operating conditions and grid disturbances [52, 53]. Fuzzy-based control is a rule-based control which deals with linguistic values rather than crisp values [54, 55]. Algorithms used in ANN provide the feature of self-learning. In case of ANFIS-based control, a set of training data trains the controller and fuzzy if–then rules are used to generate input/output (I/O) pairs [56].

The salient features of the above mentioned control techniques have also been highlighted in Table 1.

4 Conclusion

This paper covers various control strategies of droop control that have been anticipated in literature. Conventional $P–f/Q–V$ droop finds its suitability for conventional inductive grid but not for resistive microgrid. Various control techniques have been developed from the point of mitigating the shortcoming of conventional droop control. This is evident from literature that proportional power sharing, voltage stability, circulating currents, and frequency stability are major challenges that are faced during the control of parallel converters. Therefore, control methods which are robust and stable enough under steady state and transient conditions as well as capable of better voltage and frequency regulation are still an area of research.

Table 1 Features of different control techniques

Control technique	Features
$P-\omega/V-Q$ droop	*Advantages* • Implementation is easy • Suitable for inductive transmission lines and linear loads *Disadvantages* • Affected by system parameters • Poor voltage regulation • Poor frequency regulation • Slow transient response • Poor current sharing • Not suitable for nonlinear loads
Adaptive droop control	*Advantages* • Better transient response • Better performance under heavy load condition • Better voltage regulation • Better reactive power sharing *Disadvantages* • Requires system parameters to be known in advance • Implementation is complex
Virtual impedance method	*Advantages* • Implementation is easy • Not affected by system parameters • Better active and reactive power sharing • Suitable for unbalanced, linear, nonlinear loads *Disadvantages* • Poor voltage regulation • Frequency deviation can occur • Transient response is slow
Signal injection method	*Advantages* • Not affected by system parameters • Suitable for linear and nonlinear loads *Disadvantages* • Poor voltage regulation • Implementation is complex • Transient response is slow • Causes harmonic distortion of terminal voltages • Reduced power quality
Virtual frame transformation method	*Advantages* • Implementation is easy • Decoupled active and reactive power control • Suitable for linear loads *Disadvantages* • Require line impedance values to be known in advance • To ensure same transformation angle for all parallel units is difficult task • Poor voltage regulation

(continued)

Table 1 (continued)

Control technique	Features
Angle droop control	*Advantages* • Better frequency regulation *Disadvantages* • Poor power sharing
P–V/Q- ω droop	*Advantages* • Better frequency regulation • Implementation is easy *Disadvantages* • Affected by system parameters • Not suitable for nonlinear loads
H-infinity control method	*Advantages* • Not affected by system parameters • Suitable for linear, nonlinear, and unbalanced loads • Less total harmonic distortion(THD) *Disadvantages* • Slow dynamics
ANN and fuzzy control	*Advantages* • Not affected by system parameters • Suitable for large power system • Robust control performance *Disadvantages* • Slow dynamic response

References

1. J. Rocabert, A. Luna, F. Blaabjerg, P. Rodriguez, Control of power converters in AC microgrids. IEEE Trans. Power Electron. **27**(11), 4734–4749 (2012)
2. A. Molderink, V. Bakker, M.G. Bosman, J.L. Hurink, G.J. Smit, Management and control of domestic smart grid technology. IEEE Trans. Smart Grid **1**(2), 109–119 (2010)
3. R. Lasseter, A. Akhil, C. Marnay, J. Stephens, J. Dagle, R. Guttromsom, J. Eto, Integration of distributed energy resources. The CERTS Microgrid Concept (No. LBNL-50829). Lawrence Berkeley National Lab.(LBNL), Berkeley, CA (United States, 2002)
4. J. Yang, X. Jin, X. Wu, P. Acuna, R.P. Aguilera, T. Morstyn, V.G. Agelidis, Decentralised control method for DC microgrids with improved current sharing accuracy. IET Gener. Transm. Distrib. **11**(3), 696–706 (2017)
5. B.T. Irving, M.M. Jovanovic, Analysis, design, and performance evaluation of droop current-sharing method, in *APEC 2000. Fifteenth Annual IEEE Applied Power Electronics Conference and Exposition (Cat. No. 00CH37058)*, vol. 1 (IEEE, 2000) pp. 235–241
6. A. Khorsandi, M. Ashourloo, H. Mokhtari, R. Iravani, Automatic droop control for a low voltage DC microgrid. IET Gener. Transm. Distrib. **10**(1), 41–47 (2016)
7. R. Majumder, A. Ghosh, G. Ledwich, F. Zare, Angle droop versus frequency droop in a voltage source converter based autonomous microgrid, in *2009 IEEE Power & Energy Society General Meeting* (IEEE, (2009, July), pp. 1–8
8. C. Dou, Z. Zhang, D. Yue, M. Song, Improved droop control based on virtual impedance and virtual power source in low-voltage microgrid. IET Gener. Transm. Distrib. **11**(4), 1046–1054 (2017)

9. Y. Zhu, B. Liu, F. Wang, F. Zhuo, Y. Zhao, A virtual resistance based reactive power sharing strategy for networked microgrid, in *2015 9th International Conference on Power Electronics and ECCE Asia (ICPE-ECCE Asia)* (IEEE, 2015, June), pp. 1564–1572
10. P. Zhang, H. Zhao, H. Cai, J. Shi, X. He, Power decoupling strategy based on 'virtual negative resistor' for inverters in low-voltage microgrids. IET Power Electronics **9**(5), 1037–1044 (2016)
11. Y. Lin, W. Xiao, Novel piecewise linear formation of droop strategy for DC microgrid. IEEE Trans. Smart Grid **10**(6), 6747–6755 (2019)
12. T.V. Vu, D. Perkins, F. Diaz, D. Gonsoulin, C.S. Edrington, T. El-Mezyani, Robust adaptive droop control for DC microgrids. Electric Power Syst. Res. **146**, 95–106 (2017)
13. H. Mahmood, D. Michaelson, J. Jiang, Reactive power sharing in islanded microgrids using adaptive voltage droop control. IEEE Trans. Smart Grid **6**(6), 3052–3060 (2015)
14. H. Wu, H. Sun, L. Cai, X. Tao, Simulation on control strategies of grid-connected inverters. in *The 2nd International Symposium on Power Electronics for Distributed Generation Systems* (IEEE, 2010), pp. 904–908
15. J.W. Kim, H.S. Choi, B.H. Cho, A novel droop method for converter parallel operation. IEEE Trans. Power Electron. **17**(1), 25–32 (2002)
16. S. Dahale, A. Das, N.M. Pindoriya, S. Rajendran, An overview of DC-DC converter topologies and controls in DC microgrid, in *2017 7th International Conference on Power Systems (ICPS)*. (IEEE, 2017, December), pp. 410–415
17. A.K. Sahoo, K. Mahmud, M. Crittenden, J. Ravishankar, S. Padmanaban, F. Blaabjerg, Communication-less primary and secondary control in inverter-interfaced AC microgrid: an overview. IEEE J. Emerg. Select Top. Power Electron. (2020)
18. S. Luo, Z. Ye, R.L. Lin, F.C. Lee, A classification and evaluation of paralleling methods for power supply modules, in *30th Annual IEEE Power Electronics Specialists Conference Record (Cat. No. 99CH36321)*, vol. 2 (IEEE, 1999, July), pp. 901–908
19. K. De. Brabandere, B. Bolsens, J. Van den Keybus, A. Woyte, J. Driesen, R. Belmans, A voltage and frequency droop control method for parallel inverters. IEEE Trans. Power Electron. **22**(4), 1107–1115 (2007)
20. I. Batarseh, K. Siri, H. Lee, Investigation of the output droop characteristics of parallel-connnected DC-DC converters, in *Proceedings of 1994 Power Electronics Specialist Conference-PESC'94*, vol. 2. (IEEE, 1994, June), pp. 1342–1351
21. K. Siri, J. Banda, Current distribution for parallel-connected DC power sources without remote sensing, in *Proceedings of Intelec 94*. (IEEE, 1994, October), pp. 196–203
22. S.K. Sahoo, A.K. Sinha, N.K. Kishore, Control techniques in AC, DC, and hybrid AC–DC microgrid: a review. IEEE J. Emerg. Selected Top. Power Electron. **6**(2), 738–759 (2017)
23. Xu, H., Pan, Z., An adaptive droop control strategy of energy storage control for microgrid with wind power and energy storage systems, in *2017 20th International Conference on Electrical Machines and Systems (ICEMS)* (IEEE, 2017, August), pp. 1–5
24. J.M. Guerrero, J. Matas, L.G. de Vicuna, M. Castilla, J. Miret, Decentralized control for parallel operation of distributed generation inverters using resistive output impedance. IEEE Trans. Industr. Electron. **54**(2), 994–1004 (2007)
25. J. He, Y.W. Li, Analysis, design, and implementation of virtual impedance for power electronics interfaced distributed generation. IEEE Trans. Ind. Appl. **47**(6), 2525–2538 (2011)
26. J. Zhang, J. Ning, L. Huang, H. Wang, J. Shu, Adaptive droop control for accurate power sharing in islanded microgrid using virtual impedance, in *IECON 2017–43rd Annual Conference of the IEEE Industrial Electronics Society*. (IEEE, 2017, October), pp. 2383–2388
27. Q.C. Zhong, Robust droop controller for accurate proportional load sharing among inverters operated in parallel. IEEE Trans. Industr. Electron. **60**(4), 1281–1290 (2011)
28. Y.W. Li, C.N. Kao, An accurate power control strategy for power-electronics-interfaced distributed generation units operating in a low-voltage multibus microgrid. IEEE Trans. Power Electron. **24**(12), 2977–2988 (2009)
29. B. Liu, Z. Liu, J. Liu, R. An, H. Zheng, Y. Shi, An adaptive virtual impedance control scheme based on small-AC-signal injection for unbalanced and harmonic power sharing in islanded microgrids. IEEE Trans. Power Electron. **34**(12), 12333–12355 (2019)

30. P. Sreekumar, V. Khadkikar, A new virtual harmonic impedance scheme for harmonic power sharing in an islanded microgrid. IEEE Trans. Power Deliv **31**(3), 936–945 (2015)
31. A.N. Ronghui, L.I.U. Zeng, L.I.U. Jinjun, Modified adaptive virtual impedance method to compensate mismatched line impedances in microgrids, in *2019 IEEE Applied Power Electronics Conference and Exposition (APEC)*, (IEEE, 2019, March), pp. 1109–1114
32. A. Tuladhar, H. Jin, T. Unger, K. Mauch, Control of parallel inverters in distributed AC power systems with consideration of line impedance effect. IEEE Trans. Ind. Appl. **36**(1), 131–138 (2000)
33. G. Hou, F. Xing, Y. Yang, J. Zhang, Virtual negative impedance droop method for parallel inverters in microgrids, in *2015 IEEE 10th Conference on Industrial Electronics and Applications (ICIEA)*. (IEEE, 2015, June), pp. 1009–1013
34. M. Eydi, R. Ghazi, A novel communication-less control method to improve DC microgrid performance based on AC signal injection. Int. J. Electr. Power Energy Syst. **123**, 106309 (2020)
35. Y. Cao, Z. Li, Z. Lu, Z. Chen, X. Liu, A full-dq based adaptive virtual impedance compensation in virtual power frame for droop-controlled parallel inverters under unbalanced conditions, in *IECON 2017–43rd Annual Conference of the IEEE Industrial Electronics Society*. (IEEE, 2017, October), pp. 8710–8715
36. Y. Li, Y.W. Li, Decoupled power control for an inverter based low voltage microgrid in autonomous operation. in *2009 IEEE 6th International Power Electronics and Motion Control Conference*, (IEEE, 2009, May), pp. 2490–2496
37. Y. Li, Y.W. Li, Power management of inverter interfaced autonomous microgrid based on virtual frequency-voltage frame. IEEE Trans. Smart Grid **2**(1), 30–40 (2011)
38. S. Ghosh, S. Chattopadhyay, S. Samanta, A synchronous reference frame based decentralized control architecture for inverters connected to an autonomous microgrid, in *2019 IEEE Applied Power Electronics Conference and Exposition (APEC)* (IEEE, 2019, March), pp. 540–547
39. Y. Li, Y.W. Li, Virtual frequency-voltage frame control of inverter based low voltage microgrid, in *2009 IEEE Electrical Power & Energy Conference (EPEC)* (IEEE, 2009, October), pp. 1–6
40. H. Moussa, A. Shahin, J.P. Martin, S. Pierfederici, N. Moubayed, Optimal angle droop for power sharing enhancement with stability improvement in islanded microgrids. IEEE Trans. Smart Grid **9**(5), 5014–5026 (2017)
41. B. John, A. Ghosh, F. Zare, Load sharing in medium voltage islanded microgrids with advanced angle droop control. IEEE Trans. Smart Grid **9**(6), 6461–6469 (2017)
42. E. Espina, M. Espinoza, R. Cárdenas, Active power angle droop control per phase for unbalanced 4-wire microgrids, in *2017 IEEE Southern Power Electronics Conference (SPEC)* (IEEE, 2017, December), pp. 1–6
43. R.R. Kolluri, I. Mareels, T. Alpcan, M. Brazil, J. De Hoog, D. Thomas, Power sharing correction in angle droop controlled inverter interfaced microgrids. in *2015 IEEE Power & Energy Society General Meeting* (IEEE, 2015, July), pp. 1–5
44. X. Hou, Y. Sun, W. Yuan, H. Han, C. Zhong, J.M. Guerrero, Conventional P-ω/QV droop control in highly resistive line of low-voltage converter-based AC microgrid. Energies **9**(11), 943 (2016)
45. A.E.M. Bouzid, P. Sicard, A. Yamane, J.N. Paquin, Simulation of droop control strategy for parallel inverters in autonomous AC microgrids, in *2016 8th International Conference on Modelling, Identification and Control (ICMIC)* (IEEE, 2016, November), pp. 701–706
46. T.L. Vandoorn, J.D. De. Kooning, B. Meersman, J.M. Guerrero, L. Vandevelde, Automatic power-sharing modification of $ P $/$ V $ droop controllers in low-voltage resistive microgrids. IEEE Trans. Power Deliv **27**(4), 2318–2325 (2012)
47. T. Hornik, Q.C. Zhong, Voltage control of grid-connected inverters based on h∞ and repetitive control, in *2010 8th World Congress on Intelligent Control and Automation* (IEEE, 2010, July), pp. 270–275
48. A.M. Bouzid, M.S. Golsorkhi, P. Sicard, A. Chériti, H∞ structured design of a cascaded voltage/current controller for electronically interfaced distributed energy resources. in *2015 Tenth International Conference on Ecological Vehicles and Renewable Energies (EVER)* (IEEE, 2015, March), pp. 1–6.

49. T. Hornik, Q.C. Zhong, A current-control strategy for voltage-source inverters in microgrids based on H-infinity and repetitive control. IEEE Trans. Power Electron. **26**(3), 943–952 (2010)
50. A.M. Bouzid, P. Sicard, A. Chériti, M. Bouhamida, H-∞ loopshaping controller design of micro-source inverters. in *2014 IEEE 27th Canadian Conference on Electrical and Computer Engineering (CCECE)* (IEEE, 2014, May), pp. 1–6
51. T. Hornik, Q.C. Zhong, H∞ repetitive current controller for grid-connected inverters, in *2009 35th Annual Conference of IEEE Industrial Electronics*. (IEEE, 2009, November), pp. 554–559
52. T. Vigneysh, N. Kumarappan, Artificial neural network based droop-control technique for accurate power sharing in an Islanded microgrid. Int. J. Comput. Intell. Syst. **9**(5), 827–838 (2016)
53. W. Dong, S. Li, X. Fu, Artificial neural network control of a standalone DC microgrid. In *2018 Clemson University Power Systems Conference (PSC)* (IEEE, 2018, September), pp. 1–5
54. S. Ahmadi, S. Shokoohi, H. Bevrani, A fuzzy logic-based droop control for simultaneous voltage and frequency regulation in an AC microgrid. Int. J. Electr. Power Energy Syst. **64**, 148–155 (2015)
55. M.A. Hossain, H.R. Pota, W. Issa, M.J. Hossain, Overview of AC microgrid controls with inverter-interfaced generations. Energies **10**(9), 1300 (2017)
56. H. Bevrani, S. Shokoohi, An intelligent droop control for simultaneous voltage and frequency regulation in islanded microgrids. IEEE Trans. Smart Grid **4**(3), 1505–1513 (2013)

Economic Analysis of Floating Photovoltaic Plant in the Context of India

Divya Mittal and K. V. S. Rao

1 Introduction

The solar photovoltaic power plant utilizes energy available from the Sun and provides a cleaner alternative to produce electricity. Widespread solar installations are the consequence of decrement in module prices and advancement in semiconductor technology. Solar photovoltaic (SPV) power plants are the solar installations on land, and solar rooftop photovoltaic (RTPV) plants are installed on the roof of the buildings. Many measures are taken, and schemes are continuously being implemented by the Indian government for encouraging the installation of solar photovoltaic power plants in India.

Solar photovoltaic modules when installed on water bodies offer numerous advantages and are called floating photovoltaic (FPV) power plants. FPV power plant as compared to SPV power plant does not require any landholdings, and therefore, the land is conserved. FPV plants have slightly greater energy generation than SPV due to the cooling effect of water. FPV plants conserve the water by preventing evaporation. FPV plant requires additional floating and support structure than SPV, thereby increasing the cost and complexity of installation. The Energy and Resources Institute (TERI) has predicted that India has the potential of installing 280 GW of FPV plants. By 2019, floating solar power plant capacity in India has already crossed 2.7 MW, and more than 1.7 GW was under development phase [1]. MNRE aims to install 10 GW of floating solar power plants by 2022 as a part of its 227 GW renewable energy target [2].

D. Mittal (✉)
Department of Renewable Energy, Rajasthan Technical University, Kota, India

K. V. S. Rao
Gayatri Vidya Parishad College of Engineering for Women, Visakhapatnam, Andhra Pradesh, India

Sanjeevikumar et al. (eds.), *Advances in Renewable Energy and Electric Vehicles*, Lecture Notes in Electrical Engineering 767,
https://doi.org/10.1007/978-981-16-1642-6_13

In this paper, an economic analysis of installing 1 MW FPV power plant in Jaipur, Rajasthan, is done. Economic parameters for 1 MW SPV are calculated by considering different land costs. This paper illustrates the cost-effectiveness of the FPV plant in comparison with the SPV plant based on economic parameters. Economic analysis of 1 MW FPV plant is done for different values of cost–benefit of potable water, which is saved.

In this paper, Sect. 2 describes solar photovoltaic (SPV) power plants, Sect. 3 describes floating photovoltaic plants (FPV) power plants, and Sect. 4 presents a review on the installation cost of floating photovoltaic plants (FPV) power plants. Section 5 describes the economic parameters considered in the study. The methodology of the study is described in Sects. 6, and 7 presents the results from the study.

2 Solar Photovoltaic (SPV) Power Plant

Solar photovoltaic plants (SPV) convert solar energy to electricity using photovoltaic modules. Solar photovoltaic power plants can be installed on land and therefore require large landholdings. To encourage and facilitate SPV power plant installations, the Indian government is planning to create ten solar zones and 25 solar parks. The objective of solar zones and solar parks is to provide a collective land, transmission, and distribution infrastructure for minimizing the initial investment required in the SPV installations [3].

3 Floating Photovoltaic (FPV) Power Plant

Floating photovoltaic power plant (FPV) is installation of solar photovoltaic modules on the water body. Mittal et al. [4] reviewed the studies conducted on FPV system and also described the 10 kW FPV system installed in India at West Bengal, Kerala, and Chandigarh. Saving the valuable land, such installations can prevent water from getting evaporated by decreasing the exposed water surface to the Sun. Mittal et al. [5] showed the amount of water saved from being evaporated by FPV plants for different lakes of Rajasthan and concluded 64 million litres to 496 million litres water savings annually for lakes located at different places. Mittal et al. [6] estimated that the FPV plant has a 2.48% increase in annual energy generation and a 14.56% decrease in average module temperature.

These plants have an added advantage of increased energy production than SPV plants due to cooling of the back surface of PV modules by water. These plants when installed on a cooling pond of a conventional power plant could serve as an auxiliary power source, preventing evaporation of cooling water and also reducing carbon footprints of the plants. These plants when installed on a lake or pond could supply power to nearby localities with minimal transmission and distribution infrastructures.

These plants can maintain water quality by preventing algae growth. Despite numerous advantages offered by FPV power plants over SPV power plants, installations of FPV plants are globally less in number with a total capacity of 98 MW, and in India, it is a recent development with a small number of FPV plants having 10 kW rating. This is due to the enhanced capital cost of FPV plants, requiring floating structure, mooring system, and buoyancy anchors and also due to lack of knowledge on the breakeven analysis of FPV plants. This paper attempts to do an economic analysis of FPV plants.

4 Installation Cost of Floating Photovoltaic System

The floating photovoltaic system has a high installation cost. The total installation cost of FPV can be stated in two parts: the cost of installing a PV system and the cost of installing a floating system or structure. Table 1 shows the installation cost of FPV as stated by different researchers [7–12]. Sahu et al. [11] stated the FPV cost in the context of India. Therefore, the installation cost of 80 Rs/W is considered for calculating LCOE in this paper as per [11].

Table 1 Installation cost of FPV [7–12]

FPV	PV system cost (Rs/W)	Floating system cost (Rs/W)	Total installation cost (Rs/W)	References
FPV system with pontoon platform and monocrystalline PV modules	70.35	44.16	114.51	[7]
1 MW FPV system	NA	NA	175.76	[8]
Thin-film flexible floating PV array	NA	NA	733.932	[9]
100 kW FPV system with pontoon platform	148.6	52.04	200.64	[10]
1 MW FPV system	NA	NA	80	[11]
pontoon-based (poly-crystalline Si) FPV system	NA	NA	191.12	[12]
Flexible (amorphous Si) FPV system	NA	NA	132.48	[12]

NA; Not Available

5 Parameters Considered for Economic Feasibility and Financial Analysis in the Study

5.1 Levelized Cost of Electricity (LCOE)

LCOE is a parameter used for comparing the cost of energy production from different energy sources. LCOE takes into account the investment required and the revenue earned for the entire lifetime of the project as shown in (1). It indicates the price (Rs/kWh) at which the energy must be sold to completely recover the investment done for the entire life of the plant [13]. For solar projects, the high construction cost and low capacity factor produce higher LCOE [14]. Branker et al. [15] reviewed the general assumptions required for LCOE determination of solar PV projects.

Some of the shortcomings of LCOE are: (i) Financing method is generally considered the same for different technologies; (ii) The price in market is dynamic in nature, while LCOE gives a static price; (iii) LCOE is highly sensitive to the assumptions considered; and (iv) The lifetime and other parameter considered for LCOE determination mostly differ from actual value for the plant. Despite these shortcomings, LCOE provides a unique way to indicate the economic feasibility and cost-effectiveness of energy produced from different energy sources. Also, LCOE indicates the technical competitiveness of a power plant in comparison with other power plants.

$$\text{LCOE} = \frac{\text{Initial cost} + \sum_{n-1}^{N} \frac{(O\&M\ \text{Cost})+(\text{Insurance Cost})}{(1+r)^n}}{\sum_{n-1}^{N} \frac{E(1-d)^n}{(1+n)^n}} \tag{1}$$

where r is discount rate, d is degradation rate of photovoltaic modules, N is lifetime of the project, and E is the energy generation in kWh.

5.2 Net Present Value (NPV)

Net present value denotes the difference between present revenue and present investment of a project as shown in (2). Positive value of NPV indicates the net profit earned in the lifetime of the project, and therefore, the project will be economically feasible [16].

$$\text{NPN} = \sum_{n-1}^{N} \frac{(\text{Gross Revenue})}{(1+r)^n} - \sum_{n-1}^{N} \frac{(\text{Gross Cost})_n}{(1+r)^n} \tag{2}$$

where $(\text{Gross Revenue})_n = (\text{Net Revenue})_n - [t_n \times (\text{Total Investment})]$, and t_n is $O\ \&\ M$ cost as a percentage of total investment.

5.3 *Internal Rate of Return (IRR)*

Internal rate of return denotes the discount rate at which the NPV becomes zero as shown in (3). For the project to be economically feasible, the IRR should be greater than the discount rate [17].

$$\sum_{n-1}^{N} \frac{(\text{Gross Revenue})}{(1+IRR)^n} - \sum_{n-1}^{N} \frac{(\text{Investment Cost})_n}{(1+IRR)^n} = 0 \tag{3}$$

5.4 *Profitability Index (PI)*

The profitability index is the ratio of NPV to the initial cost as shown in (4). PI indicates the effectiveness of utilization of initial cost in the project [18].

$$\text{PI} = \frac{\text{NPV}}{\text{Initial Cost}} \tag{4}$$

5.5 *Discounted Payback Period (DPP)*

A discounted payback period is the period for which the NPV of the project becomes zero. If the DPP is less than the lifetime of the project, then the project is considered to be economically feasible [17].

6 Methodology

In this paper, an economic analysis of installing 1 MW floating photovoltaic plant in Jaipur is done and illustrates the cost-effectiveness of the FPV plant in comparison with the SPV plant. The PVWatts calculator is used to calculate the annual energy generation for 1 MW SPV plant in Jaipur. Firstly, the economic parameters of 1 MW SPV plant in Jaipur are calculated. The variation in economic parameters of SPV is studied by considering the following cases of A, B, C, D, E: (A) Zero land cost; (B) Land cost of Rs. 10.12 lakhs/acre [19]; (C) Land cost of Rs. 20.24 lakhs/acre; (D) Land cost of Rs. 30.36 lakhs/acre; and (E) Land cost of Rs. 40.49 lakhs/acre.

Secondly, the economic parameters of 1 MW FPV plant in Jaipur are calculated by considering a 3.6% increase in energy generation [20] due to the cooling effect of water in FPV systems. The economic parameters of FPV are studied by considering

Table 2 Assumptions considered for the Study [8, 11, 15, 19, 21]

Parameter	Value considered	References
Insurance cost	0.25% of initial cost	[21]
Lifetime	25 years	[21]
Degradation rate	1%	[15]
Discount rate	10%	[21]
Operation and maintenance cost for SPV	1% of initial cost for first year and 10% increment for each year	[21]
Operation and maintenance cost for FPV	0.69% of initial cost for first year and 10% increment for each year	[8, 21]
Initial cost of PV installations for 1 MW SPV plant	Rs. 476 lakhs	[19]
Initial cost for 1 MW FPV plant	Rs. 800 lakhs	[11]

the following cases of a, b, c, d, e, f: (a) Zero cost of water savings; (b) Rs. 0.10/l cost of water; (c) Rs. 0.20/l cost of water; (d) Rs. 0.30/l cost of water; (e) Rs. 0.40/l cost of water; and (f) Rs. 0.50/l cost of water. The LCOE, gross revenue, and NPV for FPV are calculated using (5), (6), and (2), while the IRR, PI, and DPP are calculated in the same way as calculated for SPV. The general assumptions considered for calculating economic parameters for the SPV and FPV plants are shown in Table 2 [8, 11, 15, 19, 21].

$$\text{LCOE} = \frac{\text{Initial cost} + \sum_{n-1}^{N} \frac{(O\&M\ \text{Cost})+(\text{Insurance Cost})-(\text{Cost water saved})}{(1+r)^n}}{\sum_{n-1}^{N} \frac{E(1-d)^n}{(1+r)^n}} \tag{5}$$

$$\text{Gross Revenue} = (\text{Net Revenue}) + (\text{Coast of water saved}) - [t_n \times (\text{Total Investment})] \tag{6}$$

7 Results and Discussion

7.1 Economic Analysis of 1 MW SPV Plant in Jaipur

The AC energy generation of 1 MW SPV plant in Jaipur is calculated by the PVWatts calculator as shown in Fig. 1. The annual energy production from 1 MW SPV plant will be 18, 37, 134 kWh. The cases considered for varying land costs are (A) Zero land cost; (B) Land cost of Rs. 10.12 lakhs/acre; (C) Land cost of Rs. 20.24 lakhs/acre; (D) Land cost of Rs. 30.36 lakhs/acre; and (E) Land cost of Rs. 40.49 lakhs/acre.

The variation in LCOE of SPV due to variation in land cost is shown in Fig. 2, and the variation in LCOE, NPV, IRR, PI, and DPP is shown in Table 3. It is observed

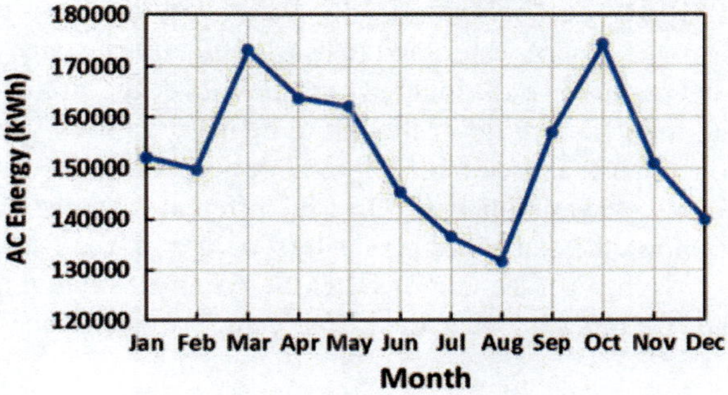

Fig. 1 AC energy generation of 1 MW SPV plant in Jaipur

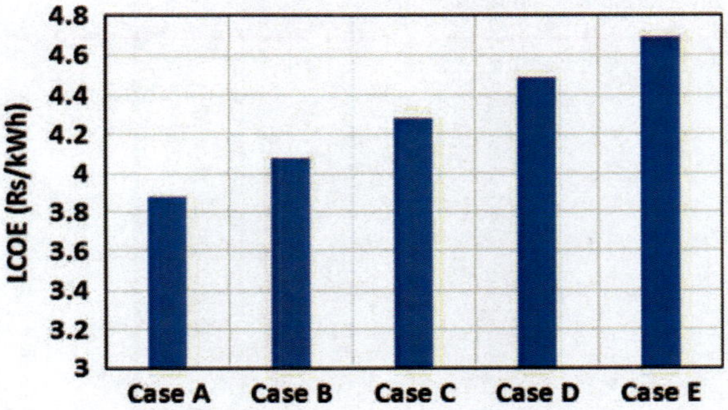

Fig. 2 Variation in LCOE of SPV due to variation in land cost

Table 3 Economic parameters of 1 MW SPV plant for different land costs

SPV land cost (Lakhs Rs. per acre)	LCOE (Rs/kWh)	NPV (Rs.)	IRR (%)	PI	DPP (Years)
0	3.88	5, 45, 23, 972	25.12	1.15	4
10.12	4.08	5, 13, 42, 241	20.83	1.02	5
20.24	4.28	4, 81, 60, 509	20.57	0.91	5
30.36	4.49	4, 49, 78, 778	20.31	0.82	6
40.49	4.69	4, 17, 97, 046	20.05	0.72	6

that LCOE and DPP increase while NPV, IRR, and PI decrease with increment in land cost for 1 MW SPV plant.

7.2 *Economic Analysis of 1 MW FPV Plant in Jaipur*

Considering a 3.6% increase in energy generation [17] due to the cooling effect of water in FPV, the annual energy generation is calculated. Economic parameters are estimated for 1 MW FPV plant by considering: (a) Zero cost of water savings; (b)

Rs. 0.10/l cost of water; (c) Rs. 0.20/l cost of water; (d) Rs. 0.30/l cost of water; (e) Rs. 40/l cost of water; and (f) Rs. 0.50/l cost of water. Figure 3 shows the cash flow and cumulative discounted cash flow over the lifetime for case a. The cumulative discounted cash flow is found to be negative till 9 years, while from the tenth year (i.e. after the discounted payback period), the value is found to be positive. Figure 4 shows the variation in LCOE with different values of cost–benefit of water, and Table 4 shows the economic parameters' values for the 1 MW FPV plant.

As the cost–benefit of water increases, it is found that LCOE for FPV decreases and becomes less than the LCOE value estimated for SPV although the initial cost of

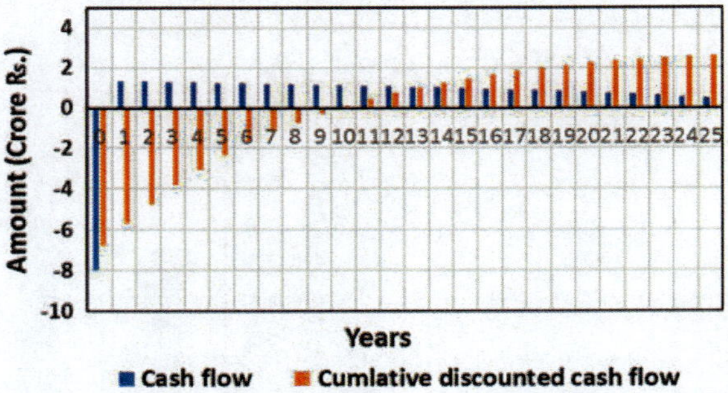

Fig. 3 Cash flow and cumulative discounted cash flow over the lifetime of 1 MW FPV plant for case a

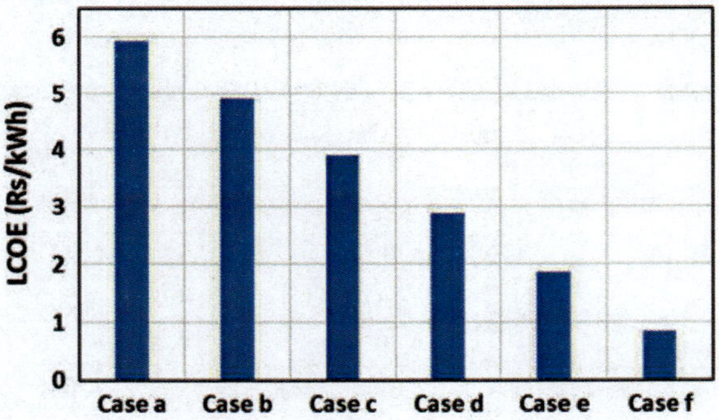

Fig. 4 Variation in LCOE of FPV due to different cost of water

Table 4 Economic parameters of 1 MW FPV plant

Case	LCOE (Rs/kWh)	NPV (Rs.)	IRR (%)	PI	DPP (years)
a	5.93	2, 48, 87, 005	10.91	0.3	10
b	4.92	3, 98, 22, 703	15.43	0.5	8
c	3.90	5, 60, 12, 924	15.91	0.7	7
d	2.88	7, 22, 03, 146	20.43	0.9	6
e	1.86	8, 83, 93, 367	20.88	1.1	5
f	0.84	10, 50, 00, 000	25.37	1.3	4

FPV is higher than SPV. The amount of water saved from being evaporated by FPV represents a large profit even at lower rates of water tariff. Therefore, such a system can ensure economic energy generation along with substantial water conservation.

7.3 Comparison of Economic Parameters for 1 MW SPV and 1 MW FPV.

1 MW photovoltaic plants when installed on land and water have different working environments as well as different types of construction. While SPV plants require large landholdings, still they have comparatively easier installations and maintenance. The FPV plants require no land, but construction and mounting of the floating system on the water body require a skilled workforce. Figure 5 shows the comparison of the total investment required over the lifetime and net present value for SPV and FPV plants. Figure 6 shows LCOE, Table 5 shows cases considered, and Table 6 shows a comparison of parameters for 1 MW SPV and FPV power plant.

1 MW FPV, with zero and Rs. 0.10/l water tariff, is found to be less economical than 1 MW SPV (all cases). While, 1 MW FPV with Rs. 0.20/l and above, water tariff is found to be more economical than 1 MW SPV (all cases). The FPV plants may seem

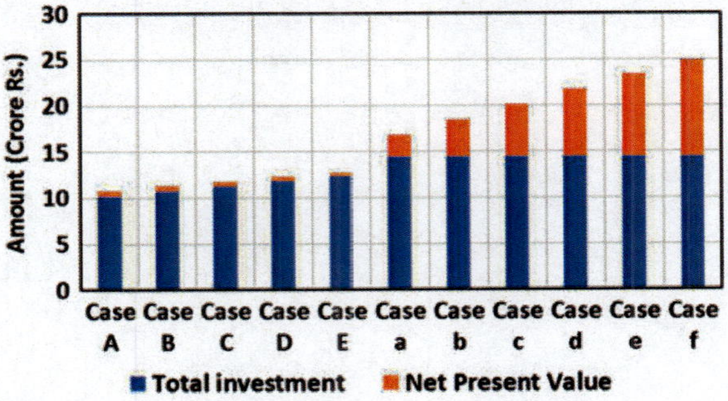

Fig. 5 Total investment required over the lifetime and net present value for 1 MW SPV and FPV plants

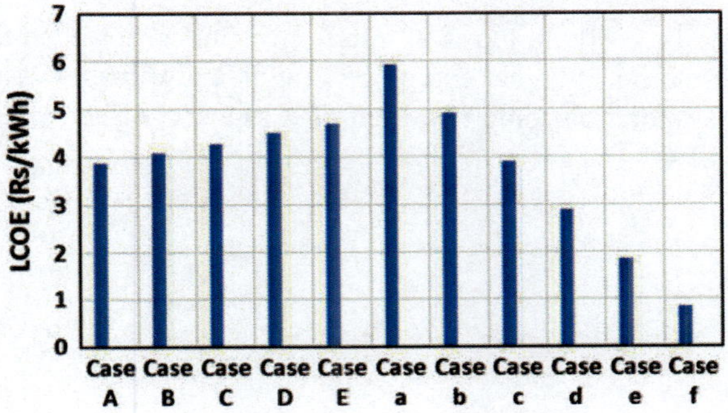

Fig. 6 LCOE for 1 MW SPV and FPV power plant

Table 5 Cases considered in the study

PV type	Case	Description
SPV	A	Zero land cost
	B	Land cost of Rs. 10.12 lakhs/acre
	C	Land cost of Rs. 20.24 lakhs/acre
	D	Land cost of Rs. 30.36 lakhs/acre
	E	Land cost of Rs. 40.49 lakhs/acre
FPV	a	Zero cost of water saving
	b	Rs. 0.10/l cost of water
	c	Rs. 0.20/l cost of water
	d	Rs. 0.30/l cost of water
	e	Rs. 0.40/l cost of water
	f	Rs. 0.50/l cost of water

Table 6 Comparison of parameters for 1 MW SPV and 1 MW FPV plant

		Annual energy generation (kWh)	Initial cost (crore Rs.)	Water saved from evaporation (l)	Annual cost of water saved (Lakh Rs.)	LCOE (Rs/kWh)	DPP (years)
SPV	*A*		4.8			3.9	4
	B	18, 37, 134	5	0	0	4.1	5
	C		5.3			4.3	5
	D		5.5			4.5	6
	E		5.8			4.7	6
FPV	*a*	19, 03, 270	8	1, 78, 36, 455	0	5.9	10
	b				18	4.9	8
	c				36	3.9	7
	d				54	2.9	6
	e				71	1.9	5
	f				89	0.8	4

to be costly in terms of initial investment; however, the long-term benefits of FPV in terms of higher energy generation and valuable water savings due to reduction of evaporation could represent FPV as a cost-effective technology in comparison with SPV.

8 Conclusions

In this paper, economic analysis of installing 1 MW FPV power plant in Jaipur is done. Economic parameters for 1 MW SPV are calculated by considering different land costs. Economic analysis of 1 MW FPV plant is done by considering different values of cost of potable water saved there by giving overall cost–benefit. Levelized cost of electricity and discounted payback period increase, while net present value, internal rate of return, and profitability index decrease with increment in land cost for 1 MW SPV plant. The LCOE value varies from 3.88 Rs/kWh (SPV with zero land cost) to 4.69 Rs/kWh (SPV with land cost of Rs. 40.49 lakhs/acre). The discounted payback period varies from 4 to 6 years.

The LCOE for 1 MW FPV plant is estimated to be 5.93 Rs/kWh for zero water tariff. However, LCOE decreased with increase in water tariff and even became close to 0.84 Rs/kWh (for Rs. 0.50/l water tariff). The discounted payback period varies from 10 to 4 years. The values of LCOE obtained indicate that even at lower tariff of electricity, the project can be economical.

FPV plants have higher initial cost than SPV plant and seem to be less economical, if water saving due to reduction in evaporation by FPV is not considered. If the water saving due to reduced evaporation by FPV is taken into account, then FPV could become meritorious and cost effective than SPV. More advancement and research on materials for floating structures can help reduce installation cost of FPV plants. The advantages offered by FPV plant over SPV plant can encourage adaption of the FPV technology.

References

1. Bridge to India, Floating solar: Opportunities and way ahead. [Online]. (2019, Jun 21) Avaliable: http://www.kwattsolutions.com/kcpst/SolarInstallNew/program1/img/Floating-Solar-Report_WEB.pdf
2. TERI, Floating solar photovoltaic (FSPV): A third pillar to solar PV sector? (2020, Nov 28.). [Online]. Available: https://www.teriin.org/sites/default/files/2020-01/floating-solar-PV-report.pdf
3. MNRE, Implementation of a scheme for development of solar zones in the country commissioning from 2016–17 and onwards (i.e. from the year 2016–17 to 2020–21) (2017, Sep. 08) [Online]. Available: http://mnre.gov.in/file-manager/UserFiles/Scheme-for-development-of-Solar-Zone.pdf
4. D. Mittal, B.K. Saxena, K.V.S. Rao, Floating solar photovoltaic systems: An overview and their feasibility at Kota in Rajasthan, in *2017 International Conference on Circuits Power and Computing Technologies* (Kollam, Kerala, 2017)
5. D. Mittal, B.K. Saxena, K.V.S. Rao, Potential of floating photovoltaic system for energy generation and reduction of water evaporation at four different lakes in Rajasthan, in *2017 International Conference on Smart Technologies for Smart Nations* (Bengaluru, Karnataka 2017).
6. D. Mittal, B.K. Saxena, K.V.S. Rao, Comparison of floating photovoltaic plant with solar photovoltaic plant for energy generation at Jodhpur in India, *in 2017 Int. Conf. Technol. Adv. Power Energy (TAP Energy)* (2017) pp. 1–6

7. S. Azami, M. Vahdaty, F. Torabi, Theoretical analysis of reservoir-based floating photovoltaic plant for 15-khordad dam in Delijan. Energy Equip. Sys. **5**(2), 211–218 (2017)
8. J. Song, Y. Choi, Analysis of the potential for use of floating photovoltaic systems on mine pit lakes: case study at the Ssangyong open-pit limestone mine in Korea. Energies **9**(102), 1–13 (2016)
9. K. Trapani, D.L. Millar, The thin film flexible floating PV (T3F-PV) array: The concept and development of the prototype. Renew. Energy **71**, 43–50 (2014)
10. C. Gisbert, J.J. Gozálvez, M. Santafé, P. Gisbert, F.J. Romero, J.B. Soler, A new photovoltaic floating cover system for water reservoirs. Renew. Energy **60**, 63–70 (2013)
11. Sahu, N. Yadav, K. Sudhakar, Floating photovoltaic power plant: a review. Renew. Sustain. Energy Rev. **66,** 815–824 (2016)
12. K. Trapani, D.L. Millar, Proposing offshore photovoltaic (PV) technology to the energy mix of the Maltese islands. Energy Convers. Manage. **67**, 18–26 (2013)
13. S.B. Darling, F. You, T. Veselka, A. Velosa, Assumptions and the levelized cost of energy for photovoltaics. Energy Environ. Sci. **4**(9), 3133–3139 (2011)
14. P. Wild, The levelised cost of energy of three solar PV technologies installed at UQ Gatton campus [Online]. (2017, Sep. 3). Available: http://gci.uq.edu.au/filething/get/12443/APSRC-PAPER-gsrf-lcoe.pdf
15. K. Branker, M.J.M. Pathak, J.M. Pearce, A review of solar photovoltaic levelized cost of electricity. Renew. Sustain. Energy Rev. **15**(9), 4470–4482 (2011)
16. K. Shukla, K. Sudhakar, P. Baredar, Design, simulation and economic analysis of standalone roof top solar PV system in India. Sol. Energy **136**, 437–449 (2016)
17. F.M. Camilo, R. Castro, M.E. Almeida, V.F. Pires, Economic assessment of residential PV systems with self-consumption and storage in Portugal. Sol. Energy **150**, 353–362 (2017)
18. M. Haegermark, P. Kovacs, J.O. Dalenbäck, Economic feasibility of solar photovoltaic rooftop systems in a complex setting: a Swedish case study. Energy **127**, 18–29 (2017)
19. CERC,. Determination of Benchmark Capital Cost Norm for Solar PV power projects and Solar Thermal power projects applicable during FY 2016–17 [Online]. (2017, Mar. 23). Available: http://www.cercind.gov.in/2016/orders/SO17.pdf
20. M.S.M. Azmia, M.Y. Othmana, M.H. Ruslanb, K. Sopianb, Z.A.A. Majidc, Study on electrical power output of floating photovoltaic and conventional photovoltaic, in*AIP Conference Proceedings*, vol. 1571, pp. 95–101 (2013)
21. J.H. Moro, J.M.M. Duart, Analytical model for solar PV and CSP electricity costs: Present LCOE values and their future evolution. Renew. Sustain. Energy Rev. **20**, 119–132 (2013)

Electrical Field and Potential Distribution Simulation of 220 kV Porcelain String Insulator Using COMSOL Multiphysics

A. M. Vasudeva and H. C. Gururaj

1 Introduction

Generation and consumption of the electrical power seldom happens at close proximity. Overhead transmission lines span over thousands of kilometres carrying the electrical power from generating station to consumer premises [1]. To minimize the transmission losses, electrical power is transmitted at higher voltage levels in the order of kilovolts [2]. Conductors carrying high voltage are attached to tower using insulators. Insulators along with providing the mechanical support also serve the purpose of electrically isolating the live conductors from the tower [3]. Both porcelain and glass insulators are being used for over hundred years [4]. Although these materials have proven themselves in resisting the environmental ageing, owing to the hydrophilic surface the pollution performance of these insulators is meagre. Lately, polymer insulators are extensively used due to their superior pollution performance [5]. The polymer insulator market is the fastest growing amongst the different types [6].

The potential distribution across the insulator string varies greatly due to presence of coupling capacitor between disc insulators and the conductor. The voltage and electric field distribution around disc close to the conductor is 3–4 times greater than the one away from it leading to corona effect. Deterioration of the insulator surface can lead to flashover affecting the safe operation of transmission lines. Hence, it is important to calculate EFPD in and around the string insulator.

A. M. Vasudeva (✉)
Department of Electrical and Electronics Engineering, UBDT College of Engineering, Davangere, Karnataka, India

H. C. Gururaj
Department of Electrical and Electronics Engineering, DRR Government Polytechnic, Davangere, Karnataka, India

Sanjeevikumar et al. (eds.), *Advances in Renewable Energy and Electric Vehicles*, Lecture Notes in Electrical Engineering 767,
https://doi.org/10.1007/978-981-16-1642-6_14

When the contaminated layer on the surface of the insulator becomes wet under fog and rainy conditions, the layer becomes conductive causing electric field enhancement and triggering of the surface discharges, which may ultimately lead to an undesirable flashover [7]. With higher voltages, the problem amplifies, resulting in damage of the equipment. The pollution flashover voltage of ceramic insulators depends not only on the conditions but also on the pollution severity. Hence calls for improving the pollution performance of insulators. Comprehensive analysis of the electric field strength intensification owing to water droplets covering the surface of the ceramic insulator due to wet conditions is imperative for the thorough comprehension of the discharge process and the mechanism of pollution flashover commencement on the ceramic insulator surface. Visible problems can be found relatively easily by inspection [8]. However, the non-visible defects inside the insulators are very dangerous. The carbonization of the rod not only reduces the insulating length of the insulators but also weakens the rod mechanically. Because the electric field strength in environs of a ceramic insulator decreases considerably in form of an internally shorted or defective insulator, the electric field strength measurement method permits the detection of this kind of non-visible defects.

The rest of the paper is organised as: Section 2 explains the need to study electric field strength distribution and methods employed for the same. Section 3 describes the modelling of the 220 kV porcelain string insulator using CATIA and its analysis using COMSOL Multiphysics. Section 4 has the results and the discussion that follows, leading to the conclusion part in Section 5.

2 Need for and Methods to Study Electric Field Strength Distribution

2.1 *Need for the Study of Electric Field Strength Distribution Along the Ceramic Insulator*

The electric field distribution on the insulator [9] is generally non-linear primarily because of the intermediate metal parts. Owing to the below mentioned reasons the electric field strength on the ceramic insulator and the associated hardware is to be regulated.

- To avert substantial discharge activity over the surface of the ceramic insulators under dry, wet conditions influencing the pollution performance of ceramic insulators.
- To avert the discharge activity on the inside of ceramic insulator (thus preventing electrical/mechanical failure).

2.2 Methods to Study of Electric Field Strength Distribution

Two methods are used to study the electric field strength distribution along the insulator namely,

- Experimental methods
- Numerical analysis methods.

In experimental methods as the name suggests physical instruments like flux metres, capacitive probes, electro-optical quartz sensors and dipole antennas are employed to analyze the electric field strength dissemination under dry/wet conditions. The numerical analysis method is further divided into,

- Domain method or Field approach
- Boundary method or Source distribution technique.

Domain method is associable with domains having finite boundaries and comprises of,

- Finite Difference Method (FDM) [10]
- Finite Element Method (FEM) [11].

Boundary method is associable with domains having open boundaries with no restrictions on the domain's geometry and comprises of,

- Charge Simulation Method (CSM) [12]
- Boundary Element Method (BEM) [13].

Differential form of the maxwell's equation is solved by the numerical method of FEM. FEM breaks down the whole problem space in conjunction with the encompassing region into a bunch of sub-regions (finite elements) which are non-separate and non-overlapping and the process is termed as meshing. Finite elements can appear in many shapes. However, typically triangles are used for 2D analysis, tetrahedral for 3D analysis. Every geometric element is articulated by polynomials having nodal values being the co-efficient. Employing weighted residual approach, partial differential equations are condensed to a few definite, positive and symmetric matrix equations.

An uninterrupted object will have boundless degree of freedom, and it is not feasible to solve problems with this configuration. FEM dwindles the degree of freedom from infinite to finite using discretization or meshing (nodes and elements). Every calculation is made at small bunch of points called nodes. Entity attaching nodes and creating a distinct profile-like quadrilateral/triangular is called as element. To obtain the variable anywhere amidst calculation points, interpolation function is employed.

3 Modelling and Analysis of String Insulator

3.1 Modelling of the Insulator Profile Using Catia V5

Distribution of electric field strength with regard to a full scale insulator amid field conditions needs to be studied. A 120 kN ceramic insulator is chosen for the purpose. Relative permittivity of the insulator is 6.3 where as of the pin and cap is 1000. Ground and the line ends of the insulator are provided with the metal fittings. The stress in the insulator under energised condition is uniformly distributed by appropriately designing head portion of cap and ball pin. The surface of the insulator and metal parts coming in contact with the cement are coated with resilient paint in order to negate the effect of unequal thermal expansion. Attention needs to be paid during designing and production of the insulator so that radio interference voltage (RIV) [14] levels are less than the standard values. Figure 1 shows disc type porcelain insulator, labelling and the type of material used is depicted in Table 1.

2D and 3D model of a Single disc porcelain insulator are done using CATIA V5 [15] software. The steps for modelling of the insulator are,

Step-1: Take the profile dimension of test insulator using a thick thread.

Step-2: Plot the profile in a graph sheet and consider the *X*, *Y* coordinates of different points as shown in Fig. 2.

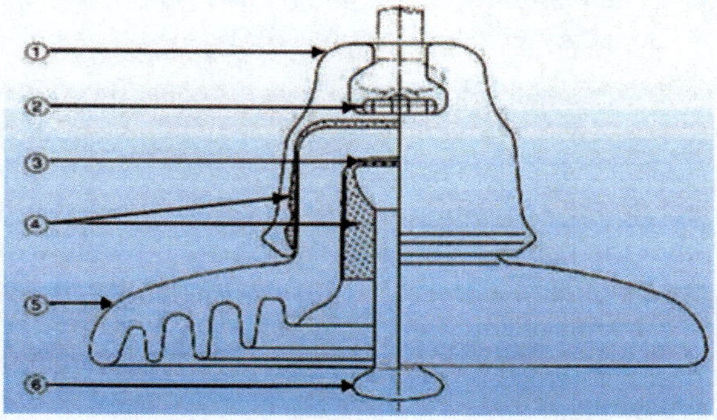

Fig. 1 Design of simulated insulator

Table 1 Porcelain disc type insulator parts and its materials

Item no	Description	Material
1	Socket cap	Spheroidal graphite iron
2	Standard split pin (security clip)	Stainless steel
3	Cushion	Cork/synthetic foam
4	Binding material	Portland cement
5	Shell	Porcelain (brown or grey)
6	Ball pin	Forged steel

Fig. 2 Profile Dimension of 120 kN disc insulator

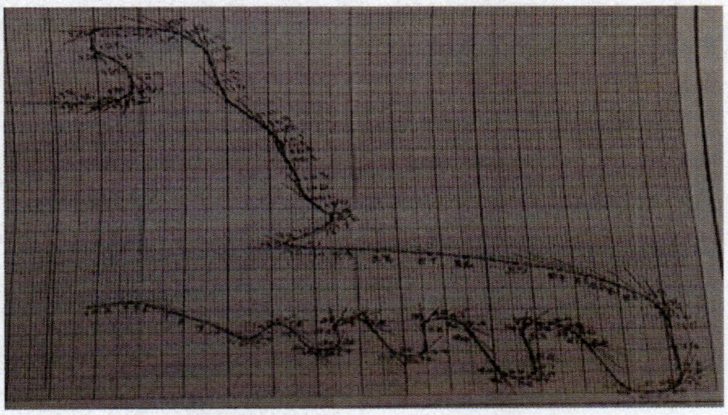

Step-3: Open CATIA software and go to "Start" → click on "Mechanical Design" → select "Part Design", and select one of plane and click on "Sketch" on the right side of the window to draw the 2D model.

Step-4: Turn on "construction /standard element" and then click on the "point by using coordinates" tool and enter the value of X,Y coordinates according to profile readings of porcelain part only.

Step-5: Using "spline" draw over the points to complete the design as shown in Fig. 3 and assure that your model is "iso-constrained" before saving the 2D model.

Step-6: Exit the workbench, and by using "shaft" tool revolve the model to 360° degrees and save the 3D model.

Step-7: Repeat step-4 to step-6 to draw different parts the insulator such as pin, cap, cement and cushion.

Step-8: Go to "start" click on "mechanical design" and select "assembly design", go to tree and right click on "product" and select "components" then select "Existing components" and load all parts one after the other by renaming them.

Step-9: Assemble the different parts by using "coincidence constraint" and "contact constraint" command tool as shown in Fig. 4.

Step-10: for the sectional view of the insulator, revolve it through an angle less than 360° (say 270°).

Fig. 3 Insulator design using SPLINE command

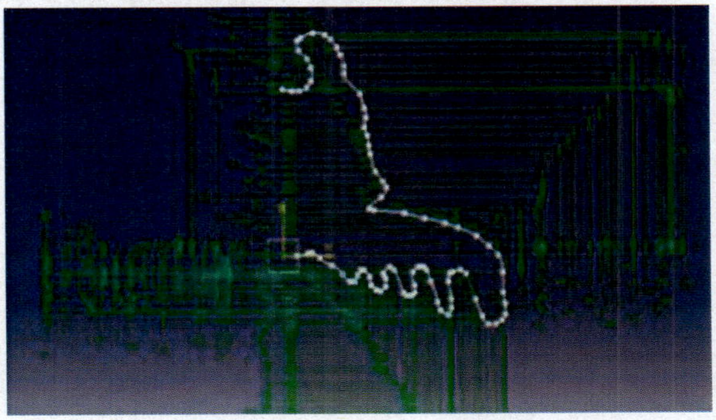

Fig. 4 Assembly of different parts using "coincidence constraint" and "contact constraint commands

Fig. 5 Half elevation CATIA Model of 120 kN standard insulator

Step-11: to apply the material to a particular part, select "Apply material" tool. Drag and place the material on that respective part and click on "Apply material", after completion of the assembly save the design.

Step-12: in order to place the insulator discs one above the other, click on rectangular pattern command; give the number, direction and distance, to obtain the insulator string.

Figure 5 shows the half elevation and Fig. 6 represents the 3D model of the single insulator whilst Fig. 7 shows the insulator string with 15 discs.

3.2 *Analysis of Insulator Using Comsol Multiphysics*

In order to analyse electric field and the voltage distribution in insulators, 15 number of 120, 220 kV porcelain disc insulators are connected in series to form a string. In order to solve MAXWELL's equation COMSOL MULTIPHYSICS [16] employs the Finite Element Method, hence it is mandatory to impose the boundary conditions.

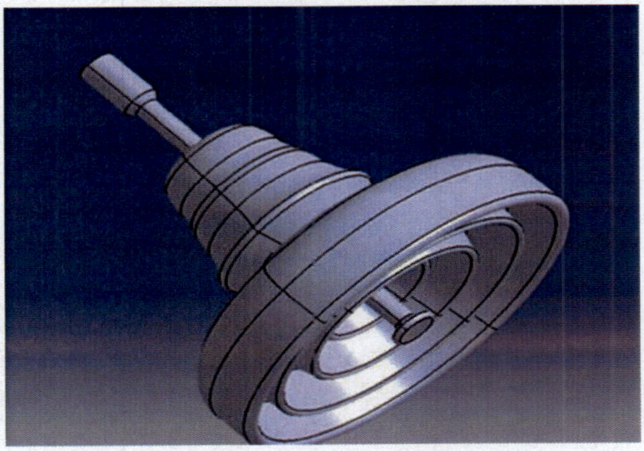

Fig. 6 The insulator 3D model designed using CATIA software

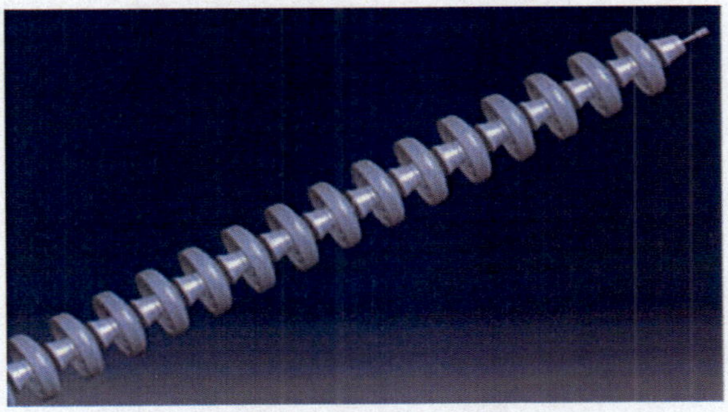

Fig. 7 Insulator string with 15 discs

The outer boundary conditions are set to zero. The following steps are carried out for the analysis,

Step-1: Run "Analysis product launcher" and click on "Run".
Step-2: Go to "stationary " settings.

- Include geometric non linearity in "off".
- In physics and variables selections prefer "Electrostatics (es)".

Step-3: import the file which is saved in "mph" format to COMSOL.
Step-4: to obtain required result, the data on geometry needs to be set.
Step-5: select the component.

- Set the unit system "SI".
- Set geometry shape order as "Automatic".

Step-6: In definition select the "Pairs".

- Select "identity pair"
- Give geometric entity level as "boundary" and select the boundary.
- Follow the same procedure for source selection and destination selection.

Step-7: Repeat the above step to select all the pairs and give the extrapolation tolerance value as 1e−4.

Step-8: Select the assembled product and then choose the option of surfaces as "off" and remove redundant edges and turn vertices "on".

Step-9: Go to the Array option and give the size value as {1, 1, 16} and displacement value as {0, 0,160.732}.

- Click on "form Assembly option" and give relative repair tolerance value as 1e−10.

Step-10: Click on the materials option then select the related materials which are already defined in the software. The required value can be defined or changed.

Step-11: Once all relative materials are selected then go for selection of terminal in the geometry.

- Click on geometry and select potential points and enter the voltage value of the selected domains, i.e. initial value as "zero" and electric potential value as 22,000.

Step-12: Click on continuity and select all domains which are available in the selection box.

- For "Meshing" click on "Mesh" select minimum element quality as 0.05978. Select Tetrahedral and triangular elements. Select the maximum and minimum size of the meshing for the domain required.

Step-13: For computation of the problem,

- Click on stationary.
- Select electrostatics (ES).

Step-14: to solve the configuration click on "study", select "study1" & study step as "stationary".

Step-15: for dependant variables define the study step as "stationary" and select electric potential and then click "Field components".

Step-16: the obtained results can be plotted by using "Plot graphs " option. Both electric potential & field distribution can be plot on 2D and 3D graphs.

4 Simulation Results Using Finite Element Analysis in Comsol

4.1 Potential Distribution on 220 kV Porcelain Disc Insulator String

Four different cases are considered,

- Case 1: insulator string under clean/dry condition

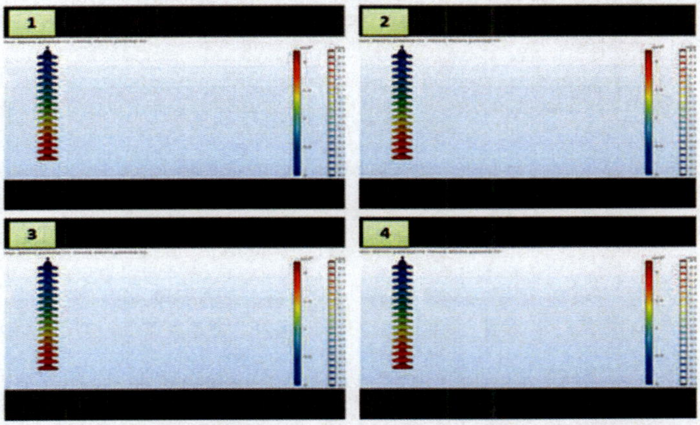

Fig. 8 A plot showing electric potential distribution of ceramic insulator string, Under Clean/ dry, lightly polluted, moderately polluted, heavily polluted condition

- Case 2: low level pollution, 25 g/L of Sodium Chloride (Nacl) is uniformly covered on the surface of each insulator in the string and the equivalent salt deposit density (ESDD) value is 0.04 mg/cm^2
- Case 3: moderate level pollution, 50 g/L of Nacl is uniformly covered on the surface of each insulator in the string and the ESDD value is 0.10 mg/cm^2
- Case 4: high level pollution, 120 g/L of Nacl is uniformly covered on the surface of each insulator in the string and the ESDD value is 0.46 mg/cm^2

Case 1: Fig. 8 (1) illustrates potential distribution along the 220 kV porcelain insulator string under clean/dry condition. It is noticed that the potential is distributed over the insulator surface from ground to live end and maximum potential observed near conductor end is 215 kV.

Case 2: Fig. 8 (2) presents potential distribution along the 220 kV porcelain insulator string under ESDD = 0.04 mg/cm^2. It is observed that the potential is distributed over the insulator surface from live to ground end and maximum potential observed near conductor end is 215 kV.

Case 3: Fig. 8 (3) shows the potential distribution along the 220 kV porcelain insulator string under ESDD = 0.10 mg/cm^2. It is observed that the potential is distributed over the insulator surface from live to ground end and maximum potential observed near conductor end is 204 kV.

Case 4: Fig. 8 (4) shows the potential distribution along the 220 kV porcelain insulator string under ESDD = 0.46 mg/cm^2. It is observed that the potential is distributed over the insulator surface from live to ground end and maximum potential observed near conductor end is 215 kV.

Table 2 shows the potential distribution across each disc in the string under clean/dry as well as different contamination conditions. The insulator disc close to the ground end is most exposed to the environmental conditions, hence chances of

Table 2 Potential distribution across each disc in the string under clean and dry and different pollution level condition

Different pollution levels →	Under clean and dry condition	ESDD (mg/cm^2) 0.04	IESDD (mg/cm^2) 0.10	ESDD (mg/cm^2) 0.46
Number of discs connected in string ↓	Potential in kV	Potential in kV	Potential in kV	Potential in kV
Unit 1 (conductor end)	215	215	204	215
Unit 2	204	193	193	204
Unit 3	193	181	181	193
Unit 4	160	160	171	160
Unit 5	149	149	160	149
Unit 6	138	126	138	126
Unit 7	126	116	126	116
Unit S	116	104	116	104
Unit 9	104	94	104	94
Unit 10	94	82	94	71
Unit 11	71	71	71	60
Unit 12	60	60	60	50
Unit 13	28	50	39	28
Unit 14	17	28	28	17
Unit 15 (ground end)	17	6	17	6

pollution accumulation is more compared to other discs in the string. The accumulation of pollutants on the surface of the insulator combined with moisture, rain or fog, results in smaller value of potential distribution for polluted condition compared to clean/dry condition. From the data in Table 2, it is clear that the potential distribution decreases as we move upwards from the conductor end towards the ground end.

The magnitude of electric potential decreases when moved from the conductor end to the ground end. Upto certain disc numbers the magnitude of electric potential decreases under both clean/dry and polluted conditions, beyond which almost similar electric potential values are obtained.

4.2 *Electric Field Distribution of 220 kV Porcelain Disc Insulator String*

In order to analysis the electric field distribution of porcelain insulator string following regions are considered.

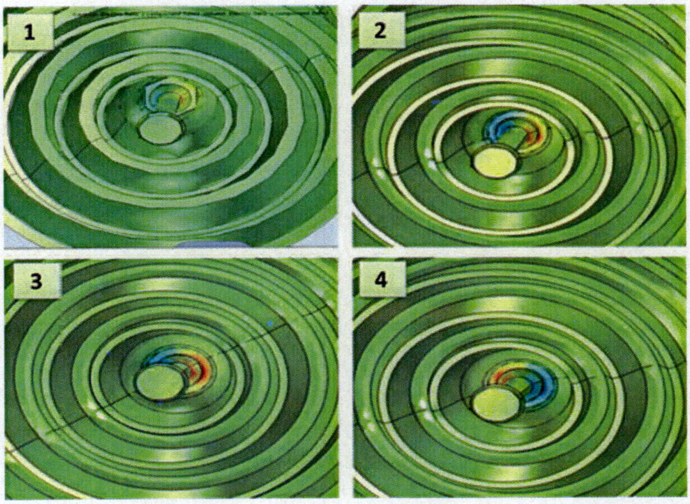

Fig. 9 A plot showing electric field distribution at conductor end disc of 220 kV insulator string under, clean and dry, lightly polluted, moderately polluted, heavily polluted condition

- Junction of pin and cement
- Junction of pin and cap.

Figure 9 shows the electric field distribution at the disc closest to the conductor under clean/dry condition and after adding the pollutants. The first condition considered is when the insulator is clean/dry, as seen in part 1 the electric field is moderate, but as the contamination is added the electric field intensity goes on increasing as indicated by the increase in concentration of the colours as we move from part 2 (lightly polluted condition) to part 3 (moderately polluted condition) towards part 4 (heavily polluted condition).

Figure 10 shows the electric field distribution at the pin and the cap region of the discs away from the conductor. Part 1 shows the electric field distribution at the cap and pin region at 4th, 5th and 6th discs under clean and dry condition, Part 2 at cap and pin region of 5th , 6th discs under lightly polluted, part 3 at cap and pin region of 6th disc under moderately polluted and finally Part 4 at cap and pin region of 2nd disc under heavily polluted condition. Under clean and dry condition maximum electric field strength observed at pin and cement region is 1.56 kV/mm and at cap and pin region 1.82 kV/mm.

From the above figures and data in Table 3, it is clear that the electric field distribution across the insulator string increases with the increase in the pollution level. The electric field distribution is more at cap and pin region compared to pin and cement region. Magnitude of the electric field is high at the energised end and reduces along the length of the insulator string as moved towards the ground end. The highest electric field of 1.68 kV/mm along the insulator string is recorded at cap and pin region.

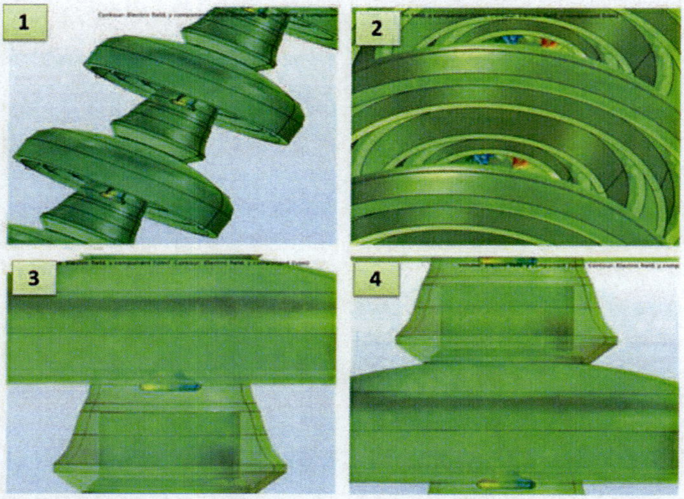

Fig. 10 A plot showing electric field distribution at cap and pin region for various discs under different pollution levels

The maximum field recorded under highly polluted condition at pin and cement junction is 1.15 kV/mm and cap and pin junction is 1.68 kV/mm.

5 Conclusion

In this study, the electric field and potential distribution in the vicinity of ceramic insulator string under clean/dry and different level of pollution have been presented. Finite element analysis based electric field analysis is carried out employing COMSOL for the calculation of EFPD. Standard 120 kN porcelain insulator string with 15 discs is selected to simulate EFPD.

The modelling and discretization of the insulator are carried by using commercially available CATIA. Based on these discussions the following conclusions are drawn,

- For the insulators near the conductor end the potential distribution is almost same for clean/dry and polluted conditions. But as we move up towards the ground end which has higher chances of pollutants settling on the surface, the value of potential distribution under polluted condition is far less compared to clean/dry condition. The difference in the value of potential distribution for clean/dry and polluted condition is highest at the ground end
- The electric field intensity increased with the increase in pollution level

Table 3 Electric field distribution across each disc in the string under Clean/dry and different pollution conditions

Pollution level →	Under clean and dry condition		ESDD (mg/cm^2) 0.04		lESDD (mg/cm^2) 0.1		ESDD (mg/cm^2) 0.46	
Creepage length in mm ↓	Maximum stress or field points considered in KV/mm		Maximum stress or field points considered in kV/mm		Maximum stress or field points considered in kV/mm		Maximum stress or field points considered in kV/mm	
	At pin and cement	At cap and pin	At pin and cement	At cap and pin	At pin and cement	At cap and pin	At pin cement	At cap and pin
450	0.78	1.3	0.81	1.5	0.81	1.5	0.98	1.68
1018	0.78	1.56	0.81	1.33	0.81	1.15	1.15	1.5
1586	1.04	1.82	0.81	0.98	0.63	0.98	0.81	1.33
2154	1.04	1.82	0.81	1.15	0.81	0.98	0.81	1.15
2722	0.78	1.56	0.98	1.51	0.98	1.33	0.98	1.5
3290	1.04	1.3	0.98	1.33	0.81	1.33	1.15	1.33
3858	1.04	1.56	0.98	1.15	0.81	1.15	0.81	1.15
4426	1.04	1.3	0.81	1.68	0.63	1.5	0.81	1.5
4994	0.52	1.82	0.98	1.5	0.98	1.5	0.81	1.33
5562	0.78	1.3	0.98	1.15	0.63	1.33	0.98	1.15
6130	1.3	1.56	1.15	1.5	0.81	0.98	0.81	1.15
6698	1.56	1.56	0.81	1.15	0.98	0.81	0.81	0.98
7266	1.04	1.82	0.98	1.5	0.98	1.33	0.98	1.5
7854	0.52	1.82	0.81	1.68	0.81	1.5	1.15	1.68

- The electric field distribution at the region of pin and cement junction, pin and cap junction increased with the increase in pollution level all along the length of the insulator string
- Pin region is identified as the most probable source for starting of partial arcs.

References

1. E. Shishkov, V. Goldstein, I. Krivihin, Open overhead transmission lines. Appl Mech Mater **792**, 293–299 (2015). https://doi.org/10.4028/www.scientific.net/AMM.792.293
2. K. Goh, Efficiency of high voltage transmission in power lines. Phys Educ **01**, 1920009 (2019). https://doi.org/10.1142/S2661339519200099
3. Jiang X-Y, Zhao W-Y (2011) The design of insulation of high voltage transmission line and lightning protection device 24:150–153+158
4. W. Shugg, Ceramic and glass insulations, pp. 417–432 (2009). https://doi.org/10.1109/9780470544778.ch15

5. M. W. Rachmawati, Suwarno, Study on leakage current characteristics and electrical equivalent circuit properties of aged polymer insulator under artificial environmental condition. in *2018 Conference on Power Engineering and Renewable Energy (ICPERE)* (Solo, Indonesia, 2018) ,pp. 1–6. https://doi.org/10.1109/ICPERE.2018.8739617
6. Electric Insulator Market by Type (Ceramic, Glass, and Composite). Accessed on Nov 29, 2020. [online]. Available : https://www.marketsandmarkets.com/Market-Reports/industrial-insulators-market-239382720.html
7. B. Sarang, V. Lakdawala, P. Basappa, Electric field calculations on a high voltage insulator under wet conditions 86–90 (2009). https://doi.org/10.1109/EIC.2009.5166327
8. Practical Guidelines for Visual Inspection & Condition Assessment of Transmission Insulation. Feb 29, 2020. Accessed on Nov 29,2020. [Online]. Available: https://www.inmr.com/practical-guidelines-visual-inspection-condition-assessment-transmission-insulation/
9. S. Zungu, G. Ininahazwe, I. Davidson, Voltage and electric field distribution on polymeric insulators (2015)
10. R. Ohnishi, D. Wu, T. Yamaguchi, S. Ohnuki, Numerical accuracy of finite-difference methods, in *2018 International Symposium on Antennas and Propagation (ISAP)* (Busan, Korea (South), 2018) pp. 1–2
11. J. Katsikadelis, The finite element method (2020). https://doi.org/10.1016/B978-0-12-818643-5.00011-X
12. D. Harimurugan, G.S. Punekar, A comparative study of field computation methods: charge simulation method and method of moments, in *2018 International Conference on Power, Signals, Control and Computation (EPSCICON)* (Thrissur, 2018), pp. 1–4. https://doi.org/10.1109/EPSCICON.2018.8379586
13. S. Chakraverty, N. Mahato, P. Karunakar, T.D. Rao, Boundary element method (2019). https://doi.org/10.1002/9781119423461.ch8
14. S. Coatu, M. Costea, D. Rucinschi, V. Cristea, T. Leonida, Measurement of radio interference voltages due to insulator sets and other line equipment (2004)
15. "CATIA v5" Accessed on Nov 29, 2020. [online]. Available: https://www.3ds.com/products-services/catia/?wockw=card_content_cta_1_url%3A%22%3A%2F%2Fblogs.3ds.com%2Fcatia%2F%22
16. *"COMSOL MULTIPHYSICS"* Accessed on Nov 29,2020. [online]. Available: https://www.comsol.co.in/

Energy Prospects for Sustainable Rural Livelihood in Vijayapur District, Karnataka India

Mukta M. Bannur and Suresh H. Jangamshetti

1 Introduction

Agriculture and its allied sectors significantly require electricity for irrigation and farm mechanization [1]. However, grid supply in rural India lacks both in quality and quantity [2]. It is hampering income generating activities. Moreover, women and children in rural area are spending 3–5 h/day on collection of basic needs. More often, they are exposing too many threats like; health hazards, assaults and snake bites, etc. Indeed, a sound technology which could provide sustainable energy for rural consumers is essential infrastructure to improve their livelihood [3].

Renewable energy-based power supply: Solar photovoltaic (SPV) and small wind turbine generator (SWTG) are gaining due importance in current scenario to supply power for remote and rural demands [4]. Energy use and its management are closely linked to climatic and socio-economic status of that place. Furthermore, designing and planning of power supply depend on the current and future programs. Thus, energy approaches for sustainable development should not only be compatible with, but also contribute towards regional development [5].

Rural energy provisions certainly necessitate sound strategy and administration judgments on three factors such as resources, economics and environment. In this regard, a study is conducted in rural areas of Vijayapur district (East latitude 15° 50′–17° 28′, North longitude 74° 59′–76° 28′) Karnataka India. Vijayapur district is one of the agricultural potential regions of Karnataka and endowed with natural

M. M. Bannur (✉)
Department of Electrical and Electronics Engineering, BLDEA's V P Dr. PG Halakatti College of Engineering and Technology, Vijayapur, Karnataka, India
e-mail: ec.mukta@bldeacet.ac.in

S. H. Jangamshetti
Department of Electrical and Electronics, Basaveshwar Engineering College, Bagalkot, Karnataka, India
e-mail: suresh.j@ieee.org

Sanjeevikumar et al. (eds.), *Advances in Renewable Energy and Electric Vehicles*, Lecture Notes in Electrical Engineering 767,
https://doi.org/10.1007/978-981-16-1642-6_15

resources. District needs attention to boost up the sustainable livelihood in the rural areas [6]. Major infrastructure essential to carry out the agri and its allied sectors is a reliable power supply. Thus, the objectives of this study is: to identify on-farm and off-farm opportunities for sustainable livelihood, to understand their energy consumption patterns, assess renewable resource potential for power generation and finally propose a sustainable energy option for rural applications.

2 Methodology

Renewable resources potential assessment and consumers' power demand with annual consumption pattern are essential aspects for designing techno-economical viable power supply. The necessary meteorological time-series data includes solar radiation, wind speed and temperature data is collected from automatic weather station at regional agricultural research station (RARS), Vijayapur, Karnataka India. Electricity consumption data composed from field survey. MATLAB is used to analyse real-time meteorological time-series data, and findings are explored according to [7]. The results of solar and wind resources analysis are used in HOMER software [8] to obtain optimal size of solar photovoltaic (SPV), small wind turbine generator (SWTG) and solar-wind hybrid energy (SWHE) systems for electrification of typical farmhouse load.

2.1 Solar Power Potential in Vijayapur

The time-series solar radiation data for past five years (2015–2019) is processed to study the solar potential in the Vijayapur location [7]. Statically processed monthly variation of temperature and mean solar potential at study location is presented in Table 1. District experiences 320–330 clear sunny days around a year and temperature variation observed between 7.1 and 41.3 °C. In brief, an average solar energy varies from 3.2 to 6.78 kWh/m^2/day, and it is available for 10–11 months over the year. Solar data analysis also depicts that, in study location, August–September is lean period for application solar potential.

Variation of average solar potential over five years is processed, and findings are presented in Table 2. Table 2 depicts that, seasonal variation of solar radiation is approximately same for most of the months over five years, except during year 2017, it is low (2.89 kWh/m^2/day) (Table 2).

Table 1 Monthly variation of solar potential and temperature at Vijayapur

Month	Solar energy potential kWh/m²/day	Temp (°C)	
		Low	High
Jan	5.29	8.2	35.2
Feb	5.81	12.9	37.1
Mar	6.51	20.5	39.6
Apr	6.64	21.6	40.7
May	6.34	21.0	41.3
Jun	4.94	20.9	34. 1
Jul	3.86	20.3	33.2
Aug	3.96	21.1	32.7
Sep	3.78	18.3	33.1
Oct	4.46	19.3	33.4
Nov	4.55	20.3	31.2
Dec	4.79	7.1	32.7

Table 2 Monthly variation of daily solar potentialand temperature at Vijayapur

S. No.	Season	Total horizontal radiation (kWh/m²/day)				
		2015	2016	2017	2018	2019
1	Summer	6.32	6.37	5.85	5.92	6.22
2	Monsoon	4.01	3.9	2.89	3.8	4.12
3	Winter	5.20	5.13	4.77	5.2	5.10

2.2 *Wind Power Potential in Vijayapur*

The statically processed wind potential characteristics of Vijayapur resource at 10 and 20 m heights are presented in Table 1. It presents monthly and annual mean wind speed, Weibull parameters and wind power density at Vijayapur resource. From Table 1, it is found that annual mean wind speed 4.55 m/s and wind power density 109 W/m^2 designate that resource is good for power applications. The monthly mean wind speeds over a year are >3.3 m/s, including all other parameters: $k > 2$, c=5 m/s, $V_{\max} = 7.1$ m/s and power density 110 W/m^2 presents that, Vijayapur resource is feasible for wind power applications. In addition, the wind regime is fair enough at 20 m height.

Table 3 Monthly and annual mean wind speed, Weibull parameters and wind power density at Vijayapur resource

Month	Mean wind speed (m/s)		k-parameter		c-parameter (m/s)		V_{max} (m/s)		WPD (W/m^2)	
	10 m	20 m	10 m	20 m	10 m	20 m	10 m	20 m	10 m	20 m
Jan	3.26	3.47	1.61	1.91	3.42	4.02	5.63	5.86	45.82	65.37
Feb	3.28	3.69	1.6	1.89	3.66	4.31	6.08	6.32	58.42	67.67
Mar	3.73	4.45	1.67	1.97	4.17	4.91	6.71	7.03	77.34	92.16
Apr	4.25	4.33	2.19	2.59	4.54	5.28	6.12	6.58	62.83	88.49
May	4.83	5.73	2.89	3.41	5.42	6.27	6.51	7.18	99.34	159.94
Jun	5.57	6.61	2.88	3.4	6.06	7.01	7.28	8.03	145.06	191.82
July	7.00	7.57	3.1	3.66	7.69	8.9	9.04	10.02	248.28	364.3
Aug	6.45	6.67	2.78	3.28	6.91	8.2	8.4	9.24	188.73	273.08
Sep	5.38	5.79	2.31	2.73	6.07	7.05	7.96	8.63	144.24	222.74
Oct	3.74	4.56	1.5	1.77	4.14	4.9	7.3	7.52	88.8	130.08
Nov	3.51	4.42	1.52	1.79	3.89	4.6	6.77	6.99	63.8	92.37
Dec	3.42	3.73	1.27	1.51	3.45	4.14	7.24	7.25	74.43	77.59
Annual	4.55	5.18	2.11	2.49	4.98	5.8	7.08	7.55	108.09	152.34

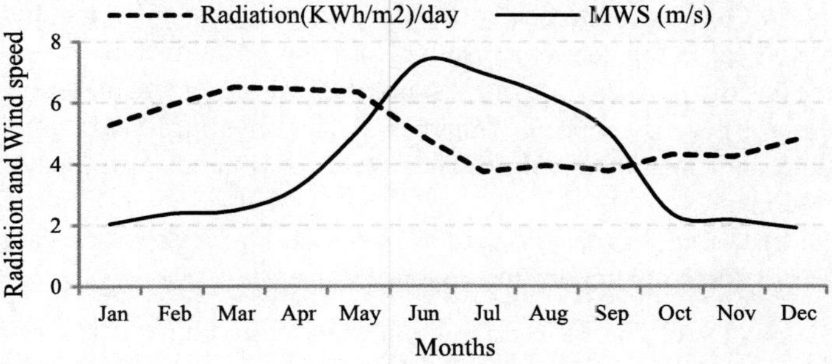

Fig.1 Complementary nature of wind and solar potential

2.3 *Wind-Solar Hybrid Energy Potential in Vijayapur*

From the wind and solar resource analysis, it is found that both the potentials are fortunately complementary in nature over the day and year. Variation of wind and solar potential is shown in Fig. 1. This kind of opportunity can be effectively utilized by integrating them in SWTG-SPV hybrid energy system. The power system can exploit inter dependent strengths of both SPV and SWTG systems [8].

2.4 *On-Farm and Off-Farm Opportunities*

In the study area, there are ample of small-scale opportunities that have potential to generate employment and sustainable livelihood for rural population. Prospectus suitable for rural public has been identified both on-farm and off-farm sectors and are listed as:

On Farm Activities: The sustainable agricultural allied activities in the study area include:

- Animal husbandry: The climatic conditions and resources are ideal for dairy development. Dairy value chain is currently not organized and underdeveloped potential in the district. The waste land development and water management programs are necessary for developing more of fodder areas to support the dairy sector in the region.
- Poultry: Resources such as waste land, barren land and meteorological conditions are in favour of poultry farming. However, poultry farming is not planned and developed in the district. Its development could bring speedy economy and social transformation among rural culture.
- Horticulture: Potential land, water supply and excellent climatic conditions are more suitable for fruits and vegetable farming. In the district, there is a scope for vineries, jam, packed fruit and vegetables.

- Floriculture: Access to the flower trading places nearby, small landholdings and water resources in the district are suitable for flower cultivation.
- Fisher: Five rivers in the district provide huge scope of inland and fresh water for fishing. There is scope for ornamental fish breeding. Sericulture: Climatic conditions and small landholdings are suitable for mulberry cultivation and silk-worm rearing.
- Food and Beverages processing: Man power and agri-resources such as crops, vegetables and fruits support this sector.

Load demand and energy consumption pattern of above-listed activities are collected during walk-in survey. The load demand and energy consumption pattern of cattle farming (animal husbandry) are discussed, and it is presented in subsection E.

Non-farm Activities: The major off-farm activities are listed with major electric fixtures required to process the activity. The average power demand for lighting and cooling fixtures is between 500 and 1000 W.

- Milling and grinding: Major electric fixture is grinder of capacity 2–5 HP.
- Tailoring and sewing: Major electric fixture is universal motor of 50 W.
- Repairing and servicing: Form requires welding and soldering machines of 1 kW.
- General shop including grocery: Arrangement requires major equipment refrigerator for cold drinks (1 kW).
- Health clinic: Essential equipment in setup is sterilizer (1 kW).
- Education and training centres: TV, computer, printer and photocopier are the necessary gadgets (2 kW) in the system.
- Home restaurant, mix-grinder and refrigerator are the major appliances (1.1 kW).
- Carpentry shop: saw, vertical saw, levelling and drilling machines (1–2.5 kW), wood and metal grinders with 1–2.5 HP motors.

The listed sectors need minimum infrastructure, manpower and financial support. The qualified unemployed youth capacity in the district is more appropriate to run the sectors. The reliable electric supply and training are bottle neck to achieve self-employment in the region.

3 Electricity Consumption Pattern

Load demand and electricity consumption patterns are paramount in designing renewable-based power systems. Representative electricity consumption pattern in cattle farmhouse located near Vijayapur is considered for the present study. Farming requires water stock of 2000–2500 L/day, and it is more during summer season. The peak load of farmhouse on the utility grid is 2 kW. The electric gadgets with power rating and working duration are presented in Table 4.

The energy consumption pattern is simple and almost fixed. Load profile at 30 min time resolution over 24 h on typical day in summer season is shown in Fig. 2. The

Table 4 Electric appliances in farmhouse with capacity and consumption over a day

Sl	Electric appliance	Quantity	Power rating	Total watt	Duration (h)
1	CFL	8	11	88	8–10
2	CFL	5	5	25	6–8
3	Fluorescent tube	4	40	160	5–7
4	Ceiling fan	5	60	300	0–12
5	Mix grander	1	175	175	20–30 min
6	Television	1	100	100	2–5
7	Refrigerator	1	100	100	50% duty cycle
8	Water pump	1	740	740	1–2
9	Other	1	100	100	2
	Total	27	–	1788	–

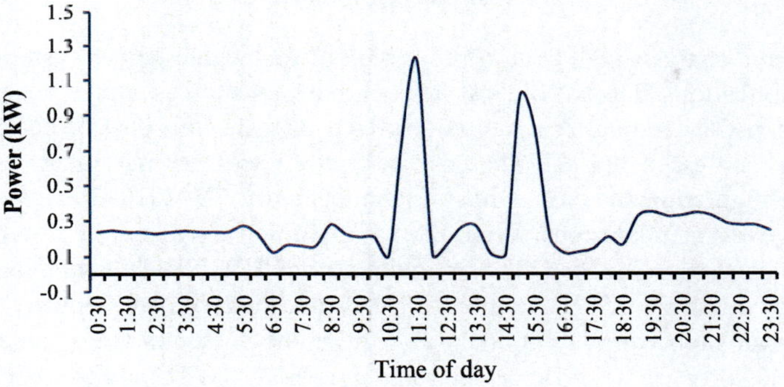

Fig. 2 Hourly load demand during typical day in summer

monthly energy consumption pattern depends on seasons around a year, and it is more during summer season. Monthly energy consumption pattern during year 2018 is shown in Fig. 3.

4 Design and Techno-Economic Analysis of Power Systems

Techno-economic analysis is paramount to promote technology for power applications. HOMER software is used to obtain optimal size of renewable energy-based standalone SPV, SWTG and wind-solar hybrid energy systems [9, 10]. The power systems are designed for the load demand of farmhouse load discussed in subsection D. The loss of power supply probability (LPSP) is estimated to analyse the technical viability of the power systems. It is evaluated using Eq. (1).

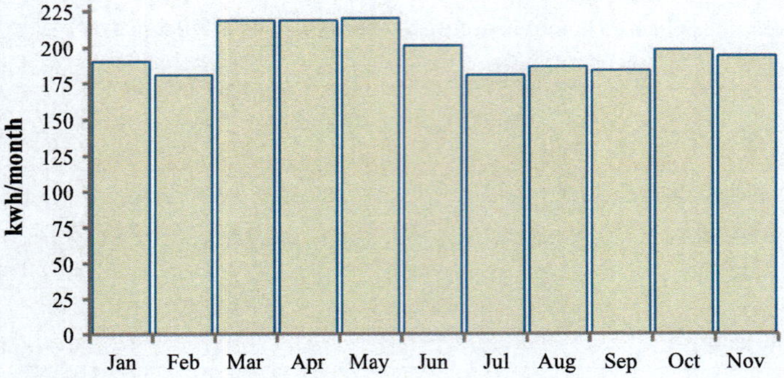

Fig.3 Monthly energy consumption at farmhouse

$$\text{LPSP} = \frac{\text{Total power dificit over a year}}{\text{Total power produced over year}} \tag{1}$$

Economic analysis is crucial in promotion of renewable power system for standalone applications. The well-known cost parameters such as capital cost (CC), life cycle cost (LCC), annualized capital cost (ALCC) and unit cost of energy (UCE) parameters are estimated [9]. The cost and other necessary parameter used in the system configuration and cost estimation model are for FY: 2018–19 and no benefits from the government are considered. The whole capital costs of systems are assumed to be furnished from the loan. The optimum size of SPV, SWTG and hybrid system in terms of number of SPV modules, SWTGs and battery bank required to satisfy loss of power supply probability (LPSP) required by costumer are obtained.

HOMER software is used to design standalone SPV, SWTG and wind-solar hybrid energy systems. The optimal system configurations having number of SPV modules (NPT), SWTG (NWT) and batteries (NBT) at desired value of LPSP are evaluated. The optimal sizing results obtained at 2% loss of power supply probability (LPSP) for standalone SPV, SWTG and hybrid power system with different cost parameters are presented in Table 5.

Table 5 Optimum sizing results for standalone SPV, SWTG and hybrid systems

System type	System configuration			Hybrid system costs				LPSP (%)
	N_{PT}	N_{WT}	N_{BT}	CC (Lakh)	LCC (Lakh)	ALCC (Lakh)	UCE (Rs./kWh)	
SPV	12	Nil	18	5.52	13.81	1.31	10.50	1.91
SWTG	Nil	10	14	9.12	15.84	1.14	26.18	1.98
Hybrid	8	2	6	3.59	6.26	0.43	9.10	1.72

5 Results Analysis and Discussion

District endowed with rich solar and wind potential. Wind and solar potential are compliment in nature. Vijayapur district has plenty of opportunities in small-scale farm and non-farm sectors. During survey, it is observed that in the district, rural families spend around 10% of their income on domestic energy need, about 80% of rural population distrust the grid supply and around 25% of them are willing to pay Rs. 6–10/kWh for reliable electricity. Ultimately, the lack of access to viable electric supply significantly diminishes the opportunities. Thus, there is a need of assured electricity for income generation to ensure sustainable livelihood and environmental security in the district.

HOMER software is used to design standalone solar photovoltaic (SPV), small wind turbine generator (SWTG) and wind-solar hybrid energy systems. The optimal system configurations having number of SPV modules (N_{PT}), SWTG (N_{WT}) and batteries (N_{BT}) at desired value of LPSP and other system constraints are obtained. The optimal sizing results obtained at 2% loss of power supply probability (LPSP) for standalone SPV, SWTG and hybrid power system are presented in Table 6.

Capital cost (CC), life cycle cost (LCC), annualized capital cost (ALCC) and unit cost of energy are estimated and presented in Table 5. It is obvious from Table 5 that both SPV and SWTG system configurations resulted in a higher annualized capital cost (ALCC) compared to the wind-solar hybrid energy system. The critical analysis made on these options highlights that the wind-solar hybrid system seems to be more feasible in terms of techno-economic performance.

The costs of electricity produced by hybrid system are relatively less (9.10 Rs./kWh) as compared to SPV and SWTG systems (Rs. 10.50/kWh and Rs. 26.18/kWh, respectively) in realizing same LPSP. Further, hybrid energy system requires smaller size of renewable generators and less battery capacity. Thus, hybrid system is techno-economically feasible for electrification of farmhouse.

In renewable energy technologies, local skills and knowledge can play essential factors to finance, install, operate, maintain and repair the equipment. It lowers the production and maintenance costs, creates prosperity within the benefited community and increase the social acceptance of the developed technology. It certainly creates self-employment and prevents the migration of youth in to nearby states. Furthermore, it helps in overcoming the cultural barriers and empowers the poor, cutting the reliance on subsidies and hand-outs from NGOs.

6 Conclusion

Electricity to ensure sustainable livelihood in rural areas is essential infrastructure. Study conducted on renewable resources at Vijayapur highlights that study area is endowed with rich wind and solar potential. Techno-economic performance analysis on SPV, SWTG and hybrid system revelled that the climatic condition is viable for promotion of SPV and wind-solar hybrid energy system for rural small-scale power applications. The proposed power systems require low/no maintenance cost, but initial investment is unaffordable for rural potential consumers. Financial support from the government for installation of power system is essential.

Acknowledgements The authors wish to thank the Authorities of Regional Agricultural Research Station at Vijayapur, HECSOM Vijayapur Karnataka, India, and rural public for providing appropriate information for the present research.

References

1. A. De1, Renewable energy sources for the development of rural India. Int. J. Eng. Sci. Invent. **3**, 2319–6734 (2014)
2. Department of Agricultural Research and Education by Ministry of Agriculture, April, 2018.
3. D.K. Subramania, T.V. Ramachandra, Aspects of Agriculture and Irrigation In Karnataka Energy Research Group Centre For Ecological Sciences, Researcg article Indian Institute of Science, Bangalore 560 012, India (2016)
4. J. Charles Rajesh Kumar, M.A. Majid, Renewable energy for sustainable development in India: Current Status, future prospects, challenges, employment, and investment opportunities. Energy Sustainab. Soc. **10**, Article number: 2 (2020)
5. T.V. Ramachandra, Ecologically sound energy planning strategies for sustainable development (IISc)", Humanity Development Library, Versio - 2.0
6. District Profile- Vijayapur, Karnataka Global Agribusiness & Food Processing Summit (2019)
7. T.V. Ramachandra, Solar energy potential assessment using GIS. Energ. Educ. Sci. Technol. **18**(2), 101–114 (2007)
8. G.L. Johnson, *Wind Energy Systems* (Prentice Hall Inc. Englewood Cliffs, N.J., 1985); J.V. Seguro, T.W. Lambert, Modern estimation of the parameters of the Weibull wind speed distribution for wind energy analysis. J. Wind Eng. Industr. Aerodynam. **85**(3):75–84 (2002)
9. M.A. Mohamed, A.M. Eltamaly, A.I. Alolah, Sizing and techno-economic analysis of stand-alone hybrid photovoltaic/wind/diesel/battery power generation systems. J. Renew. Sustainab Energy, **7**, Article ID 063128 (2015)
10. M. Kolhe, Techno-economic optimum sizing of a stand-alone solar photovoltaic system. IEEE Trans. Energy Convers. **24**(2), 511–519 (2009)

Estimation of Levelized Cost of Electricity (LCOE) of 1 MW SPV Plants Installed at 33 Different Locations in Rajasthan, India

Vineet Kumar Mahaver and K. V. S. Rao

1 Introduction

In India SPV (grid connected) installed capacity has reached 35.12 GW as on June 30, 2020 [1]. Renewable energy production is increasing in India with an aim to install 100 GW solar power by 2022. It includes grid connected SPV and concentrating solar power of 60 GW, rooftop SPV of 40 GW [2]. The state of Rajasthan has bright sunshine of 8.0 to 8.2 h daily and solar radiation of 6–7 kWh/m^2/day with approximately 300 sunny days per year [3].

Levelized cost of electricity or LCOE represents the overall functioning of the plant and financial feasibility of the power plant [4–6].

- The LCOE calculation is a fundamental method used to assess the energy-producing project.
- The LCOE can also be used to evaluate whether the project is commissionable or as a compare different energy-producing projects
- The formula to calculate the LCOE is: (Present Value of Total Cost Over the Lifetime)/(Present Value of All Electricity Generated Over the Lifetime)

In this paper, LCOE of 1 MW SPV plant if established in all 33 district headquarters of Rajasthan state in India is studied. To study LCOE, parameters like interest rate, inflation rate, plant life, and capital cost of the plant which includes cost of land, cost of solar panel, cost of inverter, operation, and maintenance cost are considered at each of the selected plant sites with electricity generation. The electricity generation data plant is simulated using PVsyst and SAM software for each location.

V. K. Mahaver (✉)
Department of Renewable Energy, Rajasthan Technical University, Kota, Rajasthan, India

K. V. S. Rao
Gayatri Vidya Parishad College of Engineering for Women, Visakhapatnam, Andhra Pradesh, India

Sanjeevikumar et al. (eds.), *Advances in Renewable Energy and Electric Vehicles*, Lecture Notes in Electrical Engineering 767,
https://doi.org/10.1007/978-981-16-1642-6_16

2 Simulated SPV Plant and Solar Radiation Data with Generation (E_{grid})

The modeling of 1 MW SPV plant and simulation is done for different selected locations using PVsyst, SAM software [7–9]. The simulation results provide plant generation data (annually), performance ratio (PR), and CUF. For simulation, required data of the number of modules, number of inverters, plant area, etc., are shown in Table 1.

SPV plant description is given in Table 1, which are the inputs to model the 1 MW SPV plant at different locations using PVsyst and SAM.

Plant annual generation (E_{grid}) data and radiation data of 1 MW SPV plant for different locations which is obtained by simulation using PVsyst and SAM software are shown in Table 2 [7]. The radiation data is assessed from Meteonorm and NASA by internal architecture of PVsyst & SAM software, respectively.

Capital cost of each of the plant is calculated using the land cost, module cost, transmission line cost, and other installation cost. This capital cost obtained after summation of these costs is shown in Table 4.

As per PVsyst and SAM software, maximum solar radiation and electricity generation are recorded at Dungarpur city.

Bano and Rao [10] calculated LCOE using generation data and capital cost of existing 1 MW SPV plant at Jaipur, which is shown in Table 3.

Table 1 Description of SPV Plant [7]

Parameters	PVsyst	SAM
PV module manufacturer	Canadian	Canadian
PV module power	315 Wp	315 Wp
Inverter	Power gate plus (500 kWac)	Satcon technology (500 kWac)
Tilt angle	30°	30°
No. of modules in series	11	12
No. of modules in parallel	289	264
Total no. of modules	3179	3168
No. of inverters	2	2
Plant area	6181 m^2	6114 m^2

Table 2 E_{grid} and radiation data from PVsyst, Sam Software for 1 MW SPV Plant [7]

Location no. and name		PVsyst		SAM	
		Radiation (kWh/m^2/year)	E-Grid (MWh/year)	Radiation (kWh/m^2/year)	E-Grid (MWh/year)
1	Ajmer	2149	1752	2064	1819
2	Alwar	2019	1663	2058	1664
3	Banswara	2201	1784	2030	1749
4	Baran	2076	1700	1684	1728
5	Barmer	2166	1760	1956	1839
6	Bharatpur	1980	1653	1919	1643
7	Bhilwara	2149	1747	1750	1791
8	Bikaner	2048	1670	2029	1791
9	Bundi	2079	1701	2005	1734
10	Chittorgarh	2147	1751	2005	1778
11	Churu	2087	1705	1983	1771
12	Dausa	2078	1702	1974	1739
13	Dhaulpur	1984	1644	2083	1671
14	Dungarpur	2248	1823	2094	1775
15	Sri Ganganagar	1957	1624	2082	1654
16	Hanumangarh	1964	1622	1922	1670
17	Jaipur	2124	1737	1942	1690
18	Jaisalmer	2156	1754	1900	1822
19	Jalore	2143	1742	2051	1861
20	Jhalawar	2079	1701	2045	1739
21	Jhunjhunun	2072	1701	1807	1760
22	Jodhpur	2145	1743	2046	1823
23	Karauli	2044	1688	2018	1718
24	Kota	2077	1697	2050	1757
25	Nagaur	2128	1731	1734	1821
26	Pali	2150	1745	2023	1830
27	Pratabgarh	1835	1521	2046	1744
28	Rajasamand	2195	1779	2046	1797
29	Sikar	2134	1744	1778	1793
30	Sirohi	2151	1745	2004	1797
31	SawaiMadhopur	2061	1702	2011	1721
32	Tonk	2087	1671	1957	1733
33	Udaipur	2204	1792	2061	1799

Table 3 Capital cost and lcoe of 1 mw spv plant at jaipur [10]

Location name	Ccap (million Rs.)	E-Grid (MWh/Year)	O&M (million Rs.)	LCOE (Rs./kWh)
Jaipur	140	1641	1.4	11.33

Table 4 Capital cost for establishment of SPV plant per MW [12]

S. No.	Entities	Lakhs per MW AC in million Rs.
1	Land	2.5
2	Module	33
3	Civil and general work	5.5
4	Mounting structures	4
5	Inverters	4
6	Electrical cables transformer, etc.	6.5
7	Grid extention and bay extention	4
8	Preliminary expenses approvals, land leveling, etc.	1.5
	Total cost	61

3 Calculation of LCOE

LCOE is the unit cost (Rs./kWh) of electricity generation at which the electricity should be sold while considering capital cost, operation, and maintenance cost of the plant [10, 11].

$$C_{\mathrm{pe}} = \left(\frac{C_{\mathrm{om}}}{f - i}\right)\left[\left(\frac{1 + f}{1 + i}\right)^{N} - 1\right] \tag{1}$$

where C_{pe} is present equivalent cost, C_{om} is operation and maintenance cost (3% of total capital cost), f is inflation rate, i is interest rate, and N is plantlife span.

$$C_{\mathrm{a}} = \left(C_{\mathrm{cap}} + C_{\mathrm{pe}}\right)\left[\frac{i(1 + i)^{N}}{(1 + i)^{N} - 1}\right] \tag{2}$$

where C_{a} is Equivalent annual cost and C_{cap} is Capital cost of the plant.

Levelized cost of electricity is given by [10, 11]:

$$\mathrm{LCOE} = \left[\frac{C_{\mathrm{a}}}{E_{\mathrm{grid}}}\right] \tag{3}$$

A. Calculation of Capital Cost

Breakup cost of commissioning 1 MW SPV plant is taken from solar Mango website which shows the capital cost of SPV plant per MW [12].

B. Comparison from LCOE of SPV Plant at Different Locations

To calculate the LCOE, O&M cost of SPV plant is considered as 3% of the total capital cost of the plant and interest rate (i) as 8%, inflation rate (f) as 6%, plant life N 20 years are assumed.

E_{grid} is annual energy produced by the plant which is taken from Table 2.

When the interest rate and inflation rate have kept at 8% and 6%, respectively, the average LCOE assessed using generation data (E_{grid}) of PVsyst and SAM software has found to be Rs. 5.33 and Rs. 5.19, respectively, from Table 5. For the actual 1 MW SPV plant situated at Jaipur, the LCOE is Rs.11.33 by using capital cost of Rs. 14 million [10, 11].

(1)*Effect of the interest rate on LCOE.*

When the inflation rate is kept constant ($f = 6\%$) and interest rate (i) is varying between 8, 10, 12%, the effect of interest rate on LCOE is studied using PVsyst and SAM generation data, and results have been shown in Table 5 for PVsyst and SAM software, respectively.

As i increases LCOE also increase which can be seen in Table 5 and Figs. 1, 2, and 3 for both PVsyst and SAM software, respectively.

As per the solar mango website capital cost of 1 MW SPV plant accrued as Rs. 61 million in Table 4. Tabular data of LCOE achieved is shown in Table 5 when $f =$ 6% and $i = 8$, 10, 12%.

LCOE calculation is also done, if capital cost of 1 MW SPV plant reduced to Rs. 50 million as per MNRE [13], while keeping other parameters as before at $f = 6\%$ and $i = 8$, 10, 12%. The LCOE achieved is shown in Table 6.

4 Results

When interest rate i is at 8% and inflation rate f is at 6%, the LCOE varies between Rs. 5/kWh (Dungarpur) to Rs. 6/kWh (Pratabgarh) for PVsyst and Rs. 4.90/kWh (Jalore) to Rs. 6.63/kWh (Bharatpur) for SAM, respectively. The average LCOE of 1 MW SPV plants for different locations is Rs. 5.33/kWh for PVsyst and Rs. 5.19/kWh for SAM.

When f is fixed at 6% and i varies as 8, 10, 12%, average LCOE varies as Rs. 5.33/kWh, Rs. 5.83/kWh, Rs. 6.37/kWh for PVsyst, and Rs. 5.19/kWh, Rs. 5.68/kWh Rs. 6.20/kWh for SAM at a capital cost of Rs. 61 million.

When f is fixed at 6% and i varies 8, 10, 12%, average LCOE also varies as Rs. 4.37/kWh, Rs. 4.78/kWh, Rs. 5.22/kWh for PVsyst and Rs. 4.25/kWh, Rs. 4.65/kWh, Rs. 5.08/kWh for SAM at a capital cost of Rs. 50 million.

Table 5 LCOE of 1 MW SPV capital cost at Rs. 61 million, $F = 6\%$, $I = 8, 10, 12\%$

S. No.	City name	LCOE					
		i (interest rate is varying %) and $f = 6\%$					
		PV syst			SAM		
		8	10	12	8	10	12
1	Ajmer	5.21	5.70	6.22	5.01	5.48	5.99
2	Alwar	5.48	6.00	6.55	5.48	6.00	6.55
3	Banswara	5.11	5.59	6.11	5.21	5.70	6.23
4	Baran	5.37	5.87	6.41	5.28	5.77	6.30
5	Barmer	5.18	5.67	6.19	4.96	5.43	5.92
6	Bharatpur	5.52	6.04	6.59	5.55	6.07	6.63
7	Bhilwara	5.22	5.71	6.23	5.09	5.57	6.08
8	Bikaner	5.46	5.97	6.52	5.09	5.57	6.08
9	Bundi	5.36	5.86	6.40	5.26	5.75	6.28
10	Chittorgarh	5.21	5.70	6.22	5.13	5.61	6.13
11	Churu	5.35	5.85	6.39	5.15	5.63	6.15
12	Dausa	5.36	5.86	6.40	5.24	5.74	6.26
13	Dhaulpur	5.55	6.07	6.63	5.46	5.97	6.52
14	Dungarpur	5.00	5.47	5.97	5.14	5.62	6.14
15	Sri Ganganagar	5.62	6.14	6.71	5.51	6.03	6.59
16	Hanumangarh	5.62	6.15	6.72	5.46	5.97	6.52
17	Jaipur	5.25	5.75	6.27	5.40	5.90	6.45
18	Jaisalmer	5.20	5.69	6.21	5.01	5.48	5.98
19	Jalore	5.23	5.73	6.25	4.90	5.36	5.85
20	Jhalawar	5.36	5.87	6.40	5.24	5.74	6.26
21	Jhunjhunun	5.36	5.87	6.40	5.18	5.67	6.19
22	Jodhpur	5.23	5.72	6.25	5.00	5.47	5.97
23	Karauli	5.40	5.91	6.45	5.31	5.81	6.34
24	Kota	5.37	5.88	6.42	5.19	5.68	6.20
25	Nagaur	5.27	5.76	6.29	5.01	5.48	5.98
26	Pali	5.23	5.72	6.24	4.98	5.45	5.95
27	Pratabgarh	6.00	6.56	7.16	5.23	5.72	6.25
28	Rajasamand	5.13	5.61	6.12	5.08	5.55	6.06
29	Sikar	5.23	5.72	6.24	5.09	5.56	6.07
30	Sirohi	5.23	5.72	6.24	4.99	5.46	5.97
31	Sawai Madhopur	5.36	5.86	6.40	5.30	5.80	6.33
32	Tonk	5.46	5.97	6.52	5.26	5.76	6.29
33	Udaipur	5.09	5.57	6.08	5.07	5.55	6.05

(continued)

Table 5 (continued)

S. No.	City name	LCOE					
		i (interest rate is varying %) and $f = 6\%$					
		PV syst			SAM		
		8	10	12	8	10	12
	MAX	6.00	6.56	7.16	5.55	6.07	6.63
	MIN	5.00	5.47	5.97	4.90	5.36	5.85
	AVG	5.33	5.83	6.37	5.19	5.68	6.20

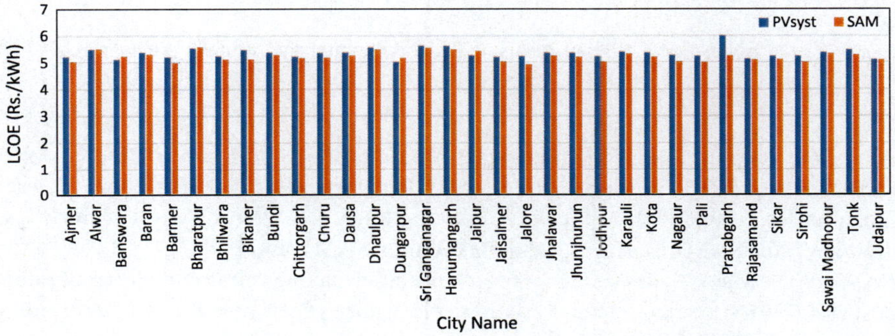

Fig. 1 LCOE of the 1 MW SPV plant while *i* and *f* are fixed at 8% and 6%, respectively

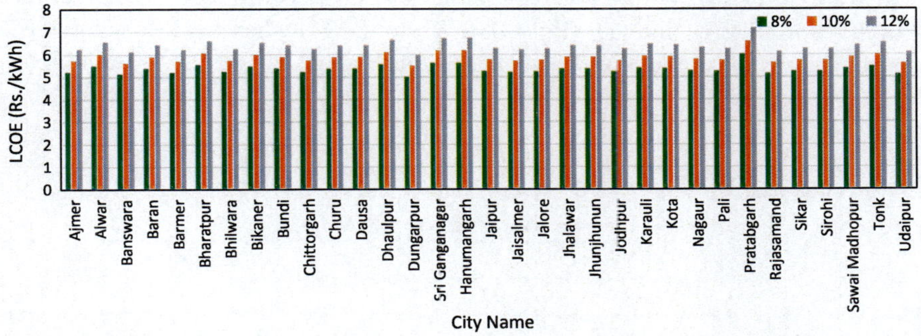

Fig. 2 LCOE of the 1 MW SPV plants using PVsyst generation data while *f* is kept fix at 6% and *i* varies 8, 10, 12%

5 Conclusions

In this paper, the study of 1 MW SPV plant is carried out to estimate the feasibility of the plant at different 33 locations in Rajasthan state for economic considerations. The results of LCOE have been figured out by considering interest rate and inflation

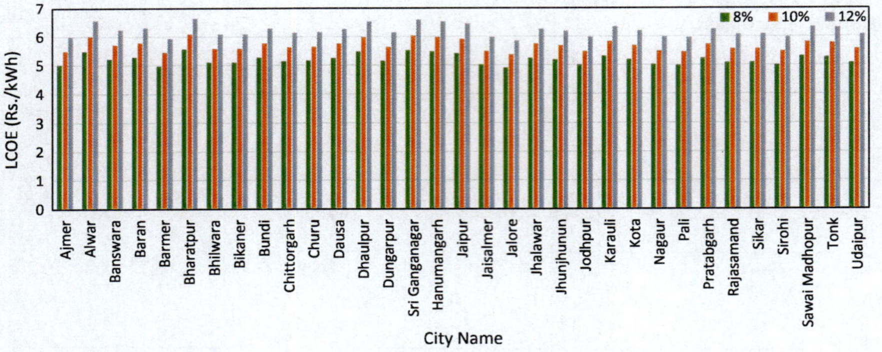

Fig. 3 LCOE of the 1 MW SPV plants using SAM generation data while*f* is kept fix at 6% and *i* varies 8, 10, 12%

rate while keeping one of them constant. The most feasible results of LCOE for a capital cost of Rs. 61 million have found for Dungarpur (Rs. 5.00/kWh for PVsyst), Udaipur (Rs. 5.09/kWh for PVsyst), Banswara (Rs. 5.11/kWh for PVsyst), Jalore (Rs. 4.90/kWh for SAM), Barmer (Rs. 4.96/kWh for SAM), and Pali (Rs. 4.98/kWh for SAM) cities when $i = 8\%$ and $f = 6\%$; these cities having minimum electricity unit cost (Rs./kWh) due to comparatively high electricity generation among the different locations of 1 MW SPV plants.

The values are obtained for LCOE when capital cost is Rs. 50 million. LCOE for Dungarpur (Rs. 4.10/kWh for PVsyst), Udaipur (Rs. 4.17/kWh for PVsyst) and Banswara (Rs. 4.19/kWh for PVsyst), Jalore (Rs. 4.02/kWh for SAM), Barmer (Rs. 4.06/kWh for SAM), and Pali (Rs. 4.08/kWh for SAM) when $i = 8\%$ and $f = 6\%$ are considered lower than LCOE calculated for a capital cost of Rs. 61 million.

Table 6 LCOE of 1 MW SPV capital cost at Rs. 50 million, $F = 6\%$, $I = 8, 10, 12\%$

S.No	City name	LCOE					
		i (interest rate is varying %) and $f = 6\%$					
		PV syst			SAM		
		8	10	12	8	10	12
1	Ajmer	4.27	4.67	5.10	4.11	4.50	4.91
2	Alwar	4.50	4.92	5.37	4.49	4.91	5.37
3	Banswara	4.19	4.58	5.00	4.27	4.68	5.10
4	Baran	4.40	4.81	5.25	4.33	4.73	5.17
5	Barmer	4.25	4.65	5.07	4.06	4.45	4.85

(continued)

Table 6 (continued)

S.No	City name	LCOE					
		i (interest rate is varying %) and $f = 6\%$					
		PV syst			SAM		
		8	10	12	8	10	12
6	Bharatpur	4.52	4.95	5.40	4.55	4.98	5.43
7	Bhilwara	4.28	4.68	5.11	4.17	4.57	4.98
8	Bikaner	4.48	4.90	5.34	4.17	4.57	4.98
9	Bundi	4.39	4.81	5.25	4.31	4.72	5.15
10	Chittorgarh	4.27	4.67	5.10	4.20	4.60	5.02
11	Churu	4.38	4.80	5.24	4.22	4.62	5.04
12	Dausa	4.39	4.80	5.25	4.30	4.70	5.13
13	Dhaulpur	4.55	4.97	5.43	4.47	4.89	5.34
14	Dungarpur	4.10	4.49	4.90	4.21	4.61	5.03
15	Sri Ganganagar	4.60	5.04	5.50	4.52	4.94	5.40
16	Hanumangarh	4.61	5.04	5.51	4.48	4.90	5.35
17	Jaipur	4.30	4.71	5.14	4.42	4.84	5.28
18	Jaisalmer	4.26	4.66	5.09	4.10	4.49	4.90
19	Jalore	4.29	4.69	5.12	4.02	4.39	4.80
20	Jhalawar	4.39	4.81	5.25	4.30	4.70	5.13
21	Jhunjhunun	4.39	4.81	5.25	4.25	4.65	5.07
22	Jodhpur	4.29	4.69	5.12	4.10	4.49	4.90
23	Karauli	4.43	4.85	5.29	4.35	4.76	5.20
24	Kota	4.40	4.82	5.26	4.25	4.65	5.08
25	Nagaur	4.32	4.73	5.16	4.11	4.49	4.90
26	Pali	4.28	4.69	5.12	4.08	4.47	4.88
27	Pratabgarh	4.91	5.38	5.87	4.29	4.69	5.12
28	Rajasamand	4.20	4.60	5.02	4.16	4.55	4.97
29	Sikar	4.29	4.69	5.12	4.17	4.56	4.98
30	Sirohi	4.28	4.69	5.12	4.09	4.48	4.89
31	Sawai Madhopur	4.39	4.81	5.25	4.34	4.75	5.19
32	Tonk	4.47	4.89	5.34	4.31	4.72	5.15
33	Udaipur	4.17	4.56	4.98	4.16	4.55	4.96
	MAX	4.91	5.38	5.87	4.55	4.98	5.43
	MIN	4.10	4.49	4.90	4.02	4.39	4.80
	AVG	4.37	4.78	5.22	4.25	4.65	5.08

References

1. *Programme/scheme wise physical progress in 2020–21 & cumulative upto june 2020*, Ministry of new and renewable energy, Sept. 2020. [Online]. Available: https://mnre.gov.in/the-ministry/physical-progress
2. *Scaling up rooftop solar power in india: the potential of solar*, Climate Policy Initiative, Sep. 2020. [Online]. Available:https://www.climatepolicyinitiative.org/publication/scaling-rooftop-solar-power-india-potential-solar-municipal-bonds/
3. *Solar radiation and sunshine availability in Rajasthan,* Rajras, Oct. 2018. [Online]. Available: https://www.rajras.in/index.php/solar-radiation-sunshine-availability-rajasthan/
4. *Three LCOE Methods and Seven Value Drivers*, Dec 2020. [Online]. Available: https://edbodmer.com/electricity-technology-comparisons/
5. *Levelized Cost of Energy (LCOE)*, CFI, Dec. 2020. [Online]. Available: https://corporatefinanceinstitute.com/resources/knowledge/finance/levelized-cost-of-energy-lcoe/
6. *Levelized Cost Of Electricity Renewable Energy Technologies*, Fraunhofer. Dec. 2020. [Online]. Available: https://www.ise.fraunhofer.de/content/dam/ise/en/documents/publications/studies/EN2018_Fraunhofer-ISE_LCOE_Renewable_Energy_Technologies.pdf
7. V.K. Mahaver, K.V.S. Rao, Solar Energy Potential of the State of Rajasthan in India. *Innovative Applications of Computational Intelligence on Power, Energy and Controls with their impact on Humanity (CIPECH-18)* (2018) pp. 56–61. https://doi.org/10.1109/CIPECH.2018.8724198
8. A.E. Baghdadi, K. Yaakoubi, Z. Attari, A. Leemrani, A. Asselman, Performance investigation of a PV system connected to the grid. Procedia Manuf. **22**, 667–674 (2018). https://doi.org/10.1016/j.promfg.2018.03.096
9. T. Bano, K.V.S. Rao, Performance analysis of 1MW grid connected photovoltaic power plant in Jaipur, India, in *2016 International Conference on Energy Efficient Technologies for Sustainability (ICEETS)* (2016), pp.165–170, https://doi.org/10.1109/ICEETS.2016.7582919
10. T. Bano, K.V.S. Rao, Levelized electricity cost of five solar photovoltaic plants of different capacities. *Procedia Technol.* **24**, 505–512, (2016). https://doi.org/10.1016/j.protcy.2016.05.086.
11. T. Bano, K.V.S. Rao, The effect of solar PV module price and capital cost on the levelized electricity cost of the solar PV power plant in the context of india, in *2016 Biennial International Conference on Power and Energy Systems:Towards Sustainable Energy (PESTS E)* (2016) https://doi.org/10.1109/PESTSE.2016.7516468
12. *What are the Initial Investment and O&M Costs Required for a MW Solar Plant in India?*, Solar Mango, Sept. 2020, [Online]. Available: http://www.solarmango.com/ask/2015/09/20/what-are-the-initial-investment-and-om-costs-required-for-a-mw-solar-plant-in-india-what-kind-of-financial-returns-can-we-expect-from-it/
13. *Twenty-Eighth Report Standing Committee On Energy (2016–2017),* Ministry of New And Renewable Energy, Sept. 2020. [Online]. Available: http://164.100.47.193/lsscommittee/Energy/16_Energy_28.pdf

Harnessing Solar Energy from Wind Farms: Case Study of Four Wind Farms

Monika Agrawal and K. V. S. Rao

1 Introduction

Large-scale solar photovoltaic (PV) and wind energy systems are developed all over the world for fulfilling the demand of electricity. Solar and wind resources have great potential for power generation due to their social, economical, and environmental benefits and are also motivated by government policies and incentives. As per the Ministry of New and Renewable Energy (MNRE) of the Government of India, the achieved total solar and wind power installed capacities are 33.730 GW and 37.505 GW, respectively, in the country till the end of December 2019 [1]. But both, solar PV and wind energy systems operate at varying times depending on their availability, and thus continuous reliable power production is not possible in separate solar PV and wind power plants.

In wind farms, a large amount of land is vacant to avoid wake losses. The better utilization of this free land in between wind turbines can be possible by the installation of solar PV modules. Therefore, additional electricity can be generated from the same wind farm [2, 3]. A combined solar PV-wind power plant is more reliable and economical than a standalone power plant. Also, the combined systems provide other advantages-like saving the land, having the same grid for both systems and greater reliability of the power system for the supply of energy to the grid. The combined solar-wind power plant becomes cost-effective as the cost of solar PV module is continuously reducing [4].

The study [5] presented a grid-connected solar-wind hybrid system performance at a remote island in India and observed a low tariff rate of electricity. The levelized cost

M. Agrawal (✉)
Centre for Energy and Environment, Malaviya National Institute of Technology, Jaipur, India

K. V. S. Rao
Gayatri Vidya Parishad College of Engineering for Women, Visakhapatnam, Andhra Pradesh, India

Sanjeevikumar et al. (eds.), *Advances in Renewable Energy and Electric Vehicles*, Lecture Notes in Electrical Engineering 767,
https://doi.org/10.1007/978-981-16-1642-6_17

of electricity was observed as 2.74 Rs/kWh as compared to supply from the grid. The study [6] discussed the performance and detailed design of a grid-connected solar-wind hybrid system at a large scale with the incorporation of control approaches. The authors illustrated the annual energy yield was 1.509 TWh/year from hybridization of 50 MW solar power plant and 200 MW wind power plant. A feasibility study of hybridization of the solar power plant in exiting wind farm of 18 wind turbines (each of 2 MW) was presented in [7]. The authors observed the annual capacity improvement was 90% and the payback period is 7 years only. The repowering of the wind farm concept not only improves the system performance but also increases the lifespan of the system too. In this context, the study [8] presented an analytical study of wind farm repowering by the integration of new wind turbines and solar photovoltaics. Using PVsyst, the authors analyzed the shaded area between wind turbines for PV module installation with an understanding of sun geometry. The wind turbines installed in 5D × 7D configuration in 6 MW wind farm at Kayathar, Tuticorin District, Tamil Nadu. They observed huge power improvement with the integration of solar power plants in vacant land.

From the motivation, the following work presents a theoretical performance study of the wind farms integrated with solar photovoltaic power plants installing in vacant land between wind turbines. As per MNRE [1], the wind and solar installation potential wise, the top seven states in India are Karnataka, Rajasthan, Tamil Nadu, Andhra Pradesh, Gujrat, Maharashtra, and Madhya Pradesh. Therefore, four different wind farms located in four states Tamil Nadu, Rajasthan, Andhra Pradesh, and Karnataka of India are considered in the present study. The wind farm layout has been considered as 5D × 7D and a comparative study of the electricity generation from solar PV plant in different cases of vacant space coverage has been presented.

The paper embodied as Sect. 2 explains the concept of combined solar PV and wind power plants. In Sect. 3, the electric energy production and performance of solar photovoltaic power plants at four wind farm sites by taking six cases of the area covered by solar PV modules are estimated. The monthly electric energy production for 5% of area coverage and annual electric energy production for 5%, 10%, 15%, 20%, 25%, and 30% of area coverage by PV modules are estimated using PVsyst software. Section 4 presents the conclusions of this paper.

2 Combined Solar PV and Wind Power Plant

The arrangement of wind turbines (micro siting) has to be done properly in a wind farm for minimizing the wind farm array losses and enhancing the performance of the wind power plant [9]. The land between wind electric generators remains unutilized and can be used for producing electric energy. The performance of separate solar PV power plant is dependent on weather conditions and hence its reliability is also less. The combined solar PV-wind power plant will be more reliable and more efficient than a separate solar PV and a separate wind power plant. The advantage of the combined system is that the same land can be used for the dual purpose of producing electricity from two different types of power plants and efficient utilization of land is possible. The spacing between wind turbines can produce additional electricity by implementing the solar photovoltaic power plant in between wind turbines. Solar photovoltaic and wind energy systems are complementary to each other and their respective power generation timing is also different for most of the time. This will improve overall power fluctuations.

3 Case Study of Four Wind Farms Located in Four Different States for Feasibility of Combined Solar PV and Wind Power Plants

In India, most of the states have the availability of solar insolation that is sufficient for energy generation by using solar PV modules for about 300 days in a year. The states like Tamil Nadu, Gujarat, Rajasthan, Madhya Pradesh, and Andhra Pradesh have sites with high wind power density for wind electricity generation and also high solar insolation for solar PV based electricity generation. In this paper, four locations namely Pratapgarh in Rajasthan, Davangere in Karnataka, Tirunelveli in Tamil Nadu, and Anantpur in Andhra Pradesh are selected for estimation of solar PV power plant electricity generation at four previously installed wind farms in these four sites. The locations of four chosen sites are shown in Fig. 1.

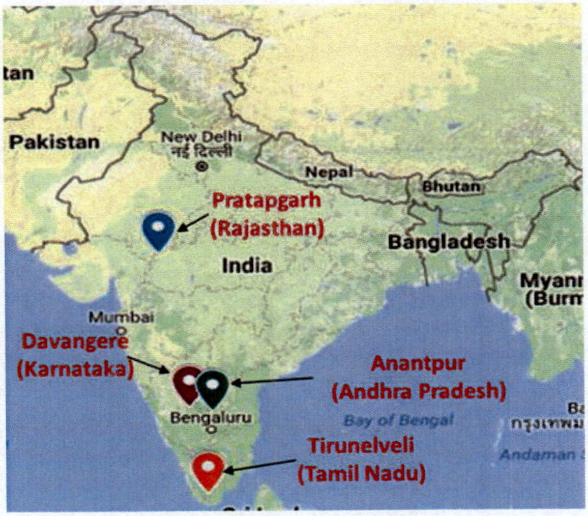

Fig. 1 Four wind farm sites are located on the India map for which solar PV power plant feasibility analysis is done

3.1 Description of the Four Working Wind Farm Sites

Site 1: wind farm of 45 MW is located at village Dalot, taluka Arnod of district Pratapgarh in Rajasthan (RJ) state of India. The geographical coordinates of the project site are latitude 23° 39' 28.7" N and longitude 74° 47' 42.9" E, respectively. May and June are the hottest months. After summer, rainfall occurs due to the southwest monsoon. The month of December and January are the coldest. The wind farm consists of thirty 1.5 MW ReGen Powertech make wind turbines and are installed by Green Infra Wind Farms Assets Limited (GIWFAL). The annual electric energy production of the wind farm is 78,840 MWh at a plant load factor (PLF) of 20% [10].

Site 2: a 29.70 MW wind power project is situated at Davangere district in Karnataka (KA) state of India. This project is spread in two villages namely Anabaru and Arasinagundi in Jagalur taluka, with Anabaru having an installed capacity of 13.20 MW and Arasinagundi having an installed capacity of 16.50 MW. These villages are approximately 250 km from Bangalore city of Karnataka. The site is situated between latitudes 14°28'–14°34' N and longitudes 76°20'–76°23' E. The altitude is 700-810 m above mean sea level. This project has eighteen 1.65 MW Vestas V82 make wind turbines. The annual electric energy production of wind farms is 94,884.72 MWh at a plant load factor of 36.47%. The owner of the project is Accion Wind Energy Pvt. Ltd. (AWEPL) [11].

Site 3: this wind farm of 33 MW is covering the four villages namely Melamaruthapapuram, Balapathiramapuram, Keelakalangal, and Ichchanda of V.K. Puthur taluka, Tirunelveli district in Tamil Nadu (TN) state of India. The project is located between latitudes 9°01'19.2"–9°03' N and longitudes 77°18'18"–77°22'24" E. The wind farm has twenty-two 1.5 MW Suzlon S82 wind turbines. The annual electric energy production of the wind farm is 86,377.104 MWh at a plant load factor of 29.88%. The project owner is Super Wind Project Private Ltd [12].

Site 4: a wind farm of 50.4 MW is situated at Anantpur district in Andhra Pradesh (AP) state of India. The site is located around the villages Gondipali, Duddebanda, Kogira, and Mustikovilla. The geographical coordinates of the site are latitude 14°10'32.3" N and longitude 77°34'15.7" E. The wind farm has sixty-three 0.80 MW Enercon E53 make wind turbines. The annual electric energy production of the wind farm is 1,12,186.166 MWh at a plant load factor of 25.41%. Tadas Wind Energy Private Limited (TWEPL) initiated this project [13].

The specifications of wind turbines installed at the above-mentioned four wind farms are given in Table 1. The estimation of electricity production by the solar photovoltaic power plant is done by using PVsyst 6.6.2 software. The global solar radiation data of chosen sites are taken from PVsyst through Meteonorm 7.1 database and is shown in Table 2 and its comparative analysis is shown in Fig. 2. The technical parameters of the PV module and inverter used in PVsyst software for designing solar PV power plants are shown in Table 3.

Table 1 Specifications of wind turbines installed at four wind farms of case study

Parameters	Site 1 (RJ)	Site 2 (KA)	Site 3 (TN)	Site 4 (AP)
Make	ReGen Powertech	Vestas	Suzlon	Enercon
Model	V82	V82	S82	E53
Rotor diameter (m)	82	82	82	53
Hub height (m)	85	78	82	75
Rated power (kW)	1500	1650	1500	800
Cut in speed (m/s)	3.5	3.5	4	3
Rated speed (m/s)	12	7.5	14	12.6
Cut out speed (m/s)	18	20	20	28

Table 2 Solar radiation data of four chosen sites

Months	Global horizontal solar irradiation (kWh/m²/day)			
	Pratapgarh (RJ)	Davangere (KA)	Tirunelveli (TN)	Anantpur (AP)
Jan	4.78	5.43	5.45	5.54
Feb	5.67	5.97	5.98	5.87
Mar	6.81	6.64	6.37	6.5
Apr	7.04	6.64	5.92	6.4
May	7.3	6.3	5.42	6.05
Jun	6.12	5.38	4.51	5.04
Jul	4.41	4.91	4.74	4.77
Aug	4.1	4.76	5.18	4.53
Sep	5.47	5.32	5.43	5.08
Oct	5.74	5.38	5.11	4.97
Nov	4.87	5.18	4.60	4.97
Dec	4.54	5.09	4.98	5.04
Annual	5.57	5.58	5.30	5.39

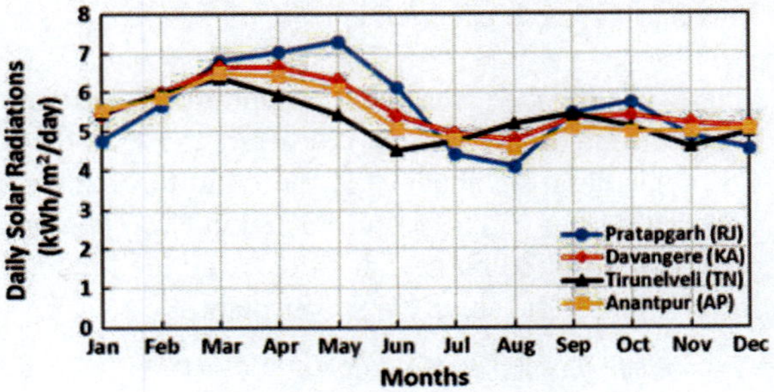

Fig. 2 Monthly average daily global horizontal solar radiation of four locations used in the case study

Table 3 Design parameters of solar PV power plant used in PV syst software for four locations

Components	Parameters	Specifications
PV module	Solar tracking mode	Fixed
	Type	Poly-Si
	Tilt angle	Latitude angle
	Surface azimuth angle	0° (South facing)
	Manufacturer	Canadian solar Inc.
	Model	CS6X—305P
	Module area	$1.92\ m^2$
	Open circuit voltage	44.8 V
	Short-circuit current	8.97 A
	Max. power point voltage	36.3 V
	Max. power point current	8.41 A
	Maximum power	305.3 W
Inverter	Manufacturer	SMA
	Model	500CP XT
	Operating voltage	430–850 V
	Nominal AC power	$500\ kW_{ac}$
	Maximum AC power	$550\ kW_{ac}$

3.2 *Methodology of Calculation*

(1) *Free area available in wind farm for solar PV plant*

In this study, it is assumed that the wind farm layout is of a 5D × 7D configuration. The free space area for installing solar photovoltaic modules in the wind farm is calculated by using (1) [2]:

$$A = (5D \times 7D) - \left(\frac{\pi(D+H)^2}{4}\right) \tag{1}$$

where A is the free area for solar photovoltaic modules installation around a single wind turbine, D is the rotor diameter, and H is the hub height of the wind turbine. For neglecting the shadow of the wind turbine on the solar photovoltaic module, the area $\pi(D + H)^2/4$ is subtracted in (1). The number of wind turbines in a wind farm is multiplied by (1) for calculating the total available area in a wind farm for installation of the solar PV power plant.

(2) *Performance Ratio (PR)*

Various researchers have analyzed the performance of grid-connected solar PV power plants [14–16]. The performance ratio is a factor by which a solar PV system's efficiency is calculated. This factor also indicates that how much electric energy is available for supply to the grid. The performance ratio is useful for comparative analysis of different PV module technologies. PR is calculated by using (2):

$$PR = \left(\frac{E_{\text{Grid}}/P_o}{G_{\text{inc}}/G_o}\right) \tag{2}$$

where, E_{grid} is AC electric energy supplied to grid, P_o is nominal power, G_{inc} is global solar radiation on tilted surface, and G_o is global solar radiations (1000 W/m^2) at Standard Test Condition (STC).

(3) *Capacity Utilization Factor (CUF)*

It is the ratio of electric energy generated from PV system to nominal power of solar PV power plant at STC in a specific time interval. It is a method for calculating solar PV plant performance. The environmental factors-like solar radiation and module degradation factor are not considered for calculation of CUF. CUF is also an indicator to check the reliability of solar PV modules [17]. The monthly CUF is calculated by using (3):

$$\text{CUF} = \frac{E_{\text{Grid}}}{n \times 24 \times P_o} \tag{3}$$

where, E_{Grid} is the AC electric energy generated from PV module and n is the number of days in specific month for calculating monthly CUF.

3.3 Electric Energy Production by Solar PV Power Plant

This paper considers six cases of 5, 10, 15, 20, 25, and 30% of total available area in the wind farm for installation of PV modules for utilization of unused land for additional energy generation. Table 4 shows the solar PV power plant installed capacity for six cases of areas as mentioned above.

By considering that 5% of the total free available area is used for the installation of the solar PV power plant, the monthly DC electric energy produced by the plant and monthly AC energy supplied by inverter to the grid are estimated by using PVsyst software. Tables 5, 6, 7 and 8 show the monthly global solar radiation on a tilted surface, monthly DC energy generated by solar PV plant, and monthly AC energy supplied to the grid by inverter for 5% of the total free area for Pratapgarh (RJ), Davangere (KA), Tirunelveli (TN), and Anantpur (AP) sites, respectively.

Table 4 Area used in wind farm for solar PV power plant installation and capacity of PV power plant

% Area of total available area covered by solar PV plant (%)	Solar PV plant capacity (MW_p) and Area Used (km^2)			
	Pratapgarh (RJ)	Davangere (KA)	Tirunelveli (TN)	Anantpur (AP)
5	45.80 (0.29 km^2)	27.71 (0.17 km^2)	33.71 (0.21 km^2)	38.51 (0.24 km^2)
10	91.60 (0.58 km^2)	55.42 (0.35 km^2)	67.42 (0.42 km^2)	77.01 (0.48 km^2)
15	137.41 (0.86 km^2)	83.14 (0.52 km^2)	101.13 (0.64 km^2)	115.52 (0.73 km^2)
20	183.21 (1.15 km^2)	110.85 (0.70 km^2)	134.84 (0.85 km^2)	154.03 (0.97 km^2)
25	229.01 (1.44 km^2)	138.56 (0.87 km^2)	168.55 (1.06 km^2)	192.54 (1.21 km^2)
30	274.81 (1.73 km^2)	166.28 (1.05 km^2)	202.27 (1.27 km^2)	231.04 (1.45 km^2)

Table 5 Monthly electric energy generated and supplied by solar PV power plant at pratapgarh (RJ) by covering 5% Area

Months	Global solar radiation on tilted surface [G_{inc}] (kWh/m^2)	DC energy from solar PV modules [E_a] (MWh)	AC energy supplied to grid by inverter [E_{Grid}] (MWh)
Jan	200.3	7568	7379
Feb	197.4	7279	7102
Mar	236.6	8419	8223
Apr	212.2	7478	7295
May	209.9	7418	7239
Jun	165.7	6075	5921
Jul	126.1	4774	4636
Aug	122.2	4667	4532
Sep	172.7	6393	6232
Oct	211.3	7672	7491
Nov	192.6	7166	6991
Dec	195.3	7339	7152
Annual	2242.3	82,248	80,193

3.4 Results and Discussion

The annual electric energy productions of solar PV power plants for six cases of available area in four wind farms are shown in Table 9. Figure 3 shows the annual electric

Table 6 Monthly electric energy generated and supplied by solar PV power plant at davangere (KA) by covering 5% area

Months	Global solar radiation on tilted surface [G_{inc}] (kWh/m^2)	DC energy from solar PV modules [E_a] (MWh)	AC energy supplied to grid by inverter [E_{Grid}] (MWh)
Jan	197.5	4414	4323
Feb	185.3	4061	3981
Mar	215.3	4597	4509
Apr	196.7	4190	4104
May	183.5	3988	3903
Jun	149.4	3343	3263
Jul	142.8	3204	3124
Aug	143.2	3216	3140
Sep	161.8	3601	3519
Oct	178.9	3984	3893
Nov	176.1	3952	3865
Dec	186.6	4196	4104
Annual	2117.8	46,746	45,728

Table 7 Monthly electric energy generated and supplied by solar PV power plant at Tirunelveli (TN) by covering 5% Area

Months	Global solar radiation on tilted surface [G_{inc}] (kWh/m^2)	DC energy from Solar PV Modules [E_a] (MWh)	AC energy supplied to Grid by Inverter [E_{Grid}] (MWh)
Jan	183.4	4955	4852
Feb	176.8	4759	4663
Mar	201.4	5361	5253
Apr	175.1	4728	4625
May	161.8	4403	4303
Jun	129.9	3615	3527
Jul	141.2	3925	3830
Aug	157.3	4327	4227
Sep	163.6	4483	4380
Oct	164.1	4471	4368
Nov	146.4	4021	3929
Dec	167.6	4576	4477
Annual	1968.6	53,624	52,434

Table 8 Monthly electric energy generated and supplied by solar PV power plant at Anantpur (AP) by covering 5% Area

Months	Global solar radiation on tilted surface [G_{inc}] (kWh/m^2)	DC energy generated from solar PV modules [E_a] (MWh)	AC energy supplied to grid by inverter [E_{Grid}] (MWh)
Jan	200.1	6222	6067
Feb	181.7	5540	5404
Mar	209.4	6243	6092
Apr	189.8	5671	5524
May	177.2	5352	5209
Jun	141.0	4398	4268
Jul	139.5	4387	4255
Aug	136.0	4292	4164
Sep	153.9	4818	4683
Oct	163.3	5095	4953
Nov	166.7	5257	5116
Dec	181.7	5718	5568
Annual	2040.3	62,993	61,303

Table 9 Annual electric energy produced by solar PV power plants for six cases of areas at four locations

% Area of total available area covered by solar PV plant (%)	Pratapgarh (RJ) (MWh)	Davangere (KA) (MWh)	Tirunelveli (TN) (MWh)	Anantpur (AP) (MWh)
5	80,193	45,728	52,434	61,303
10	160,347	91,181	104,874	121,214
15	240,874	137,190	157,008	179,184
20	320,121	182,454	209,346	245,227
25	402,059	228,654	262,178	307,393
30	481,752	273,988	314,016	368,294

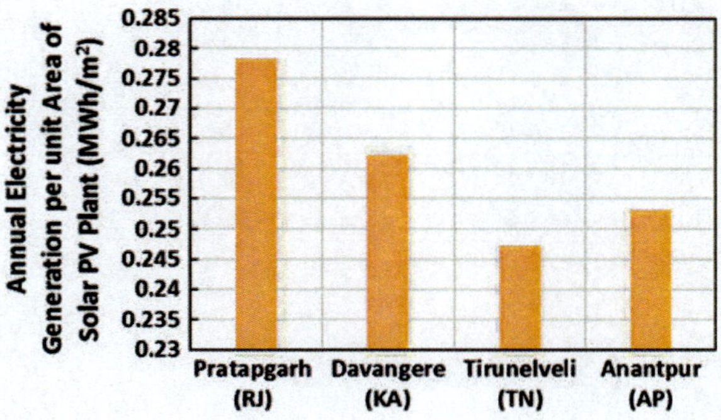

Fig. 3 Annual electric energy generation per unit area by solar PV power plant at four location

energy generation per unit area (MWh/m^2) by solar PV power plant at four locations. It is observed that solar PV electricity generation per unit area for Pratapgarh (RJ), Davangere (KA), Anantpur (AP), and Tirunelveli (TN) are 0.278 MWh/m^2, 0.262 MWh/m^2, 0.247 MWh/m^2, and 0.253 MWh/m^2 respectively. The month wise variation of performance ratio of solar PV power plants at four sites are shown in Fig. 4 and month wise variation of capacity utilization factor is shown in Fig. 5 for the case of 5% of available area of wind farm.

At Pratapgarh (RJ) the PR is maximum in the month of August (81%) and minimum in April (75.1%). At Tirunelveli (TN), the PR is maximum in month of June (80.5%) and minimum in March (77.3%). Monthly PR at Davangere (KA) and Anantpur (AP) are varying similarly and the annual PR at these two sites is calculated as 77.9% and 78.1%, respectively. The CUF of all four solar PV power plants is highest in the month of March. The annual CUF are calculated as 20%, 18.9%, 17.8%, and 18.2% for Pratapgarh (RJ), Davangere (KA), Tirunelveli (TN), and Anantpur (AP), respectively.

Table 10 gives the details of annual solar PV electricity generation, wind power plant electricity generation, and combined power plant electricity generation for per unit installed capacity (MWh/MW), respectively. Figure 6 shows a comparison of

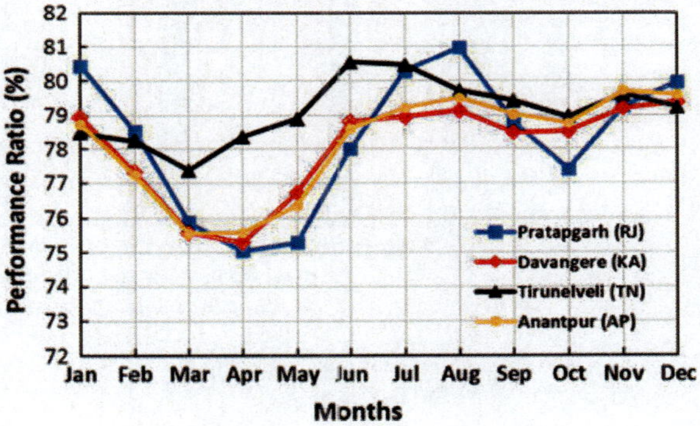

Fig. 4 Monthly variation of performance ratio of solar PV power plants at four locations for the case of 5% of available area of wind farm

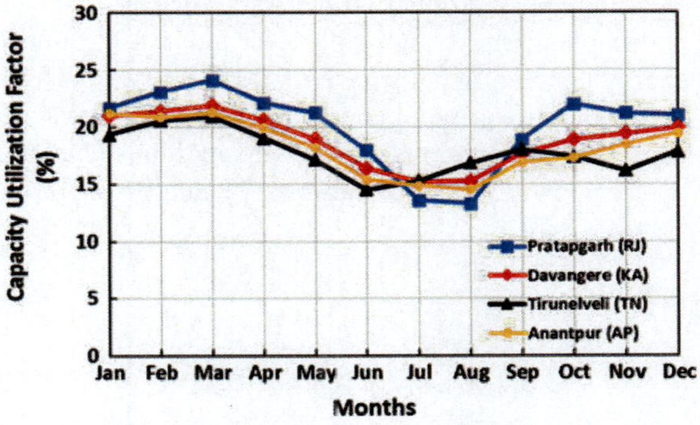

Fig. 5 Monthly variation of capacity utilization factor of solar PV power plants at four locations for the case of 5% of available area of wind farm

Table 10 Annual electricity production per unit installed capacity by covering 5% of the available area

Locations	Annual solar PV electricity (MWh/MW_p)	Annual wind electricity (MWh/MW)	Annual combined electricity (MWh/MW)
Pratapgarh (RJ)	1750.98	1752.00	1751
Davangere (KA)	1650.18	3194.77	2449.23
Tirunelveli (TN)	1555.49	2617.49	2080.85
Anantpur (AP)	1591.99	2225.92	1951.36

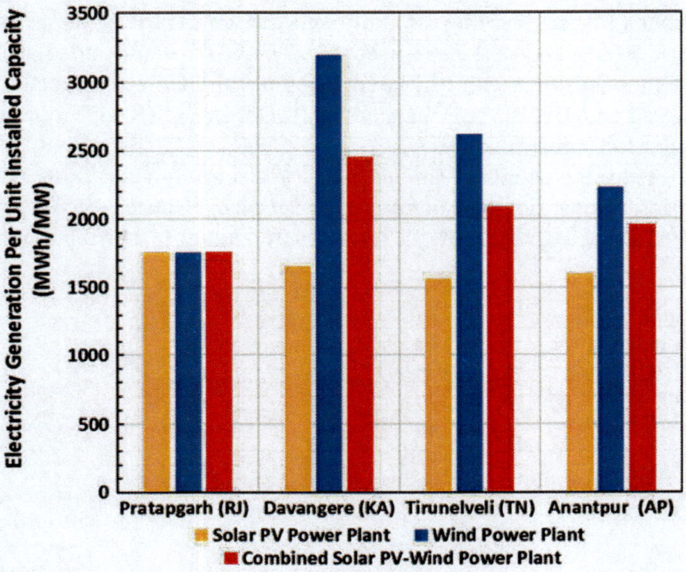

Fig. 6 Electricity generation (MWh/MW) by solar PV power plant, wind power plant, and combined solar and wind power plant

electric energy production per unit capacity (MWh/MW) of solar PV plant, wind power plant, and combined power plant at the four locations.It is noted that solar energy generation per square meter or per MW installed capacity in Pratapgarh (RJ) is maximum and wind energy generation at Davangere (KA) is maximum. So, the combined maximum electricity generation is obtained at Davangere (KA), followed by Tirunelveli (TN), Anantpur (AP), and Pratapgarh (RJ), respectively.

4 Conclusions

In this paper, power generation from the solar plant installation at vacant land of existing wind farms has been investigated. It is seen that solar PV electricity generation potential by covering 5% area of vacant land in wind farm by solar PV modules is maximum at Pratapgarh (1750.98 MWh/MW), followed by Davangere (1650.18 MWh/MW), Anantpur (1591.99 MWh/MW), and Tirunelveli (1555.49 MWh/MW). It is also noted that in Davangere, solar PV production per unit area is more than Anantpur, irrespective of the fact that available land area and plant installed capacity both are more at Anantpur than in Davengere. It may be due to slightly more daily solar insolation at Davengere than at Anantpur. The annual average performance ratio and annual average capacity utilization factor of solar power plants are varying from 77%–79% to 18%–20%, respectively, for the four locations. The annual wind electricity generation per unit installed capacity of the wind farm is highest at Davangere (3,194.77 MWh/MW) and is, followed by Tirunelveli (2,617.49 MWh/MW), Anantpur (2,225.92 MWh/MW), and Pratapgarh (1,752 MWh/MW). The combined electricity generation of solar PV and wind power plants are found to be the highest at Davangere in Karnataka and lowest at Pratapgarh in Rajasthan.

References

1. MNRE (2020) Ministry of new and renewable energy annual report 2019–20. MNRE, 2020. (Online). Available: https://mnre.gov.in/knowledge-center/publication. Accessed 29 Nov 2020
2. Agrawal M, Saxena BK, Rao KVS (2017) Feasibility of establishing solar photovoltaic power plants at existing wind farms. In: 2017 International conference on smart technologies for smart nation (SmartTechCon), pp 251–256. https://doi.org/10.1109/smarttechcon.2017.8358378
3. M. Agrawal, B.K. Saxena, K.V.S. Rao, *Techno-economic analysis of a grid-connected hybrid solar-wind energy system* (Springer, Singapore, 2019), pp. 81–92
4. S. Bhattacharjee, S. Acharya, PV-wind hybrid power option for a low wind topography. Energy Convers Manage **89**, 942–954 (2015). https://doi.org/10.1016/j.enconman.2014.10.065
5. Goswami A, Sadhu P, Sadhu PK (2020) Development of a grid connected solar-wind hybrid system with reduction in levelized tariff for a remote island in India. J Sol Energy Eng Trans ASME 142(4). https://doi.org/10.1115/1.4046147
6. A.F. Tazay, A.M.A. Ibrahim, O. Noureldeen, I. Hamdan, Modeling, control, and performance evaluation of grid-tied hybrid pv/wind power generation system: case study of gabel el-zeit region, egypt. IEEE Access **8**, 96528–96542 (2020). https://doi.org/10.1109/ACCESS.2020.2993919
7. R.S. Yendaluru, G. Karthikeyan, A. Jaishankar, S. Babu, Techno-economic feasibility analysis of integrating grid-tied solar PV plant in a wind farm at Harapanahalli, India. Environ Prog Sustain Energy **39**(3), 1–10 (2020). https://doi.org/10.1002/ep.13374
8. K. Boopathi et al., Optimization of the wind farm layout by repowering the old wind farm and integrating solar power plants: a case study. Int J Renew Energy Res **10**(3), 1287–1301 (2020)
9. R.M.A. Feroz, A. Javed, A.H. Syed, S.A.A. Kazmi, E. Uddin, Wind speed and power forecasting of a utility-scale wind farm with inter-farm wake interference and seasonal variation. Sustain Energy Technol Assessments **42**, (2020). https://doi.org/10.1016/j.seta.2020.100882

10. Clean Development Mechanism (2017) 45 MW Wind energy based power generation in Pratapgarh, Rajasthan. (Online). Available: https://cdm.unfccc.int/Projects/Validation/DB/SSCJKI8JQBKLQ4FLF0YARBAWFWZX1Z/view.html
11. Clean Development Mechanism (2017) 29.7 MW Wind project in Karnataka. (Online). Available: https://cdm.unfccc.int/Projects/DB/DNV-CUK1216117082.43/view
12. Clean Development Mechanism (2017) Grid connected wind energy project in Tamil Nadu by Super Wind Project Private Ltd. (Online). Available: https://cdm.unfccc.int/Projects/DB/DNV-CUK1280379317.22/view
13. Clean Development Mechanism (2017) Nallakonda wind farm in Andhra Pradesh. (Online). Available: https://cdm.unfccc.int/Projects/DB/LRQA%20Ltd1355495522.4/view
14. E. Kymakis, S. Kalykakis, T.M. Papazoglou, Performance analysis of a grid connected photovoltaic park on the island of crete. Energy Convers Manage **50**(3), 433–438 (2009)
15. R. Dabou, F. Bouchafaa, A.H. Arab, A. Bouraiou, M.D. Draou, A. Necaibia, M. Mostefaoui, Monitoring and performance analysis of grid connected photovoltaic under different climatic conditions in south Algeria. Energy Convers Manage **130**, 200–206 (2016)
16. K. Attari, A. Elyaakoubi, A. Asselman, Performance analysis and investigation of a grid-connected photovoltaic installation in Morocco. Energy Reports **2**, 261–266 (2016)
17. M.E. Basoglu, A. Kazdaloglu, T. Erfidan, M.Z. Bilgin, B. Cakir, Performance analyzes of different photovoltaic module technologies under Izmit, Kocaeli climatic conditions. Renew Sustain Energy Rev **52**, 357–365 (2015)

HVDC Fault Analysis and Protection Scheme

Sankarshan Durgaprasad, Shreya Nagaraja, and Sangeeta Modi

1 Introduction

High-voltage direct current (HVDC) transmission is regarded as a popular medium for bulk power transmission over long distances. This transmission system involves the conversion of power to high voltage DC at the generating side and is converted back to the desired form of Energy at the consumer side. Advancements in the field of power electronics and improved flexibility of power control have enabled HVDC transmission to become a desirable choice despite the dynamic behavior of the grid. An HVDC link avoids various problems present in HVAC transmission. Comparisons between the two systems are made based on the economics of transmission, reliability and technical performance. Figure 1 shows the cost (relative) between AC and DC transmission lines versus distance. The initial cost of investment is more in the case of HVDC systems. However, this breaks even relative to HVAC systems and is more economically beneficial for distances greater than break-even distance. An HVDC system requires one or two conductors, whereas a three-phase HVAC system requires three conductors. Bipolar DC lines and double circuit AC lines are equally reliable, due to its ability to transfer energy during the event of failure or fault at one of the poles, the other pole can supply at least 50% energy with a ground return. DC links have complete autonomy over the power transmission in both directions, and fast control enables DC lines to limit fault currents. Theoretically, power transmitted by a bipolar line is equal to that transmitted by a three-phase system. This is a technical

S. Durgaprasad (✉) · S. Nagaraja · S. Modi
Department Electrical and Electronics, PES University, Bangalore, India
e-mail: sankarshandurgaprasad@pesu.pes.edu

S. Nagaraja
e-mail: shreyan@pesu.pes.edu

S. Modi
e-mail: smodi@pes.edu

Sanjeevikumar et al. (eds.), *Advances in Renewable Energy and Electric Vehicles*, Lecture Notes in Electrical Engineering 767,
https://doi.org/10.1007/978-981-16-1642-6_18

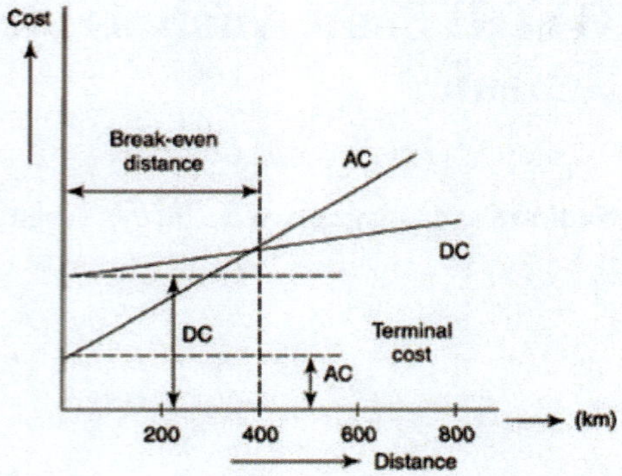

Fig. 1 Relative cost of AC and DC transmission lines vs distance [3]

and economic superiority of HVDC systems over HVAC systems [1]. The drawbacks of HVDC transmission are mainly the generation of harmonics due to the operation of converters, mandatory use of static var compensators, the futility of transformers in a DC system.

HVDC systems are branched into three most common schemes based on the number of conductors in operation, flexibility in load demand, reliability concerns and economic feasibility. Following three are the most common schemes [2]. Monopolar: two converters are linked by a single line where either a negative or positive voltage is used with the ground as the return path for transmission of voltage [2]. Bipolar: usually a combination of two monopolar systems, i.e., two conductors of opposite polarity are used. Homopolar: two conductors or more that are of the same polarity are used with the ground acting as the return path.

Protection schemes for HVDC are designed to isolate the faults quickly and protect the operator

Protection Requirements

- Fault restoration ought to be fast and accurate.
- Sensors of high sensitivity are to detect all faults
- Protection method ought to be robust and eliminate redundancies.
- Operating conditions should be restored within critical time.

In order to achieve faster response against faults and timely restoration, DIR algorithm has been developed in this paper.

2 System Components of HVDC

Figure 2 shows the basic system components of HVDC which can be categorized into,

- Converters
- Transformers

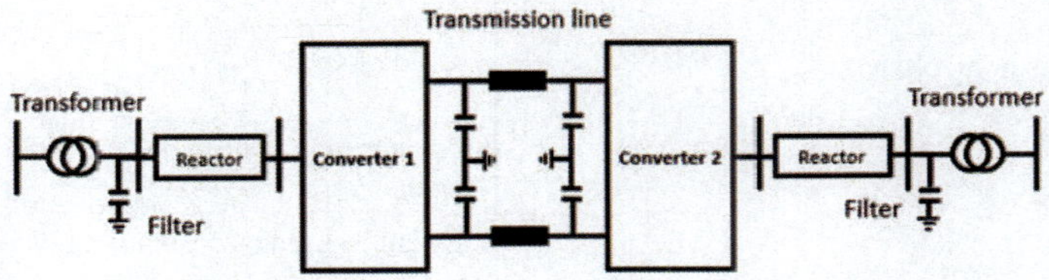

Fig. 2 Basic components of HVDC system

- Filters
- Transmission Line.

Converters at the sending end side convert AC to DC and at the receiving end side of the HVDC system convert DC to AC. These converters are connected to their respective sides through a transformer that steps up/down voltage levels to desired levels. Filters are of two types, AC side filters and DC side filters. AC side filters limit the harmonic currents produced and compensate the reactive power consumed by HVDC converters. DC side filters are used to suppress the voltage ripples that are caused by the converters. Considering the environmental impacts of HVDC transmission, overhead transmission lines are preferred [4].

3 Faults in HVDC System

Circuit breakers are used to disconnect the line during the occurrence of faults and restored once the fault has been cleared. Deadtime is shorter for DC line as DC fault current is lesser than AC fault current. Commonly, observed faults in HVDC transmission systems are shown in Table 1 [5–7].

4 Protection Schemes and Power States for HVDC Systems

4.1 Fault Clearing Strategies

Line Protection: the time of fault clearing is limited to several seconds by using fast DC breakers and can be increased using additional current limiters. Another method is to use fault blocking converters like IGBT, which terminate the faulted line and limit the current to be interrupted hence reducing the time constraint [8].

Open Grid Strategy: along with the strategy used above, the line is put out of order for repair, and the healthy line is immediately reclosed. Since it involves switching between two lines, the stability concern increases and hence the time constraint is more stringent [8].

Table 1 IEE 1159-1995 [10]—measured voltage events classification

Types of faults	
DC faults	
L-G	Most common fault in underground HVDC systems
L-L	Rare and caused by insulation failure between two conductors
AC faults	
L-G	Comprises of 65–70% of the faults
L-L/L-L-G	Less frequent
L-L-L-G/L-L-L	Severe, rare and also known as balanced faults
Malfunction of controller and valves	
Arc back	Non-self-clearing faults which are random in nature.
Arc through	Occurs when the faulted valve experiences a firing delay
Misfire	Occurs when the valve at the incoming end is unable to fire

Low-Speed HVDC Protection - this is ideally useful in smaller DC networks where switching from DC to AC is not an issue. Fault current is limited, which prevents complete shutdown of the HVDC system hence enabling faster recovery [8].

According to [9, 10] IEEE 1159-1995 Power Quality Standards, variations in the power system are classified as depicted in Table 1. For simulation purposes, momentary, temporary, long duration interruptions are simulated. Depending on the severity of the fault causing the interruption circuit breaker operations are performed. Operations of the circuit breaker are performed based on the above-specified power system states. According to [11], power system states can be classified into the following five states as depicted in Fig. 3.

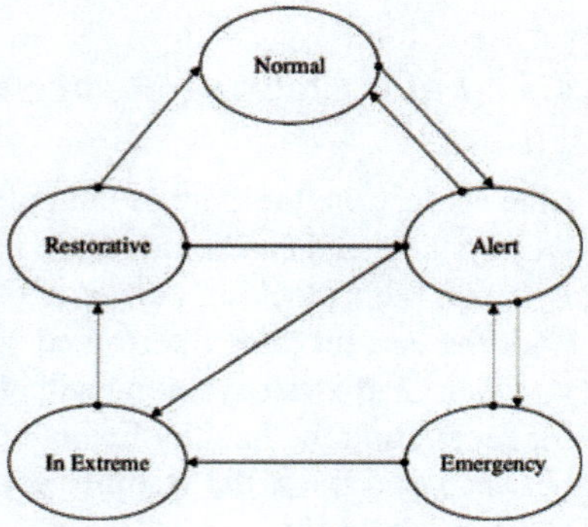

Fig. 3 Power system operational states [11]

5 Simulation Using the Proposed Algorithm

5.1 Simulation Model Under Consideration

This model consists of two utility grids, 500 kV 5000 MVA at sending end and 345 kV 10,000 MVA at receiving end. A 500 km HVDC line links the two. Twelve pulse converters are used on either side. Base current and voltage are 2000 A and 500 kV, respectively. Figure 4 shows the pu representation of the system. Figure 5 shows the model of the system under consideration for fault analysis and restoration.

The output voltage at the DC end of a HVDC transmission line is calculated as,

$$V_{dc} = \frac{3 * \sqrt{3} * \sqrt{2} * V\_\mathrm{rms}}{\pi} \cos\alpha \tag{1}$$

where α is the firing delay angle of the rectifier.

The expression gives DC line current,

$$I_d = \frac{\Delta \mathrm{Vacross\ line}}{Rdc} \tag{2}$$

Power transported to across the end of the line is given by

$$P_d = V_{\mathrm{line}} \frac{\Delta \mathrm{Vacross\ line}}{Rdc} \tag{3}$$

where *Rdc* is the dc resistance for positive transportation line conductor [13], Fig. 5 shows the model for the DC fault Simulation.

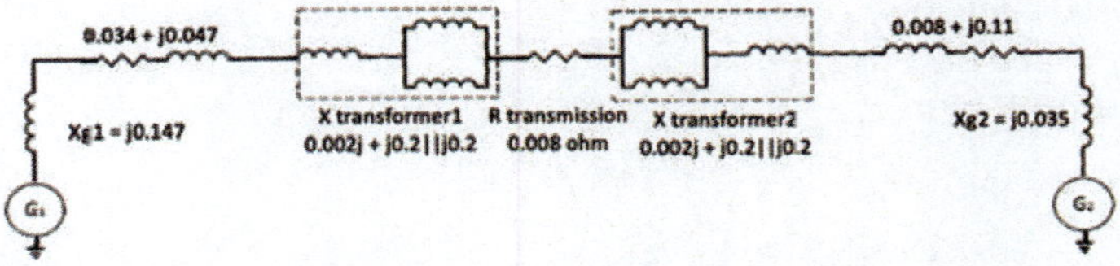

Fig. 4 pu Representation of the system

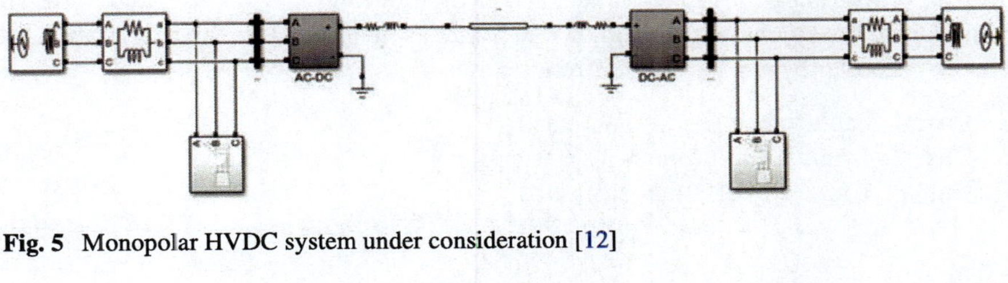

Fig. 5 Monopolar HVDC system under consideration [12]

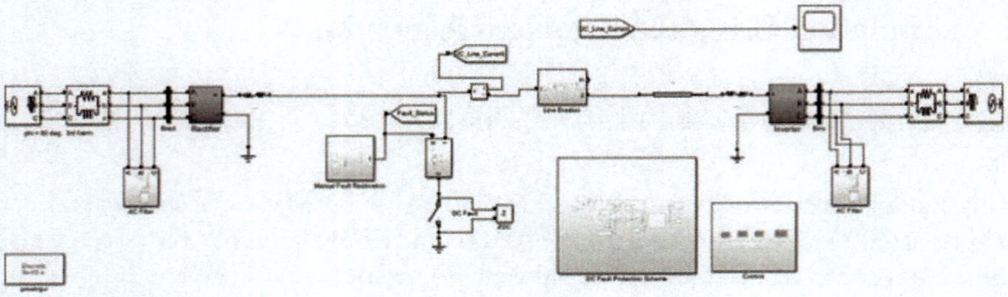

Fig. 6 DC fault simulation

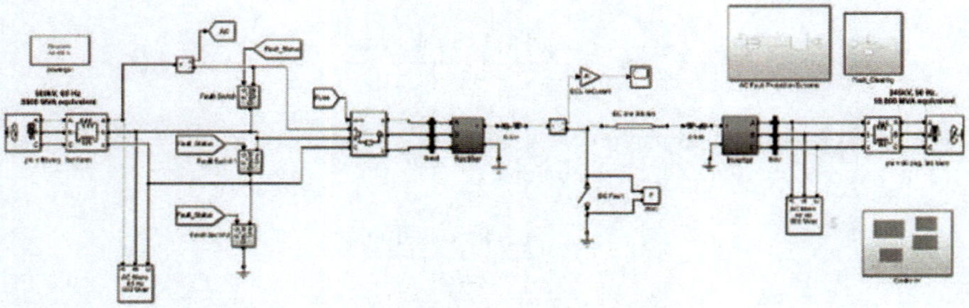

Fig. 7 AC fault simulation

$$I_{\text{fault}} = \frac{V_{\text{fault}}}{R_{\text{Transmissionlne}}} \tag{4}$$

Figure 6 and 7 show the DC & AC fault simulated circuit, respectively.
$I_{\text{base}} = 2000$ A; $V_{\text{pu}} = 1 = 500$ kV;
$Z_{\text{pu}} = 0.04 + 0.194\text{j}$
$I_{\text{fault pu}} = 5\text{pu}$ 78.3 deg lag
$I_{\text{actual}} = 10$ KA

5.2 *Proposed DIR Algorithm*

DIR algorithm has been implemented on L-G DC fault and L-L-L-G AC fault through simulation. Based on the duration of fault time, as mentioned in Table 2, simulation has been implemented for three different cases.

Case 1: Fault time < 3 s (momentary).
Case 2: 1 min > Fault time > 3 s (temporary).
Case 3: 1 min < Fault time (sustained).

Table 2 IEE 1159-1995 [10]—measured voltage events classification

Category	Duration	Voltage magnitude	Number of events
Short duration variation			
Instantaneous			
Sag	10–500 ms	10–90%	528
Swell		110–180%	9
Momentary			
Interruption	10 ms–3 s	<10%	9
Temporary			
Interruption	3 s–1 min	< 10%	1
Short duration variation			
Sustained interruption	>1 min	0%	8

Figure 8 shows the DIR algorithm for the separation of the faulty portion from the healthy portion of the system by isolating the faulty portion. The system is restored to the original state upon clearance of the fault.

For currents greater than 1.1 pu, the fault is detected, and the state of the system is alert/emergency. Increased scanning time (fault time) results in the changing of the system state. For temporary—emergency and sustained—emergency/extreme (extremis) faults the line is disconnected to prevent the flow of fault current; the line

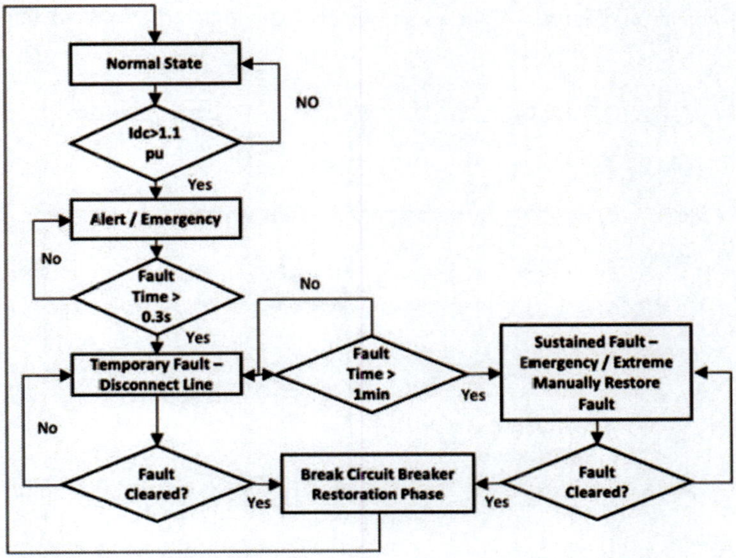

Fig. 8 Detection—isolation—restoration algorithm

is reinstalled (connected) once the fault has been cleared or isolated manually and cleared as required in the latter.

6 Result and Analysis

For simulation purposes, 1 s simulation time = 5 s real time. During fault, all observations are done on the DC line.

6.1 DC Fault Analysis

Case 1: Fig. 9 shows the result for simulated momentary fault (at $t = 3$ s). This fault is self-healing hence does not require any operation

Case 2: Fig. 10 shows the result for simulated temporary fault (at $t = 1$ s) without line disconnection. Figure 15 shows the result for simulated temporary fault (at $t = 1$ s) with line disconnection. Since the scan time exceeds 3 s (0.6 s simulated time) the state of the system moves to emergency state. Once the fault is cleared (region 4) the system state is in restorative state. In Fig. 11 the line has been disconnected in region 3 where the scan time (fault time) exceeds momentary fault condition.

Fault current of 95 Amps is disconnected.

Residual Fault current is 4 A.

Case 3: Fig. 12 shows the result for simulated sustained fault (at $t = 1$ s) without line disconnection. Figure 13 shows the result for simulated sustained fault (at t

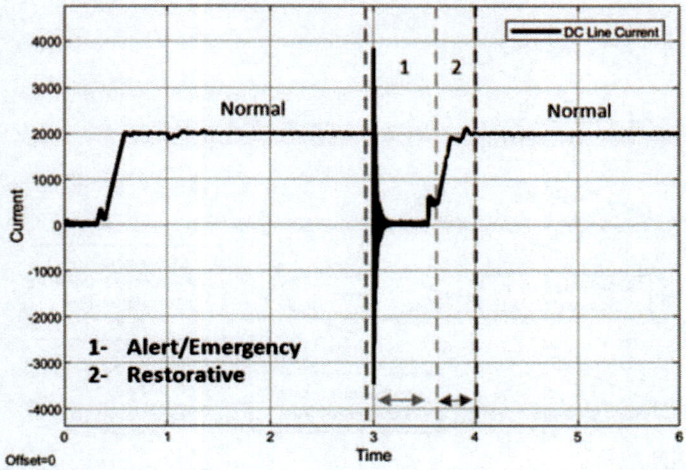

Fig. 9 Momentary DC fault-DC line current (A)

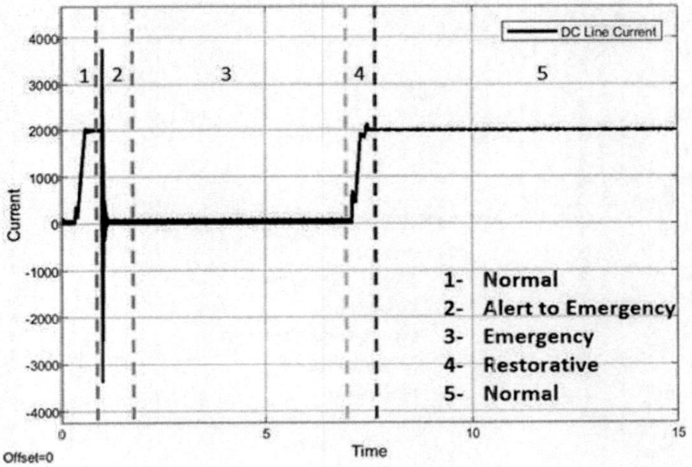

Fig. 10 Temporary DC fault-DC line current (A) without line disconnection

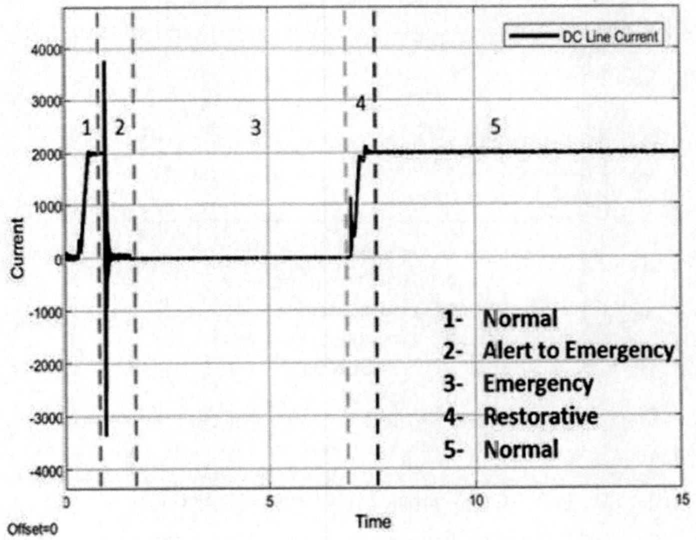

Fig. 11 Temporary DC fault-DC line current (A) with line disconnection

= 1 s) with line disconnection. The fault persists beyond momentary & temporary conditions and becomes sustained fault. Hence manual fault clearance/isolation is necessary. Once the fault has been cleared, the line is connected back to the system. This comprises of the restorative phase as shown in region 5. Region 6 is normal state. Circulating fault current is avoided.

Fault current = 95A

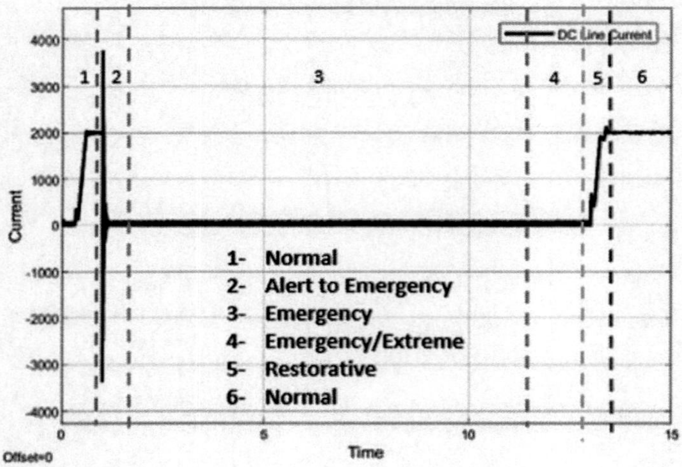

Fig. 12 Sustained DC fault-DC line current (A) without line disconnection

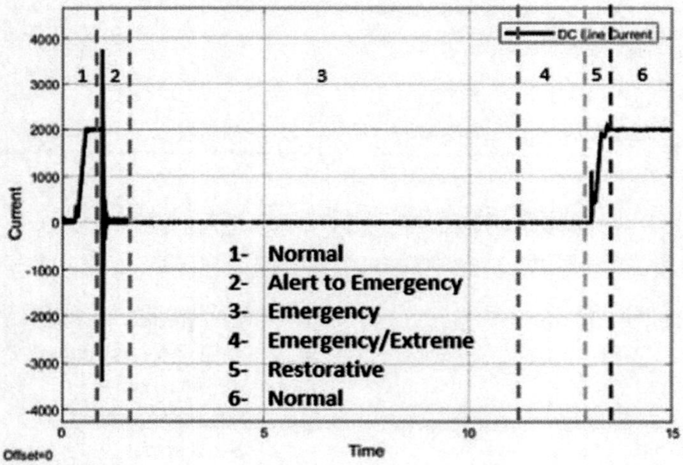

Fig. 13 Sustained DC fault-DC line current (A) with line disconnection

Residual fault current = 4A

6.2 AC Fault Analysis

Case 1: Fig. 14 shows the result for simulated momentary fault (t = 5–5.4 s). Fault duration is 2 s in real time (0.4 simulated time). This is a self-healing fault and hence

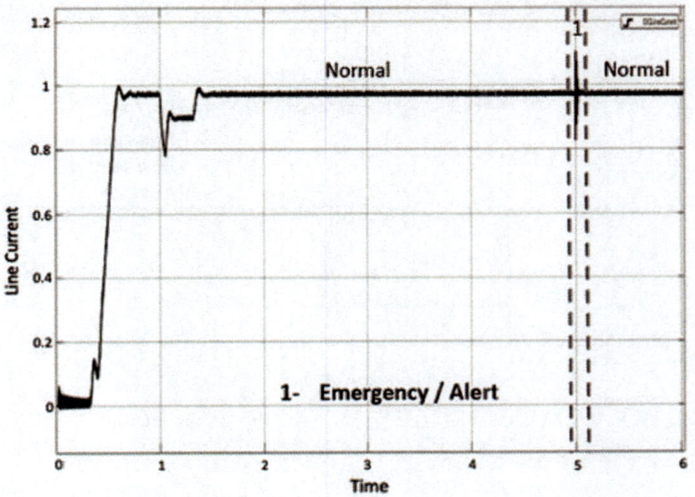

Fig. 14 Momentary AC fault-DC line current (p.u.)

no circuit breaker operation is required. The system is put to alert state at the time of fault.

Case 2: Fig. 15 shows the result for simulated temporary fault (at $t = 3$ s) without line disconnection. Figure 16 shows the result for simulated temporary fault (at $t = 3$ s) with line disconnection. The scanned time in Figs. 15 and 16 is greater than 3 s real time (0.6 s simulated time) this region is classified as Emergency state. At T = 8 s the fault is cleared and the restorative state begins. In region 2 of Fig. 16, fault

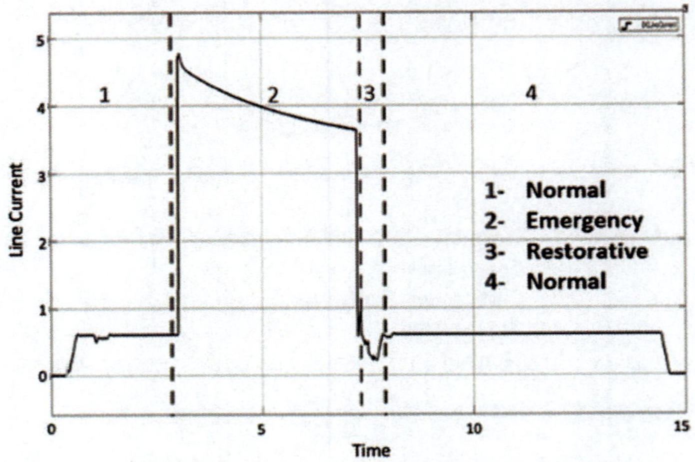

Fig. 15 Temporary AC fault-DC line current (p.u.) without line disconnection

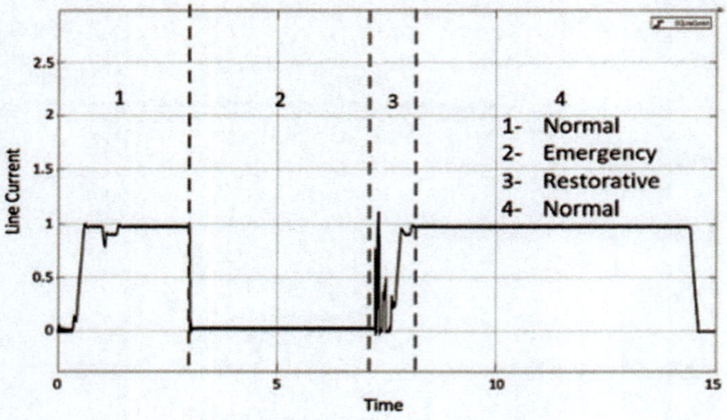

Fig. 16 Temporary AC fault-DC line current (p.u.) with line disconnection

current is not observed (close to 0 pu) because the protection scheme disconnects the line.

Case 3: Fig. 17 shows the result for simulated sustained fault (at t = 1 s) without line disconnection. Figure 18 shows the result for simulated sustained fault (at $t =$ 1 s) with line disconnection. Since the fault lasts more than 1 min (12 s simulated time) the fault is now regarded as sustained AC fault and the power system enters emergency/extremis state (regions 3). This requires manual intervention to isolate and clear the fault. Figure 18 shows At $T = 14$ s the fault is cleared and restoration state beings. In Fig. 18, the circuit breakers are used to cut the line and protect the system from fault current in regions 2 and 3 hence we do not observe any current.

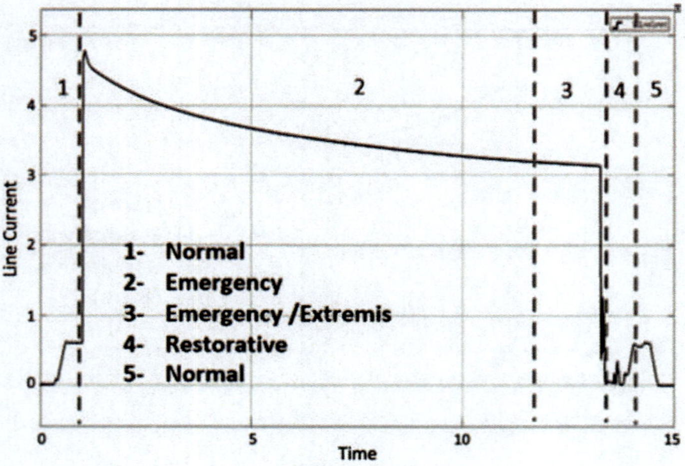

Fig. 17 Sustained AC fault-DC line current (p.u.) without line disconnection

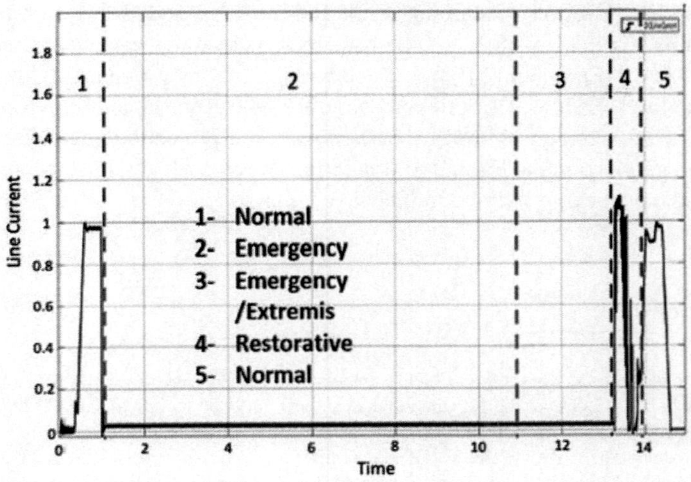

Fig. 18 Sustained AC fault-DC line current (p.u.) with line disconnection

Table 3 Summary of the simulated results

Types of faults		
DC faults	D.C line current without disconnection	D.C line current with DIR
Case 1	90A	Not-applicable
Case 2	95A	4A
Case 3	95A	4A
AC faults	D.C line current without disconnection	D.C line current with DIR
Case 1	10kA	Not-applicable
Case 2	9.6kA	3.5A
Case 3	9.6kA	3.5A

Table 3 summarizes the results obtained for the various cases. These results have been verified mathematically

7 Conclusion

In this paper, symmetrical faults on the AC and Pole-Ground fault-DC side of the Monopolar HVDC system has been analyzed through simulation. Detection-isolation-restoration (DIR) algorithm has been presented and successfully implemented employing pole controllers and circuit breakers. Thus, eliminating the fault current in the system. This method takes into account the different fault states-like momentary interruptions, temporary fault and the sustained fault which have been identified by IEEE 1159-1995 Standards. The system states have been classified as Normal, Alert & Emergency, Extreme and Restorative states depending on the duration of the fault. Each of them demands a different response and the same is taken care. This work can be used further in the process of protection gear selection. The results and graphs obtained from the simulations were verified with expected manual results and have been presented in this paper.

References

1. S. Kamakshaiah. V. Kamaraju, HVDC transmission, (TATA McGraw-Hill, 2011), pp. 22–28
2. K.S. Lakshmi, G. Sravanthi, L. Ramadevi, A review paper on technical data of present HVDC links in India. Int. J. Recent Innov. Trends Comput. Commun. (IJRITCC)
3. S. Kamakshaiah, V. Kamaraju, HVDC Transmission, (TATA McGraw-Hill, 2011). Fig 1.8(a)
4. B. Taormina, J. BaldJuan, A. Want, T. Gerard, L. Morgane, D. Nicolas, C. Antoine, A review of potential impacts of submarine power cables on the marine environment: knowledge gaps, recommendations and future directions, Renew. Sustain. Energy Rev. **96**, 380–391 (2018)
5. K. Khaimar, P.J. Shah, Study of various types of faults in HVDC transmission system, in *2016 International Conference on Global Trends in Signal Processing, Information Computing and Communication (ICGTSPICC)*, Jalgaon, 2016, pp. 480–484. https://doi.org/10.1109/icgtspicc.2016.7955349
6. A.K. Khairnar, P.J. Shah, Study of Various Types of Faults in HVDC Transmission System, in *International Conference on Global Trends in Signal Processing, Information Computing and Communication ICSPICC (Technically Sponsored by IEEE Bombay Section)*, Proceeding, (2016)
7. H. Batra, R. Khanna, Study of various types of converters station faults, Int. J. Eng. Res. Technol. (IJERT) **2**(6), 3288–3293, (2013)
8. W. Leterme, D. Hertem, *Classification of fault clearing strategies for HVDC grids* (Lund, Cigre, 2015)
9. IEEE standard 141-1993: Recommended practice for power distribution in industrial plants
10. IEEE Standard 1159-1995: Power quality monitoring standards
11. P. Kundur, *Power system stability and control* (McGraw-Hill, New York, 1994)
12. Silvano Casoria, "Discrete HVDC Controller", Power Systems Simulation Laboratory, IREQ, Hydro-Quebec
13. A. Usman, M. Kutay, E. Tuncay, MATLAB/SIMULINK model for HVDC fault calculations, (2019). https://doi.org/10.1109/acemp-optim44294.2019.9007154

Integration of Solar Photovoltaic Generation in a Practical Distribution System for Loss Minimization and Voltage Stability Improvement

S. J. Rudresha, Shekhappa G. Ankaliki, T. Ananthapadmanabha, and V. Girish

1 Introduction

The power system includes generation, transmission and distribution stations. The power is supplied to consumers by distribution system, and this control and operation of distribution system are becoming more complicated due to drastic variation of load on the system. In general, distribution system is connected in radial configuration. The R/X ratio is more in radial system which leads to more voltage drop and significant power loss. Due to continuous increase in load in the system, the source has to supply this load current which leads to more drops in voltage and increase in losses. This increase in voltage drop and system losses reduces distribution system performance [1].

Currently, distributed generation (DG) technologies are attracting the researchers as an alternative solution for the conventional power supply from the central grid in order to reduce voltage drop and minimize the losses. Distribution generation (DG) is also called as dispersed generation which is small-scale generation and mainly connected at the consumer end in the distribution system [2]. This DG technology can supply electricity from sources which includes renewable and non-renewable energy sources. The renewable energy sources are small hydro, geothermal, wind, biomass, solar and cogeneration. The non-renewable energy sources are gas turbines, fuel cell,

S. J. Rudresha (✉)
PESITM, Shivamogga, India
e-mail: rudreshsj@pestrust.edu.in

S. G. Ankaliki
SDMCET, Dharwad, India

T. Ananthapadmanabha
NIEIT, Mysore, India

V. Girish
HESCOM, Hubli, India

Sanjeevikumar et al. (eds.), *Advances in Renewable Energy and Electric Vehicles*, Lecture Notes in Electrical Engineering 767,
https://doi.org/10.1007/978-981-16-1642-6_19

reciprocating engines and micro-turbines. The emission from most of these sources is less so it is good for environment and reduces the voltage drop and minimize losses [3]. The more benefits from these DGs can be obtained by finding the optimal placement and sizing of DG in the distribution network which includes increase in voltage stability and reduction in losses, power quality improvement, reduced transmission and distribution congestion which intern reduces the overall cost. Different factors have to be considered for finding the optimal size and location of the DG such as DG penetration level, its position ambiguity and changing output from DGs [4]. Out of all the renewable sources, the use of solar photovoltaic generation (PV) is increasing day by day due to many advantages such as its environmental friendly, available in abundant, its operating and maintenance costs are low and cost of solar panels are decreasing day by day compared to costs of other renewable energy sources. The government is promoting this use of solar energy by providing financial support, and installation of residential solar panels is simple on rooftops or on the ground without any intervention to human lifestyle [5, 6].

The one of the realistic distribution feeder is Shivamogga, Karnataka, India. The DGs are grouped into four types depending on its delivering capacity of real and reactive power into system.

"Type1: DG capable of delivering both active and reactive power (e.g. synchronous machine).

Type2: This type of DG is capable of delivering only active power (e.g. module, micro-turbines and fuel cells).

Type3: DG capable of delivering only reactive power (e.g. capacitor banks).

Type4: DG capable of delivering active power but consuming reactive power (e.g. induction generators, which are used in wind farms)" [7].

Numerous methods have been found in order to determine optimal DG position by means of either analytical or heuristic optimization techniques. In [8], optimal DG positions are found by determining critical buses using power stability index. In [9], modified voltage index is used to find and increase voltage stability margin in the system, and optimization difficulty is solved by mixed-integer nonlinear programming technique. In [10], the particle swarm optimization (PSO) which is heuristic optimization techniques is used to find the size and position of multi DGs by taking into account multi-objective index. In [11] the cuckoo search algorithm and in [12] GA algorithm are used as heuristic optimization techniques in order to solve the optimal DG placement difficulty.

Many researchers have done work for determining best possible placement and sizing of PV generation by considering power loss reduction and voltage stability improvement. In [13], in order to determine optimum placement of PV and wind turbine, PSO technique was used. In [14], VSI method is used to find optimal location of DG, and artificial bee colony optimization method is proposed to find optimal sizing of hybrid PV/wind turbine/fuel cell.

The main purpose of this work is to study this distribution feeder is having low-voltage profile, high energy losses, and more power interruptions and congestion in the feeder. The best possible position and sizing of DG decrease the system losses

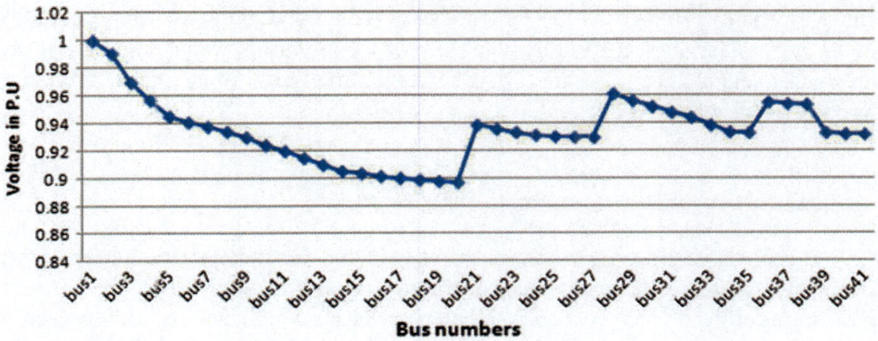

Fig. 1 Voltage profile of practical Devikathikoppa 41-bus distribution feeder

and pick up the voltage profile to stay within the suitable limits and thereby maintain voltage stability in the system.

2 Statement of the Problem

This manuscript analyses a practical distribution feeder called "Devikathikoppa feeder" emanating from 110/11 kV Alkola substation, situated in Shivamogga, Karnataka, India. The feeder has the peak load of 1.977 MW and 0.617 MVAR with 41 distribution transformers (DTC). The load on the distribution system increases with time due to rise in population. As load demand increases, there is an increase in the system losses and decrease in the voltage profile of the system. The consumer at the last bus will face the voltage stability problem, i.e. voltage profile is not within the acceptable limits; therefore, there is a decrease in performance of distribution system. Distribution generation (DG) placed at the consumer end will improve the voltage profile at the buses and decrease the total loss in the system. This paper analyses the main problems faced by the practical distribution consumers, i.e. reduced supply voltage and high system losses. The voltage profile of the existing "Devikathikoppa feeder" is shown in Fig. 1. During analysis, as per the standard we considered 6% variable voltage as acceptable stable voltage limit, i.e. $V_{\min} = 0.94$ p.u and $V_{\max} = 1.06$ p.u. In the subsequent part, we will show how optimum size and position of DG impact on voltage level of interconnecting buses.

3 Proposed Analysis Method

In our analysis, sensitivity factor method is used to calculate best possible size and position of DG using Power World Simulator software in order to decrease the losses and progress the voltages at the various buses which maintain the voltage stability in the system. For a given system, when size of DG is varied from P_{DG1} to P_{DG2} and the consequent change in power loss is correspondingly P_{L1} to P_{L2}, then formula to find the sensitivity factor is

$$\frac{dP_L}{dP_i} = \frac{P_{L1} - P_{L2}}{P_{DG1} - P_{DG2}}.$$

In this analysis, in order to reduce the search space in the system, the sensitivity factors are calculated for every bus using above equation. Out of which the bus with maximum sensitivity is recognized and all other buses which have sensitivity factors very near to the maximum value are selected for analysis. Then at all these selected buses, the DG size varies in large step value to find power loss. The bus which gives minimum loss for various DG size is best position, and corresponding generation is the optimum DG size [15].

4 Steps Used in PWS Software to Carry Out Simulation

The main aim of this sensitivity factor method is to discover best place and size of a DG unit to decrease the power losses and get better the voltage profile in the system. In a system, there might be several best possible locations and sizes for a DG unit, out of which any one solution is the best one.

"The following steps are carried out to model the test system in the Power World Simulator

1. Draw the buses and enter the data.
2. Draw the transmission lines, generators, load and enter the data as given in the test system.
3. Now run the model and study the voltage at all the buses and total losses in the system without DG.
4. Determine the sensitivity of each bus with small penetration of DG and list the most sensitive buses.
5. Select a bus from the list and calculate power loss for large variation of DG size.
6. Make sure whether all sensitive buses have been analysed.
7. Find the bus which has minimum power loss and its corresponding DG size.
8. Find the voltages at all the buses with optimum DG size and location.
9. Analyse the voltage stability of the system.
10. If the voltage stability is not maintained at all the buses. then increase the DG size at a optimum location until the voltage stability is maintained" [15].

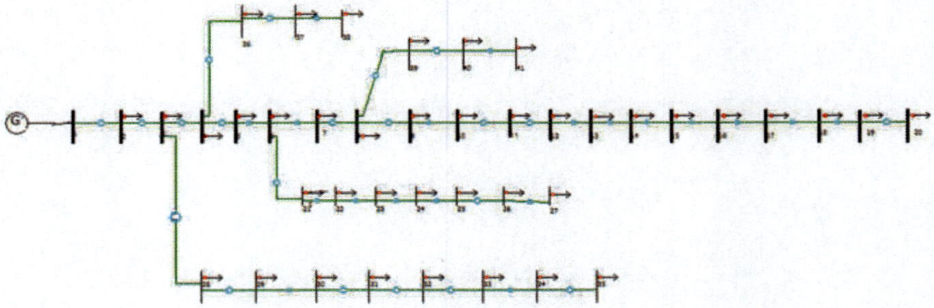

Fig. 2 41-bus practical Devikathikoppa distribution feeder

5 Simulation Results

5.1 Devikathikoppa 41-Bus Practical Distribution System, Shivamogga

The practical "Devikathikoppa feeder" is shown in Fig. 2 which is a 11 kV distribution feeder having total 41-bus with bus-1 is the slack bus and is modelled and simulated in Power World Simulator software, and voltage stability and system losses are analysed with and without placement of DG. Without DG real power loss of the system is 0.1334 MW, reactive power loss is 0.096 MVAR and minimum bus voltage is 0.8967 p.u at peak load.

The following different cases are considered as below:

Case-1: Integration of only DG units.
Case-2: Integration of only capacitor.
Case-3: Incorporation of DG and capacitor simultaneously.

Case 1: Integration of only DG (solar PV module) unit

The simulation diagram of Devikathikoppa feeder in PWS is as shown in Fig. 3. The proposed loss sensitivity factor technique is used to find the optimal placement and sizing of DG. The optimal placement obtained using this method is at bus 18 and size of DG is 30% of the total generation obtained without DG from central grid. The optimal DG location obtained from this method is at bus-18 with DG size of 0.633 MW.

The results, before and after DG placement, are shown in Table 1. After DG placement, the real power loss is reduced to 0.0783 MW from 0.1334 MW, reactive power loss is reduced to 0.0589MVAR from 0.096 MVAR, and lowest bus voltage is also improved to 0.9401 p u from 0.8967 p u.

By placing DG, the system real and reactive power losses are reduced to 41.30% and 38.54, respectively, and voltage profiles at all the buses are within the acceptable limits (0.94 < Vi < 1.06) which in turn improves the voltage stability of the system.

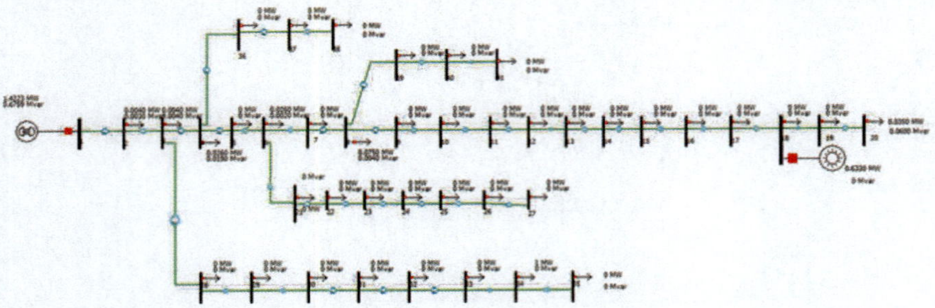

Fig. 3 Devikathikoppa feeder is modelled in PWSS with optimal placement and size of DG

Table 1 Loss reduction analysis

Cases	With out DG	DG only	Capacitor only	DG and capacitor
DG location at bus	–	18	18	18 and 16
Active power supplied by DG in MW	–	0.633	0	0.633
Reactive power supplied by DG in MVAr	–	0	0.214	0.214
Active power loss in MW	0.1334	0.0783	0.1188	0.0709
Reactive power loss in MVAr	0.096	0.0589	0.088	0.0498
P loss reduction in %	–	41.30	10.94	46.85
Q loss reduction in %	–	38.64	8.33	48.12

Case 2: Integration of only capacitor

The optimal size and location of shunt capacitor unit for 41-bus system are calculated by proposed technique, and it is 0.214 MVAR at bus-18. After capacitor placement, the real power loss is reduced to 0.1188 MW from 0.1334 MW, reactive power loss is reduced to 0.088 MVAR from 0.096 MVAR, and minimum bus voltage is also improved to 0.9166 p u from 0.8967 p u. The results are shown in Table 1.

Case 3: Integration of DG and capacitor simultaneously.

In this case, both DG and capacitor are located simultaneously, the optimal size of DG is 0.633 MW at bus-18, and optimal size of capacitor is 0.214 MVAR at bus-16. From Table 1, it can be seen that after DG and capacitor placement, the percentage real and reactive power loss drop of the practical system are 46.85% and 48.12%, respectively. The least bus voltage is also improved to 0.9431 pu from 0.8657 pu.

Comparison of voltage profile for 41-bus practical distribution system for various cases is shown in Fig. 4. It can be seen from Fig. 4 that after simultaneous position of DG and capacitor the voltage profile of each bus is within the allowable limit.

The real power loss for different cases is shown in Fig. 5, and reactive power loss for different cases is shown in Fig. 6. The figure shows that more reduction in real

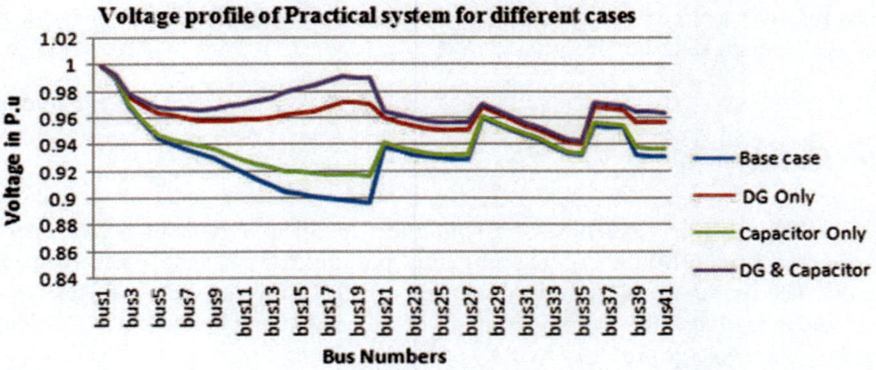

Fig. 4 Voltage profile of practical 41-bus distribution system for different cases

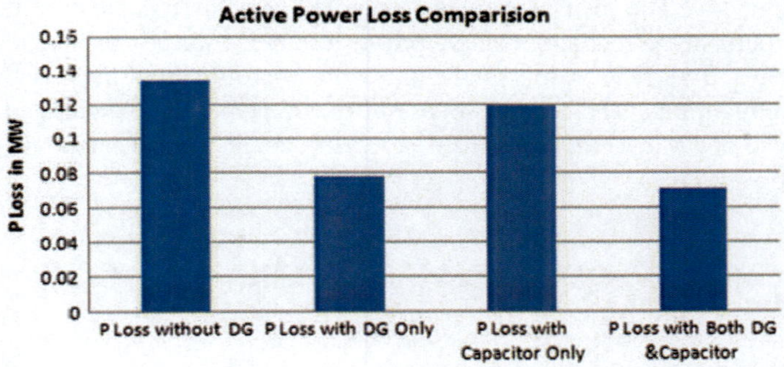

Fig. 5 Real power loss of practical 41-bus distribution system with different cases

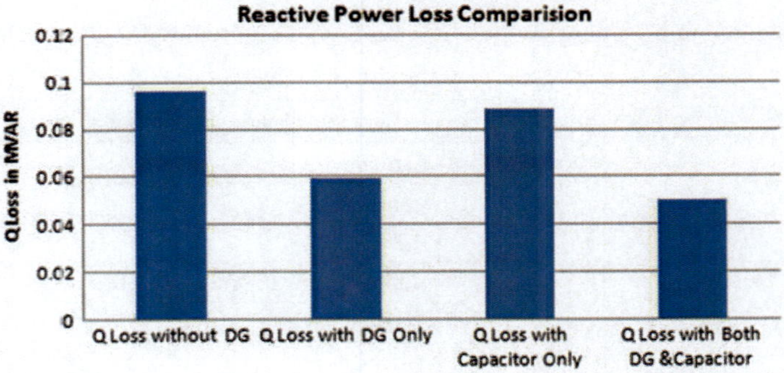

Fig. 6 Reactive power loss of practical 41-bus distribution system for different cases

and reactive power losses occurs when we place DG and combination of DG and capacitor in the distribution system.

6 Conclusion

This work analyses a practical distribution feeder called "Devikathikoppa feeder" emanating from 110/11 kV Alkola substation, positioned in Shivamogga, Karnataka, India. The feeder has 41-bus with peak load of 1.977 MW and 0.617 MVAR. New technique based on loss sensitivity factor is implemented to minimize total power loss and improve voltage profile of RDS by optimal position and sizing of DG and capacitor by considering various operating conditions. To test the effects of the proposed method, it is tested on 41-bus real distribution system of Devikathikoppa feeder in Shivamogga city. The result shows that once DG and shunt capacitor allotment is done, the percentage real and reactive power loss reduction of the practical system are 46.85% and 48.12%, respectively, and voltage profile at all the buses is within the acceptable limits. The consumers or utility companies are encouraged to set up a rooftop solar PV system at the load side after analysing the results of practical distribution system.

Appendix

Table 2 shows the geographical site condition of Shivamogga, and Fig. 7 shows the monthly solar radiation details of Shivamogga city.

Table 2 Geographical site condition of Shivamogga

S. No.	Geographical site	Technical value	Unit
1	Latitude	14	°N
2	Longitude	75	°E
3	Elevation	632	m
4	Ambient temperature	20–30	°C
5	Relative humidity	55–75	%
6	Tilt angle	7–20	Degree

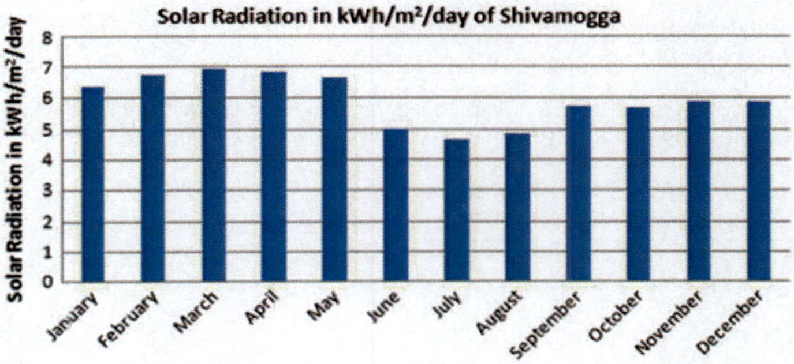

Fig. 7 Monthly solar radiation details of Shivamogga city

References

1. A.M. Eltamaly, Y.S. Mohamed, Impact of distributed generation (dg) on the distribution system network. Int. J. Eng. Sci. (2019)
2. B. Singh, J. Sharma, A review on distributed generation planning. Renew. Sustain. Energy Rev. **76**, 529–544 (2017)
3. T. Ackermann, G. Andersson, L. Soder, Distributed generation: a definition. Electric Power Syst. Res. **57**, 195–204 (2000)
4. V.S.N. Arava, V.R. Shesmatti, Voltage profile and loss analysis of radial distribution system in presence of embedded generator with a case study in MI-power, (IEEE 2013)
5. A.T. Davda, B. Azzopardi, B.R. Parekh, M. Desai, Dispersed generation enable loss reduction and voltage profile improvement in distribution network—a case study, Gujarat, India. IEEE Trans. Power Syst. **9**(4), (2013)
6. P. Dondi, D. Bayoumi, C. Haederli, D. Julian, M. Suter, Network integration of distributed power generation. J. Power Sources **106**, 1–9 (2002)
7. M. Kashyap, S. Kansal, R. Kansal, Optimal placement of multiple type dgs in radial distribution system using sensitivity based approach. Int. J. Electr. Eng. Technol. (IJEET) **9**(3), 192–198 (2018)
8. M.M. Aman, G.B. Jasmon, H. Mokhlis, A.H.A. Bakar, Optimal placement and sizing of DG based on a new power stability index and line losses. Electr. Power Energy Syst. **43**, 1296–1304 (2012)
9. R.S. Al Abri, F. El-Saadany Ehab, Y.M. Atwa, Optimal placement and sizing method to improve the voltage stability margin in a distribution system using distributed generation. IEEE Trans. Power Syst. **28**(1), 326–334 (2013)
10. A. Kaviani-Arani, Optimal placement and sizing of distributed generation units using co-evolutionary particle swarm optimization algorithms. Indonesian J. Electr. Eng. **13**(2), 247–256 (2015)
11. Z. Moravej, A. Akhlaghi, A novel approach based on cuckoo search for DG allocation in distribution network. Electr. Power Energy Syst. **44**, 672–679 (2013)
12. K.M. Shebl, M.E. Khazendar, A.E. Husseiny, Genetic algorithm for optimum siting and sizing of distributed generation, in *Proceedings of the 14th International Middle East Power Systems Conference*, Cairo University, Egypt., (2010), pp. 433–440
13. P. Kayal, C.K. Chanda, Placement of wind and solar based DGs in distribution system for power loss minimization and voltage stability improvement. J. Electr. Power Energy Syst. **53**, 795–809 (2013)

14. H. Nasiraghdam, S. Jadid, Optimal hybrid PV/WT/FC sizing and distribution system reconfiguration using multi-objective artificial bee colony algorithm. J. Solar Energy **86**, 3057–3071 (2012)
15. S.J. Rudresha, S.G. Ankaliki, T. Ananthapadmanabha, Voltage stability analysis and loss minimization with integration of different types of DGs into the distribution system. IJSDR, **4**(12), 185–189 (2019)

IoT-Based Patient Health Monitoring System Using STM32F103C8T6

K. R. Nishitha and M. Vittal Bhat

1 Introduction

IoT, i.e., Internet of Things, is a fast-growing worldwide network of interconnected variety of objects that support many input–output devices, sensors and actuators based on standard communication protocol [1]. It uses "Smart" devices that make use of different actuators as well as sensors aimed at carrying out numerous activities. Progression happening in inventive technology with IoT brings about a significant impact on human wellbeing. Human wellbeing ought to be conservation as well as advancement of healthiness through identifying, diagnosing, providing cure, and avoiding illnesses and infection in addition to some bodily and psychological harm happening within people. Human fitness is able to contribute a significant role from nation's economy point of view. Nonetheless divided behaviour involved, further exacerbated due to deficiency in devices aimed at interaction in the middle of experts, arouses to have importance in useful interaction to enhance synergy.

A significant viewpoint in the healthcare system is observing the vigorous gestures like temperature level, blood pressure values as well as pulse. Several apparatus showing indispensable gestures of one who is suffering are generally found inside serious care section within operating theatres. Nevertheless, situations sometimes arouse as medic is not being alarmed during a phase of crisis in spite of monitoring frequently. Additionally, information is not possible to remain circulated amongst another expert found in same sector as well amongst relatives. Expertise supporting entirely these deeds does not remain reachable. Visiting medics now and then is

K. R. Nishitha (✉) · M. Vittal Bhat
P. A. College of Engineering, Mangaluru, India

M. Vittal Bhat
e-mail: vittalbhatm@gmail.com

Sanjeevikumar et al. (eds.), *Advances in Renewable Energy and Electric Vehicles*,
Lecture Notes in Electrical Engineering 767,
https://doi.org/10.1007/978-981-16-1642-6_20

not economical for a number of individuals in rising countries. Thus, issue is overwhelmed via utilizing IoT. Hence, IoT for health care will become a stable solution for the healthcare system [1].

In health care, IoT empowers sensor information to be intelligent via inserting those to form a framework. It further processes whatever values are sensed as well yields results successfully and then communicates the data to the assigned IP address through a Wi-Fi transceiver which can connect to a current existing internet source when configured as an access point. The outcomes can be then accessed through a web application to monitor the bioclinical readings through a cell phone or remote computer having Internet access. In this way, doctor stays updated with the patient's health status [2].

2 Related Work

Ashwini Gutte et al. implemented an IoT-based health monitoring system using Raspberry Pi to check the heartbeat and body temperature of elderly persons. This framework could send SMS as well as email alerts to desired users in case specific sensor value exceeds the definite range because of which medics got to know about the health condition and successfully take proper care [3]. An IoT-based patient health monitoring model using ESP-8266 as well as Arduino was able to monitor the pulse rate and surrounding temperature constantly which was further updated to ThingSpeak which is an open source of IoT application [4]. A framework for healthcare monitoring using IoT was introduced in [5] which had embedded wearable sensors and observed several wellbeing constraints like temperature, heart rate and also blood pressure. The procured data was then transferred to Raspberry Pi to process as well as analyse the records. Finally, during an emergency situation, medics could receive the results of analysis. An IoT fitness device using Arduino Uno microcontroller was implemented to observe heart rate and temperature of a person and communicate the real-time data via ESP-8266 to ThingSpeak that is an IoT website [6]. Vijay Kumar et al. presented an architecture using 8051 microcontroller. This was implemented to constantly monitor temperature as well as heartbeat using sensors and further communicated to respective users. The key importance was to advise how and also why IoT system is essential. It conveyed the patient all cautious paces likely to be faced by them [7]. Narendra Swaroop et al. introduced a real-time monitoring scheme that stored vital signs of patient and directed them to the caretakers. It mainly improved wellbeing information conveyance via BLE, GSM and Wi-Fi [8]. An IoT-based patient monitoring system using PIC16F877A microcontroller was proposed in which health parameters were transmitted to user via ESP-8266 to the IoT Gecko [9]. Jayapradha et al. introduced an IoT-based human healthcare system using Arduino Uno board which emphasized on improving the monitoring of human health using IoT and cloud combined technology [10]. Khan et al. proposed an IoT-based intelligent health monitoring system to monitor BP, heartbeat and ECG, and data was monitored by Arduino Uno and Pi-camera attached to Raspberry Pi for video [11]. Krishna and Sampath designed a framework to fetch patient's body temperature, oxygen saturation percentage as well as heart rate using Arduino board, Raspberry Pi board and cloud computing. By determining pattern of parameters, the nature of disease could be predicted [12].

3 System Architecture

3.1 Block Diagram

Figure 1 features a transmitting portion, likewise a receiving portion. Here transmitting portion comprises of sensors coupled with STM32F103C8T6 microcontroller performing information attainment plus processing as well as storage of processed information on cloud. In the receiver part, the data is monitored by using ESP-8266 module which connects entire system to Wi-Fi network.

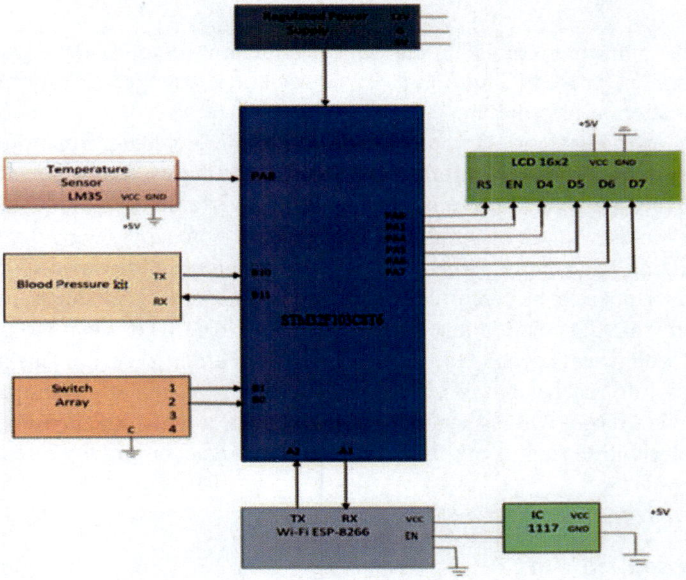

Fig. 1 Block diagram of the system

3.2 *Working*

Blood pressure kit incorporated with pulse sensor, LCD, temperature sensor and ESP-8266 unit is linked to a microcontroller as shown in the block diagram referring to STM32F103C8T6 pin diagram. Regulated power supply is given as per the hardware requirements. Program the microcontroller using Arduino IDE software using serial module. Now using the blood pressure kit, the systolic, diastolic and pulse values are sent to microcontroller. The values are displayed on the LCD. The obtained parameters can be transmitted using Wi-Fi ESP-8266 module which is programmed by using suitable network identity credentials. It can connect to available network, and hence, data can be monitored from a mobile phone or PC in the same network in case of local area network configurations or if the data is broadcasted via a static IP address over Internet, then info happens to be received and examined anywhere around the planet by just utilizing the corresponding IP address.

3.3 *Projected Framework Methodology*

Projected framework entails a number of junctures comprising of attainment of information, processing the attained information, storage of information and communication of information.

According to Fig. 2, primary job comprises of amassing the information from sick person through the sensors. Information occurs to be the wellbeing constraints like blood pressure, pulse as well as temperature. Projected framework employs sensors of low energy intake. These assemble info from sick person happening within frequent base. Utilizing recurrently gained info, wellbeing circumstance in case of specific sick person happens to be perceived so as to suggest mandatory remedies. Gained info is revealed via LCD coupled together with STM32F103C8T6. This info processing occurs as well as gets promoted to IoT computer in order to enable stockpiling the same. Now info reaches out to expected regulars via specific applications. All info procured via Web of Things server happens to be accumulated, aimed at denoting every nobles within framework furthermore conveyed according to prerequisites.

3.4 *Structural Plan*

Below stated scheme as shown in Fig. 3 presents in what manner info procured from sick person occurs to be directed towards STM32F103C8T6 via respective sensors and also displayed on LCD. The respective user thereafter examines info and also suggests remedies instantly. So health condition is supervised successfully. Info transference as well as connectivity gets enabled via Transmission Control Protocol. STM32F103C8T6 gets programmed via suitable code. Power the board from USB

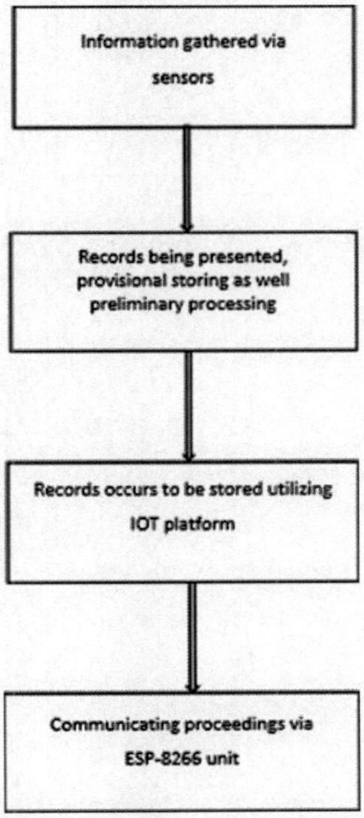

Fig. 2 Flowchart of the framework

input, and board executes the uploaded code. Liquid crystal display gets initialized displaying the project name. Later, BP kit is initialized in order to successfully communicating records involving systolic, diastolic and pulse values serially as per set variables within location. The data appears on LCD. The final output on the phone via ESP-8266 unit is procured. Meanwhile, medics make available medicaments to sick person.

4 Circuit Connection

4.1 Hardware Requirements

(1) *STM32F103C8T6 ARM Cortex M3 Microcontroller*

As shown in Fig. 4, microcontroller unit over the board occurs to be STM32F103C8T6, i.e., a specific series belonging to STMicroelectronics. Microcontroller unit does operate utilizing 3.3 V. Despite a fact that the unit of microcontroller works mainly utilizing 3.3 V, the majority general-purpose input/output pins can withstand more voltage up to even 5 V. This one additionally devises pins,

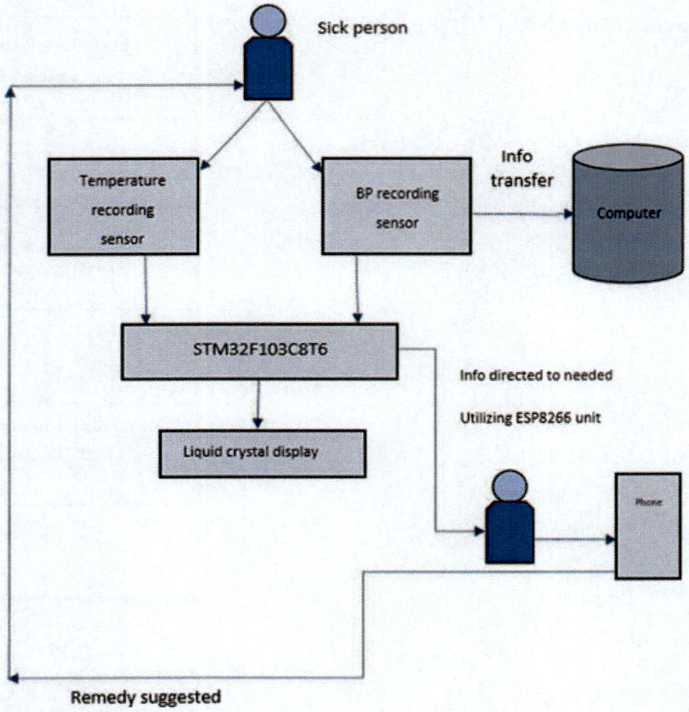

Fig. 3 Overall idea for projected scheme

Fig. 4 STM32F103C8T6 microcontroller board

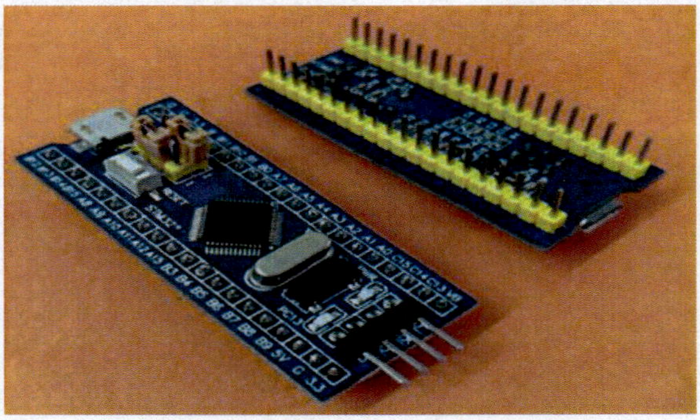

named header, which can be utilized for jumping amid program as well as operating mode. It is more stable, quick in operation and consumes less power when compared to traditional boards.

(2) *Temperature Sensor (LM35)*

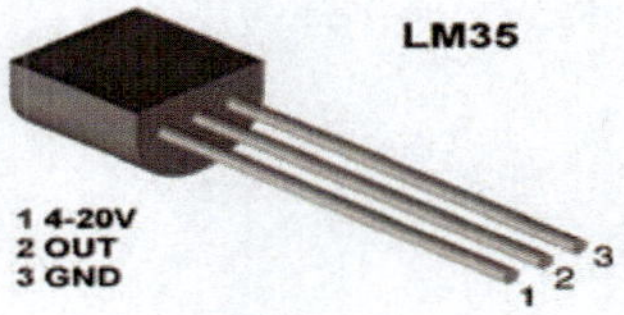

Fig. 5 Sensor utilized for temperature check

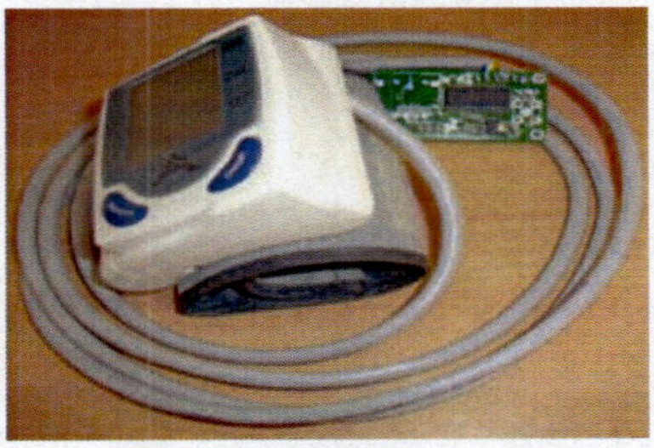

Fig. 6 Blood pressure monitoring kit

The sensor as shown in Fig. 5 is employed towards temperature checking. LM35 range is highly accurate as well as utilized in integrated circuitry. The voltage yielded from sensor remains straightly correlated towards Centigrade measurement. It can be easily interfaced with microcontroller. It is helpful for checking one's surrounding hotness that lies within the scale of −55 to 150 °C.

(3) *Blood Pressure and Pulse Sensor*

A blood pressure kit incorporated with pulse sensor is shown in Fig. 6, a smart apparatus which provides pulse and blood pressure information. This one is compressed in composition and avoids pumping process.

(4) *16*2 LCD*

As shown in Fig. 7, LCD has 224 characters as well as symbols and is called so because there are 16 characters in each line out of two such lines. This one comprises of two registers, one command likewise the other is data. First type provisions different commands directed to display, and second type stores info for being displayed.

(5) *ESP-8266 Wi-Fi Module*

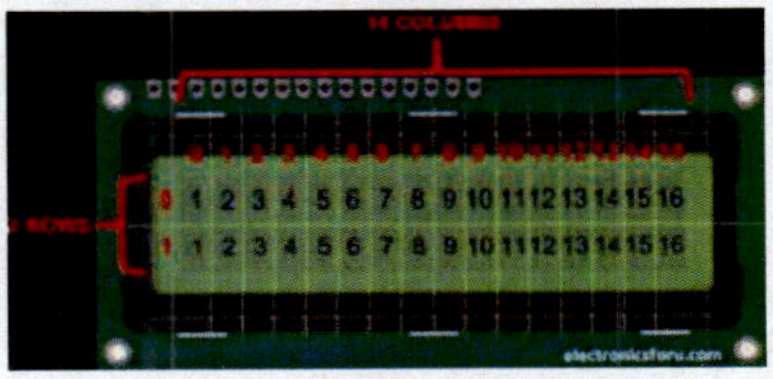

Fig. 7 16*2 LCD

Fig. 8 Image showing ESP-8266

Fig. 9 LM1117 voltage regulator

ESP-8266 is shown in Fig. 8. It occurs to be an independent SOC along integrated TCP/IP protocol pile which provides any MCU, admittance towards Wi-Fi network. It can be incorporated with sensors as well as application-specified gadgets via GPIOs because of its on-board processing in addition to storage proficiency.

(6) *LM1117*

LM1117 is shown in Fig. 9. It is a three-terminal linear voltage regulator. It provides a static potential yield of 1.8 V, 2.5 V, 3.3 V or else 5 V as per requirement.

(7) *USB to TTL serial converter module*

Figure 10 shows a universal asynchronous receiver–transmitter board, utilized for TTL serial communication. It is a breakout board for the FTDI FT232R chip along with USB interface which can use 3.3 or 5 V DC and comprises of Tx/Rx as well other breakout points. It also consists of an integrated EEPROM and optional clock generator output. The module has mini USB port used to connect to an USB port of

Fig. 10 USB to TTL serial module

interfaced device. The FTDI USB to TTL serial converter module has breakout pins, i.e. Tx, Rx, Vcc and GND connected to corresponding pins of microcontroller.

4.2 Software Requirements

(1) *Arduino IDE*

Integrated Development Environment contains a text editor enabling writing the code, a toolbar with buttons for collective functions, a message area, a sequence of menus as well as a text console. This one links to Arduino hardware for uploading code as well as communicating amongst them.

(2) *TCP UDP Terminal Application*

It is a mobile application utilized for linking your phone towards Wi-Fi unit and gets data by logging into the respective IP address. In the proposed model, we use TCP to guarantee the connection between two points..

4.3 Overall System Setup

Procedure to interface temperature sensor, blood pressure kit, LCD and Wi-Fi module to STM32F103C8T6 microcontroller is done by referring to the pin diagram. Primarily, it is essential to design the system as shown in Fig. 11.

Fig. 11 Image for presented framework

5 Implementation

5.1 Data Collection

- Install the required packages using Arduino IDE and by using a serial module connect the STM board. Upload the code by getting STM to programming mode.
- BP kit extracts pulse, systolic and diastolic values, temperature sensor measures the temperature, and these values are sent to the microcontroller and gets processed.
- Congregated data is further presented on LCD.

5.2 Receiving Data on Phone

- The entire system is interfaced with a Wi-Fi transceiver (ESP-8266) module which can be configured as an access point and also as a Wi-Fi device. It has to be programmed with its network identity credentials such as IP address, MAC address and port number. Once configuration is done, it can connect to any available network with Internet, hence successfully broadcasting the data from the sensor networks.
- Open the TCP UDP terminal application and primarily set the IP address and port number which provides the final output on the phone via Wi-Fi.

6 Results and Discussion

6.1 Result

- The obtained data from temperature sensor and blood pressure sensor is directed to STM32 as well as in addition presented on 16*2 LCD as shown in Fig. 12.

The final output on the phone via Wi-Fi is obtained utilizing respective IP via TCP UDP terminal application as shown in Fig. 13 from which medic checks info recurrently.

Fig. 12 Data displayed on the LCD screen

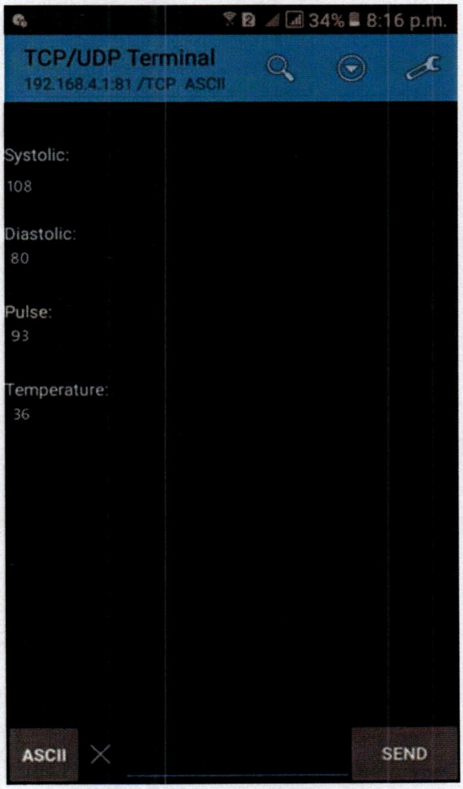

Fig. 13 Output of the data obtained via Wi-Fi

6.2 General Discussion

Typical temperature which occurs to be in regular physique in case of individual remains 37 °C. It is measured constantly using LM35. Acceptable assortment of same is depicted below in Tables 1 and 2 correspondingly [7].

Table 1 General BP records

Age category	Gender	mmHG range
Less than eighteen	Men	80:120
Eighteen to twenty	Men	80:125
Twenty-one to forty	Men	80:135
Above forty	Men	80:135
Less than twenty	Women	80:123
Twenty-one to forty	Women	80:135

Table 2 General pulsing level records

Condition	Beats per minute
Calm	60:100
Sleep	40:50
Cardiac infarction	Beyond 100

Table 3 Collected temperature data of four persons

Sick person identity	Temperature in °C	Conclusion
SP 1	36	Normal
SP 2	38	Abnormal
SP 3	38.7	Abnormal
SP 4	35.4	Normal

Convenient TCP UPD terminal application occurs to be prearranged for physicians via IP, likewise port. Medics get enabled for supervising each single delicate constraints linked towards sick person. Info equivalent comprising of physical temperature, BP in addition to pulse, all are procured serially utilizing web of things. Serially, info reaches towards medic on mobile screen or PC. Adaptable applications successively specify procured info. Reliability of projected skills was patterned via realizing definite individuals as well as chronicled in Tables 3, 4 and 5 separately.

All the constraints were examined in a frequent manner and recorded, and in case of abnormal circumstances, doctors or caretakers could take necessary action.

Table 4 Collected BP data of four persons

Sick person identity	Age	Gender	Mmhg level	Conclusion
SP 1	19	Women	80:108	Normal
SP 2	26	Women	80:134	Normal
SP 3	57	Men	80:140	Abnormal
SP 4	22	Men	80:135	Normal

Table 5 Collected pulse data of four persons

Sick person identity	Beats per minute	Conclusion
SP 1	93	Normal
SP 2	95	Normal
SP 3	112	Abnormal
SP 4	98	Normal

7 Conclusion and Future Enhancement

A model was fabricated by interfacing the sensors with STM32F103C8T6 microcontroller and is programmed using Arduino IDE software. The obtained health parameters such as blood pressure, pulse and temperature were sent serially to computer and were displayed on LCD screen. Meanwhile, the data is transmitted by using a Wi-Fi unit, i.e. ESP-8266. The data monitored is then displayed easily on mobile screen using TCP UDP terminal application.

Since medical supervisions have got vital role in humanity, automating this moderates burden over mankind and eases entire evaluating technique. Moreover, ease in admittance to framework assists sick person for depending on the same. The health of patients can be monitored regularly on a real-time basis and in turn avoids unnecessary doctor visits. Caregivers can keep a check on all the data and reach out to doctors only when needed. The goal of creating proposed model helps in dropping treatment charges via lessening doctor appointments, tedious check-up procedures and hospitalizations.

Lot of extra enhancements could be carried in presented framework. Creating the same in accurate and adaptable manner could be done by inserting many sensors of propelled kind. Framework is projected for tracing real-time data via several sensors. It aids in heightening behaviour existing in medicinal attention. The upcoming days for IoT occur to have countless chances for undergoing progressive impression happening in medicinal field by utilizing rapid broadband swiftness, technological improvements, enhanced inquiring as well as extra contenders on planetary. Another very important skill backing mainly for upcoming improvements in IoT that too in medical field happens to be presenting 5G networking. This delivers lot more quicker speediness to undergo, in comparison with old-style 4G networking one generally use. Internet of Things guarantees connection for transfer. Rapid information transmission via phone arranges for great IoT rigidity, pertaining with bulks of transmission which occurs to be happening with lot more quickness, and all the data could be stored on cloud permanently.

References

1. M.S. Islam, M.T. Islam, A.F. Almutairi, G.K. Beng, N. Misran, N. Amin, Monitoring of the human body signal through the Internet of Things (IoT) based LoRa wireless network system. Appl. Sci. **9**(9), 1884 (2019)
2. A. Mumrez, H. Tariq, U. Ajmal, M. Abrar, IOT-based framework for e-health monitoring system, in *2019 International Conference on Green and Human Information Technology (ICGHIT)*, Kuala Lumpur, Malaysia, 2019, pp. 113–115
3. A. Gutte, R. Vadali, IoT based health monitoring system using raspberry Pi, in *2018 Fourth International Conference on Computing Communication Control and Automation (ICCUBEA)*, Pune, India, 2018, pp. 1–5
4. S.K. Dharmoji, A. Anigolkar, M. Shraddha, IoT based patient health monitoring using ESP8266. Int. Res. J. Eng. Technol. (IRJET) **07**(03), 3619–3624 (2020)

5. S. Shaikh, V. Chitre, Healthcare monitoring system using IoT, in *2017 International Conference on Trends in Electronics and Informatics (ICEI)*, Tirunelveli, 2017, pp. 374–377
6. T.H.Y. Ling, L.J. Wong, R. Moore, IoT fitness device with real time health assessment and cloud storage, in *2019 7th International Conference on Smart Computing & Communications (ICSCC)*, Sarawak, Malaysia, Malaysia, 2019, pp. 1–5
7. G.V. Kumar, A. Bharadwaja, N.N. Sai, Temperature and heart beat monitoring system using IOT, in *2017 International Conference on Trends in Electronics and Informatics (ICEI)*, Tirunelveli, 2017, pp. 692–695
8. K. Narendra Swaroop, K. Chandu, R. Gorrepotu, S. Deb, A health monitoring system for vital signs using IoT, Internet of Things **5**, 116–129 (2019)
9. V.P. Jagtap, P.R. Sarode, T.B. Sonawane, M.C. Toke, IOT based patient monitoring system. Int. J. Emerg. Trends Technol. Comput. Sci. (IJETTCS), **7**(1), 036–040 (2018)
10. S. Jayapradha, P.M.D.R. Vincent, An IOT based human healthcare system using Arduino uno board, *2017 International Conference on Intelligent Computing, Instrumentation and Control Technologies (ICICICT)*, Kannur, 2017, pp. 880–885
11. I. khan et al., Healthcare monitoring system and transforming monitored data into real time clinical feedback based on IoT using raspberry Pi, in *2019 2nd International Conference on Computing, Mathematics and Engineering Technologies (iCoMET)*, Sukkur, Pakistan, 2019, pp. 1–6
12. C.S. Krishna, N. Sampath, Healthcare monitoring system based on IoT, in *2017 2nd International Conference on Computational Systems and Information Technology for Sustainable Solution (CSITSS)*, Bangalore, 2017, pp. 1–5

Islanding Detection of Grid-Connected Photovoltaic Systems Using Active Disturbance-Based Techniques

Prajwal Puranik, Bharath Prabhu, Anantha Saligram, and Suryanarayana K.

1 Introduction

Depleting fossil fuel-based energy sources, reduction in cost of PV modules and increase in their efficiency have been the primary factors for increase in dependency on PV-based systems for clean energy. Moreover, support from local political bodies towards the establishment of cleaner energy, increased environmental awareness among general public, continuous refinements in the field of power electronics have further fuelled this growth [1]. The global PV capacity which was 1.2GW in 1992 stands to be at least 627GW in 2019, out of which 115GW were installed in 2019 alone and thereby breaking the 100GW threshold for the third time in a row. In this way, PV systems are now contributing close to 3% of electricity demand in the world and reducing 5.3% of electricity generation related CO_2 emissions [2].

The substantial growth in PV systems has provoked a move from conventional inverter topologies to more complex ones with increased efficiency and reliability, with better control strategies. Review as well as studies on different single- and three-phase grid tie inverter topologies are given in [1, 3, 4]. Elimination of line frequency transformer, reduction of ground leakage current, improvement in grid current total

P. Puranik (✉) · B. Prabhu · A. Saligram
Hexmoto Controls Pvt. Ltd, Mysuru, India
e-mail: prajwal@hexmoto.com

B. Prabhu
e-mail: bharathprabhu@hexmoto.com

A. Saligram
e-mail: ananthasaligram@hexmoto.com

Suryanarayana K.
Department of Electrical and, Electronics Engineering, N.M.A.M. Institute of Technology, Nitte, India
e-mail: suryanarayana@nitte.edu.in

Sanjeevikumar et al. (eds.), *Advances in Renewable Energy and Electric Vehicles*,
Lecture Notes in Electrical Engineering 767,
https://doi.org/10.1007/978-981-16-1642-6_21

harmonic distortion and conversion efficiency are where the research is majorly focussed [5, 6]. The safety and reliability requirements have also accelerated and the PV grid-connected systems are now required to satisfy more stringent standards. The standard IEC 60,755 defines the requirements for protection against leakage currents and this makes the use of leakage current protection devices a compulsion or rather inevitable [3]. A serious and challenging risk to the safety of grid is posed by a condition called "islanding" of grid-connected PV systems. Such a condition causes the unintentionally islanded PV systems to continue to energise the grid that has been intended to be de-energised. This puts the entire grid system as well as the personnel and general public at risk and also the PV system itself as it now operates in undesirable operating conditions. The risk of islanding becomes particularly serious when the installed PV capacity is large enough such that its penetration level balances the local peak load [7]. The standard IS 16169 defines the safety requirements against islanding, that a PV grid-connected system must comply.

Preventing unintentional islanding of PV grid-connected systems is of great importance and techniques used for this purpose are called anti-islanding techniques or islanding detection techniques. These techniques may be of remote or local types [8]. Remote methods use communication strategies such as SCADA, power line carrier communication (PLCC) which have extremely small or rather no non-detection zones (NDZ) [9, 10]. NDZ is defined by a certain range of load wherein an islanding detection technique fails. Local techniques consists of passive and active types. Passive methods use the information available by measurement of voltages, currents, frequency, and phase at point of common coupling (PCC) for islanding detection [11] and have large NDZ with higher possibilities of false tripping. Active methods operate by introducing small disturbances at PCC and observing the changes in system parameters, thereby directly interacting with the grid. The introduced disturbances force any of the system parameters to reach fault detection threshold when the event of islanding occurs. Active methods have small NDZ and reduced possibility of false tripping when compared to passive methods. But since disturbances are introduced, these methods reduce the power quality and efficiency to a small extent and also increase the complexity of the controller [12]. There are several active methods such as Sandia voltage shift (SVS), Sandia frequency shift (SFS), slip mode frequency shift (SMS), reactive power disturbance and active frequency drift (AFD).

A two-stage inverter system and circuitry along with conditions for its anti-islanding test is discussed briefly in Sects. 2 and 3. Among the active methods, Sandia voltage shift (SVS) and reactive power disturbance-based methods are analysed in Sects. 4 and 5, and their simulation results are provided in Sect. 6.

2 Two-Stage Grid Tie Inverter System

A two-stage grid tie inverter system implies that the system comprises of two power conversion stages, as shown in Fig. 1. The first stage is a DC–DC boost converter

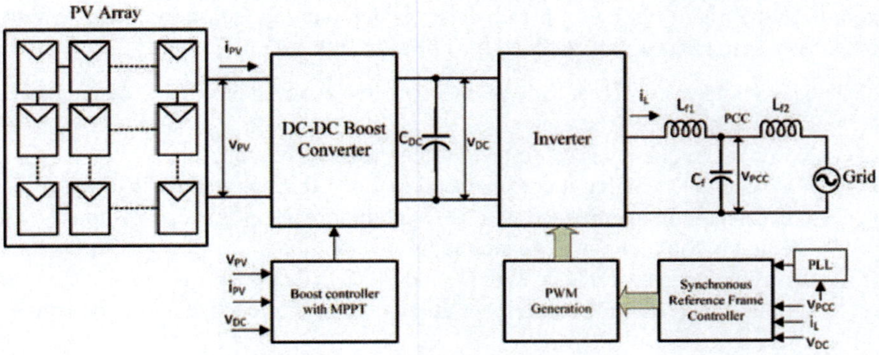

Fig. 1 Typical two-stage single-phase grid tie inverter system

that increases the PV array voltage to a level that is sufficient for the subsequent single/three-phase inverter stage to feed power to the utility grid. The first stage performs maximum power point tracking (MPPT) to extract maximum power available from the PV array and injects it into DC link capacitor C_{dc} which increases DC voltage (V_{dc}). The inverter stage uses DC bus voltage as the reference for feeding the available power to the grid. Hence, the inverter output increases current fed to the grid when the first stage increases the DC bus voltage, thereby stabilising the DC bus voltage. The inverter stage is usually controlled in synchronous reference frame wherein grid voltage angle is used to transform from stationary to synchronous reference frame [13]. A software phase locked loop (PLL) is used for grid synchronisation purpose [14].

3 Test Conditions for Islanding Detection

The test circuit and conditions for anti-islanding are defined in IS 16169:2014 [15], and the circuit is shown in Fig. 2. The voltage and frequency of the utility system are

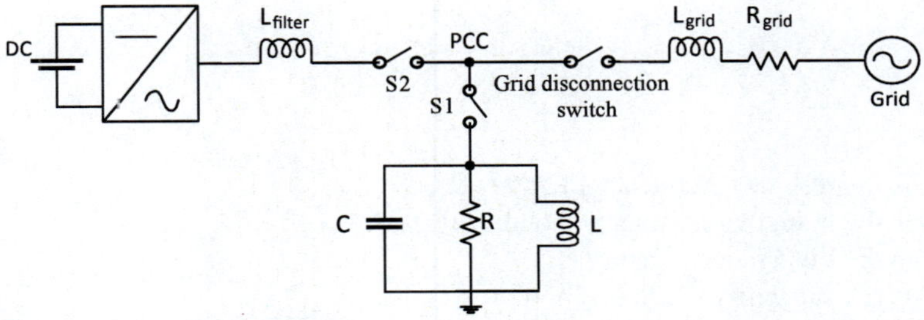

Fig. 2 Circuit for islanding detection test

kept near the middle of their operating range, without any variation during testing. RLC load used for testing is defined as follows.

(1) The quality factor is set to 1, i.e. reactive power stored is equal to active power dissipated
(2) Resonant frequency is set to be the same as grid frequency
(3) *R* is selected such that it consumes all real power generated by inverter.
(4) *L* is selected in accordance with (1) such that reactive power consumed by it is equal to the real power generated by inverter.
(5) *C* is selected in accordance with (1) and (4) such that the sum of reactive power supplied by it as well as the inverter is equal to the reactive power consumed by *L*.

The worst and challenging condition for islanding detection occurs when the inverter is generating 100% real power at unity power factor. This strikes a balance between load and generated power, resulting in zero current being drawn from the grid. In the analysis presented in this paper, only unity power factor condition is considered, and hence, mathematically *R*, *L*, and *C* could be written as:

$$f = \frac{1}{2 \times \pi \sqrt{L \times C}} \tag{1}$$

$$X_{\mathrm{L}} = \frac{V^2}{Q_{\mathrm{f}} \times P} \tag{2}$$

$$L = \frac{X_{\mathrm{L}}}{2 \times \pi \times f} \tag{3}$$

for unity pf operation,

$$X_{\mathrm{C}} = \frac{V^2}{Q_{\mathrm{f}} \times P} \tag{4}$$

$$C = \frac{1}{2 \times \pi \times f \times X_{\mathrm{C}}} \tag{5}$$

$$R = \frac{V^2}{P} \tag{6}$$

where

f is the resonant frequency in hertz (Hz)
X_{L} is the inductive reactance in ohm (Ω)
L is the inductance in henry (H)
X_{c} is the capacitive reactance in ohm (Ω)
C is the capacitance in farad (F)
R is the resistance in ohm (Ω)

P is the inverter real power in watts (W)
Q_f is the quality factor
V is the voltage at PCC in volts (V)

In the islanded situation, the PV inverter supplies its rated real power to the parallel *RLC* load. The load is represented by the transfer function $G(s)$.

$$G(s) = \frac{V_{PCC}(s)}{I_{load}(s)}$$

$$I_{load}(s) = I_R(s) + I_L(s) + I_C(s)$$

$$I_R(s) = \frac{V_{PCC}(s)}{R},\ I_L(s) = \frac{V_{PCC}(s)}{sL},\ I_C(s) = V_{PCC}(s) \times sC$$

$$G(s) = \frac{sLR}{s^2RLC + sL + R}$$

4 Sandia Voltage Shift Method

The Sandia voltage shift (SVS) is an active method that uses voltage feedback at PCC [16] for islanding detection. The PV inverter periodically reduces (or increases) its output active power by a small value. When the grid is present, the current fed by the inverter changes and there is no difference in voltage at PCC, whereas absence of grid changes voltage at PCC in order to change the output current, as governed by Ohm's law. This change in voltage is sensed and the feedback action of the SVS method in the PV inverter system is designed to further changes the voltage at PCC. Eventually under or over voltage protection is activated, which shuts down the PV inverter system and prevents the system from islanding. Usually, reduction in power is the periodic disturbance applied, since it results in under voltage protection trip and is not destructive as compared to increasing the power. Also, the RMS value of voltage at PCC is usually used as feedback. Figure 3 shows the control loop block diagram of SVS method.

In the analysis, it is assumed that the current controller perfectly regulates inverter current to be equal to the reference I_{o*} in the absence of SVS or any other islanding

Fig. 3 Control loop block diagram of SVS method

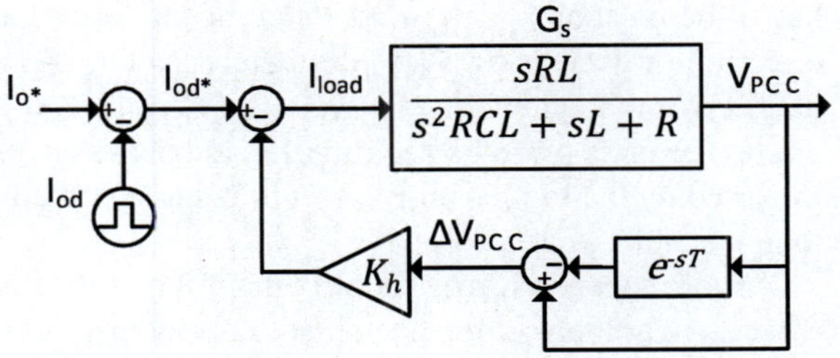

Table 1 SVS parameters

Parameter	Value
Real power, P	2200 W
Grid frequency, f (nominal)	50 Hz
Grid voltage, V (nominal)	230 V RMS
Under voltage trip level	85% of nominal RMS value = 195.5 V
Quality factor, Q_f	1

detection technique. In SVS method, I_{o*} is periodically reduced by a small value, I_{od}. RMS value of Voltage at PCC, V_{PCC} is taken as feedback and its previous value is subtracted from the present value to obtain difference in V_{PCC} as ΔV_{PCC}. The period in which this difference (T) is calculated may be more than or equal to the time period of grid voltage itself. ΔV_{PCC} multiplied by a feedback factor K_h and is fed back to the disturbed reference I_{od*}.

Analysing $G(s)$, K_h may be selected such that it is large enough to trip the system by reducing V_{PCC} to under voltage trip point within the specified time limits. The parameters for which SVS is designed and simulated are shown in Table 1. For the parameters P, f, V, and Q_f specified in Table 1, R, L, and C are obtained from (1) to (6) as:

$R = 24.045\ \Omega$, $\mathrm{L} = 76.539$ mH and $C = 132.378\ \mu$F.

Loop gain $T(s)$ of the SVS control loop is:

$$T(s) = G(s) \times \left(1 - \mathrm{e}^{-sT}\right) \times K_h \tag{7}$$

Using the third-order Padé Approximation [17], the transport delay e^{-sT} is approximated as $P(s)$.

$$P(s) = \frac{120 - 60(Ts) + 12(Ts)^2 - (Ts)^3}{120 + 60(Ts) + 12(Ts)^2 + (Ts)^3}$$

Hence,

$$T(s) = G(s) \times (1 - P(s)) \times K_h \tag{8}$$

For the PV inverter system to trip under islanded condition, the SVS control loop has to be unstable. A negative value of feedback factor K_h drives the loop in the direction of decreasing V_{PCC}, ultimately leading to an under voltage trip. The bode diagram of loop gain $T(s)$ for $K_h = -3.25$ and $T = 100$ ms is shown in Fig. 4. The phase margin is set to be negative, i.e. the phase lag provided by the loop when gain imparted by the loop is unity is set to be greater than 180° to make the SVS control loop unstable.

A more superior form of control that follows the same method may be constructed when synchronous reference frame controller is used [18]. Instead of monitoring

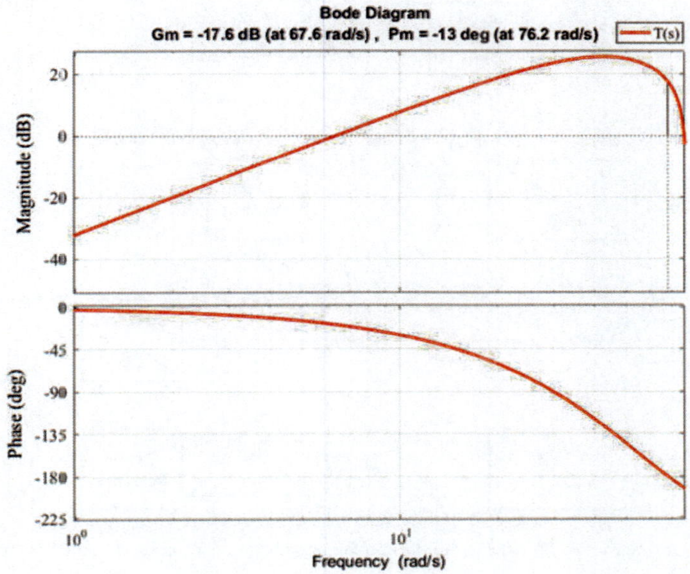

Fig. 4 Bode diagram of loop gain $\boldsymbol{T(s)}$

change in RMS value of V_{PCC}, the direct axis voltage V_d is monitored and decrease in V_d is positively fed to direct axis current reference I_{dref}. The difference block $(1 - P(s))$ may also be replaced by a band pass filter that extracts ΔV_d, as shown in Fig. 5. The pass band may be set from one-tenth of the grid frequency to twice the grid frequency. A synchronous reference frame controller may be used for single-phase inverter as well by emulating a quadrature axis from 90° phase shifted signal [19]. The block diagram shown in Fig. 5 represents the control of a three-phase system.

5 Reactive Power Disturbance Method

The synchronous reference frame controller provides the flexibility of being able to individually control active as well as reactive power being injected into the grid. This method is based on reactive power control, where the reactive power is disturbed periodically and corresponding changes in the frequency at PCC is monitored. Presence of grid results in the disturbed reactive power reference appear at the load as a change in reactive power, whereas absence of grid results in a change in frequency at PCC. This method drives the inverter frequency towards threshold levels that trigger under or over frequency trip [20].

Reactive power absorbed by the load is given by (9)

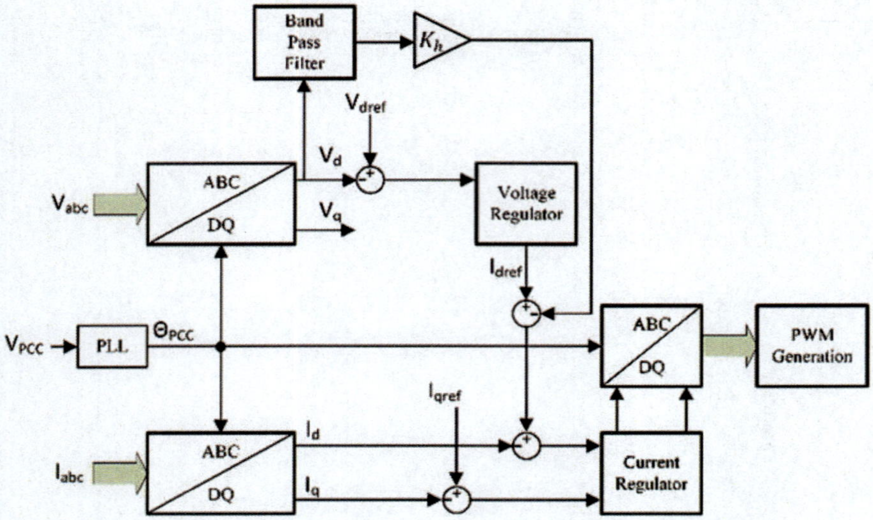

Fig. 5 Block diagram of SVS-based method using synchronous reference frame controller

$$Q_o = P_o \times Q_f\left(\frac{f_o}{f} - \frac{f}{f_o}\right) \tag{9}$$

where

P_o is the active power consumed by the load
Q_f is the quality factor of the load
f_o is the resonant frequency of the load
f is the frequency at PCC

If the resonant frequency of the load is equal to the frequency at PCC, the inverter operates at unity power factor and the load does not consume any reactive power. Under islanded condition, in order to force the frequency to deviate sufficiently up to the threshold that causes under or over frequency trip, an additional amount of reactive power ΔQ_o has to be generated, given by (10).

$$\begin{aligned}\Delta Q_o &= Q_{th} - Q_o\\ \Delta Q_o &= Q_{th} - 0\\ \Delta Q_o &= P_o \times Q_f\left(\frac{f_o}{f_{th}} - \frac{f_{th}}{f_o}\right)\end{aligned} \tag{10}$$

where Q_{th} is the reactive power at threshold frequency f_{th}.

The graph of ΔQ_o versus f_{th} which is the same as frequency at PCC under islanded condition is shown in Fig. 6, which gives the required additional reactive power.

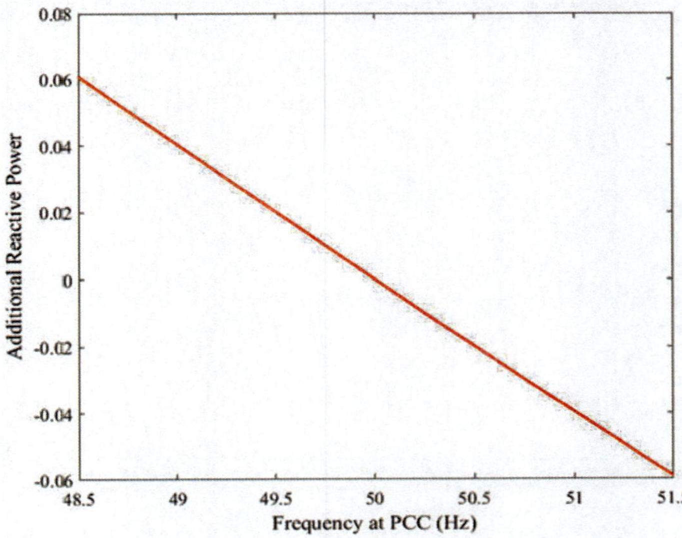

Fig. 6 Graph of ΔQ_o versus f_{th}

It is assumed that the inverter was supplying rated active power prior to entering the islanded condition, as discussed in section 3, i.e. $P_o = 1$ and also according to the testing conditions $Q_f = 1$. The threshold frequency levels according to IS 16169:2014 are around 50 Hz $+ / -$ 1.5 Hz. From the graph, it can be observed that only around 6% of additional reactive power is sufficient to reach this threshold (Fig. 6).

Since only 6% disturbance in reactive power is sufficient to reach the threshold frequency levels, a triangular periodic disturbance of this magnitude with frequency equal to one-tenth the grid frequency is sufficient and no positive feedback is required, which makes this method simple as well as efficient. The disturbance waveform is shown in Fig. 7.

The reactive power disturbance is injected by directly adding it to the actual quadrature axis reference current, as shown in the block diagram in Fig. 8. In the analysis, inverter is supplying only active power and therefore $I_{qref} = 0$ and only the reactive power disturbance forms the resulting quadrature axis current reference.

6 Simulation Results

The entire two-stage PV grid-connected inverter system is simulated in MATLAB Simulink and the above discussed islanding detection techniques are verified. The "SimPowerSystems" library [21] is used to simulate the inverter bridge along with the synchronous reference frame controller. A H6.5 configuration [22] of inverter

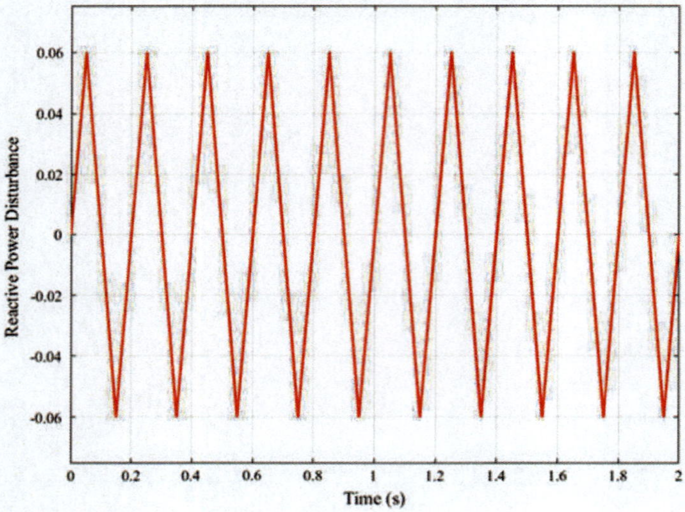

Fig. 7 Reactive power disturbance being injected

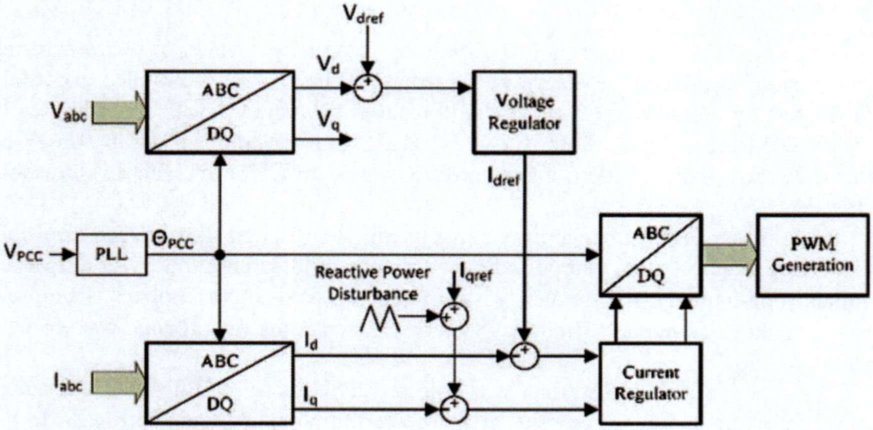

Fig. 8 Block diagram of reactive power disturbance method

bridge is selected for the simulation of single-phase system due to the advantages imparted by it with respect to reduction of common mode leakage current. A three-phase system is simulated using 3 level inverter bridge of both Neutral Point Clamped and T-type configuration. The system parameters are specified in Table 2.

Table 2 PV grid-connected system parameters

Parameter	Value
PV Input power at 1000 W/m^2	2200 W
Grid Frequency, f (nominal)	50 Hz
Line-neutral grid voltage, V (nominal)	230 V RMS
Under voltage trip level	85% of nominal RMS value = 195.5 V
Over voltage trip level	115% of nominal RMS value = 264.5 V
Under frequency trip level	48.5 Hz
Over frequency trip level	51.5 Hz
Quality factor of the load, Q_f	1

6.1 *Sandia Voltage Shift*

Sandia Voltage Shift method has been simulated for single-phase as well as three-phase systems. From Fig. 9, it can be observed that the system becomes unstable after being islanded, eventually leading to Under Voltage Trip while remaining stable under normal operating conditions.

6.2 *Reactive Power Disturbance*

It can be seen from Fig. 10 that the reactive power disturbance appears as a change in frequency as soon as the PV grid-connected system is islanded. The disturbance ultimately results in under frequency trip in this particular case.

7 Conclusions

The Sandia voltage shift and reactive power disturbance methods of islanding detection are analysed in this paper. These active methods are simulated and verified for a two-stage PV grid-connected inverter system. The discussed active methods prevent the system from supplying the load in the absence of grid, while maintaining stable operation otherwise. In Sandia voltage shift method, periodic reduction in active power reduces voltage at PCC in the absence of grid, which is monitored and the system is driven further in the same direction leading to under voltage trip. In reactive power disturbance method, the disturbance injected in reactive power output appears as a change in frequency in the absence of grid, leading to under or over frequency trip. In this way, these active islanding detection techniques ensure the safety of

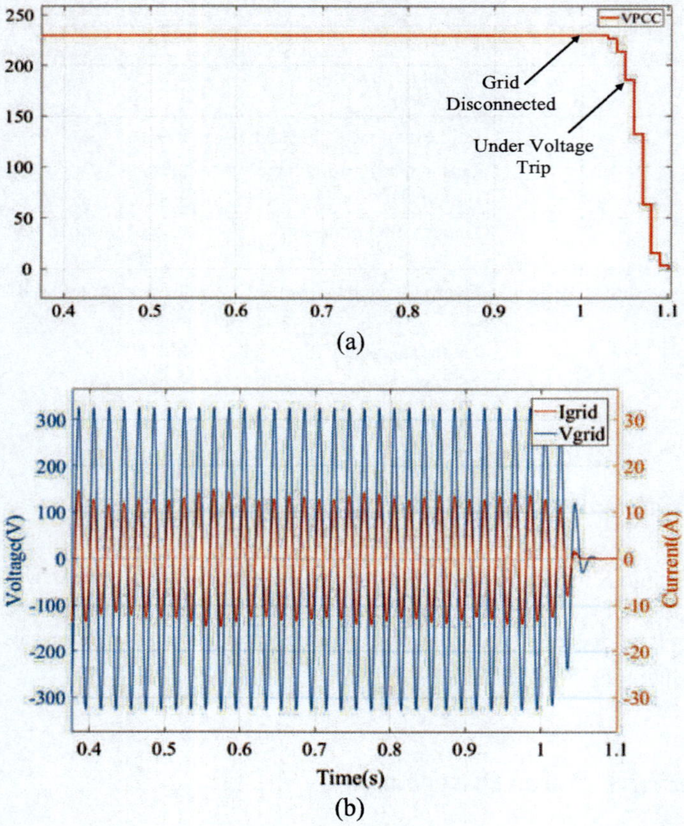

Fig. 9 **a** RMS value of voltage at PCC and, **b** grid voltage and current before and after islanding

grid and the overall system while ensuring stability with minimal interference to the actual working of the system.

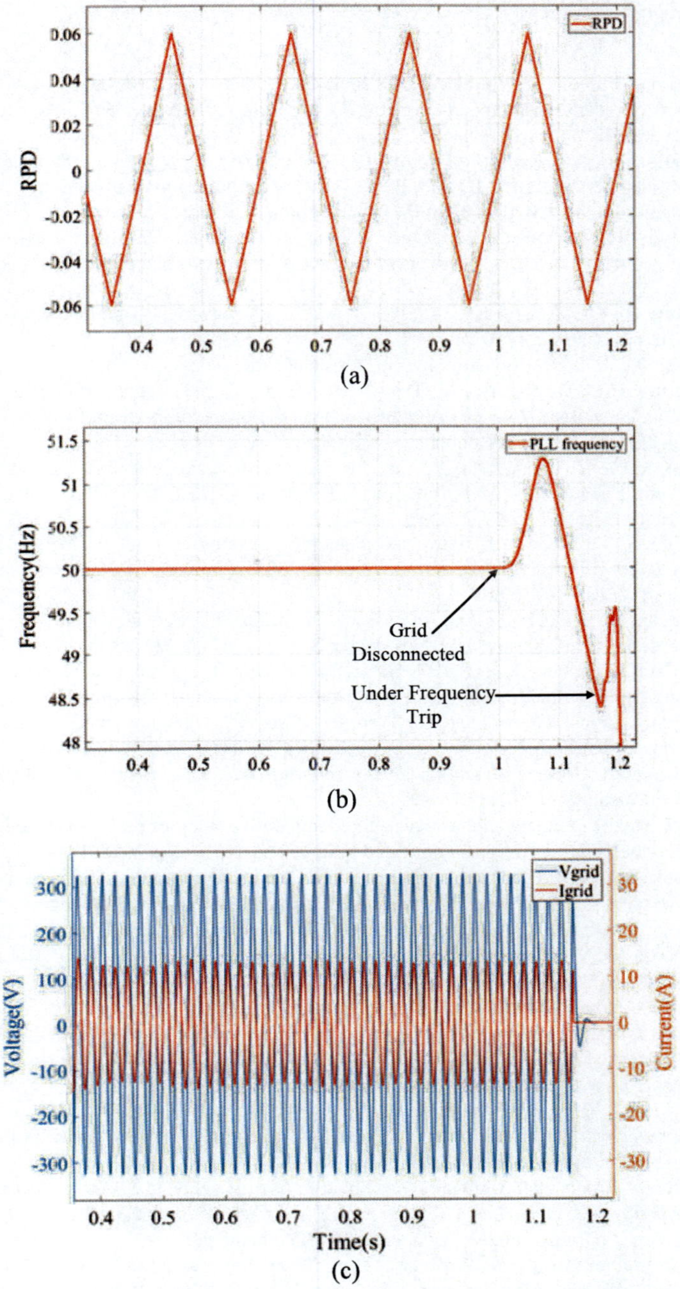

Fig. 10 **a** Reactive power disturbance injected, **b** grid frequency and, **c** grid voltage and current before and after islanding

References

1. S. Kouro, J.I. Leon, D. Vinnikov, L.G. Franquelo, Grid-connected photovoltaic systems: an overview of recent research and emerging PV converter technology. IEEE Ind. Electron. Mag. **9**(1), 47–61 (2015)
2. International Energy Agency—Photovoltaic Power Systems Programme, "Snapshots of Global PV Markets—2020, Report IEA-PVPS T1–37: 2020. (Online). Available: https://iea-pvps.org/wp-content/uploads/2020/04/IEA_PVPS_Snapshot_2020.pdf
3. A. Estévez-Bén, A. Alvarez-Diazcomas, G. Macias-Bobadilla, J. Rodríguez-Reséndiz, Leakage current reduction in single-phase grid-connected inverters—a review. Appl. Sci. **10**, 2384 (2020)
4. J. Zhang, S. Xu, Z. Din, X. Hu, Hybrid multilevel converters: topologies, evolutions and verifications. Energies **12**, 615 (2019)
5. T. LaBella, A high-efficiency hybrid resonant microconverter for photovoltaic generation systems, Ph.D. dissertation, Dept. Elect. Eng., Virginia State Univ., Blacksburg, VA, USA, 2014. (Online). Available:https://vtechworks.lib.vt.edu/bitstream/handle/10919/50526/LaBella_TM_D_2014.pdf?sequence=1&isAllowed=y
6. X. Guo, R. He, J. Jian, Z. Lu, X. Sun, J.M. Guerrero, Leakage current elimination of four-leg inverter for transformerless three-phase PV systems. IEEE Trans. Power Electron. **31**(3), 1841–1846 (2016)
7. N. Miller, Z. Ye. Report on distributed generation penetration study, in *National Renewable Energy Laboratory* (Golden, Colorado, August 2003). (Online). Available: https://www.nrel.gov/docs/fy03osti/34715.pdf
8. A. Etxegarai, P. Eguía, I. Zamora, Analysis of remote islanding detection methods for distributed resources. Int. Conf. Renew. Energies Power Q **1**(9), 1142–1147 (2011)
9. M.-S. Kim, R. Haider, G.-J. Cho, C.-H. Kim, C.-Y. Won, J.-S. Chai, Comprehensive review of islanding detection methods for distributed generation systems. Energies **12**, 837 (2019)
10. S.P. Chowdhury, S. Chowdhury, C.F. Ten, P.A. Crossley, Islanding protection of distribution systems with distributed generators—a comprehensive survey report, in *2008 IEEE Power and Energy Society General Meeting - Conversion and Delivery of Electrical Energy in the 21st Century* (Pittsburgh, PA, 2008), pp. 1–8
11. A.G. Abokhalil, A.B. Awan, Al-Qawasmi, Comparative study of passive and active islanding detection methods for PV grid-connected systems, *Sustainability* **2018**, 10 (1798)
12. E.J. Estébanez, V.M. Moreno, A. Pigazo, M. Liserre, A. Dell'Aquila, Performance evaluation of active islanding-detection algorithms in distributed-generation photovoltaic systems: two inverters case. IEEE Trans. Industr. Electron. **58**(4), 1185–1193 (2011)
13. R. Teodorescu, M. Liserre, P. Rodrguez, *Grid Converters For Photovoltaic And Wind Power Systems* (John, Chichester, West Sussex, UK, 2011).
14. Application Note: SPRABT4A, *Software Phase Locked Loop Design Using C2000™ Microcontrollers for Three Phase Grid Connected Applications* (Texas Instruments, Dallas, 2013)
15. *Test Procedure Of Islanding Prevention Measures For Utility-Interconnected Photovoltaic Inverters,* Bureau of Indian Standards IS 16169:2014
16. W. Bower, M. Ropp, Evaluation of islanding detection methods for photovoltaic utilityinteractive power systems, in *Report IEA PVPS T5–09*: 2002 (Online). Available: https://iea-pvps.org/wp-content/uploads/2020/01/rep5_09.pdf
17. V. Hanta, A. Procházka, Rational approximation of time delay, in *Institute of Chemical Technology in Prague, Department of Computing and Control Engineering* (Online). Available: https://www2.humusoft.cz/www/papers/tcp09/035_hanta.pdf
18. Z. Ye, R. Walling, L. Garces, R. Zhou, L. Li, T. Wang, Study and development of anti-islanding control for grid-connected inverters, in *General Electric Global Research Center Niskayuna* (New York, 2004) (Online). Available: https://www.nrel.gov/docs/fy04osti/36243.pdf
19. S. Sharma, Single Phase d-q transformation using as indirect control method for shunt active power filter. Int. J. Eng. Res. General Sci. **2**(3) (2014)

20. C. Xiaolong, L. Yongli, A nondestructive islanding detection method based on adaptive and periodic disturbance on reactive power output of inverter-based distributed generation, in *Journal of Applied Mathematics* (Hindawi Publishing Corporation, 2014)
21. SimPowerSystems—for use with Simulink, (The MathWorks, Inc., Natick, 2002)
22. B. Özbakir, Three-level topology for single-phase solar applications, Vincotech GmbH, Unterhaching (Germany), (Online). Available: https://www.vincotech.com/fileadmin/user_upload/content_media/documents/pdf/support-documents/technical-papers/Vincotech_TP_2016-12-001-v01_Three-Level-Topologies-for-Solar-Application_H6_5.pdf

Low-Cost Image-Based Occupancy Sensor Using Deep Learning

T. M. Sanjeev Kumar, Susan G. Varghese, Ciji Pearl Kurian, and Chandra Mouli

1 Introduction

Conventional electrical schemes rely on manual control rather than a designated rule set based on which devices can turn on or off. Human error is a factor which hinders the effectiveness of reducing energy consumption. To achieve energy savings, experts and researchers in the field have come up with various novel solutions to tackle the issue. These include use of sensors and embedded system interfacing which would allow integrating ruling algorithms to control plug-loads, lighting and mechanical loads such as heating, ventilation and air conditioning (HVAC) for thermal and visual comfort. Daylight artificial light integration scheme is one that helps to improve the occupant's comfort and reduce energy consumption [1–6]. There has been plenty of research carried out to maximize the efficiency of the system throughput using conventional sensors [7–9]. There has been steady development in image-based detection [10] for the same. Conventional schemes such as passive IR, ultrasound and dual technology suffer from long calibration time, inaccuracies due to idle occupants and lack of detections in large rooms due to range factor, leading to the requirement of multiple sensors. The process involves replacement of conventional sensors in a system with an image-based equivalent which sends signals to the hub of the control system. The Raspi-Camera Module of the Raspberry Pi acts as a sensor and processes the image under an occupancy detection algorithm which outputs the result

T. M. Sanjeev Kumar (✉) · Chandra Mouli
Manipal Academy of Higher Education, Manipal 576104, India

S. G. Varghese · C. P. Kurian
E&E, MIT, Manipal Academy of Higher Education, Manipal 576104, India
e-mail: susan.varghese@manipal.edu

C. P. Kurian
e-mail: ciji.pearl@manipal.edu

Sanjeevikumar et al. (eds.), *Advances in Renewable Energy and Electric Vehicles*, Lecture Notes in Electrical Engineering 767,
https://doi.org/10.1007/978-981-16-1642-6_22

to an existing lighting control system. For the system to be ideal, the detection needs to near-instantaneous to achieve appropriate energy efficiency.

Object detection is an active and highly researched domain under which several proposals have been made over the past decade. Haar Cascades, proposed by Viola and Jones [11], was a successful technique. Although made primarily for face detection, it worked circumstantially for other feature extractions such as upper, lower and full-body detection but the results obtained for occupancy detection was not ideal due to false detections when used in real time. Background subtraction (BS) has been successfully used for motion detection [12, 13] since it performs identically to the various motion sensors in commercial domains. For real-time occupancy detection test cases, BS has been utilized for comparison to our final chosen detection algorithm to compare the stress on the hub and the accuracy of the detection. However, it had several shortcomings which fail to produce ideal results for occupancy detection regardless of the variations to the algorithm, since the algorithm is not meant to identify occupants exclusively and their various states—in motion/idle. We aim to compare two object detector algorithms and determine which is most suitable for the application of HVAC and daylight artificial light integration scheme, provided the resources. Performance of deep learning improves with the increase in the amount of training data fed. Using deep learning to identify occupants is a logical approach which also yields the most accurate results [14]. Tiny-YOLO—a variation of you only look once (YOLO) by Redmon and Farhadi [15] and MobileNet SSD [16]—is as discussed, going to be the two primary object detectors being compared for a daylight artificial light integration scheme. In the developed system, the aim is to apply the algorithms in an embedded system format in real time while not clogging the CPU usage of the Raspberry Pi, so as to allow other functions to take place in the system hub simultaneously. The occupancy detection system is implemented in a way that it performs zone-wise occupancy detection for any room with minimal initialization steps involved. The purpose of this is to allow individual lighting and HVAC control for each pre-set zone defined as per the user-defined conditions. The immediate application is a project involving daylight artificial light integration for a test room where we aim to achieve thermal and visual comfort for an occupant based on set rules. The rest of the paper is structured as follows.

Section 2 presents information with regards to the hardware in use and the flow of operation of each node in the system. Section 3 explains the algorithms implemented in the system, Sect. 4 presents the results obtained and an inference is made and Sect. 5 provides a conclusion to the work.

2 System Description

The program was developed with the intention of making a standalone system with the aid of a Raspberry Pi and Raspi-Cam. Key Features of the system include remote access to the occupancy feed, remote hub control using mobile phones and PC, zone-wise occupancy detection in a room allowing for conditional control of lights based on occupancy status of a zone and the presence of a universal code which works identically for most cases, excluding abnormal lighting conditions. Raspi-Cam v2 is used as the primary sensor for zone-wise occupancy detection and calculation of window luminance which helps to find the dominant daylight glare position [5, 6] to achieve indoor visual comfort by considering blind control. Raspi-Cam is a low power draw accessory (200 mA) while having adequate resolution for object detection. Raspbian is a lightweight operating system intended for the Raspberry Pi. Most of the tasks performed on the Raspberry Pi is performed from the terminal with the exception of the hardware pulse width modulation (PWM) signals produced from the general purpose input/output (GPIO) pins. The flow of the system starts with the camera-based zone-wise occupancy detection followed by the sensor's data used by another program on the hub which handles the visual and thermal comfort and achieves them by following predetermined rules based on the time of the day and brightness of the lighting on the windows by calculating the luminance. The value obtained from this is used to configure the venation blind position with respect to the windows and simultaneous control of LED Luminaires in the room for optimal lighting along and maximizing energy efficiency. Figure 1 shows the hardware embedded system interface and flow to control blinds and LED luminaires, the number indicates the order in which the nodes perform their task.

Table 1 shows the specification of R-Pi camera used.

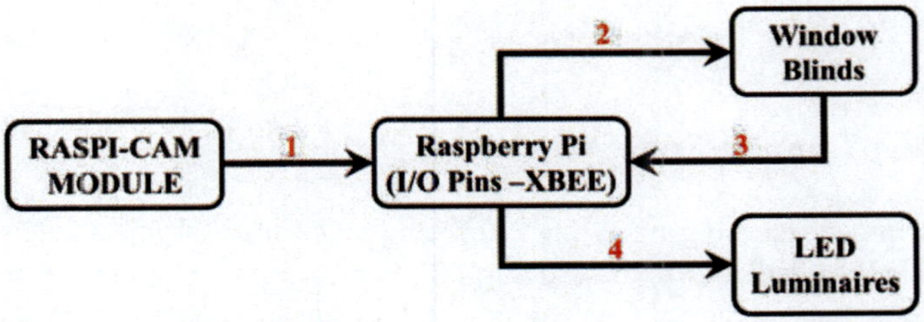

Fig. 1 Hardware network diagram

Table 1 Specifications of R-Pi camera

Image sensor	Sony IMX219
Camera resolution	8 MP
Captured picture resolution	3280 × 2464

3 Experimental Details

3.1 Occupancy Detection

3.1.1 Background Subtraction

Background subtraction [18] is a widely used technique for motion detection due to simplicity of the algorithm. It generates a foreground mask (namely a binary image containing the pixels belonging to moving objects in the scene) by using static cameras. BS with Gaussian kernel for image smoothing [17] calculates the foreground mask performing a subtraction between the current frame and a background model which contains the static part of the scene. The two main steps followed during the background modelling are: Background initialization—an initial model for the background is computed and background update—the model is updated in order to adapt to possible changes in the scene. The primitive version of occupancy detection for the test rooms was attempted using background subtraction, the code is able to pick up the initial movement by the occupant if he/she moves into the room and can use the reference background image to even identify a static occupant. The primary issue faced was false occupancy states were brought about due to the variations in lighting conditions in the test room and the background, while the blind control was being attempted or the LED luminaire's were switching between their several output levels.

3.1.2 Deep Learning Framework

The need for deep learning arises due to the simplistic nature of object detection using the previously stated algorithm. Although BS is not strenuous on the system's CPU and memory, the accuracy of the detection suffers due to false detection of non-static objects in a scene and the need for a static background and almost constant lighting condition which if not maintained can cause system to perform in a manner which was not intended. We aim to produce a multi-window zone-wise occupancy sensor while using a deep learning object detector. Deep learning techniques perform better when trained using a large number of data and suffers otherwise. Deep learning differs from traditional machine learning such that while training, the user is not required to select the features for identification of the specific image class, whereas domain knowledge is required for extracting features while using machine learning. Object detection is essentially image localization and classification done simultaneously.

3.1.3 Single Shot Multi-box Detector

SSD-MobileNet is a detector which uses MobileNet network as its feature extractor added with convolution layers and convolution filter to make a conclusion about the

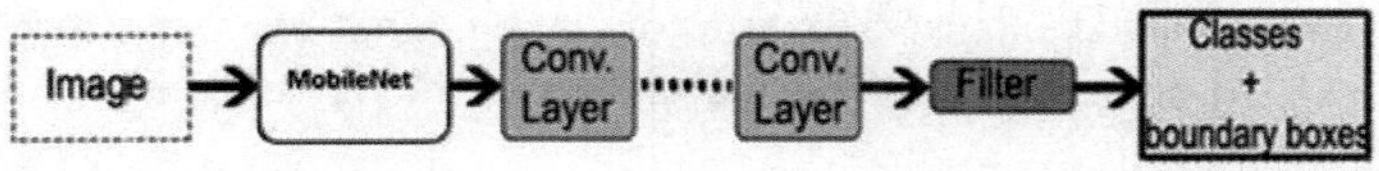

Fig. 2 Block diagram of single shot prediction flow

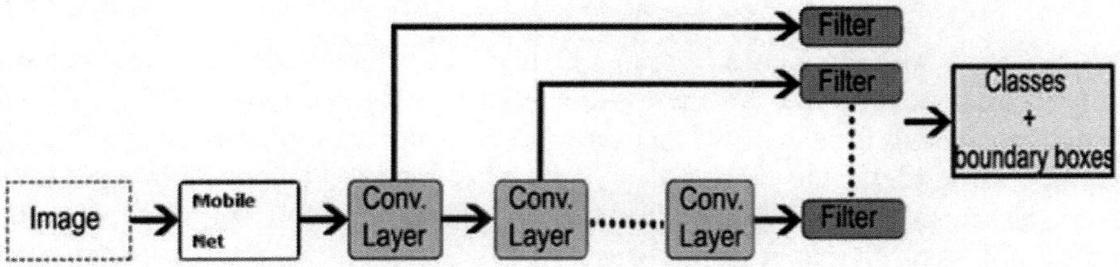

Fig. 3 Visualization of multi-scale feature maps

object class with different feature maps for separate object detections at all scaling for a given object, so as to not restrict positives for a single size.

The SSD object detection is split into two operations:

- Extraction of the feature maps.
- Application of the convolution kernels to detect objects.

SSD's architecture flow as seen in Fig. 2 shows that the object classification and localization is done in a single forward pass of the network. The prediction made for each locale uses 4 predictions of different aspect ratios, and each of them consists of a boundary and scores for each class with the highest prediction score as the dominant bounding box, hence the name assigned as such. The use of smaller, low resolution feature maps is used to detect larger-scale objects and vice versa. The detection of objects occurs by deeper layers rather than the bottom layers in convolution network; this is due to a drop in performance of the system. Hence, the resolution of the image must be kept high since the region of interest's spatial resolution might be potentially too low to detect smaller objects of interest. An adverse effect of this would mean in real-world usage, far off objects or points of interest show worse prediction trends as opposed to closer objects. In an ideal case, objects of all scale can be localized due to the multi-scale feature maps as seen in Fig. 3.

3.1.4 Tiny-YOLO

Tiny-YOLO detector divides an image into cells each of which is responsible for 'x' bounding boxes. The bounding box essentially describes the rectangle which actually enclose an object. It also predicts a confidence value for that enclosed object. Rather than using classifications and localization, YOLO makes use of a single neural network for the entire image and it divides it into regions for detection of objects.

Tiny-YOLO is not as intensive in terms of computations compared to YOLO and can be utilized in embedded applications such as the being implemented on R-Pi.

3.2 Occupancy Specifications and Scheduling

Raspi-Cam is wall mounted at a point with maximum field of view and least vulnerable to distortions to the camera's vision from lighting and movable objects. The program scheduling is done so that the system can capture images every 30 secs, while erasing previously analysed images. After image capture, the image file is fed as the input to the program and each zone is identified and zone-specific occupancy is tested. Output from the system—0/1 based on the occupancy for each zone is produced and is stored in an xls and csv data format in the local Raspberry Pi storage and cloud. The flow continues on the basis of devices to be controlled from the system hub. The program runtime is approximately 6–7 secs or lesser depending on the detector used. MobileNet SSD occupancy fetching takes a few seconds longer compared to Tiny-YOLO. Other operations like end device control and PWM generation are required, all in real time between the scripts' operational period. The operations listed are performed with no monitoring but can be viewed remotely using virtual network computing (VNC) on any mobile devices or computer, hence enabling wireless manual control. Figure 4 shows the occupancy detection and hardware interface.

VNC viewer/server is a graphical desktop sharing system that allows you to remotely control the desktop interface of one computer (running VNC Server) from another computer or mobile device (running VNC Viewer). VNC viewer transmits the keyboard and mouse or touch events to VNC server and receives updates to the screen.

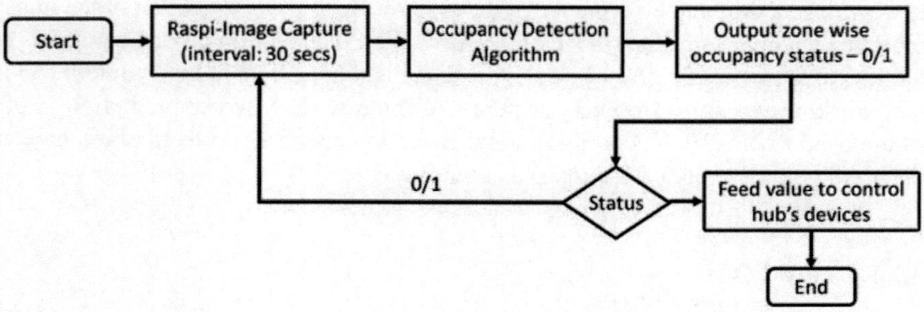

Fig. 4 Occupancy code interface

3.3 *Design of Deep Learning-Based Occupancy Detection Flow*

The algorithm initializes by loading the model and configuration prototxt. Image is read from local storage after being captured in periodic intervals and split into separate zones. Persons class label initialized for detection; 4 variables are to be predicted: 1. Persons/occupants, 2. Bounding box—top left x coordinate, 3. Bounding box—top left y coordinate, 4. Box width and height (Fig. 5).

Forward pass is computed (output values calculated from the input data) and stored as positive detection, extract the confidence values attributed for each detection and eradicate sub-par confidence values, i.e., less than 0.2 and for the class name persons, calculate the bounding box variables. In the case of SSD, non-maximal suppression (technique of suppressing several overlapping bounding boxes) is performed internally, whereas it is called explicitly in the case of Tiny-YOLO. Display the output for each zone if a bounding box is detected (room/zone is occupied) else display negative condition and store unoccupied status flag from the output. Figure 6 shows the flow chart for occupancy detection.

The code run periodically every 'x' seconds and uses still images as its arguments because running the script using live video feed rather than taking images and processing throughout the day in real time is expensive. The hardware limitation on the Raspberry Pi is the reason for using stills over a live feed. Although this might not be as instantaneous as real-time video capture, it allows the R-Pi to perform other operations such as:

- Send PWM signals to external electronic hardware control circuit due to occupant's presence.
- Use algorithms to compute ideal lighting conditions on the basis of time of day, luminance and perform window blind control.
- Log the data obtained from the occupancy feed into a cloud platform if required.
- Stream the Raspberry Pi's processes using virtual network computing based on user request.

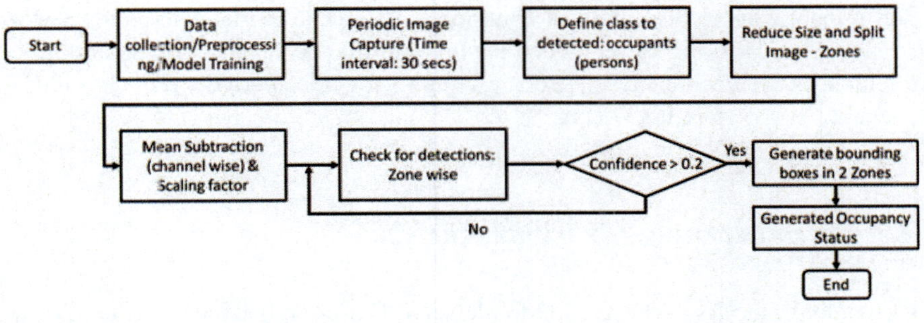

Fig. 5 Flow chart of occupancy detection algorithm

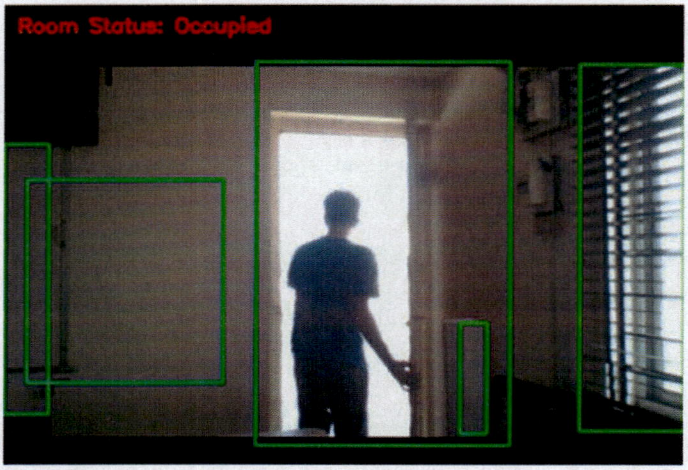

Fig. 6 False positives due to luminance variation

4 Results

In this section, conditions for optimal occupancy detection on a portable standalone system for indoor rooms and window luminance estimation to find the dominant glare position on window have been discussed.

4.1 Background Subtraction Analysis

Background subtraction needs a largely static background for the system to behave as an ideal occupancy sensor, largely due to the limitations of the algorithm as shown in Fig. 6. The algorithm is computationally inexpensive compared to deep learning detectors and its derivatives with regards to CPU demand. Zone-wise occupancy is not probable due to the nature of continuous frame extraction from a single video stream. The algorithm suffers when having multiple occupants. On the contrary if a simple motion/occupancy sensor is required for brief periods, this algorithm is adequate as shown in Fig. 7 (Fig. 8).

4.2 Deep Learning Algorithm Analysis

Utilization of the Tiny-YOLO and MobileNet SSD in an embedded form factor allows us to utilize the system resource while being able to perform other operations such as produce output based on defined fuzzy rule sets. Initial tests were made in small

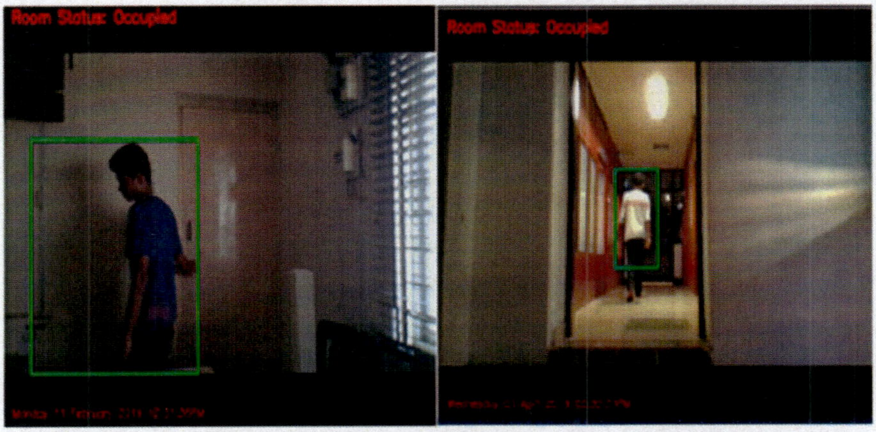

Fig. 7 Positive results obtained using background subtraction

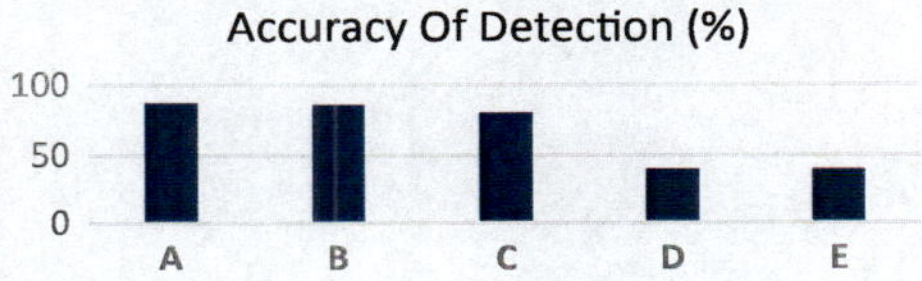

Fig. 8 Accuracy of SSD occupancy detection among five test cases

to medium-sized rooms to test the general accuracy of the R-Pi occupancy sensor as shown in Figs. 9 and 10. Once the positives were obtained from the small test rooms, the robustness and range of the occupancy detector were verified by testing in larger rooms such as the MIT Library Hall. This gave a realistic look at a high traffic occupant scenario and the maximum potential output of the sensor, as seen in Fig. 11. Occupancy detection was performed in 6 test case rooms with 30 input images. When feeding the test cases for occupancy testing on the R-Pi using both the algorithms, it is observed that the Tiny-YOLO detector-based program's runtime is a few seconds faster than the MobileNet SSD but does not produce accurate results even in well-lit scenarios, both the algorithms suffered when there was intrusive light. It was possible to increase the occupancy accuracy in both cases by modifying the exposure value of the captured image based on the luminance measured. The accuracy of positives obtained was much higher in the case of MobileNet SSD (~70% on average) when compared to Tiny-YOLO's implementation as shown in Table 2.

Comparing the results on SSD's implementation between the various room sizes: A. Primary test room; B. Lab research room 1,2; C. Department library; D. Central library; E. Office space as seen in the Fig. 8. The lower rate of detections in cases D and E where the occupants are placed fay away is due to the Raspi-Cam Sensor limitations (capture resolution) due to which the occupant features are not being identified.

Fig. 9 Zone-wise with and without occupancy detection for two different type of rooms

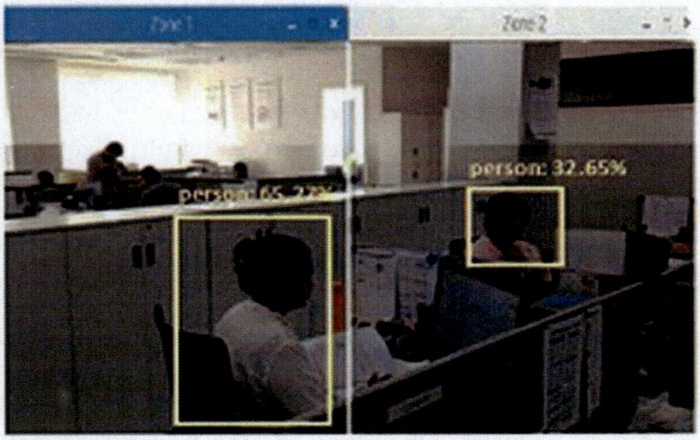

Fig. 10 Occupancy detection test in a small office space

Fig. 11 Occupancy detection in college library involving multiple occupants

Table 2 Model comparison

Algorithm used	Tiny-YOLO	Mobile net SSD
Input images tested	30	30
Accuracy obtained (before):	23.3%	56.66%
Accuracy achievable:	33.33%	70%

4.3 Limitations of the Algorithms

The algorithm ceases to produce positive occupancy if the room is ill-lit (< 300 lx), since the edges or the defining features of the occupants are not being captured. If the occupant is positioned exceptionally far away in the captured scene, the algorithm would not be able to detect the presence due to limitations in the resolution of the camera and the algorithm as shown in Fig. 12. If an occupant seated in the room has been facing away or is not seated in a posture identifiable as a human according to the training data, the occupancy status would show false negative.

4.4 Hardware Runtime and Power Consumption

Table 3 shows the time taken for the images to be processed, and the occupancy states to be stored on the local storage. Upon boot up, for the first image fed, the computation time is the longest, but as more images are captured throughout the day, the runtime decreases down to a minimum of 7.3 s.

Table 4 shows the power consumed by the Raspberry Pi 3B + during operation on load, and idle condition on LXDE (Lightweight X11 Desktop Environment) is observed. At the same time, theoretically, it is possible to reduce the LXDE idle to a

Fig. 12 Limitation of the algorithm—distinctive features are not identifiable for far off occupants

Table 3 Program runtime delay

Computational order:	Time taken (MobileNet SSD)	Time taken (Tiny-YOLO)
Iteration-1	7.48	5.62
Iteration-2	7.39	5.30
Iteration-3	7.31	5.30
Iteration-4	7.31	5.30

Table 4 Instantaneous power consumption of R-Pi

Mode of operation	Instantaneous current (ma)
Sleep	20–30 mA
Idle (desktop environment)	310 mA
CPU under load during operation	890–930 mA
Disabling peripheral (USB 2.0 and HDMI) on load	~800 mA

minimum of 200 mA; Wi-Fi is necessary to transmit the system output via VNC in real time to the user for remote-monitoring purpose.

5 Conclusions and Discussions

As per the assigned objectives of the research, we can produce a functional replacement to a conventional sensor-based lighting controlled system. It is possible to achieve a flexible and versatile occupancy sensor with zone-wise detection and which was relatively robust with minimal limitations with the use of MobileNet SSD. Utilization of the said algorithm in an embedded form factor allows us to capture the count of occupants in a given frame. Initial tests were made in small to medium-sized rooms to test the general accuracy of the R-Pi occupancy sensor, then tested for scaling in larger rooms with larger occupants upon the positive responses obtained while using the room. The algorithm used can capture and distinguish between each zone and decipher occupancy independently in either zone. Finally, in this work, we are able to produce zone-wise occupancy detection on the Raspberry Pi in real time while simultaneously being able to produce the luminance measurement required for room visual and thermal comfort and deliver output as a PWM signal (LEDs) and generate output for the modern Venetian blinds as per fuzzy rules [5, 6]. The limitations of the sensor are when an occupant is placed at a position with intrusive light or when the object to be detected is extremely far off preventing the localization and classification of the object in frame. In a bigger scale, this application can be used for the purpose of saving electricity. Further improvements can be made to increase the detection rate of far off occupants and decreasing the runtime for the program on the Raspberry Pi, further tuning with regards to obtaining luminance due variations in real time can be achieved. Mostly, training data specifically needed for this application can be done to maximize the positives obtained.

Acknowledgements The authors would like to acknowledge the sponsorship provided by MAHE, Manipal.

References

1. M.G. Kent, S. Altomonte, R. Wilson, P.R. Tregenza, Temporal effects on glare response from daylight. Build. Environ. **113**, 49–64 (2017)
2. J. Wienold, J. Christoffersen, Evaluation methods and development of a new glare prediction model for daylight environments with the use of CCD cameras. Energy Build. **38**(7), 743–757 (2006)
3. J.A. Jakubiec, C.F. Reinhart, The adaptive zone—concept for assessing discomfort glare throughout daylit spaces," light. Res. Technol. **44**, 149–170 (2012)
4. M. Bodart, C. Cauwerts, Assessing daylight luminance values and daylight glare probability in scale models. Build. Environ **113**, 210–219 (2017)
5. S.G. Varghese, C.P. Kurian, V.I. George, T.S.S. Kumar, Daylight-artificial light integrated scheme based on digital camera and wireless networked sensing actuation system. IEEE Trans. Consum. Electron. 1–1 (2019)
6. S.G. Varghese, C.P. Kurian, V.I. George, T.S. Kumar, Control and evaluation of room interior lighting using digital camera as the sensor. Int. J. Eng. Technol. (UAE) **7**(2), 99–105 (2018)

7. Y. Agarwal, B. Balaji, R. Gupta, J. Lyles, M. Wei, T. Weng, Occupancy-driven energy management for smart building automation, in *Proceedings of the 2nd ACM Workshop on Embedded Sensing Systems for Energy-Efficiency in Building* (ACM, 2010), pp. 1–6
8. D. Caicedo, A. Pandharipande, Daylight and occupancy adaptive lighting control system: an iterative optimization approach," light. Res. Technol. **48**(6), 661–675 (2016)
9. S.J. Choi, D.S. Lee, J.H. Jo, Lighting and cooling energy assessment of multi-purpose control strategies for external movable shading devices by using shaded fraction. Energy Build. **150**, 328–338 (2017)
10. V.L. Erickson, Y. Lin, A. Kamthe, R. Brahme, A. Surana, A.E. Cerpa, M.D. Sohn, S. Narayanan, Energy efficient building environment control strategies using real-time occupancy measurements, in *Proceedings of the First ACM Workshop on Embedded Sensing Systems for Energy-Efficiency in Buildings* (ACM, 2009) , pp. 19–24
11. P. Viola, M.J. Jones, Robust real-time face detection. Int. J. Comput. Vision, **57**(2), 137–154 (2004)
12. M. Piccardi, Background subtraction techniques: a review, in *2004 IEEE International Conference on Systems, Man and Cybernetics (*IEEE Cat. No. 04CH37583) (IEEE, 2004), vol. 4, pp. 3099–3104
13. H.-C. Shih, A robust occupancy detection and tracking algorithm for the automatic monitoring and commissioning of a building. Energy Build. **77**, 270–280 (2014)
14. A.R. Pathak, M. Pandey, S. Rautaray, Application of deep learning for object detection. Procedia Comput. Sci. **132**, 1706–1717 (2018)
15. J. Redmon, S. Divvala, R. Girshick, A. Farhadi, You only look once: Unified, real-time object detection, in *Proceedings of the IEEE Conference on Computer Vision and Pattern Recognition* (2016), pp. 779–788
16. A.G. Howard, M. Zhu, B. Chen, D. Kalenichenko, W. Wang, T. Weyand, M. Andreetto, MobileNets: efficient convolutional neural networks for mobile vision applications, in *Computer Vision and Pattern Recognition* (2009). https://doi.org/10.1016/S1507-1367(10)60022-3
17. P.E. Debevec, J. Malik, Recovering high dynamic range radiance maps from photographs, in *ACM SIGGRAPH 2008 Classes* (ACM, 2008), p. 31
18. J.E. Santoyo-Morales, R. Hasimoto-Beltran, Video background subtraction in complex environments. J. Appl. Res. Technol. **12**(3), 527–535 (2014)

Modeling and Analysis of 1.2 kW, 36–375 V, Push–Pull Converter

Raksha Adappa and Suryanarayana K.

1 Introduction

The demand for electrical energy is increasing at a rapid rate in India for its economic development [1]. The growing concern regarding the availability of non-renewable resources and their impact on nature has increased the demand for renewable sources of energy. Thus, renewable source of power such as solar and wind are gaining importance. Solar is a fast-growing source of renewable energy as it is clean, abundant and pollution free [2]. A PV system consisting of PV cells and DC–DC converter is required to convert voltage of PV cells to desired value for domestic and industrial applications [3]. A DC–DC converter is the main component in a PV system to transfer power to the load along with providing a constant regulated voltage. The voltage available at the output of panel employs a DC–DC converter for converting the voltage to a desired value to meet the desired application. Isolated DC–DC converters are preferred over non-isolated DC–DC converters as the transformer provides galvanic isolation between input and output side. Push–pull, forward, flyback and full-bridge are the widely used isolated converter topologies [4]. Among these topologies, push–pull converter is a preferred topology as it has isolation between input and output and better transformer core utilization [5]. As the switches in primary of push–pull converter share the current, it is suitable for high-power applications over other single-ended converters.

In this paper a 1.2 kW, 36–375 V, push–pull topology-based DC–DC converter suitable for domestic and industrial applications like electric vehicles, battery

R. Adappa (✉) · Suryanarayana K.
Department of Electrical and Electronics Engineering, NMAM Institute of Technology, Karkala, India
e-mail: raksha@nitte.edu.in

Suryanarayana K.
e-mail: suryanarayana@nitte.edu.in

Sanjeevikumar et al. (eds.), *Advances in Renewable Energy and Electric Vehicles*,
Lecture Notes in Electrical Engineering 767,
https://doi.org/10.1007/978-981-16-1642-6_23

management systems and motor drives is designed and investigated. The emphasis is to obtain a ripple free output for stable operation of the converter.

Section I presents a brief introduction to the need of DC–DC converters. The basic operation of a push–pull converter is discussed in Sect. 2. Specifications and design considerations are discussed in Sect. 3, Sect. 4 discusses the modeling aspects of the converter, Sect. 5 presents closed loop controller design for the converter, Sect. 6 gives the simulation results and observations, and Sect. 7 presents the conclusion.

2 Operation of Push Pull Converter

The push–pull converter is an isolated converter, derived from forward converter whose windings are made from a center tapped transformer. The push–pull converter better utilizes the transformer core as it is magnetized in both directions. The circuit diagram of the converter is as in Fig. 1. It consists of two active switches (Q_1& Q_2), two diodes (D_1&D_2), center tapped transformer (N_1:N_2 turns ratio) and passive elements capacitor (C) and inductor (L). The input voltage is (V_g), output voltage is ($V(t)$), R is load resistance, and (R_c) is equivalent series resistance of the capacitor. Two forward converters are operating back to back in push pull converter. During one half of the cycle, one forward converter is operating to transfer power to the load, and during the second half of the cycle, the second forward converter is operating to transfer power to the load [6].

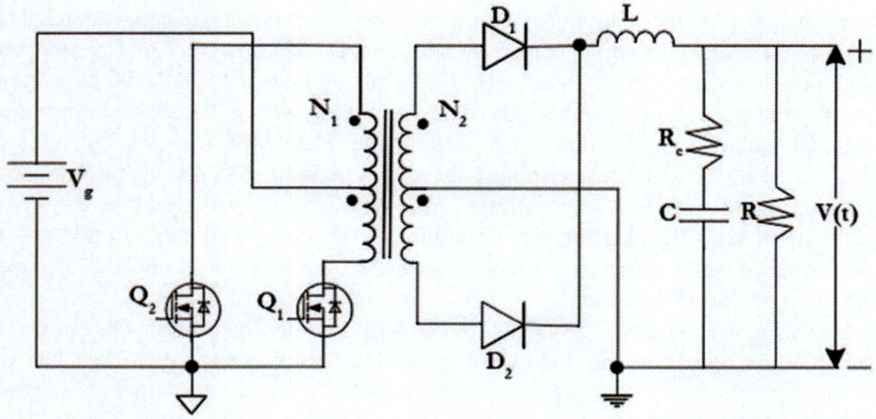

Fig. 1 Circuit diagram of push–pull converter

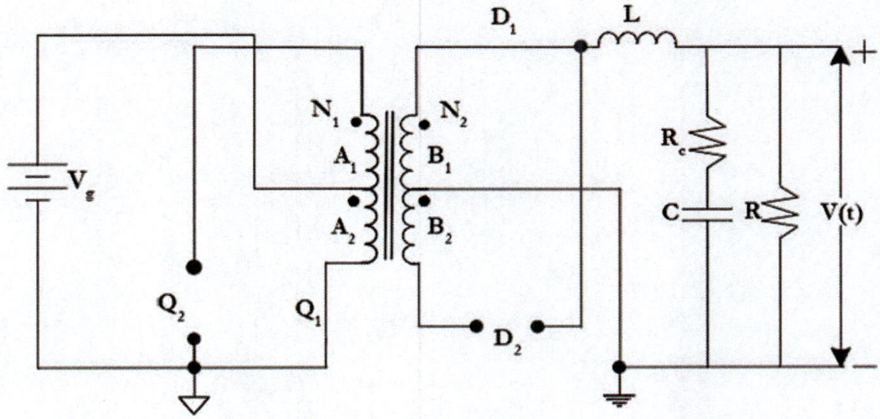

Fig. 2 Circuit diagram when Q_1 is ON and Q_2 is OFF

2.1 When Q_1 is ON and Q_2 is OFF

In this mode V_g, A_2 and Q_1 form the primary of the converter. Due to magnetic coupling, the polarity at the dot end is positive, and as a result, diode D_1 is forward biased and D_2 is reverse biased. The active secondary winding B_1 and D_1 delivers power to the load through the inductor and capacitor as shown in Fig. 2 [6].

2.2 When Q_2 is ON and Q_1 is OFF

In this mode V_g, A_1 and Q_2 form the primary of the converter. The polarity at the dot end is negative, and as a result, diode D_2 is forward biased and D_1 is reverse biased. Through the active secondary winding B_2 and D_2 power is transferred to the load through the inductor and capacitor as shown in Fig. 3 [6].

2.3 When Both Q_1 and Q_2 Are OFF

When both the primary switches are OFF, there is no transfer of power from primary to secondary. In the secondary, the inductor freewheels the current by sharing current through diodes D_1 and D_2 as in Fig. 4. The voltage across all the transformer winding is zero [6].

The waveforms of the parameters of push–pull converter are shown in Fig. 5 [6].

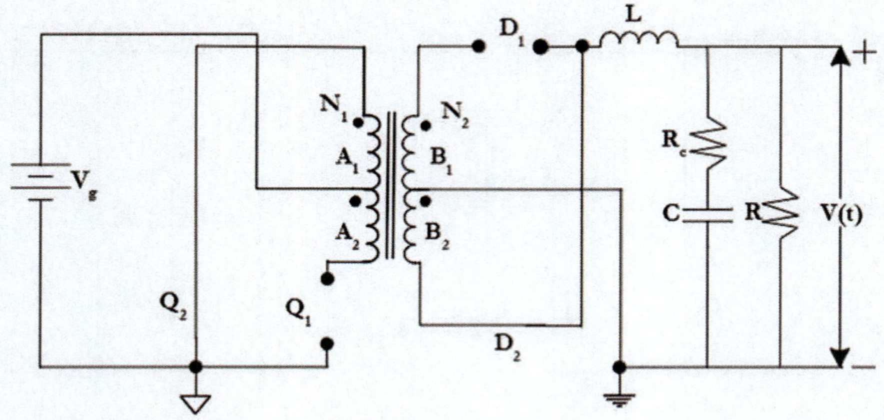

Fig. 3 Circuit diagram when Q_2 is ON and Q_1 is OFF

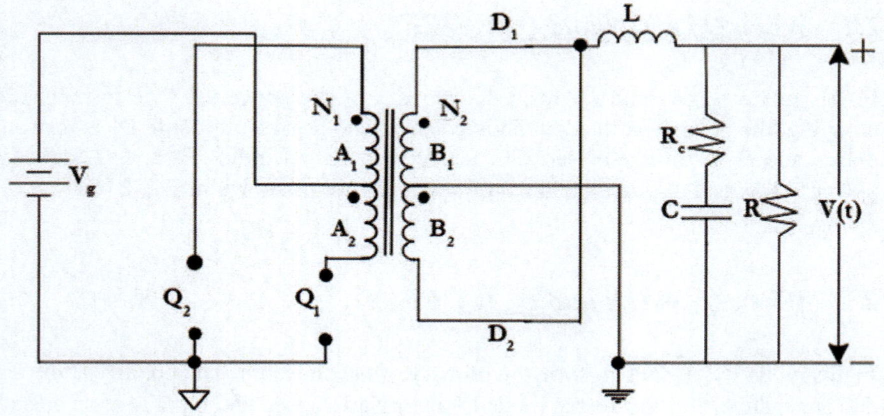

Fig. 4 Circuit diagram when both Q_1 and Q_2 is OFF

3 Specification and Design Calculations

The design specifications of the converter is listed in Table 1.

The calculated values of the system is shown in Table 2: [6].

4 Modeling of Converter

Modeling of a converter is defined as developing a set of mathematical equations relating to the behavior of the converter. The actual converter is modeled as equations

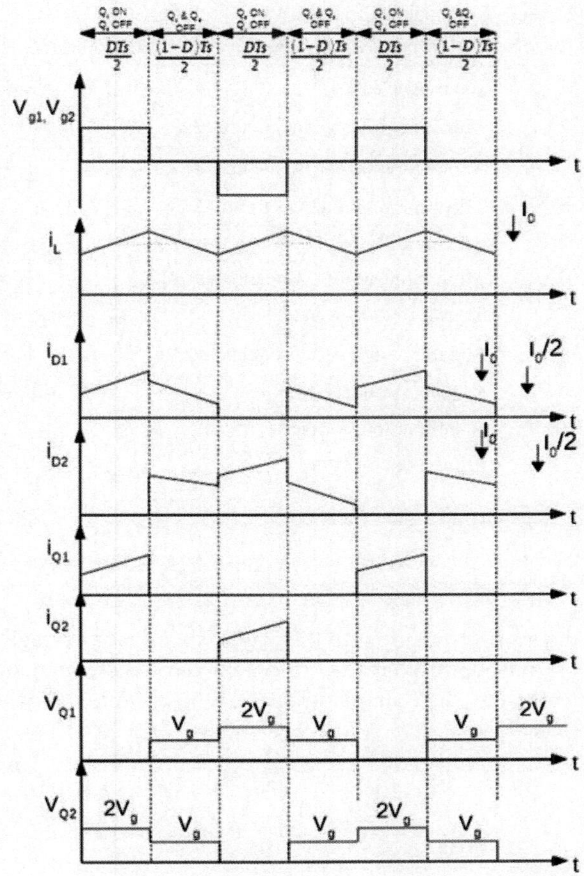

Fig. 5 Waveforms of the push–pull converter

Table 1 Specifications of converter

S. no	Description	Value
1	Input voltage	36 V
2	Power output	1.2 kW
3	Output voltage	375 V
4	Switching frequency	100 kHz
5	Maximum duty cycle	0.45
6	Converter efficiency (assumed)	90%

representing the relation between its input and output variables. The converters are modeled as a set of nonlinear differential equations to derive the small signal model and transfer function of the converter [7]. Techniques like state space averaging, current injection, circuit averaging are the methods to model the system and derive

Table 2 Component/parameter values

S. no	Component/parameter	Formula	Value
1	Input power	$P_{in} = \frac{P_{out}}{0.9}$	1.33 kW
2	Maximum average input current	$I_{in} = \frac{P_{in}}{V_{in}}$	37.03 A
3	Maximum average input current	$I_{pft} = \frac{I_{in}}{2 \times D_{max}}$	41.15 A
4	Maximum input RMS current	$I_{in(RMS)} = I_{pft}\sqrt{2 \times D_{max}}$	39.03 A
5	Maximum MOSFET RMS current	$I_{MO(RMS)} = I_{pft}\sqrt{D_{max}}$	21.18 A
6	Minimum MOSFET breakdown voltage	$V_{bk(MOS)} = 1.3 \times 2 \times V_{in(MAX)}$	93.6 A
7	Transformer turns ratio	$N = \frac{V_{out}}{2 \times D_{min} X V_{in}}$	12
8	Maximum average output current	$I_{out} = \frac{P_{out}}{V_{out}}$	3.2 A
9	Secondary RMS current	$I_{sec(RMS)} = I_{out}\sqrt{D_{max}}$	2.14 A
10	Rectifier diode voltage	$V_{diode} = N.V_{in}$	432 V
11	Filter inductor value	$L \geq \left(\frac{N_2}{N_1} V_{in} - V_{out}\right)\frac{t_{on}}{\Delta I}$	570mH
12	Output filter capacitor	$C = \frac{1}{8}\frac{\Delta I}{\Delta V} T$	1.6 μF

where ΔI and ΔV are current and voltage ripple, respectively

the transfer function. In this paper, state space averaging technique is used to model the push–pull converter to obtain the transfer functions of the converter. In this technique, the dynamic model of the converter is represented by equations of the form:

$$
\begin{aligned}
K\dot{x} &= Ax + Bu \\
y &= Cx + Eu
\end{aligned}
\tag{1}
$$

A is state matrix, B is input matrix, C is output matrix, E is direct transition matrix, and x and y are input and output vector, respectively. The above equations are state and output equation of the model. The current flowing through the inductor (i_L) and voltage across the capacitor (v_c) is taken as state variables. The output voltage (v) and input current (i_g) are considered as output variables. The two distinct modes of operation in the converter for the purpose of modeling are:

Mode 1: When one of the primary switches Q_1 or Q_2 is turned ON, the transfer of energy from primary to secondary circuit takes place and inductor current increases.

Mode 2: When both primary switches are turned off and energy stored in the inductor freewheels through secondary diodes.

A push–pull converter consisting of switches, diodes, parasitic elements, inductor and capacitor is considered for modeling the converter. The equivalent series resistance of the capacitor (R_C) is considered when modeling the converter. The inclusion of parasitic elements along with their loss components results in a precise model with improved system performance.

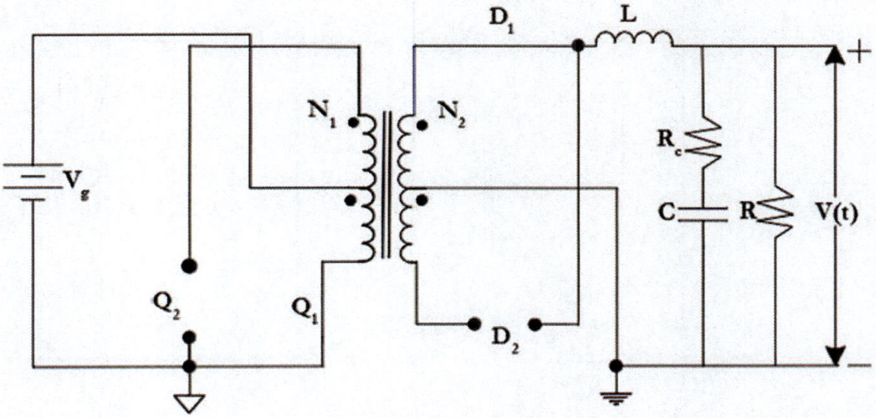

Fig. 6 Equivalent circuit when one switch is conducting

Mode 1: When Switch Q_1 is turned ON and Q_2 is OFF for the duration dTs/2, the equivalent circuit of the converter is as in Fig. 6.

The differential equation governing this section are:

$$\frac{L\mathrm{d}i_{\mathrm{L}}}{\mathrm{d}t} = \frac{N_2}{N_1}V_{\mathrm{g}} - \frac{RR_{\mathrm{c}}}{R+R_{\mathrm{c}}}i_{\mathrm{L}} - \frac{R}{R+R_{\mathrm{c}}}v_{\mathrm{c}} \tag{2}$$

$$\frac{C\mathrm{d}v_{\mathrm{c}}}{\mathrm{d}t} = -\frac{1}{R+R_{\mathrm{C}}}v_{\mathrm{c}} + \frac{R}{R+R_{\mathrm{C}}}i_{\mathrm{L}} \tag{3}$$

$$v = \frac{R}{R+R_{\mathrm{C}}}v_{\mathrm{c}} + \frac{RR_{\mathrm{C}}}{R+R_{\mathrm{C}}}i_{\mathrm{L}} \tag{4}$$

$$i_{\mathrm{g}} = \frac{N_2}{N_1}i_{\mathrm{L}} \tag{5}$$

The state equation and output equation are:

$$\begin{bmatrix} L & 0 \\ 0 & C \end{bmatrix}\begin{bmatrix} \frac{\mathrm{d}i_{\mathrm{L}}}{\mathrm{d}t} \\ \frac{\mathrm{d}v_{\mathrm{c}}}{\mathrm{d}t} \end{bmatrix} = \begin{bmatrix} -\frac{RR_{\mathrm{c}}}{R+R_{\mathrm{c}}} & -\frac{1}{R+R_{\mathrm{c}}} \\ \frac{1}{R+R_{\mathrm{c}}} & -\frac{1}{R+R_{\mathrm{c}}} \end{bmatrix}\begin{bmatrix} i_{\mathrm{L}} \\ v_{\mathrm{c}} \end{bmatrix} + \begin{bmatrix} \frac{N_2}{N_1} \\ 0 \end{bmatrix}[V_{\mathrm{g}}] \tag{6}$$

$$\begin{bmatrix} v \\ i_{\mathrm{g}} \end{bmatrix} = \begin{bmatrix} \frac{RR_{\mathrm{c}}}{R+R_{\mathrm{C}}} & \frac{R}{R+R_{\mathrm{C}}} \\ \frac{N_2}{N_1} & 0 \end{bmatrix}\begin{bmatrix} i_{\mathrm{L}} \\ v_{\mathrm{c}} \end{bmatrix} \tag{7}$$

Mode 2: When both switch Q_1 and Q_2 is turned OFF, for the duration $(1 - d)Ts/2$, the equivalent circuit of the converter is as in Fig. 7.

The differential equations governing this section are:

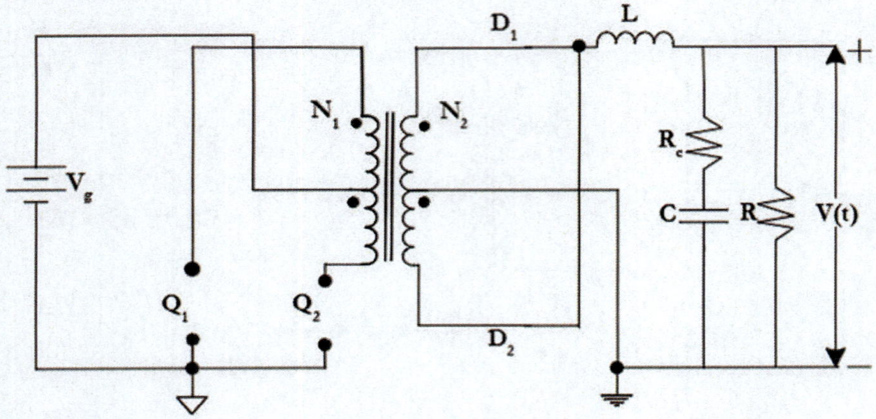

Fig. 7 Equivalent circuit in freewheeling mode

$$\frac{L\mathrm{d}i_L}{\mathrm{d}t} = -\frac{RR_\mathrm{c}}{R+R_\mathrm{c}}i_\mathrm{L} - \frac{R}{R+R_\mathrm{c}}v_\mathrm{c} \tag{8}$$

$$\frac{C\mathrm{d}v_\mathrm{c}}{\mathrm{d}t} = -\frac{1}{R+R_\mathrm{C}}v_\mathrm{c} + \frac{R}{R+R_\mathrm{C}}i_\mathrm{L} \tag{9}$$

$$v = \frac{R}{R+R_\mathrm{C}}v_\mathrm{c} + \frac{RR_\mathrm{C}}{R+R_\mathrm{C}}i_\mathrm{L} \tag{10}$$

$$i_\mathrm{g} = \frac{N_2}{N_1}i_\mathrm{L} \tag{11}$$

The state equation and output equation are:

$$\begin{bmatrix} L & 0 \\ 0 & C \end{bmatrix}\begin{bmatrix} \frac{\mathrm{d}i_\mathrm{L}}{\mathrm{d}t} \\ \frac{\mathrm{d}v_\mathrm{c}}{\mathrm{d}t} \end{bmatrix} = \begin{bmatrix} -\frac{RR_\mathrm{c}}{R+R_\mathrm{c}} & -\frac{1}{R+R_\mathrm{c}} \\ \frac{1}{R+R_\mathrm{c}} & -\frac{1}{R+R_\mathrm{c}} \end{bmatrix}\begin{bmatrix} i_\mathrm{L} \\ v_\mathrm{c} \end{bmatrix} \tag{12}$$

$$\begin{bmatrix} v \\ i_\mathrm{g} \end{bmatrix} = \begin{bmatrix} \frac{RR_\mathrm{c}}{R+R_\mathrm{C}} & \frac{R}{R+R_\mathrm{C}} \\ \frac{N_2}{N_1} & 0 \end{bmatrix}\begin{bmatrix} i_\mathrm{L} \\ v_\mathrm{c} \end{bmatrix} \tag{13}$$

Similar modeling can be arrived when switch Q_2 is ON and Q_1 is OFF. Thus, the ON and OFF period of the converter is $2\mathrm{d}T_\mathrm{s}$ and $2\mathrm{d}'T_\mathrm{s}$, respectively.

The averaged model of mode 1 and mode 2 can be written by using the equations [8, 9]:

$$A = A_1 d + A_2(1-d)$$
$$B = B_1 d + B_2(1-d)$$
$$C = C_1 d + C_2(1-d)$$

$$E = E_1 d + E_2(1 - d) \tag{14}$$

Thus, the averaged model of the push–pull converter is:

$$A = \begin{bmatrix} -\frac{RR_c}{R+R_c} & -\frac{1}{R+R_c} \\ \frac{1}{R+R_c} & -\frac{1}{R+R_c} \end{bmatrix} B = \begin{bmatrix} \frac{2dN_2}{N_1} \\ 0 \end{bmatrix} \tag{15}$$

$$C = \begin{bmatrix} \frac{RR_c}{R+R_C} & \frac{R}{R+R_C} \\ \frac{N_2}{N_1} & 0 \end{bmatrix} E = [0] \tag{16}$$

The steady-state representation of the averaged system are nonlinear. The matrices A, B, C are embedded functions of time. To linearize the system and obtain the small signal model of the converter, consider that input $vg(t)$ and duty d are varying around some quiescent operating points D and V_g. Hence,

$$\begin{aligned} v_g &= V_g + v'_g \\ d &= D + d' \end{aligned} \tag{17}$$

With

$$\frac{v'_g}{V_g} \ll 1 \text{and} \frac{d'}{D} \ll 1 \tag{18}$$

Due to above time varying perturbations, the dynamic variables x, y and u will change as

$$\begin{aligned} x &= X + x' \\ y &= Y + y' \\ u &= U + u' \end{aligned} \tag{19}$$

Using the above equations, the small signal model of the converter can be obtained

$$K\frac{\mathrm{d}X'}{\mathrm{d}t} = Ax' + Bu' + [(A_1 - A_2)X + (B_1 - B_2)U]d' \tag{20}$$

$$y' = Cx' + Eu' + [(C_1 - C_2)X + (E_1 - E_2)U]d' \tag{21}$$

For the above small signal models, the transfer functions output voltage to input voltage $G_{vg}(s)$ is obtained as [8, 9]:

$$\frac{v(s)}{v_g(s)} = C(\mathrm{SI} - A)^{-1}B \tag{22}$$

$$\frac{v(s)}{v_g(s)} = \frac{2dR\frac{N_2}{N_1}(1 + sCR_C)}{s^2LC(R + R_C) + s(L + RR_CC) + R} \tag{23}$$

The transfer function output voltage to duty $G_{vd}(s)$ is obtained as [8, 9]:

$$\frac{v(s)}{d(s)} = C(\mathrm{SI} - A)^{-1}f$$

where

$$f = [(A_1 - A_2)X + (B_1 - B_2)V_g;$$
$$X = A^{-1}BV_g \tag{24}$$

$$\frac{v(s)}{d(s)} = \frac{\frac{N_2}{N_1}RV_g(1 + sCR_C)}{s^2LC(R + R_C) + s(L + RR_CC) + R} \tag{25}$$

5 Closed Loop Controller Design

The basic control scheme for any power electronics converter is as in Fig. 8.

The plant is a power electronic system in which a specific parameter (output voltage, input current) is to be controlled. This parameter is sensed and compared with a pre-determined reference (V_{ref}). This generated error signal (error) is given as input to the controller which will generate a control parameter to obtain the desired output/response. In SMPS-based applications, the plant is controlled mostly by duty cycle; hence, a voltage to time converter to map controller input to output will be required.

Transfer function approach and state space approach are the techniques to design a controller to meet the required performances of stability, speed and accuracy. In transfer function-based method, plant is modeled as an LTI system having single input and single output (SISO) using block diagram approach. Nyquist, Bode plot, root locus and Nichols chart are approaches based on transfer function technique. In state space-based method, plant is modeled as multiple input multiple output

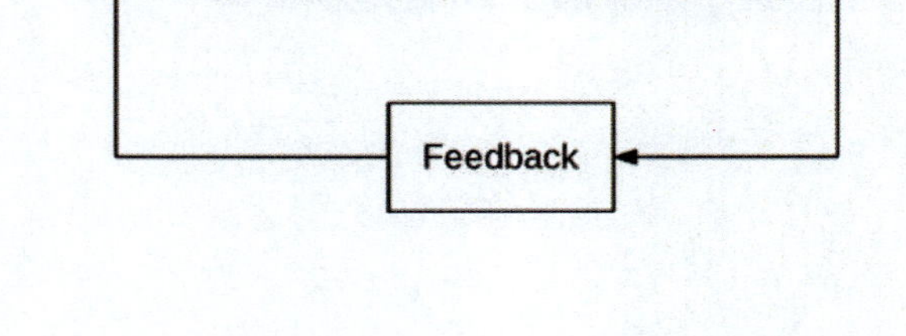

Fig. 8 Block diagram of control scheme of converter [6]

system (MIMO) using the state equations. This approach includes methods like full state feedback, output feedback, estimator, robust controller, self-tuning and adaptive controllers [6].

The main objective of this paper is to design a closed loop system to obtain the desired output and system response. Hence, Bode plot of the transfer function of converters is used to design the controller for the converter. Based on phase margin and gain margin of the open loop system obtained from the plot, a Proportional (*P*), Integral (*I*) and Derivative (*D*)-based controllers are employed to improve the performance of the system. In this paper, a PI controller with simple structure, low cost and wide industrial applications is used to obtain desired output and performance parameters [10].

The proportional-integral controller combines the advantages of proportional and integral controller and has the transfer function as:

$$\frac{U(s)}{E(s)} = K_{\mathrm{p}} + \frac{K_{\mathrm{I}}}{s} \tag{26}$$

The integral action reduces the speed, and the addition of proportional action contributes to improve the speed. Thus, the use of a PI controller improves transient response along with making the steady-state error close to zero. This controller also removes high-frequency noise along with improving phase margin and gain margin. [11].

6 Simulation, Results and Observations

The Bode plot of output voltage to duty transfer function of open loop converter in Eq. (25) is shown in Fig. 9, and the open loop Bode plot has a phase margin (PM) of 7.78.

To improve the performance parameters of the system, a PI compensator with transfer function as in Eq. (26) is added to the system, a pole is placed 10,000 Hz and a zero is placed origin, and the response of the system is shown in Fig. 10. Due to addition of the compensator, the system has a gain margin of 33.3 db phase margin of 78.8 degree at a frequency of 6.77 Hz.

The step response obtained from the system shown in Fig. 11 has a settling time of 90 ms.

Simulink model of the system with the design parameters is shown in Fig. 12. The desired output voltage of 375 V with a settling time of 0.07 ms and output current of 5.15 A is shown in Figs. 13 and 14, respectively.

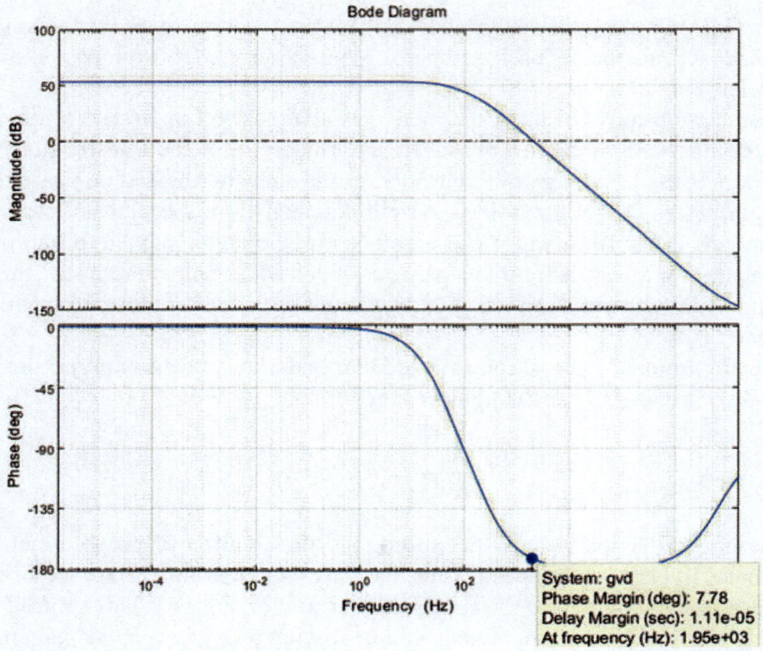

Fig. 9 Open loop Bode plot of the converter

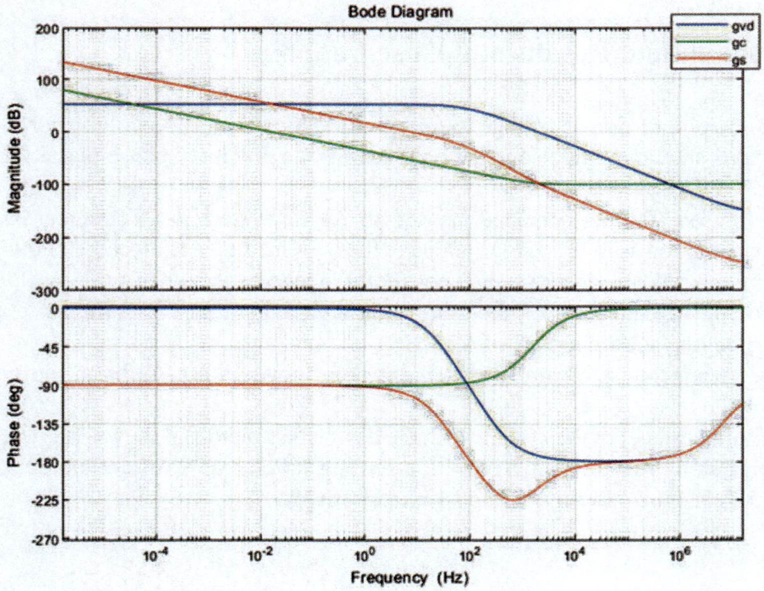

Fig. 10 Bode plot of the closed loop system

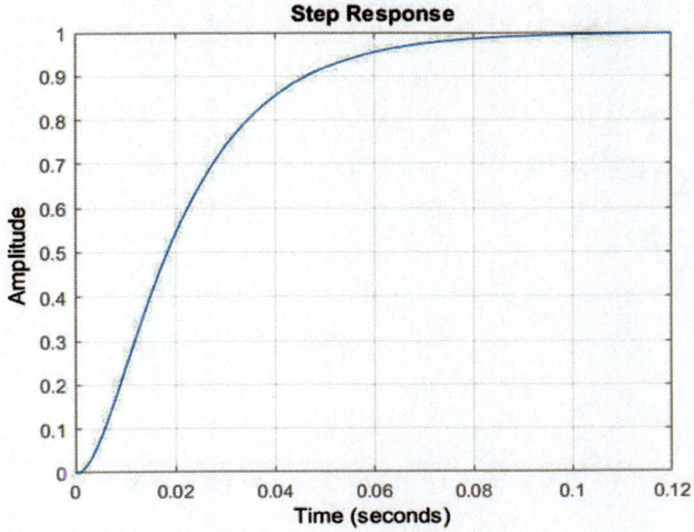

Fig. 11 Step response of closed loop system

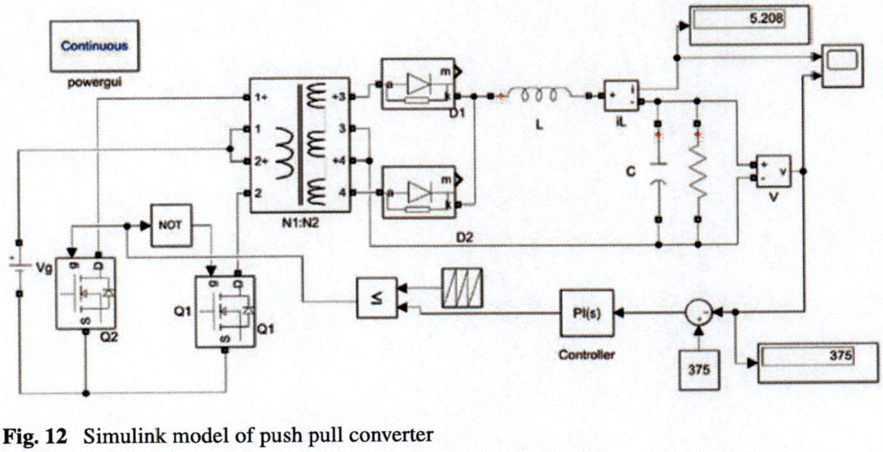

Fig. 12 Simulink model of push pull converter

7 Conclusion

A 1.2 kW, 36-375 V push–pull converter is modeled using state space average technique. The inclusion of PI controller makes the converter stable and improves the settling time of the controller. The designed converter response is verified in MATLAB Simulink. The proposed converter with galvanic isolation and higher transformation ratio is suitable for domestic and industrial applications.

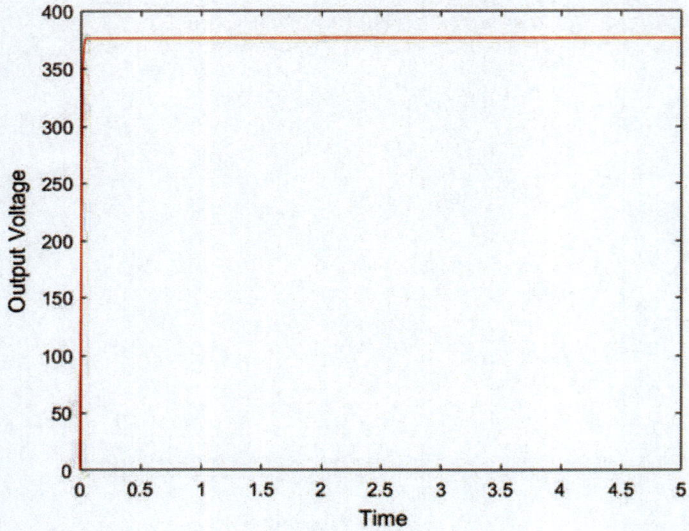

Fig. 13 Output voltage waveform

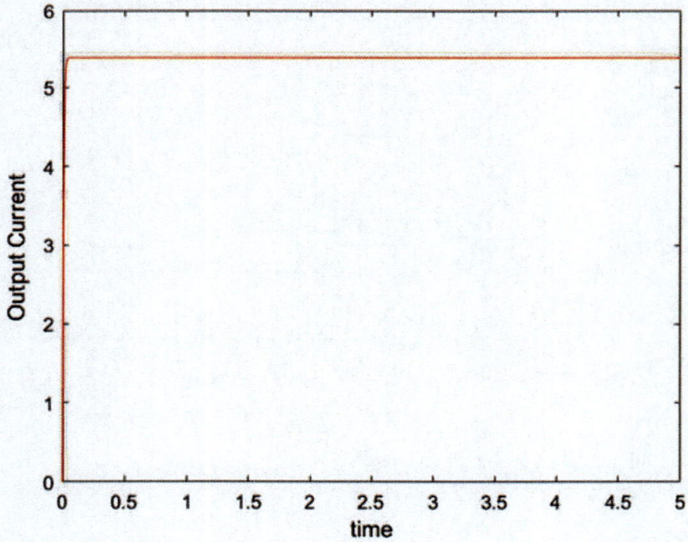

Fig. 14 Output current waveform

References

1. G.D. Kamalapur, R.Y. Udaykumar, Electrical energy conservation in India—Challenges and achievements, in 2009 International Conference on Control, Automation, Communication and Energy Conservation (Perundurai, Tamilnadu, 2009), pp. 1–5
2. M. Darameičikas, F. Muhammad-Sukki , S.H. Abu-Bakar, N. Sellami, N.A. Bani, A. Abubakar Mas'ud, J.A. Ardila-Rey, Design of a DC-DC converter in residential solar photovoltaic system, in 2nd International Conference on Women in Science, Engineering and Technology (2017), pp 1–11
3. S. Bansal, L.M. Saini, D. Joshi, Design of a DC-DC converter for photovoltaic solar system, in 2012 IEEE 5th India International Conference on Power Electronics (IICPE) (Delhi, 2012), pp. 1–5
4. M.P. Manasa, S. Oommen, Simulation analysis of closed loop dual inductor current-fed push-pull converter by using soft switching. J. Res. **2**, 75–83 (2016)
5. I.U.C. Zamora, A.A.E. Bén, A.A. Díaz-Comas, J.J.E.R. Segura, Construction of a current-fed push-pull converter: practical considerations. Pistas Educativas **130**, 54–69 (2018)
6. L. Umanand, *Power Electronics Essentials & Applications, New Delhi* (Wiley India Pvt. Ltd., India, 2009).
7. D.S.L. Simonetti, J.L.F. Viera, G.C.D. Sousa, Modeling of the high-power-factor discontinuous boost rectifiers. IEEE Trans. Indus. Electron. **46**(4), 788–795 (1999). https://doi.org/10.1109/41.778242
8. M.B. Patil, V. Ramanarayanan, V.T. Ranganathan, *Simulatioon Oof Power Electronic Circuits* (Narosa Publishing House Pvt. Ltd., New Delhi, 2013).
9. R.W. Erickson, D. Maksimovic, Fundamentals of Power Electronics (Kluwer Academic/Plenum Publishers, New York, 2001)
10. D.E. Oku, E.P. Obot, Comparative study Of PD, PI And PID controllers for control of a single joint system in robots. Int. J. Eng. Sci. **7**(6), 51–54 (2018). doi: https://doi.org/10.9790/1813-0709025154
11. K. Smriti Rao, R. Mishra, Comparative study of P, PI and PID controller for speed control of VSI-fed induction motor. Int. J. Eng. Dev. Res. **2**(2), 2740–2744

Modeling and Analysis of GaN-Based Buck Converter

H. Swathi Hatwar, Ravikiran Rao M, and Suryanarayana K.

1 Introduction

DC-DC converter plays a major role in the design of all electronic systems to achieve various voltage and current levels because of its high efficiency, compact size, and low weight. A non-isolated buck converter is one of the widely used DC-DC converters in the industry. This converts higher input voltage into lower output with high efficiency over a wide load range. When practical application is considered, efficiency drops due to inductor copper loss, losses associated with switch, and loss tangent of capacitor. With Si-based converters, due to high on-state resistance of Si switch $R_{\text{ds(on)}}$, gate charge requirement Q_{G}, losses associated with reverse recovery charge Q_{rr}, efficiency of the converter is found in the range 80–87% [1, 2]. Using WBG devices like SiC and GaN in place of Si-based switches in DC-DC converters enable a smaller and more efficient converter due to low $R_{\text{ds(on)}}$, Q_{G} and Q_{rr}. Also the devices could be operated at high frequencies which reduce volume of the converter and help in achieving high power density [3, 4]. However, use of WBG devices poses challenge for the designers in terms of switching loss, increased EMI if layout is not carefully considered [5, 6]. Replacing Si devices with GaN gives an efficiency in the range 94–99% [1, 7]. To analyze the system, study of mentioned parameters in designing a converter will lead to a fruitful result with reduced efforts in the implementation stage.

To design any system accurately, system behavior to variation in load, effect of duty cycle to output voltage, sensitivity to disturbances in input voltage must be modeled. The modeling process considers the dominant system behavior parameters

H. Swathi Hatwar (✉) · Ravikiran Rao M · Suryanarayana K.
Department of Electrical and Electronics Engineering, NMAM Institute of Technology, Nitte, Karkala 574110, India
e-mail: hswathihatwar@nitte.edu.in

Ravikiran Rao M
e-mail: ravikiranraom@nitte.edu.in

Suryanarayana K.
e-mail: suryanarayana@nitte.edu.in

Sanjeevikumar et al. (eds.), *Advances in Renewable Energy and Electric Vehicles*,
Lecture Notes in Electrical Engineering 767,
https://doi.org/10.1007/978-981-16-1642-6_24

and neglects the insignificant ones leading to approximation behavioral model [8]. The dynamic model and small-signal transfer function predicts the system behavior to the perturbations like duty cycle, load, and disturbances. From the small-signal model of non-ideal converter, the behavior of actual circuit can be analyzed to design the compensation network for stable operation of the system and meet the required dynamic performance [9].

Bode plot is one of the techniques to analyze the converter transfer functions derived using small signal equivalent circuit model. The method is preferred as it gives a quick and effective assessment of circuit performance. From the plot, system behavior is measured for the uncompensated network. Using the knowledge of system parameters, feedback mechanism to achieve desired loop gain, crossover frequency, gain, and phase margin is deduced [10].

The compensator design to achieve regulated output voltage from buck converter involves selection of PID controller and tuning of the parameters K_P, K_I and K_D. The selection of values considers a trade off between response time, overshoot, oscillations, settling time, and steady-state error [11]. The choice of proper algorithm depends on the uncompensated network bode plot. From the plot, observations on system behavior are seen, and algorithm which gives less oscillations, steady-state error, overshoot, and a better response is chosen. The tuning method involves setting K_I and K_D to zero initially, and K_P is varied and system behavior is observed as per requirement. Once K_P is fixed, K_I is varied to achieve system stability. If desired response is not achieved, K_D constant is introduced and value is varied such that the system reaches the desired reference value faster [12].

In this paper, mathematical model of an non-ideal synchronous buck converter is derived to design closed-loop system. To achieve better accuracy, parasitic resistance of all elements is considered. Using state-space averaging technique, duty to output voltage transfer function is derived. Voltage regulation is achieved through PI compensator implementation.

The paper is organized as follows: Sect. 1 discusses the need of mathematical modeling and compensator design. In Sect. 2, working principle of buck converter is discussed. The mathematical modeling and transfer function of a non-ideal buck converter is presented in Sect. 3. The bode plot and its aspects are highlighted in Sects. 4 and 5 and discuss the simulation and hardware results of GaN-based buck converter. Concluding remarks are provided in Sect. 6.

2 Theoretical Background

2.1 Ideal Buck Converter

The ideal buck converter consists of switch, inductor, capacitor and a load connected to a DC voltage as in Fig. 1. The converter does not take into consideration the losses associated with switch, inductor, capacitor, gate charge requirements, etc. In

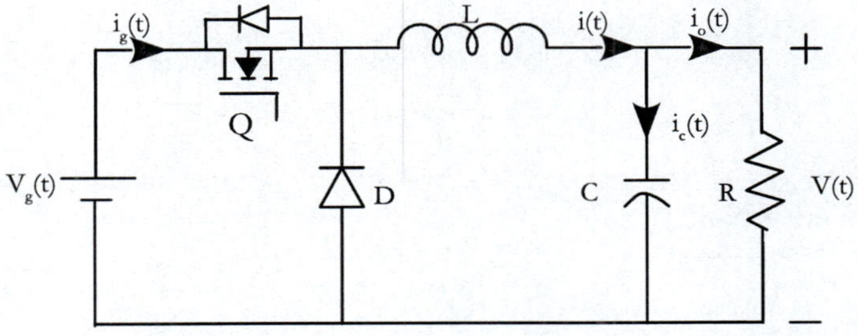

Fig. 1 Ideal buck converter

a non-synchronous buck converter, the bottom switch is a diode and top switch is a MOSFET or transistor. This paper deals with a synchronous buck converter where the bottom switch is also a MOSFET. The use of low-side MOSFET improves conversion efficiency as the on-state resistance drop is less than diode drop reducing losses. The converter operates in two modes based on the current flow through inductor: continuous conduction mode (CCM) if current through inductor never becomes zero; & discontinuous conduction mode (DCM) if current through inductor becomes zero every cycle. In this paper, CCM mode of operation is considered for analysis.

Based on the PWM pulse given to the switches, when the top switch is ON bottom switch is ensured to be kept in OFF condition in a synchronous converter. During this condition, the current flow path is from source to load through the inductor. When the bottom switch is ON and top switch is kept OFF, there is no connection between load and source & current flows through inductor, capacitor, and through the bottom switch. This ON and OFF of switches done at regular intervals based on switching frequency, and the output voltage is an average value lesser than the source voltage in CCM operation. In order to obtain desired fixed output voltage, duty cycle needs to be controlled [13]. If the duty cycle is not controlled, output might vary due to load or source variations.

Small-signal analysis is one of the methods which could be used for compensator design of the switching converter. The aim of analysis is to observe the AC behavior of the converter around a fixed operating point. Based on the requirement of the designer, transfer functions such as control to output, input admittance, output impedance, audio susceptibility, and many more could be derived. From the transfer function characteristics, a suitable compensator need to be designed to achieve the desired goals.

To represent small-signal model using state-space averaging technique, current through inductor $i_L(t)$ and voltage across capacitor $v_c(t)$ are considered as the state variables, and $v(t)$ is taken as the output variable.

In the ON state, state-space representation of state vector X and output vector Y is written in the form:

$$\begin{aligned} K\dot{X} &= A_1X + B_1U \\ Y &= C_1X + E_1U \end{aligned} \tag{1}$$

and in the OFF state by:

$$\begin{aligned} K\dot{X} &= A_2X + B_2U \\ Y &= C_2X + E_2U \end{aligned} \tag{2}$$

The proportionality matrices A, B, C, E are the average values over one period and can be written as:

$$\begin{aligned} A &= A_1D + A_2D' \\ B &= B_1D + B_2D' \\ C &= C_1D + C_2D' \\ E &= E_1D + E_2D' \end{aligned} \tag{3}$$

where D is duty ratio, U is input vector, and K is a constant matrix representing energy storage elements (L & C)

Considering the perturbations as $x\hat{(t)}$, $y\hat{(t)}$, $u\hat{(t)}$, and $d\hat{(t)}$, small signal model can be written as

$$\begin{aligned} K\frac{\mathrm{d}x\hat{(t)}}{\mathrm{d}t} &= Ax\hat{(t)} + Bu\hat{(t)} + ((A_1 - A_2)X + (B_1 - B_2)U)d\hat{(t)} \\ y\hat{(t)} &= Cx\hat{(t)} + Eu\hat{(t)} + ((C_1 - C_2)X + (E_1 - E_2)U)d\hat{(t)} \end{aligned}$$

3 Mathematical Modeling of Converter

A non-ideal synchronous buck converter given in Fig. 2 includes on-state resistance of the top and bottom switches $R_{\mathrm{ds(on)}}$, inductor loss component R_{L}, and loss tangent of the capacitor R_{C}.

The switches Q_1 and Q_2 are turned ON and OFF at regular intervals with a switching frequency f_{s} such that when Q_1 is ON Q_2 is kept OFF and vice versa. In the ON state operation when the switch Q_1 is ON, on-state resistance of the bottom switch Q_2 has no effect. The equivalent circuit in this state is shown in Fig. 3. In

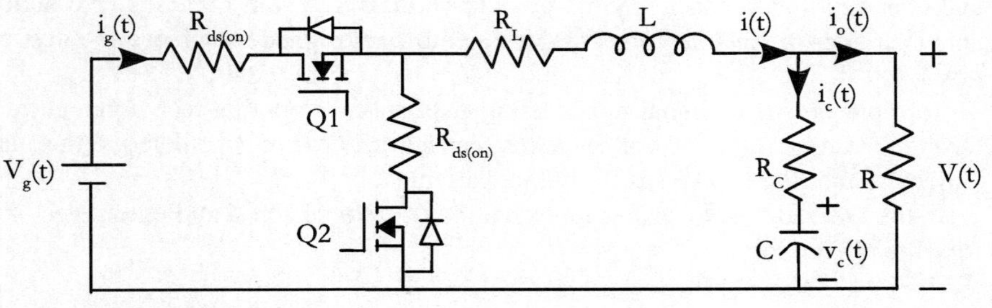

Fig. 2 Non-ideal buck converter

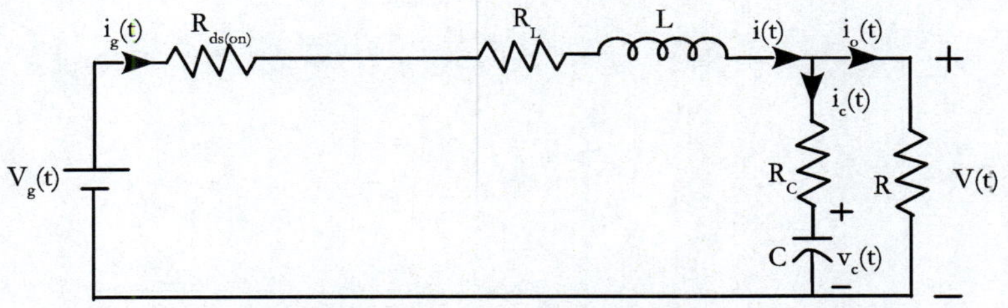

Fig. 3 Non-ideal buck converter-ON state

the ON state, on-state resistance of Q_1 switch and inductor copper loss components affects the circuit behavior.

The inductor loop equation in the ON state is: 4.

$$V_g(t) = (R_{\text{ds on}} + R_L)i(t) + L\frac{di(t)}{dt} + v(t)$$

$$L\frac{di(t)}{dt} = V_g(t) - (R_{\text{ds on}} + R_L)i(t) - v(t) \tag{4}$$

where,

$$v(t) = (\frac{R}{R + R_c})v_c(t) + (\frac{RR_c}{R + R_c})i(t) \tag{5}$$

Therefore, substituting (5) in (4), the voltage drop across inductor could be rewritten as

$$L\frac{di(t)}{dt} = V_g(t) - (R_{\text{ds on}} + R_L + \frac{RR_c}{R + R_c})i(t) - (\frac{R}{R + R_c})v_c(t) \tag{6}$$

The capacitor voltage equation in the ON state can be written as

$$C\frac{dv_c(t)}{dt} = (\frac{R}{R + R_c})i(t) - (\frac{1}{R + R_c})v_c(t) \tag{7}$$

Thus, state-space representation equation in the ON state is given as

$$\begin{bmatrix} L & 0 \\ 0 & C \end{bmatrix}\begin{bmatrix} \frac{di(t)}{dt} \\ \frac{dv_c(t)}{dt} \end{bmatrix} = \begin{bmatrix} -(R_{\text{ds on}} + R_L + \frac{RR_c}{R+R_c}) & -\frac{R}{R+R_c} \\ \frac{R}{R+R_c} & -\frac{1}{R+R_c} \end{bmatrix}$$

$$\begin{bmatrix} i(t) \\ v_c(t) \end{bmatrix} + \begin{bmatrix} 1 \\ 0 \end{bmatrix} v_g(t)$$

$$\begin{bmatrix} v(t) \end{bmatrix} = \begin{bmatrix} \frac{RR_c}{R+R_c} & \frac{R}{R+R_c} \end{bmatrix}\begin{bmatrix} i(t) \\ v_c(t) \end{bmatrix} + \begin{bmatrix} 0 \end{bmatrix} v_g(t)$$

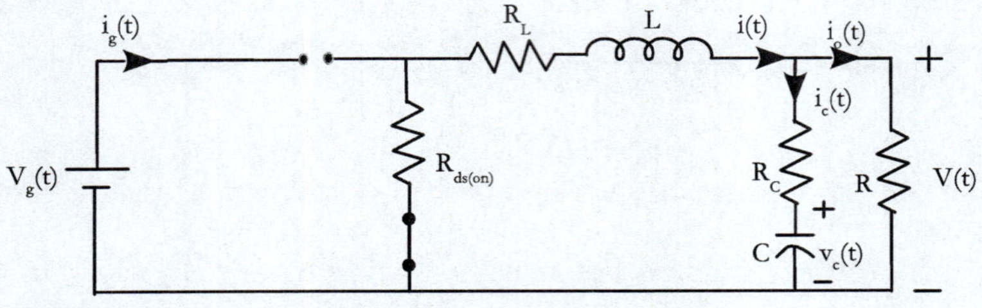

Fig. 4 Non-ideal buck converter-OFF state

In the OFF state operation, Q_1 is OFF and Q_2 is ON, and $R_{ds(on)}$ of Q_2 needs to be considered. The equivalent circuit in this state is shown in Fig. 4.

$$L\frac{\mathrm{d}i(t)}{\mathrm{d}t} = -R_{\mathrm{L}}i(t) - R_{\mathrm{c}}i_{\mathrm{c}}(t) - v_{\mathrm{c}}(t) - R_{\mathrm{ds\ on}}i(t) \tag{8}$$

where

$$i_{\mathrm{c}}(t) = i(t)(\frac{R}{R+R_{\mathrm{c}}}) - (\frac{1}{R+R_{\mathrm{c}}})v_{\mathrm{c}}(t) \tag{9}$$

$$v(t) = v_{\mathrm{c}}(t)(\frac{R}{R+R_{\mathrm{c}}}) + (\frac{RR_{\mathrm{c}}}{R+R_{\mathrm{c}}})i(t) \tag{10}$$

Thus, (8) can be rewritten as

$$L\frac{\mathrm{d}i(t)}{\mathrm{d}t} = i(t)(-R_{\mathrm{L}} - R_{\mathrm{ds\ on}} - \frac{RR_{\mathrm{c}}}{R+R_{\mathrm{c}}}) + v_{\mathrm{c}}(t)(\frac{-R}{R+R_{\mathrm{c}}}) \tag{11}$$

Capacitor loop equation can be written as

$$C\frac{\mathrm{d}v_{\mathrm{c}}(t)}{\mathrm{d}t} = i(t)(\frac{R}{R+R_{\mathrm{c}}}) - (\frac{1}{R+R_{\mathrm{c}}})v_{\mathrm{c}}(t) \tag{12}$$

OFF state proportionality matrices can be written as

$$\begin{bmatrix} L & 0 \\ 0 & C \end{bmatrix}\begin{bmatrix} \frac{\mathrm{d}i(t)}{\mathrm{d}t} \\ \frac{\mathrm{d}v_{\mathrm{c}}(t)}{\mathrm{d}t} \end{bmatrix} = \begin{bmatrix} -R_{\mathrm{L}} - R_{\mathrm{ds\ on}} - \frac{RR_{\mathrm{c}}}{R+R_{\mathrm{c}}} & -\frac{R}{R+R_{\mathrm{c}}} \\ \frac{R}{R+R_{\mathrm{c}}} & -\frac{1}{R+R_{\mathrm{c}}} \end{bmatrix}\begin{bmatrix} i(t) \\ v_{\mathrm{c}}(t) \end{bmatrix} + \begin{bmatrix} 0 \\ 0 \end{bmatrix} v_{\mathrm{g}}(t)$$

$$\begin{bmatrix} v(t) \end{bmatrix} = \begin{bmatrix} \frac{RR_{\mathrm{c}}}{R+R_{\mathrm{c}}} & \frac{R}{R+R_{\mathrm{c}}} \end{bmatrix}\begin{bmatrix} i(t) \\ v_{\mathrm{c}}(t) \end{bmatrix} + \begin{bmatrix} 0 \end{bmatrix} v_{\mathrm{g}}(t)$$

Small signal model for non-ideal buck converter can be written as

$$\begin{bmatrix} L & 0 \\ 0 & C \end{bmatrix}\begin{bmatrix} \frac{d\hat{i(t)}}{dt} \\ \frac{d\hat{v_c(t)}}{dt} \end{bmatrix} = \begin{bmatrix} -R_{\text{ds on}} - R_L - \frac{RR_c}{R+R_c} & \frac{-R}{R+R_c} \\ \frac{R}{R+R_c} & \frac{-1}{R+R_c} \end{bmatrix}$$

$$\begin{bmatrix} \hat{i(t)} \\ \hat{v_c(t)} \end{bmatrix} + \begin{bmatrix} D \\ 0 \end{bmatrix} \hat{v_g(t)} + \begin{bmatrix} V_g \end{bmatrix} \hat{d(t)}$$

$$\begin{bmatrix} \hat{v(t)} \end{bmatrix} = \begin{bmatrix} \frac{RR_c}{R+R_c} & \frac{1}{R+R_c} \end{bmatrix}\begin{bmatrix} \hat{i(t)} \\ \hat{v_c(t)} \end{bmatrix}$$

Using Laplace transform, the output to control transfer function can be derived from (13).

$$\hat{y(s)} = C((sI - K^{-1}A)^{-1}K^{-1}F + G)\hat{d(s)} \tag{13}$$

such that

$$F = (A1 - A2)X + (B1 - B2)U$$
$$G = (C1 - C2)X + (E1 - E2)U$$
$$X = -A^{-1}BU$$

The transfer function (TF) of synchronous buck converter considering all loss components and assuming identical top and bottom switches which leads to same value of $R_{\text{ds on}}$ is given as

$$G_{vd} = \frac{\widehat{v(s)}}{\widehat{d(s)}} = \frac{\text{Num}}{\text{Den}} \tag{14}$$

$$\text{Num} = \frac{R^3V_g + R^2R_cV_g(CR_{\text{ds on}} + CR_cs + 1)}{(R + R_c)(R + R_{\text{ds on}} + R_L)}$$
$$\text{Den} = s^2LC(R + R_c) + s(L + C(RR_c + RR_{\text{ds on}}+$$
$$RR_L + R_LR_c + R_{\text{ds on}}R_c)) + R_L + R_{\text{ds on}} + R$$

If all the loss components are equated to zero in the above TF, the ideal buck converter can be realized and the corresponding output to duty TF is given in (15).

$$\frac{v(s)}{d(s)} = \frac{RV_g}{RLCs^2 + Ls + R} \tag{15}$$

4 Control Loop Aspects

To achieve the desired system response like stability and steady-state accuracy, accurate controller design is of importance. One of the design methods for a LTI single-input single-output (SISO) is a traditional technique transfer function-based bode diagram method [14]. The bode plot of the control to output transfer function obtained using state space averaging technique (14) is analyzed. The frequency response of uncompensated buck converter with values as in Table 1 is simulated and plotted using MATLAB.

From the bode diagram of the uncompensated buck converter Gvd(s) Fig. 5, it is observed the system has low frequency gain of 28.4dB which is not preferred. A stable system should have a high initial gain at low frequency preferably -20dB per decade slope.

To achieve desired performance, addition of poles and zeros at certain frequencies plays a major role. The placement of poles and zeros is called as compensator network. The design requires to increase DC gain at low frequency. The design here uses a PI (proportional plus integral) controller often called lag compensator. The objective of this compensator is to increase the loop gain at low frequency to regulate better and have more loop gain. The transfer function of this type of compensator adds a inverted zero at low frequency of w_l and a gain of Gco and is expressed as in (16).

$$C1 = Gco * (1 + wl/s) \tag{16}$$

With the introduction of a PI compensator, bode plot of the compensated buck converter T(s) in Fig. 5 shows a high DC gain at low frequency.

5 Simulation and Results

The electrical model of a closed-loop non-ideal buck converter is realized using MATLAB/Simulink. The model includes capacitor with ESR, inductor with loss component, GaN MOSFETs with on-state resistance. For the system analysis, the specifications of GaN based buck converter are given in Table 2.

Table 1 Simulation parameters

S. No.	Description	Data
1	On-state resistance of MOSFET	7 mΩ
2	Inductor loss component	90 mΩ
3	Loss tangent of capacitor	6 mΩ
4	Inductor	6.03 μH
5	Capacitor	200 μF
6	Input DC voltage	36 V

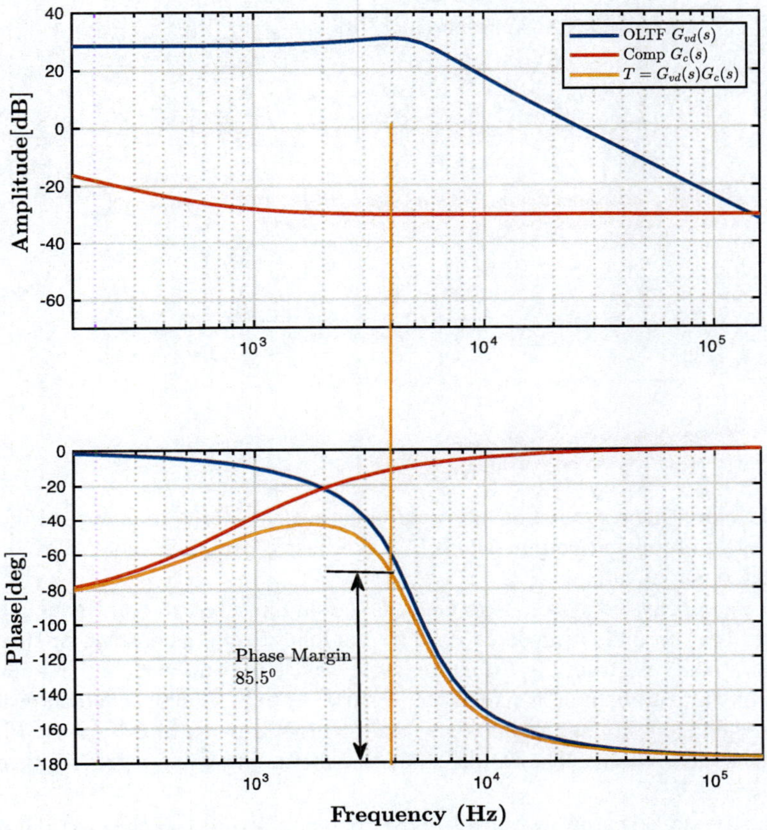

Fig. 5 Bode plot of non-ideal buck converter

Table 2 Buck converter parameters

S. No.	Description	Abbreviation	Data
1	Output power	P_o	300 W
2	Input voltage	V_g	36 V
3	Output voltage	V	12 V
4	Switching frequency	f_s	100 kHz
5	Inductor	L	6 μH
6	Capacitor	C	10 μF

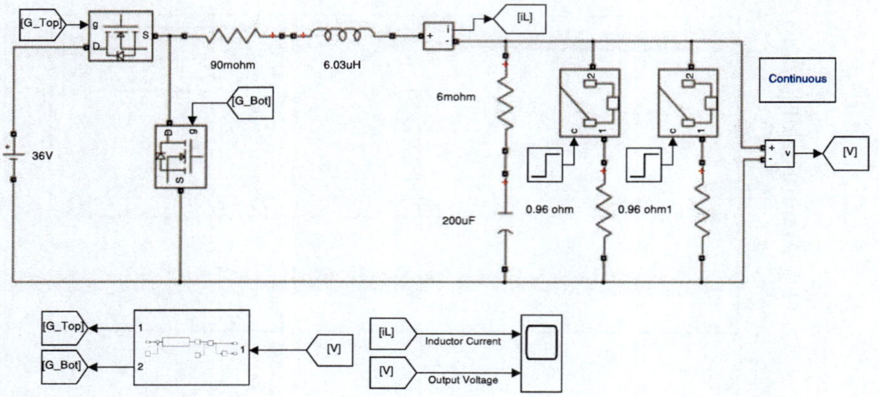

Fig. 6 Closed-loop buck converter

The closed-loop simulation of synchronous buck converter is shown in Fig. 6. The duty cycle of top and bottom MOSFETs is adjusted using a PI controller to obtain regulated output voltage.

The compensated system with Gco of 0.03 and inverted zero at frequency w_l of 5000 rad per second is simulated for different loads keeping constant input voltage, and it is observed from Fig. 7 that the system is stable and gives a fixed output of 12 V. In simulation, input is fixed at $V_g = 36$ V. At time $t = 6$ms, the load resistance is changed from no load to 0.96ohm, and at 12ms it is doubled by decreasing the resistance to 0.48ohm. The results show that the output voltage gets regulated and settles to 12V.

The developed hardware of buck converter using WBG GaN devices to step down 36–12 V is shown in Fig. 8. The switching of these devices is controlled using an in-house developed control card employing NXP make MC56F84789 microcontroller [15]. The control card generates PWM signals for atop and bottom switches, monitors status of current & voltages at input and output side; and secure switches against faulty conditions. To develop the algorithm and implement necessary control loop of the system, CodeWarrior-V10.4 software IDE is used. The compiled code is transferred to microcontroller by OnCE mode USB Tap JTAG.

The developed system is loaded with various loading conditions. Figure 9 shows the input, output voltage, and current waveforms. CH1 and CH2 represent output voltage and inductor current waveforms with values 12 V and 22 A, respectively. CH3 and CH4 represent input voltage and current waveforms with values 34 V and 8 A, respectively.

Figure 10 shows sudden application of load from 12.5 to 25 A. The PI compensator values obtained from simulation are implemented on MC56F84789 controller in discrete domain. It is shown that system gets regulated to 12 V after sudden change in the load current. Also, the simulation and experimental results match.

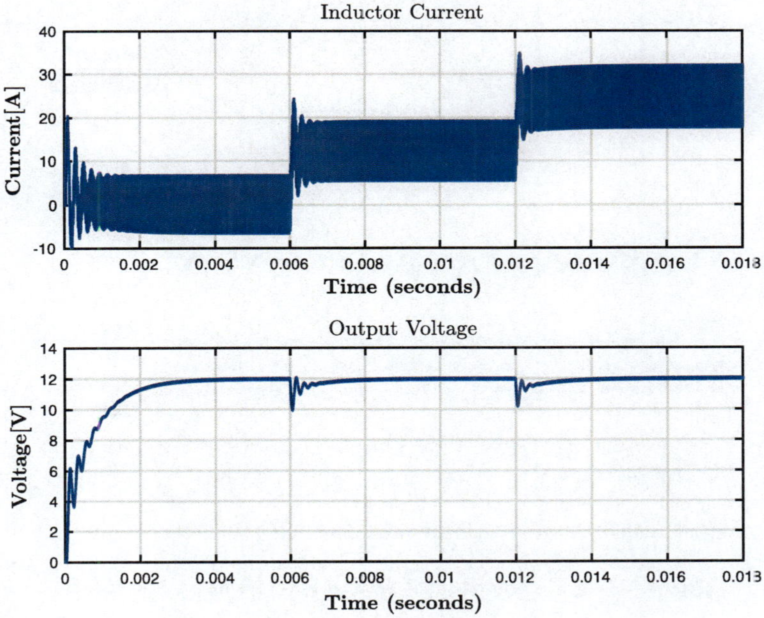

Fig. 7 Inductor current and output voltage

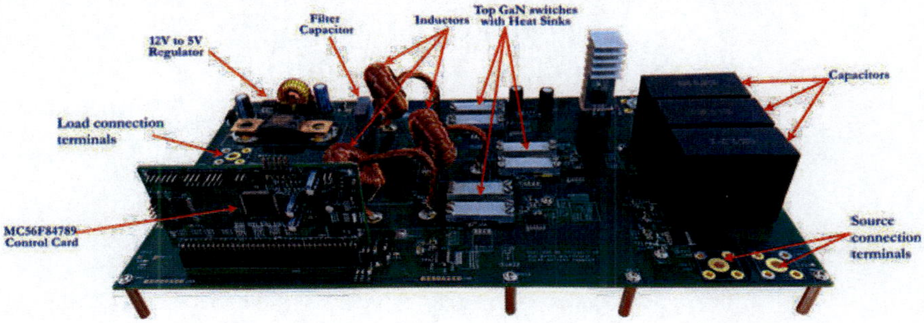

Fig. 8 Hardware setup of developed converter using GaN switches

Table 3 shows different loading values, and in each case, efficiency is calculated by measuring input, output voltage and current. From the obtained result, efficiency against load applied to the system is plotted as in Fig. 11

From the plot, following observations could be made

1. Efficiency η of the converter is found in the range of 95–98%
2. Employing WBG GaN devices, a higher efficiency is achieved in comparison with a conventional Si-based converters

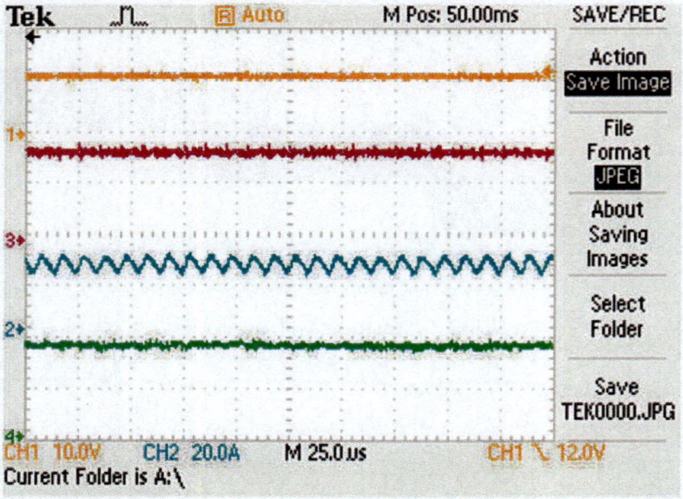

Fig. 9 Input and output voltage and current waveforms

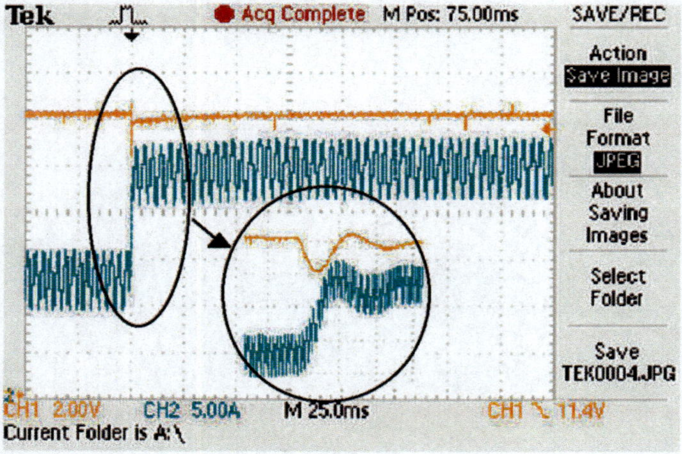

Fig. 10 Step load from 12.5 to 25 A

6 Conclusion

In this paper, mathematical model of synchronous buck converter considering all parasitics is derived. Using state-space averaging technique, transfer function of control to output voltage of non-ideal buck converter is developed. The uncompensated bode plot using MATLAB is plotted and PI compensator is designed to obtain the necessary performance. Using the coefficients obtained, a closed-loop buck converter

Table 3 Experimental readings

S. No.	V_{in} (V)	I_{in} (A)	V_{out} (V)	I_{out} (A)	Efficiency (%)
1	35.4	1.76	12	5	96.30
2	35.2	2.69	12	7.6	96.32
3	35.1	3.57	12	10.2	97.68
4	34.9	4.41	12	12.6	98.24
5	34.8	5.3	12	15	97.59
6	34.6	6.33	12	17.6	96.43
7	34.3	7.34	12	20.2	96.28
8	34.2	8.23	12	22.4	95.50
9	33.9	9.5	12	25.6	95.39

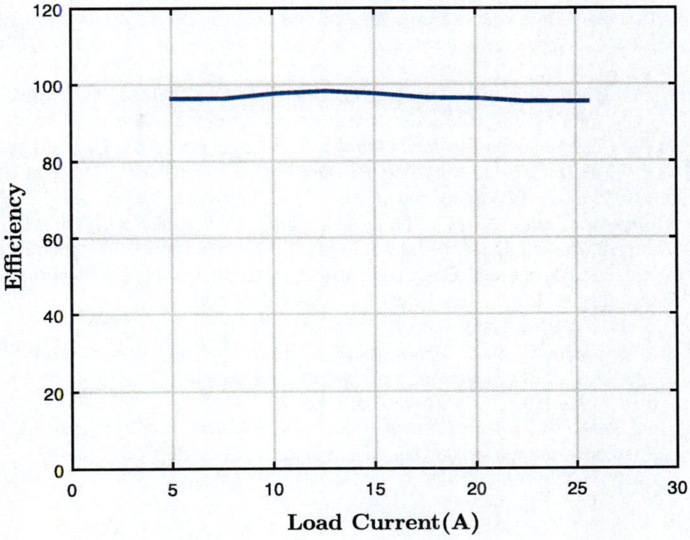

Fig. 11 Efficiency versus load current

is simulated using Simulink. The simulation results are verified by implementing the same on a GaN-based 300 W buck converter. The developed board is tested for different load, and it is observed that the GaN-based converter gives an efficiency in the range 95% to 98% which is much higher than conventional Si-based converters.

Acknowledgements The authors would like to thank VGST for funding the project under RGS/F scheme and the management of NMAM Institute of Technology, Nitte, Karkala, for providing a platform to carry out work in Research & Innovation Center. A special thanks to Dr. Niranjan N Chiplunkar, Principal, and Dr. Nagesh Prabhu, HoD, EEE, who helped throughout the journey. Authors would also like to thank Mr. Anup Shetty for validating the results.

References

1. R.L. Kini, A.J. Sellers, M.R. Hontz, M.R. Kabir, R. Khanna, Comparison of gan and si-based photovoltaic power conversion circuits using various maximum power point tracking algorithms. IEEE Appl. Power Electron. Conf. Exposition (APEC) **2017**, 2977–2982 (2017)
2. A.M.S. Al-bayati, S.S. Alharbi, S.S. Alharbi, M. Matin, A comparative design and performance study of a non-isolated dc-dc buck converter based on si-mosfet/si-diode, sic-jfet/sic-schottky diode, and gan-transistor/sic-schottky diode power devices. North Am. Power Symp. (NAPS) **2017**, 1–6 (2017)
3. M. Lenzhofer and A. Frank, Efficiency and near-field emission comparisons of a si- and gan based buck converter topology," in *2018 IEEE 18th International Power Electronics and Motion Control Conference (PEMC)*, 2018, pp. 818–823
4. F. Gamand, M.D. Li, C. Gaquiere, A 10-mhz gan hemt dc/dc boost converter for power amplifier applications. IEEE Trans. Circ. Syst. II Express Briefs **59**(11), 776–779 (2012)
5. Alex Lidow, Johan Strydom, *White Paper: eGaN FET drivers and layout considerations* (Tech. Rep, EPC, 2016)
6. D. Kim, D. Joo, B.Lee, and J. Kim, "Design and analysis of gan fet-based resonant dc-dc converter," in *2015 9th International Conference on Power Electronics and ECCE Asia (ICPE-ECCE Asia)*, 2015, pp. 2650–2655
7. H. Chong, H. Chen, K. Sun, J. Lin, S. Mu, and Y. Zhou, Efficiency evaluation of si-based and gan-based interleaved buck/boost converters for energy storage systems, in *2019 22nd International Conference on Electrical Machines and Systems (ICEMS)*, 2019, pp. 1–5
8. D.W. Robert Erickson, *Fundamentals of Power Electronics*, 2nd edn. (Springer, U.S., 2001)
9. H. Xu, Y. Liu, Z. Wang, Modeling and control of peak current mode non-ideal buck converter, in *IOP Conference Series: Earth and Environmental Science*, vol. 300, p. 042119, 08 2019
10. B. Dan Mitchell, Designing stable control loops, Texas Instruments, Technical Report (2017)
11. A.M. Rahimi, P. Parto, P. Asadi, Compensator design procedure for buck converter with voltage-mode error-amplifier, International Rectifier, Technical Report
12. M.F.B.U. Baki, Modelling and control of dc to dc converter (buck) (2008)
13. S. Ghosh, S. Satpathy, S. Das, S. Debbarma, B.K. Bhattacharyya, Different controlling method of closed loop dc-dc buck converter: a review, in *International Conference on Smart Systems and Inventive Technology (ICSSIT)* vol. 2018, pp. 29–33 (2018)
14. L. Umanand, *Power Electronics Essentials and Applications*, 1st edn. (Wiley, New Delhi, 2009)
15. Freescale Semiconductor, *Mc56f847xx technical data* (Tech. Rep, NXP, 2014)

Modeling, Simulation and Analysis of Static Synchronous Compensator Using OpenModelica

K. Navaneeth, Harshita M. Bharadwaj, Aakash, R. Shreya, and Apoorva Gopal

1 Introduction

Enhancing the power transfer capability of long transmission lines is effectively carried out by decreasing the effective line reactance and by providing dynamic voltage control with the help of a static synchronous compensator (STATCOM). In this paper, an STATCOM placed at the electrical center of the transmission line is chosen for analysis with a 12-pulse, two-level VSC with a type 2 reactive current controller. The modeling and analysis are done with the help of OpenModelica.

OpenModelica (https://www.openmodelica.org/) is an open-source Modelica-based modeling and simulation environment intended for both industrial and academic use. Modelica is a declarative, object-oriented language with general class concept which supports multi-domain modeling. The OpenModelica Connection Editor (OMEdit), an advanced open-source user-friendly graphical user interface is used to help develop the model easily.

The paper is organized as follows. Section 2 describes the modeling of STATCOM. The implementation is discussed in Sect. 3. Section 4 presents the evaluation of the performance of OpenModelica implemented STATCOM model. The major conclusions of the paper are presented in Sect. 5.

K. Navaneeth (✉) · H. M. Bharadwaj · Aakash · R. Shreya · A. Gopal
Department of Electrical and Electronics Engineering, N.M.A.M Institute of Technology, Nitte, India

Sanjeevikumar et al. (eds.), *Advances in Renewable Energy and Electric Vehicles*, Lecture Notes in Electrical Engineering 767,
https://doi.org/10.1007/978-981-16-1642-6_25

2 Modeling of Twelve Pulse, Two-Level VSC-Based STSTCOM

The power circuit of a STATCOM comprises multi-pulse and/or multilevel converter configurations.

STATCOM with 12-pulse, two-level configuration is used for analysis in this paper.

The complete details of the modeling of two-level VSC are discussed in [1], and we shall not discuss them here.

2.1 Mathematical Model of STATCOM in D-Q Frame of [1–2]

The switching functions are estimated by their fundamental frequency components with their harmonics being neglected, and by the transformation of the voltages and currents to D-Q variables with the help of Kron's transformation, a STATCOM can be modeled. The STATCOM can be represented as shown in Fig. 1.

The differential equations of the STATCOM, in the D-Q frame of reference, are as follows:

$$\frac{dI_{sD}}{dt} = -\frac{R_s\omega_B}{X_s}I_{sD} - \omega_0 I_{sQ} + \frac{\omega_B}{X_s}\left[V_{sD} - V_{sD}^i\right] \tag{1}$$

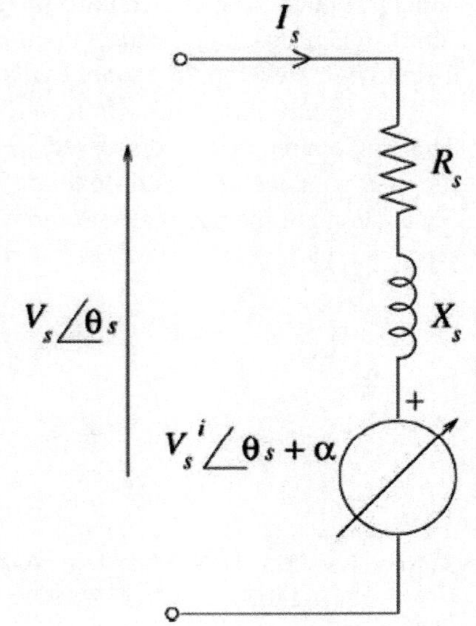

Fig. 1 Functional diagram of STATCOM shunt FACTS controller

$$\frac{dI_{sQ}}{dt} = \omega_0 I_{sD} - \frac{R_s\omega_B}{X_s} I_{sQ} - + \frac{\omega_B}{X_s}\left[V_{sQ} - V_{sQ}^i\right] \tag{2}$$

$$\frac{dV_{dc}}{dt} = -\frac{\omega_B}{b_c} i_{dc} - \frac{\omega_B}{b_c R_p} V_{dc} \tag{3}$$

where

$$i_{dc} = -\left[k\ \sin(\theta_s + \alpha) I_{sD} + k\ \cos(\theta_s + \alpha) I_{sQ}\right] \tag{4}$$

The components of STATCOM current in the D-Q frame of reference are I_{sD}, I_{sQ}.

k is known as the modulation index, and for a 12-pulse converter, it is given by $k = \frac{2\sqrt{6}}{\pi}$.

In the D-Q frame of reference, the converter output voltage V_s^i can be represented as

$$V_{sD}^i = V_s^i \sin(\theta_s + \alpha) \tag{5}$$

$$V_{sQ}^i = V_s^i \cos(\theta_s + \alpha) \tag{6}$$

The magnitude of V_s^i, $\left|V_s^i\right| = kV_{dc}$

The phase angle of V_s^i is $\theta_s + \alpha$

The phase angle of the bus voltage is θ_s.

α is the angle between the converter voltage and the STATCOM bus voltage V_s, wherein the converter voltage leads V_s.

The real and reactive currents are defined as

$$I_P = I_{sD} \sin(\theta_s) + I_{sQ}\cos(\theta_s) \tag{7}$$

$$I_R = -I_{sD} \cos(\theta_s) + I_{sQ}\sin(\theta_s) \tag{8}$$

Here the following conventions are used regarding the reactive current. Negative and positive I_R indicate the STATCOM is operating in the inductive region and capacitive region, respectively [2].

2.2 *STATCOM Current Control (Two-Level) [2]*

By changing the value of α alone, the control of reactive current can be established in a two-level VSC. The modulation index k is kept constant.

3 Implementation in OpenModelica

The STATCOM, along with the reactive current controller, has been modeled using basic Modelica blocks available preloaded in OMEdit.

Figure 2 shows the high-level overview of the STATCOM model along with reactive current controller.

The STATCOM block is designed using the differential equations associated with it.

As Modelica is a declarative language, the equations are entered as such, with the specified keywords for the respective operations. This helps in simplifying the development.

As the quantities in the differential equations are real numbers, the variables in the equations are declared as such using a keyword "real". The quantities that do not vary during the operation of the STATCOM are treated as constants by declaring them as such using the additional keyword "parameter".

The quantities which are treated as inputs or outputs, to or from the STATCOM blocks are realized using the classes "Modelica.Blocks.Interfaces.RealInput" and "Modelica.Blocks.Interfaces.RealOutput", respectively.

The equations listed below show the equations declared in OpenModelica which define the STATCOM model.

The reactive current controller is shown in Fig. 3. The reactive current controller contains a proportional–integral controller and a nonlinear feedback. It is found that in the frequency response of the system, conventional PI controller used as reactive current control causes instability in the inductive region as there is little phase margin near the system resonant frequency in the inductive region [1]. Therefore, a nonlinear feedback is used to improve the damping in the inductive region.

The controller is developed with the help of graphical connection editor tab of OMEdit. It is shown in Fig. 4.

The block named "control" is defined such that it passes values that are only greater than or equal zero.

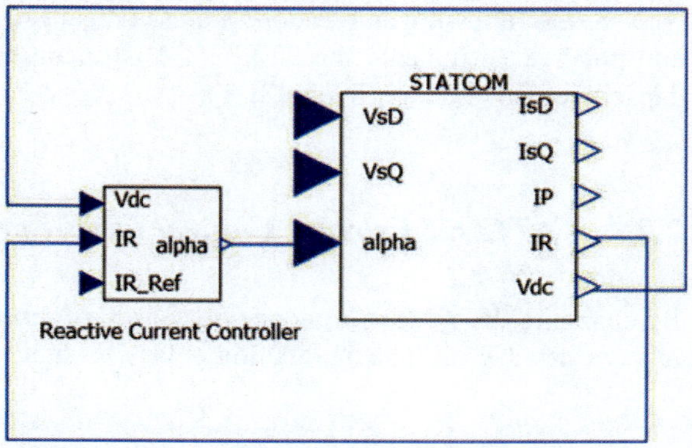

Fig. 2 OpenModelica diagram representing STATCOM and reactive current controller

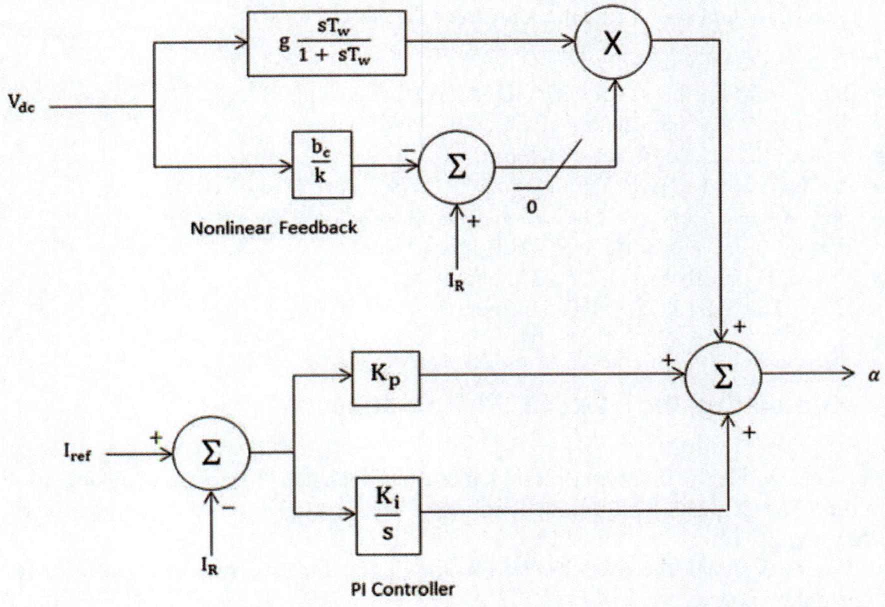

Fig. 3 Reactive current controller

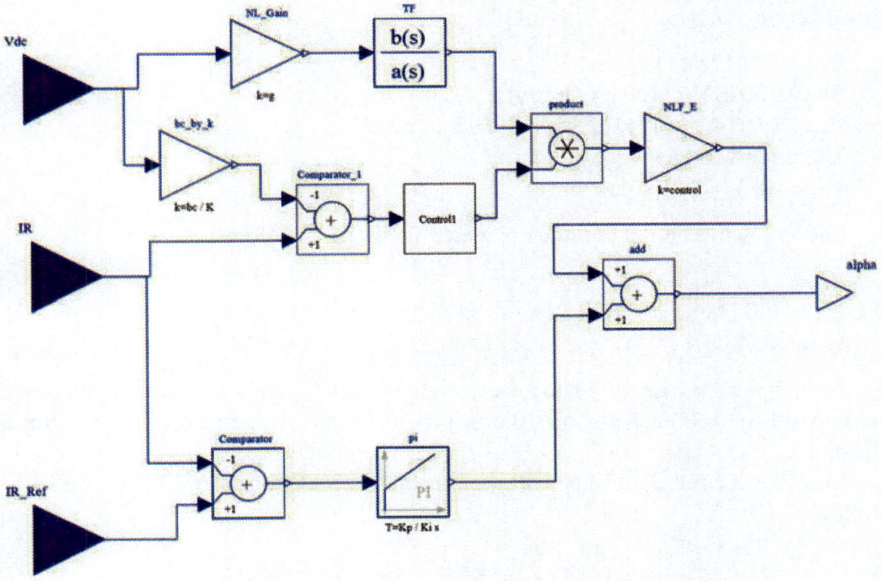

Fig. 4 OpenModelica diagram representing the reactive current controller

The Modelica code implementation for the STATCOM:

- thetas = atan(VsD/VsQ);
- Idc = - (k*sin(thetas+alpha)*IsD+k*cos(thetas+alpha)*IsQ);
- VisD = k*Vdc*sin(thetas+alpha);
- VisQ = k*Vdc*cos(thetas+Alpha);
- der(IsD) = (-Rs*(wB/Xs)*IsD)-w0*IsQ+(wB/Xs)*(VsD-VisD);
- der(IsQ) = (-Rs*(wB/Xs)*IsQ)+w0*IsD+(wB/Xs)*(VsQ-VisQ);
- der(Vdc) = (-wB/bc)*Idc - (wB/Rp*bc))*Vdc;
- IP = IsD*sin(thetas)+IsQ*cos(thetas);
- IR = -IsD*cos(thetas)+IsQ*sin(thetas);

4 Simulating the STATCOM Operation

The OpenModelica diagram descried above has been used to simulate the operation of the STATCOM under different conditions. The simulation was done using discrete time step (20 μs).

Figure 5 shows the model used for simulation for step response and also for constant response.

The input voltages to the STATCOM model are in the D-Q frame of reference, and all electrical quantities are expressed in the per unit system, with the STACOM itself serving as base.

STATCOM Data [1]
Base: 300 MVA, 400 kV, 12 pulse
$R_s = 0.01$, $X_s = 0.15$, $b_c = 1.136$, $R_p = 78.7$
The input voltages were
$V_{sD} = 0$, $V_{sQ} = 1$ (p.u)

The following model parameters were initialized as follows:

$I_{sD} = -1$
$I_{sQ} = 0.0138$
$V_{dc} = 0.7$

The response of the model to a step change in the I_{ref} without nonlinear feedback is shown Fig. 6. Figure 7 shows the response to step change in I_{ref} with nonlinear feedback.

Reactive current controller with nonlinear feedback (on STATCOM base) is as follows: [1]

$$k_p = 0.33,\ k_i = 3.33,\ g = 2.0,\ T_w = 0.01$$

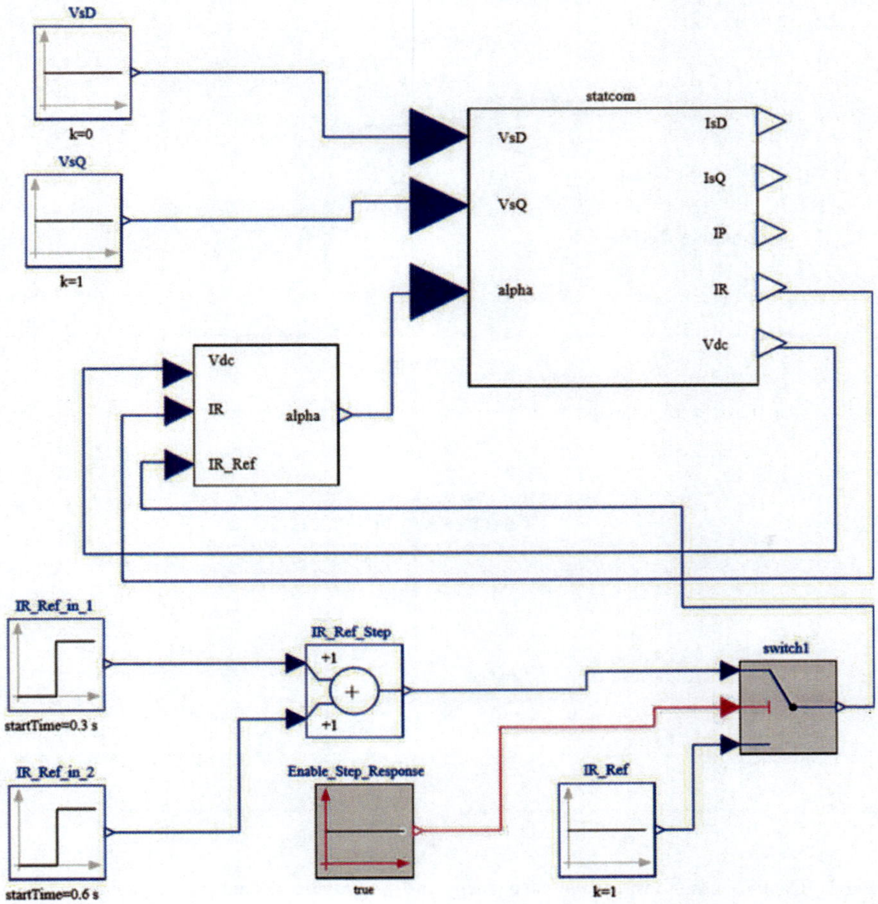

Fig. 5 Model used for simulation which includes options for both. **a** step response. **b** constant response

4.1 Eigenvalue Analysis

In this analysis, the STATCOM model along with the reactive current controller is linearized at their operating point, and the eigenvalues of the system matrix are computed. The stability of the system is determined by the location of the eigenvalues of system matrix. The system is stable if the eigenvalues have negative real parts [2].

Table 1 contains the eigenvalues of the system matrix that was obtained from the linearized model of the STATCOM without nonlinear feedback, while Table 2 contains the eigenvalues of the system matrix that was obtained from the linearized model of the STATCOM with nonlinear feedback.

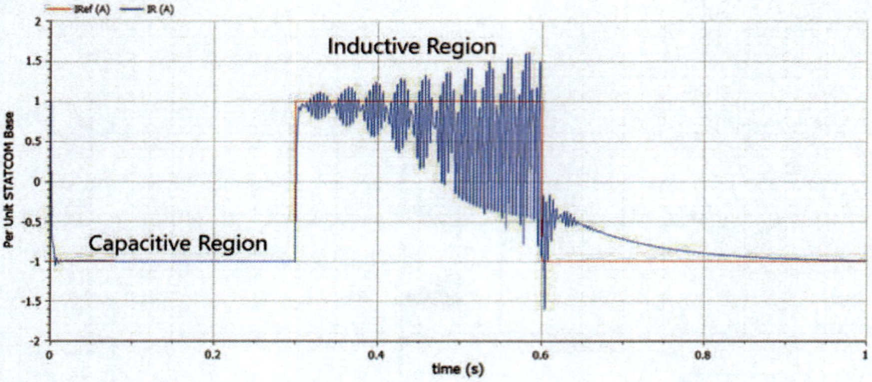

Fig. 6 Instability in the inductive region

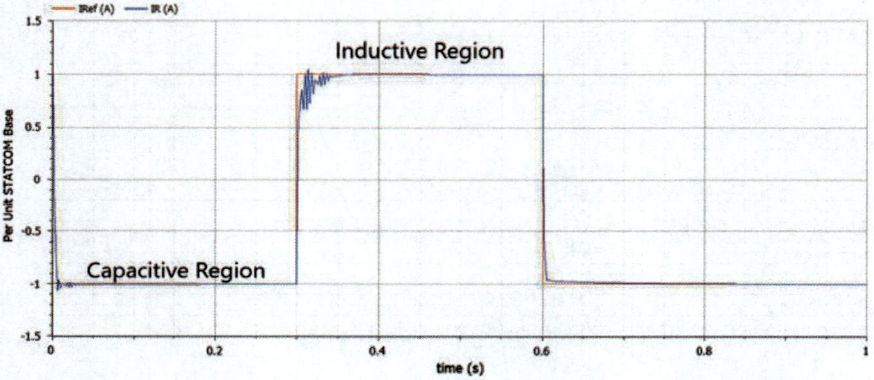

Fig.7 Reactive current response for step change in I_{ref} in nonlinear feedback controller

Table 1 Eigenvalues of the system matrix

Eigenvalues without nonlinear feedback	
Capacitive region $I_R = -1$	Inductive region $I_R = 1$
−100	−100
−9.91373	−9.91374
−834.578	−779.207
−81.8162 ± j1429.82	4.70569 ± j1482.24

5 Conclusion

An attempt at developing a STATCOM model with reactive current control using the OpenModelica platform has been presented.

Table 2 Eigenvalues of the system matrix

Eigenvalues with nonlinear feedback	
Capacitive region $I_R = -1$	Inductive region $I_R = 1$
−100	−1210.96
−9.91373	−96.3886 ± j1395.41
−834.578	−69.3418
−81.8162 ± j1429.82	−10.3303

The model developed was tested as a stand-alone system, with inputs given with the help of external sources.

The STATCOM equations are nonlinear; a linearized model is used for analysis.

It is observed that without the use of a nonlinear feedback, the system is highly unstable in the inductive region of operation when the model was simulated. It is also verified with the help of eigenvalue analysis.

The obtained simulation results have demonstrated the validity of the developed model.

Acknowledgements We thank Dr. Nagesh Prabhu, Professor and HOD, and Mr. Dinesh Shetty, Assistant Professor, Department of Electrical and Electronics Engineering, N.M.A.M Institute of Technology, Nitte, for their academic assistance.

References

1. K.R. Padiyar, A.M. Kulkarni, Design of reactive current and voltage controller of static condenser. Int. J. Electr. Power Eng. Syst. **19**(6), 397–410 (1997). ISSN 0142–0615, https://doi.org/10.1016/S0142-0615(97)00010-0
2. K.R. Padiyar, N. Prabhu, Design and performance evaluation of subsynchronous damping controller with STATCOM. IEEE Trans. Power Delivery **21**(3), 1398–1405 (2006). https://doi.org/10.1109/TPWRD.2005.861332

Modeling and Real-Time Simulation of Photovoltaic Plant Using Typhoon HIL

Minal Salunke and Diksha Tiwari

1 Introduction

Renewable sources of energy have great importance in advancing technologies and power applications. Clean power through renewable sources of energy is of great demand in industries. These renewable sources of power can be in the form of wind, turbines, solar modules, fuel cells, and many more that are handed down as distributed energy sources in the implementation of micro-grids and smart-grids. In the mentioned alternate renewable sources of energy, photovoltaic is the most efficient and widely used renewable energy source. The photovoltaic source of power is the cheapest source of energy where various photovoltaic panels are combined as an array to supply maximum electrical power. The electrical power generated is pollution free as there is no greenhouse gas emission during its conversion [1].

Due to the extensive applications of photovoltaic plants and to explore the benefits of these, separate experiments are under the research process to increase the overall efficiency, accuracy, cost, integration, and performance. Over many years, PV plant design and simulation are studied using various control strategies to know the existing power system. For the best conduction of the photovoltaic plant, several methods drifted to achieve maximum power. In this paper, a detailed design and modeling of a photovoltaic plant are considered that produces electrical capabilities in a single phase. The entire idea of photovoltaic plants in a single-phases system is realized using the MPPT algorithm. There are several traditional mathematical models to know the existing power; however due to their limited controlled strategies, it is challenging to get the most accurate results.

Hardware in the loop (HIL) system is one of the proceeding methods that enable real-time emulation of a power converter or small electrical applications in software

M. Salunke (✉) · D. Tiwari
Department of Electrical and Electronics, KLE Technological University, Hubballi, India
e-mail: minal@kletech.ac.in

Sanjeevikumar et al. (eds.), *Advances in Renewable Energy and Electric Vehicles*,
Lecture Notes in Electrical Engineering 767,
https://doi.org/10.1007/978-981-16-1642-6_26

but can easily interface with real-world hardware [2]. It is an aim for the designers to make the embedded system work suitably through HIL systems and help by creating a virtual environment to carry out simulations. It is perhaps the most crucial aspect as a system can be modeled and designed on software and can interface with real-time laboratories. In this paper, a CHIL simulation demonstrates to generate electric power. The simulation is performed through Typhoon HIL (version 402) [3].

2 HIL Testing for Photovoltaic Plant

HIL test methodology is one of the most powerful tools used for testing embedded software. In this test, one part of the system gets replaced with a model simulated in real time but the other remains as real. HIL is an approach in which the most common is CHIL that is a controlled hardware in the loop system that uses a real-time controller. Supervisory control and data acquisition (SCADA) system uses real-time inputs and outputs in HIL simulations.

HIL simulation includes the testing, verification, and development of photovoltaic plants in both real-time hardware and software embedded systems.

2.1 Block Diagram

To simulate, the photovoltaic plant representation is discussed below. SCADA inputs and outputs are to be real-time inputs and outputs and hence do not contribute to simulation. These SCADA inputs and outputs connect through various protocols that can be either Modbus or USB interface or Ethernet. SCADA interfaces with a photovoltaic plant that serves as a distributed generator (DG) as an application of micro-grid and smart grid. With this, the complicated system also provides practical results when stimulated with the help of CHIL [4]. In Fig. 1, general block diagram

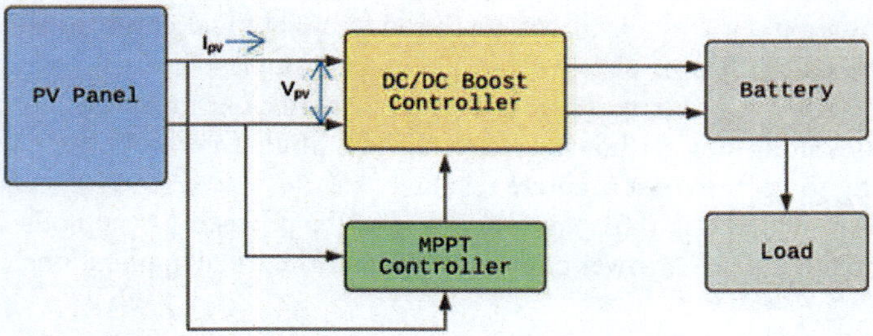

Fig. 1 Block diagram of HIL simulation of photovoltaic plant

of the photovoltaic plant is represented. The block diagram is composed of a photovoltaic panel and various controllers further composed of switches, flip-flops, and passive components.

3 PV Plant Model

The PV plant is designed and modeled to produce electrical power in a single phase. The model is represented below.

3.1 PV Panel (Photovoltaic Panel)

The photovoltaic panel is modeled as voltage-controlled current source IPV with module capacitance CPV connected in parallel. The current source controls through voltage VPV across the PV panel. HIL simulations represent a single photovoltaic panel that consists of arrays of PV cells modeled as shown in Fig. 2.

Once the modeling for the PV panel completes, the maximum power point tracking (MPPT) algorithm extracts the maximum power. In the MPPT algorithm, the variable factors are irradiation and temperature, always take care to produce the maximal power. In MPPT, the curve is plotted between voltage and current to find the maximum capabilities of the PV plant. Over many years, several MPPT algorithms are widely adopted and developed. Perturb and observe technique (P&O) is used here for tracking the maximum power. The P&O technique is an iterative algorithm to track the MPPT continuously through voltage and current measurement of the PV module. It is easy to implement because the oscillation problem is unavoidable when compared to other MPPT algorithms. Figure 3 represents the perturb and observe technique flowchart.

Following the steps involved in the P&O algorithm for tracking the MPPT:

Step 1: Start.
Step 2: Measurement of PV panel voltage and current.
Step 3: Measurement of power through the multiplication of PV panel voltage and current.

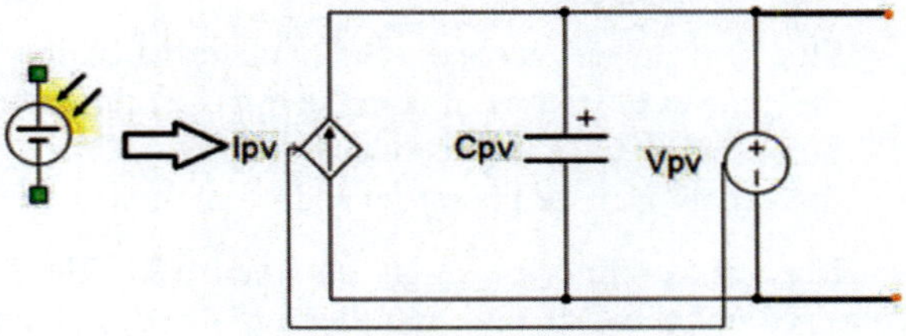

Fig. 2 Modeling of PV panel

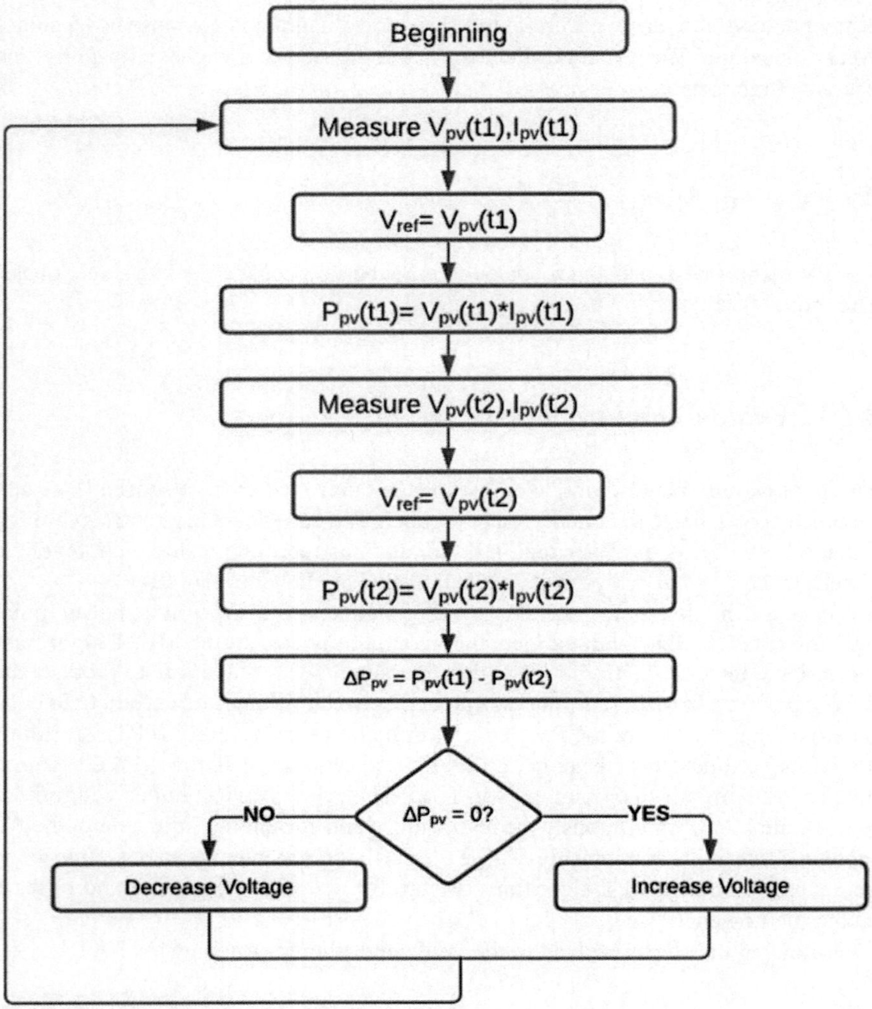

Fig. 3 Flowchart of P&O technique

Step 4: Panel voltage is perturbed (increased or decreased), and power is compared with the disturbance.
Step 5: If power increases due to disturbance, the effect is in the same direction, or if power decreases, it is in the opposite direction.
Step 6: With the above perturbation process, the system operates closer to MPPT for variations in temperature and irradiation.

Maximum output power is obtained using the MPPT algorithm (perturb and observe technique) in the controller part. In the HIL simulation [4], the P&O method implements the C function block used for comparison purposes. Comparison by the industry set datasheets and further perturbation (increment or decrement) is considered.

3.2 Boost Converter

The boost converter is the DC–DC converter used for boosting the output power of the PV plant. In HIL simulation, boost controller is used for boosting the power according to the load requirements. The load considered here is a complete resistive load. The boost controller can be controlled either by digital inputs, internal modulators, or through modeling [5]. Here internal modulator is considered for the operation.

3.3 Battery

Boost converters boost the photovoltaic plant output power such that output power handles effectively. Different types of batteries are available but according to the requirement here, a lead-acid battery is considered. For charging and discharging of a battery, comparator is used with SR flip-flop to enable the inputs accordingly. SR flip-flop sets only when one of the outputs from both the comparator is high. Analysis of working of SR flip-flop is detailed in Table 1. Battery nominal voltage considered here is 48 V such that maximum and minimum constant values for comparison are considered to be 49 and 52 V. The battery specifications in the HIL simulation of the PV plant can be listed in Tables 1 and 2.

Table 1 Truth table of SR flip-flop

S	R	Q_{N+1}	Q_{N+1}'	State
0	0	Q_N	Q_N'	Hold
0	1	0	1	Reset
1	0	1	0	Set
1	1	0	0	Invalid

Table 2 Battery specification

Parameters	Values
Nominal voltage	48 V
Capacity	10 AH
Initial SOC	100%

4 Simulation Result

HIL simulation of PV plant modeling is performed using the MPPT algorithm, boost controller, battery, and passive components. Each block has varying operations though performance counts as each of them has a crucial contribution. In Fig. 4, a complete simulation of the model is represented.

In the real-time simulation of the photovoltaic plant, irradiation is a variable parameter, and the temperature is constant. The photovoltaic plant model uses perturb and observe technique to track the MPPT in the system. Output across the PV panel checks the maximum power output obtained through the PV panel. The output power generated through the PV plant is applied to boost the converter to achieve boosted output voltage. On increment of output power, it is utilized either by a battery or the load. Contactors are for feeding the output to a battery or load according to the requirements. SCADA panel utilization is to verify the results of the photovoltaic plant. This system enables real-time emulation for a given embedded system [3]. It works basically on Python programming language and can be interfaced through real-time inputs and outputs using the USB interface. For a given set of data values, a virtual test bed is tested, verified, and then applied to a real-time hardware system. Results are analyzed using these simulations.

Simulation results are depicted in Figs. 5, 6, 7, 8, and 9. Output power and efficiency are verified using load contactor and discharge contactor. The contactors are varied accordingly at the same temperature and irradiance.

In Fig. 5, output power of the photovoltaic plant is attained after the introduction of the MPPT algorithm. Irradiation and temperature changed with the help of irradiation and temperature knob. Keeping temperature constant and varying the irradiation output power is known, when irradiation is 1000 W/m^2 and temperature of 25 °C,

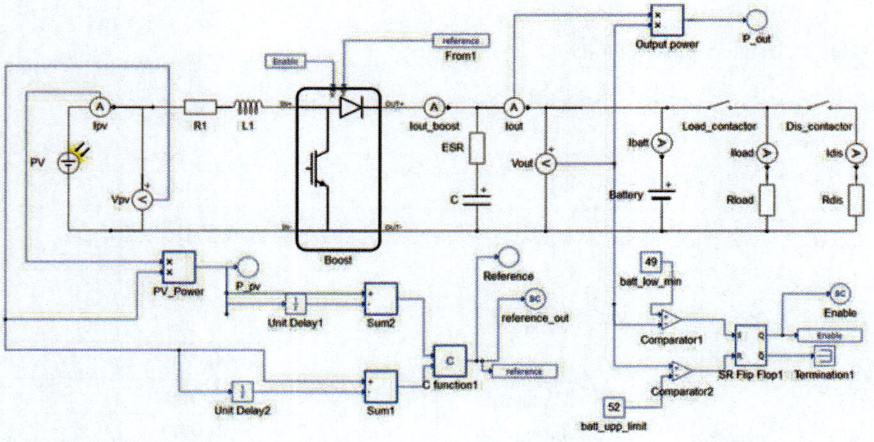

Fig. 4 HIL simulation of photovoltaic plant

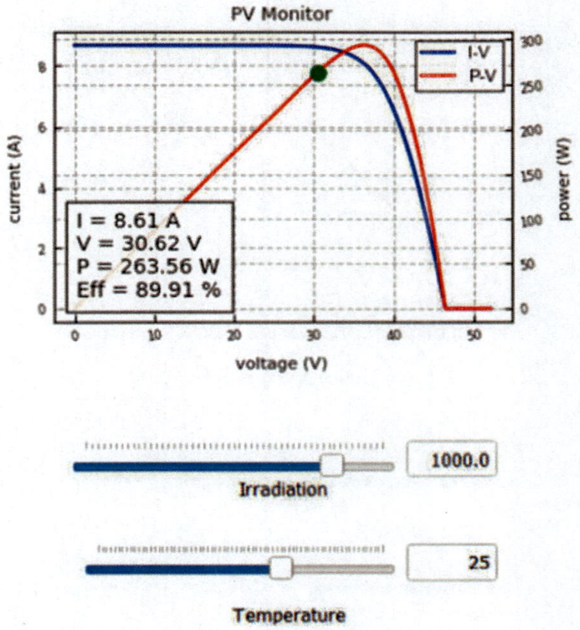

Fig. 5 Output power of PV plant with both contactors ON

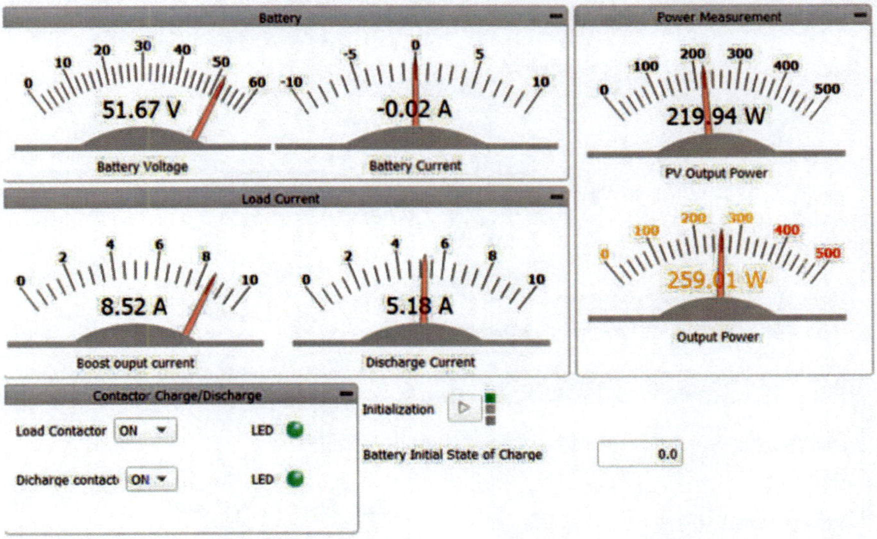

Fig. 6 SCADA panels representing different outputs with both contactors ON

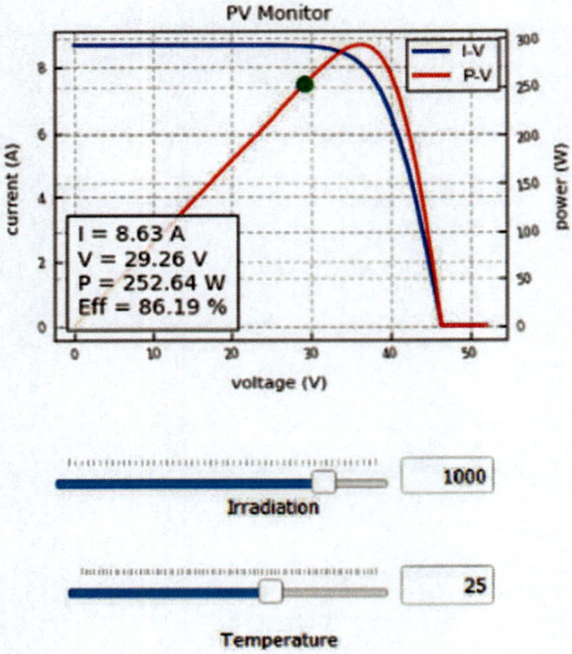

Fig. 7 Output power of PV plant with one contactor kept OFF

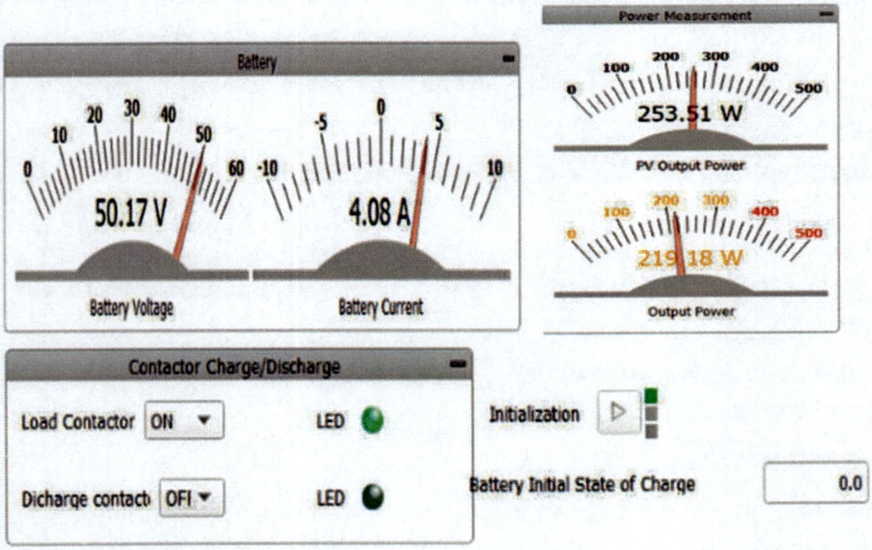

Fig. 8 SCADA panels representing different outputs with one contactor kept OFF

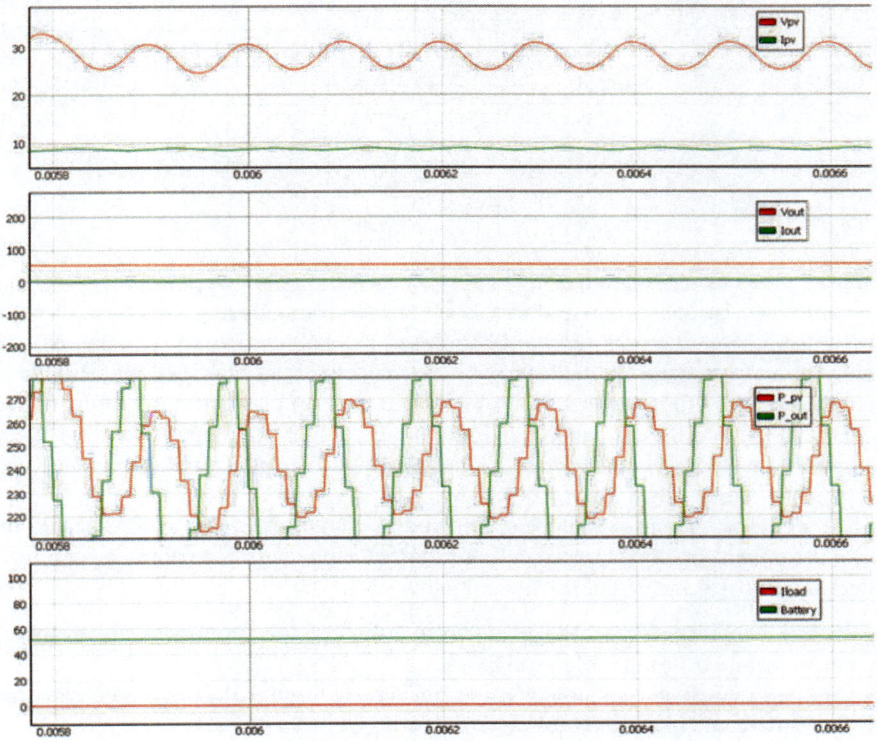

Fig.9 Output waveforms

output power attained is 263.56 W across the PV plant. The efficiency of the PV panel turns out to be 89.91%.

In Fig. 6, various groups and panels are created for analyzing the power through boost converter, charging, and discharging of the battery, battery voltage, battery current, and discharge current. Here, both load contactor and discharge contactor are in ON state that draws negative current in the battery such that it gets discharged. Load utilizes the output power in this case. Three different panels present output voltages and current. In the battery panel, battery voltage and current are distinguished. The battery current is negative here. In load, panel boost output and discharge current are distinguished that are positive here because both contactors are in ON state. The power measurement panel represents output power that is 259.01 W.

In Fig. 7, output power of the photovoltaic plant is attained after the introduction of the MPPT algorithm. Keeping temperature constant and varying the irradiation output power is known, when irradiation is 1000 W/m^2 and temperature of 25 °C, output power attained is 252.64 W across the PV plant. The efficiency of the PV panel turns out to be 86.19%.

Table 3 Interpreted PV power and output power

Irradiation (W/M^2)	Output power (W)	PV power (W)	Battery voltage (V)
500	227.03	201.18	51.72
700	241.58	207.63	51.73
900	254.04	218.03	51.65
1000	259.01	219.94	51.67
1200	289.01	254.46	51.73

In Fig. 8, various groups and panels are created for analyzing the power through boost converter, charging, and discharging of the battery, battery voltage, battery current, and discharge current. Here, load contactor is in ON state and discharge contactor is in OFF state that draws positive current in the battery such that it gets charged. Three different panels present output voltages and current. In the battery panel, battery voltage and current are distinguished. The power measurement panel represents output power that is 219.18 W.

In Fig. 9, different waveforms obtained as output are represented. There are four different scopes each representing different parameters and captured for a time duration of 0.1 s and sampling rate around 1 MPS.

- Scope 1 represents the voltage and current through the photovoltaic plant that is positive and is around 32 V and 8.2 A.
- Scope 2 represents the output current and voltage through the boost converter that is positive and is around 50 V and 0.11 A.
- Scope 3 represents output power through the boost converter and photovoltaic power that is around 259.01 and 279.20 W.
- Scope 4 represents the load current and battery voltage flowing through pure resistive loads that are around 0.1 A and 51.2 V.

In the above modeling, the simulations are performed for different values of irradiation and temperature accordingly. In Table 3, different output values are analyzed on a variation of irradiance and keeping the temperature constant around 25 °C.

From the above table, the PV power and output power are interpreted. The output attained is through a real-time simulator that are SCADA inputs and outputs, and it varies on changing the irradiance keeping the temperature constant around 25 °C.

5 Conclusion

The photovoltaic plant design and modeling are discussed in the proposed paper. The photovoltaic plant considered is composed of a controller, passive components, a boost controller, an MPPT algorithm, flip-flops, and switches. Real-time simulation results are carried out in different scenarios. The photovoltaic plant simulated using HIL simulation is flexible, accurate, and efficient from an industrial point of view. The advantage of using a HIL in the development is a substantial decrease in development time and cost of the controls. Once the controllers have been thoroughly tested and verified with the HIL, it can prefer over traditional methods due to their accuracy. External controllers and SCADA are added on high flexibility and ductility in the system.

References

1. M.S. Thomas, A. Nisar, Data-driven modeling and simulation of PV array, in *International Conference on Computing for Sustainable Global Development (INDIACom)* (2015)
2. A. Genić, P. Gartner, D. Medjo, M. Dinić, Multi-layer hardware-in-the-loop testbed for microgrids, in *International Conference on Smart System and Technologies (SST)* (2016)
3. Typhoon HIL, Inc., *Typhoon HIL Software Manual* (Online) (2016)
4. Typhoon HIL, Inc., *Typhoon Schematic Library User Guide* (Online) (2016)
5. M. Salunke, J. Patel, H Praveenkumar, K. Bellad, S.T. Umarani, Design and simulation of boost derived hybrid converter for nano grid applications. Asian J. Convergence in Technol. **3**(2) (2017)

Performance Evaluation of Knowledge-Based Reactive Current Controllers for STATCOM

Dinesh Shetty and Nagesh Prabhu

1 Introduction

The voltage source converter (VSC)-based flexible AC transmission system (FACTS) devices are employed in transmission system for compensation of various parameters in the power system. Static synchronous compensator (STATCOM) is a shunt FACTs device generally deployed for reactive power compensation. The connection between the VSC and the system bus is linked through the coupling transformer [1–3]. To improve the capacity of the transmission system for transferring the power, the optimal place is to include STATCOM somewhere middle of long transmission line to control the voltage. The compactness, the reliable reactive current injection capacity and ability to supply active power to grid in the presence of renewable energy sources are the important features of STATCOM over static VAR compensator (SVC). The STATCOM can be more efficiently used in the presence of highly reliable reactive current controller.

In this paper, two-flevel, twelve-pulse type 2 controller is used for reactive current controller for STATCOM [1–4]. PI controller based reactive current control causes the oscillations, and hence the instability occurs in the inductive region of operation as described in [4] which introduces unacceptable oscillations and instability. Therefore to ensure the stability during inductive mode of operation of STATCOM, the nonlinear feedback-based controller is proposed by Schauer and Mehta [2]. The performance enhancement of nonlinear feedback controllers has been proposed and

D. Shetty (✉) · N. Prabhu
Department of Electrical and Electronics Engineering, NMAM Institute of Technology, Nitte, India
e-mail: dinesh1383@nitte.edu.in

N. Prabhu
e-mail: prabhunagesh@nitte.edu.in

Sanjeevikumar et al. (eds.), *Advances in Renewable Energy and Electric Vehicles*, Lecture Notes in Electrical Engineering 767,
https://doi.org/10.1007/978-981-16-1642-6_27

implemented by Janaki et al. [5], wherein the controller parameters are tuned using genetic algorithm techniques to ensure the fast transient response.

The objective of this paper is to compare the performances of a type 1 Mamdani fuzzy logic controller and ANFIS fuzzy controllers. This paper is organized as follows. Section 2 describes the modeling of STATCOM. The design of different reactive current controller is explained in Sect. 3. The type 1 fuzzy logic controller is described in Sect. 4, ANFIS is described in Sects. 5 and 6, and Sect. 7 describes the results and discussion and conclusions.

2 Mathematical Modeling of STATCOM in D-Q Frame of Reference

The transformation of AC voltage and current into synchronously rotating frame of reference is carried out using Kron's transformation. This transformation is power invariant. Therefore, the STATCOM equations are derived in D-Q frame of reference [6] provided that the switching functions are approximated by their fundamental frequency components neglecting the harmonics. Figure 1 show the STATCOM functional diagram. V_{C}^{m} is the output voltage of the converter, and in type 2 STATCOM converter output voltage is dependent on DC Vdc [2, 4]. V_{s} is the voltage at STATCOM bus. The converter output voltage is given by.

$$V_{\mathrm{C}}^{m} = \sqrt{V_{\mathrm{CD}}^{m^2} + V_{\mathrm{CQ}}^{m^2}} \tag{1}$$

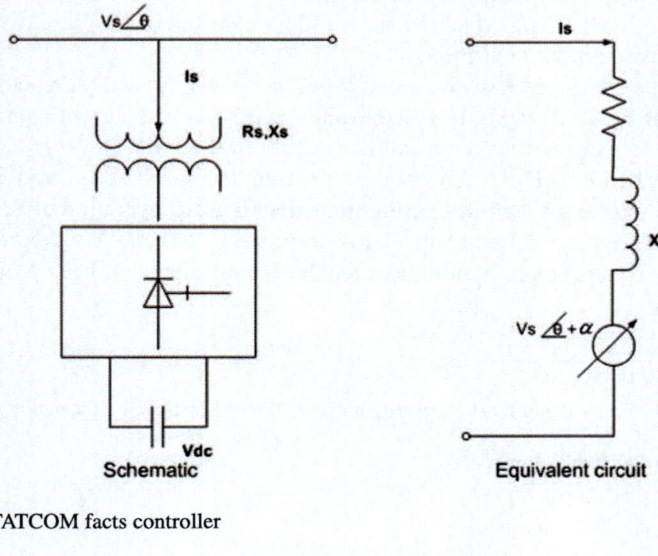

Fig. 1 STATCOM facts controller

$$V_{CD}^{m} = kV_{dc}\sin(\theta_s + \alpha), \tag{2}$$

$$V_{CQ}^{m} = kV_{dc}\cos(\theta_s + \alpha) \tag{3}$$

where θ_s is the phase angle of bus voltage, and α is the relative angle supply and injected voltage to the bus under consideration. The modulation index k for 12-pulse converter is $k = \frac{2\sqrt{6}}{\pi}$.

The following equations in the D-Q variables can be given for describing STATCOM

$$\frac{dI_{sD}}{dt} = -\frac{R_s w_B}{X_s} i_{sD} - w_0 i_{sQ} + \frac{w_B}{X_s}\left[V_{sD} - V_{CD}^{m}\right] \tag{4}$$

$$\frac{dI_{sQ}}{dt} = -\frac{R_s w_B}{X_s} i_{sQ} + w_0 i_{sD} + \frac{w_B}{X_s}\left[V_{sQ} - V_{CQ}^{m}\right] \tag{5}$$

$$\frac{dV_{dc}}{dt} = \frac{w_B}{b_c} i_{dc} - \frac{w_B}{X_s b_c} V_{dc} \tag{6}$$

where

$$i_{dc} = -\left[k\sin(\theta_s + \alpha)i_{sD} + k\cos(\theta_s + \alpha)i_{sQ}\right] \tag{7}$$

i_{sD}, i_{sD} are the D-Q components of STATCOM current, $i_P = i_{sD}\sin(\theta_s) + i_{sQ}\cos(\theta_s)$,

$$i_R = -i_{sD}\cos(\theta_s) + i_{sQ}\sin(\theta_s), \tag{8}$$

where i_P and i_R are the STATCOM real and reactive current.

3 Reactive Current Control Design for Type 2 STATCOM

In the type 2 STATCOM, the reactive current control can be achieved only by controlling the firing angle of the converter The PI controller-based reactive current controller proposed in [3] shows that the converters achieve stable operation in capacitive mode but its oscillatory behavior leads to instability in inductive mode of operation as shown in Fig. 2. Hence to overcome this, Schauder and Mehta [3] implemented a nonlinear feedback controller as shown in Fig. 3. For type 2 STATCOM, the condition for stability is found to be $i_R < \frac{wC}{k}V_{dc}$.Therefore, the nonlinear controller will get active only when $i_R > \frac{wC}{k}V_{dc}$, and hence it ensures the stability in both capacitive and inductive modes. It is observed in [3] that the transient response

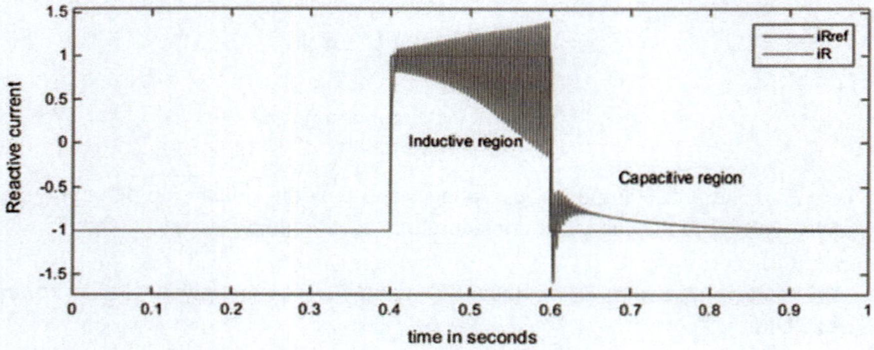

Fig. 2 Response to PI controller [3]

Fig. 3 Nonlinear feedback controller [3]

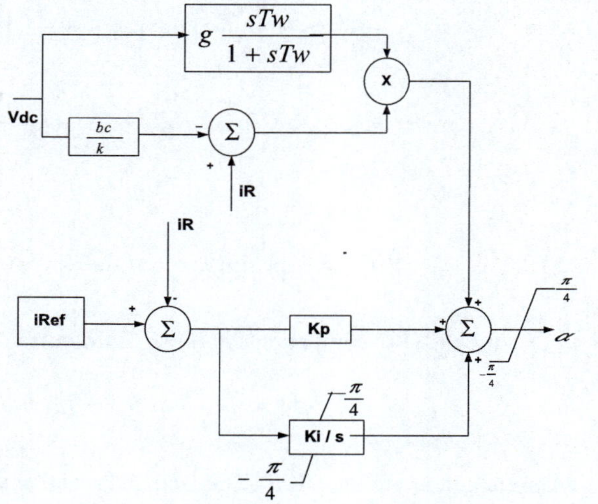

of the type 2 STATCOM with nonlinear feedback is slow and efforts are made to enhance the transient response of the system by optimizing the control parameters [5] to avoid the mathematical complexity in designing the nonlinear feedback. In this paper, we propose an alternative mathematical model-free approach for the design of controllers using knowledge-based controllers which ensures the stability of all modes with enhanced transient response.

4 Type 1 Mamdani Fuzzy Logic System

In recent days, the intelligent control techniques are preferred over traditional control techniques because of their modernized approach of addressing the problem in controlling of dynamic systems. Fuzzy logic is one of those control techniques which works efficiently by collecting complete information about the system and creates the knowledge-based control. Basic block diagram of fuzzy logic control is shown in Fig. 4. Fuzzifier forms the fuzzy sets for input and output variables which are crisp in nature. Fuzzy sets consist of degree of membership for each members contained in the fuzzy sets. A number of linguistic variables and assignment of membership functions are the key factors for the better performance of fuzzy systems [7]. The inference engine of the fuzzy controller rules build the relation between the input and output fuzzy sets using IF-THEN-ELSE rules. There Mamdani and Tagaki-Sugeno are the two types of rule base systems available for the fuzzy logic system. Defuzzification is the process of converting fuzzified variables to crisp. There are several defuzzification approaches available [7, 8] in literature. In the proposed work, centroid method of defuzzification is adapted for its degree of accuracy compared to other methods [9].

The proposed MAMDANI type 1 fuzzy-based system for type 2 STATCOM takes the inputs as error and change in error. The implementation of the model is done in MATLAB/Simulink platform. Figure 5 shows the MATLAB/Simulink model of type 2 STATCOM with reactive current controller. The fuzzy-based reactive current controller is implemented as shown in Fig. 6.

The input variables for the fuzzy systems are error and change error signals. The error signal is generated by comparing reactive current reference (iRef) with actual reactive current (iR). The linguistic variables chosen for nine rule system are positive, zero and negative. The rule base is constructed for 9, 25 and 49 Mamdani inference system as shown in Tables 1, 2 and 3, respectively.

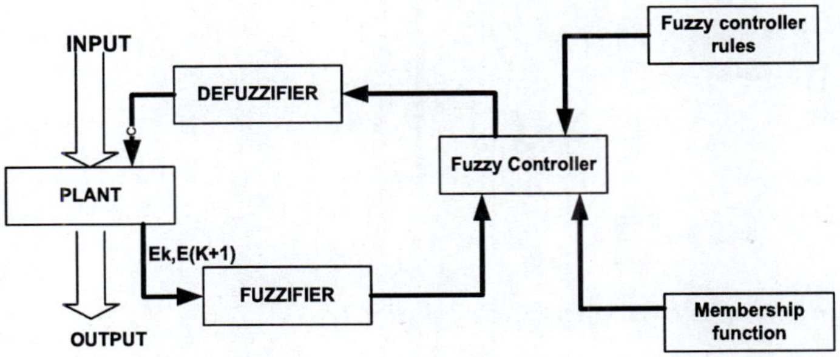

Fig. 4 Fuzzy logic controller

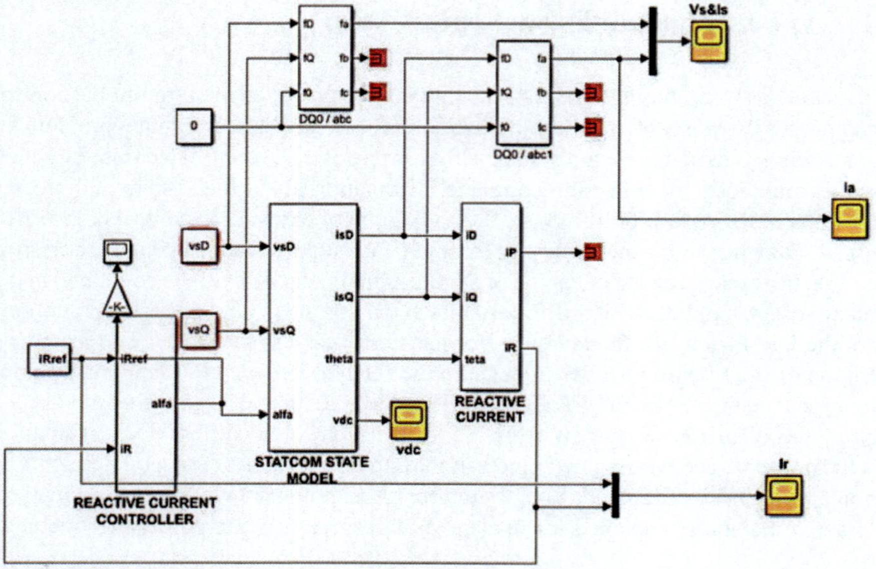

Fig. 5 MATLAB/Simulink implementation of type 2 STATCOM

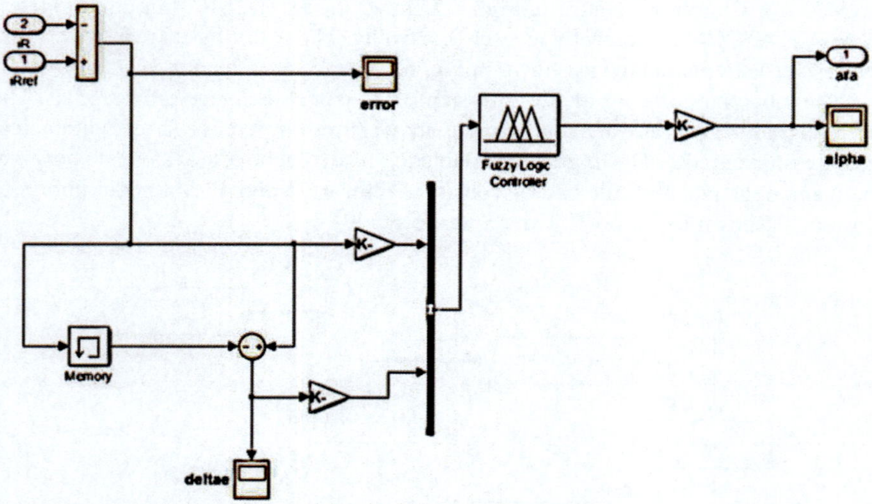

Fig. 6 Type 1 fuzzy-based reactive current controller

Table 1 Mamdani rule base for nine rules

Cee	N	Z	P
N	P	N	Z
Z	P	Z	N
P	Z	N	N

Table 2 Mamdani rule base for 25 rules

Ce					
e	NL	NS	Z	PS	PL
NL	PL	PL	PS	Z	Z
NS	PL	PS	Z	NS	NL
Z	PL	PS	Z	NS	NL
PS	PS	Z	NS	NL	NL

Table 3 Mamdani rule base for 49 rules

Cee	NL	NM	NS	Z	PS	PM	PL
NL	PL	PL	PL	NM	PM	Z	Z
NM	PL	PL	PL	PS	PS	Z	NS
NS	PL	PM	PM	PS	Z	NS	NM
Z	PL	PM	PS	Z	NS	NM	NL
PS	PM	PS	Z	NS	NM	NL	NL
PM	PS	Z	NS	NS	NL	NL	NL
PL	Z	NS	NM	NM	NL	NL	NL

5 ANFIS-Based Reactive Current Controller

ANFIS system contains Tagaki-Sugeno-based inference engine. The ANFIS is data-based modeling. It requires training data, test data and checking data to validate the system performance. It has an inherent intelligence to generate the inference system by using grid partitioning or sub-clustering methods observing the data pattern [10, 11]. The ANFIS has a capability of tuning FIS using optimization techniques. In the proposed work, MATLAB/Simulink model of the STATCOM-reactive current controller using PI is used to generate the data. The performance of the system is verified using MATLAB ANFIS tool. The error and change in error are chosen as input, and reactive current is chosen as output variable. The error signal is generated by comparing reactive current reference with actual current. Total of 1000 data samples are generated from the simulation. The 7500 samples are used as training data, and 2500 samples for testing 0.1000 random samples from training data are used as checking data. The FIS is generated using grid partitioning techniques which comprises nine rules. FIS is trained using hybrid optimization technique with error tolerance 0 and epochs = 500. In hybrid optimization, the input memberships are tuned using backpropagation method, and output membership functions are obtained using least square estimation [11]. The trained FIS is tested against trained data, and very accurate data fitting is observed with root mean square error (RMSE) of 0.00094. The checking data is also tested against trained FIS, and error is found to be 0.00094396 as shown in Fig. 7. The performance of the ANFIS controller is

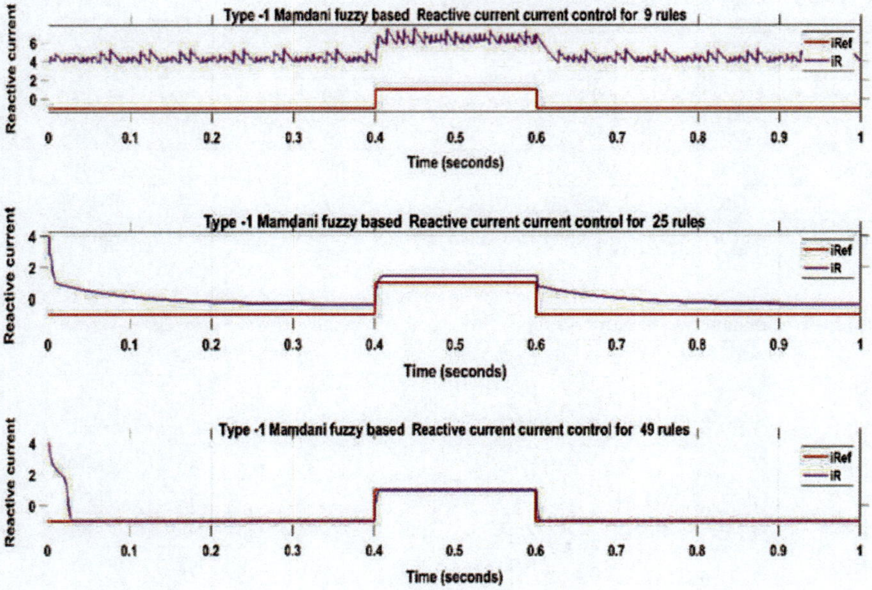

Fig. 7 Trained FIS parameters associated with the minimum checking error

evaluated by exporting the generated nine rules Tagaki-Sugeno inference system into STATCOM system.

6 Result and Discussion

The reactive current control of for type 2 STATCOM using type 1 fuzzy logic controller is developed using MATLAB/Simulink platform. Figure 8 shows the response of reactive current for different set of fuzzy rules. It can be observed that the approximation to the reference current becomes closer as the set of Mamdani rule increases.

For fuzzy system with nine rules, the poor control action is observed. The performance of the controller enhances as the rule base increases. As the set of rules increases, the memory requirement for the computation of control action increases.

In contrast, the ANFIS controllers converge to the reference current at very faster rate. Figure 9 shows the response of the reactive current of ANFIS-based controller. The transient response is improved and able to achieve the fast settling time using nine rules of Tagaki-Sugeno model developed using ANFIS grid partition algorithm and hybrid optimization method of tuning FIS.

The comparison of performances of Mamdani-based type 1 fuzzy controller of 49 rules and Tagaki-Sugeno-based ANFIS controller of nine rules is shown in Fig. 10.

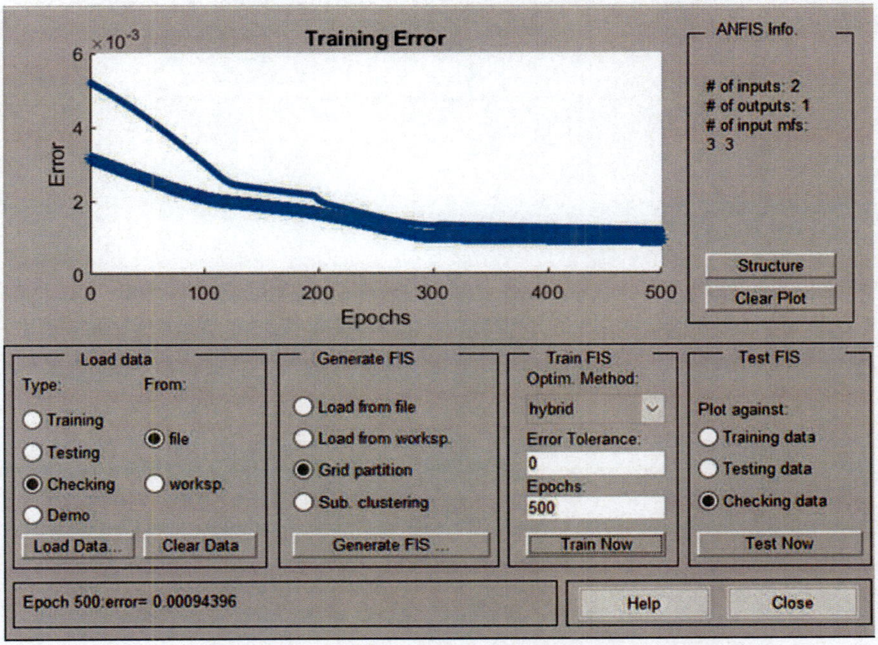

Fig. 8 Response of reactive current for different set of fuzzy rules

Both of these methods ensure the stability in the inductive region. The transient response parameters are tabulated in Table 4 which shows that ANFIS controllers give better performance. It is observed that, with less number of rules with adequate tuning of FIS, the ANFIS controller has better response compared to fuzzy logic controller which has certain noise in the inductive mode of operation though it maintains the stability.

The comparison of error and change of error signals for type 1 fuzzy and ANFIS controller shown in Fig. 11 depicts that during inductive operation (0.4–0.6 s) the type 1 fuzzy has noisy pattern, and it demands more inference rules, but in ANFIS controller the response is smooth even for less number of rules. Therefore, implementation of type 1 fuzzy recommends more memory than ANFIS system.

7 Conclusion

The type 1 fuzzy and ANFIS-based reactive current controller is implemented as an alternative to the traditional nonlinear feedback controller. The performance of type 1 fuzzy-based reactive current controller is evaluated for different set of rules, and it is observed that the approximation on reactive current reference increases

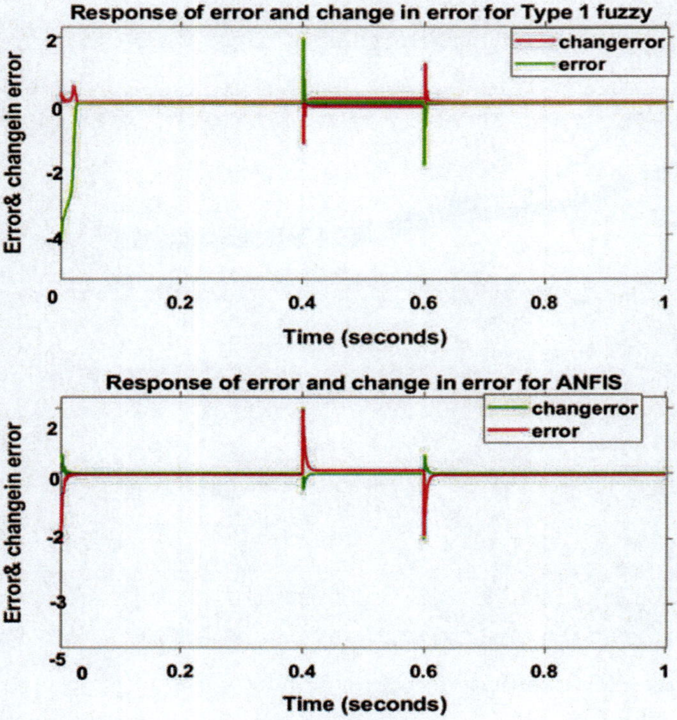

Fig. 9 Response of reactive current for ANFIS controller

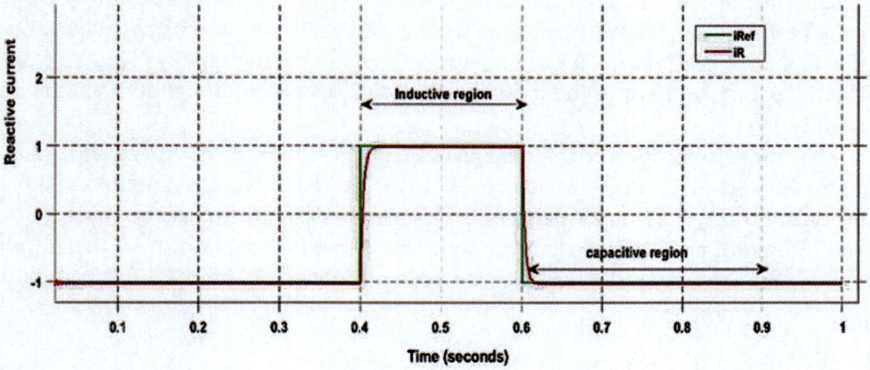

Fig. 10 Comparison of reactive current for type 1 fuzzy and ANFIS controller

Table 4 Transient response of system

Transient response of system	Overshoot	Steady-state error
Type 1 fuzzy controller for 49 rules	4.2%	0.057
ANFIS controller for nine rules	0	−0.0144

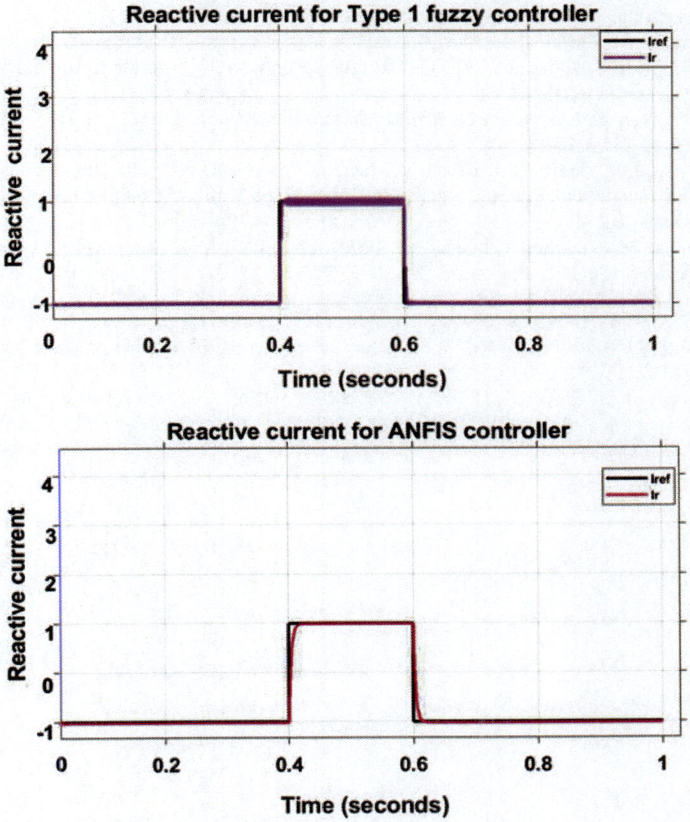

Fig. 11 Comparison of error and change of error signals for type 1 fuzzy and ANFIS controller

with increase in fuzzy rules. ANFIS-based reactive current controller with hybrid optimization techniques gives better performance with less number of rules. It is observed from the results that both type 1 and ANFIS-based controller ensure the stability in inductive mode but type 1 fuzzy controller requires large memory as it demands large number of rules to provide proper control action, and ANFIS offers the better control action with less number of rules.

References

1. N.G. Hingorani, Flexible AC transmission. IEEE Spectr. **30**(4), 40–45 (1993)
2. K.R. Padiyar, *FACTS Controllers in Power Transmission and Distribution, New Age International (P)* (New Delhi, India, 2007)
3. C. Schauder, H. Mehta, Vector analysis and control of advanced static VAR compensators. IEE Proceedings C **140**(4), 299–306 (1993) (K. K. Elissa, "Title of paper if known," unpublished)
4. K.R. Padiyar, A.M. Kulkarni, Design of reactive current and voltage controller of static condenser. Int. J. Electr. Power Energy Syst. **19**(6), 397–410 (1997)
5. M. Janaki, R. Thirumalaivasan, N. Prabhu, Design of robust current controller for two-level 12-pulse VSC-based STATCOM. Hindawi Publishing Corporation Adv. Power Electron **2011**(912749), 7 (2011)
6. K.R. Padiyar, *Power System Dynamics: Stability and Control*, 2nd edn. (B.S. Publications, Hyderabad, India, 2000)
7. B. Sumantri, E. Hnefri, R. Rokhana, I. Mandala, Fuzzy PID controller for an energy efficient personal vehicle: two wheeled electric skateboard. Int. J. Electr Comput. Eng. (IJECE). **9**(6), 5304–5311 (2019)
8. Z. Kamis, M. Ruddhin, Ghani et al., Fuzzy controlled SVC for power system damping. Indonasian J. Electr. Eng. Comput. Sci. (IJEECS) **18**(3), 1673–1678 (2020)
9. D. Shetty, N. Prabhu, Analysis and performance evaluation of type 1 fuzzy reactive current controller with STATCOM. Energy Procedia **117**, 551–558 (2017)
10. G. Joshi, A.J. Pinto Pius, ANFIS controller for vector control of three phase Induction Motor **19**(3), 1177–1185 (2020)
11. H.V. Ngyyen, H Nguyen, K.H. Le, *ANFIS and Fuzzy Tuning of PID Controller for STATCOM to Enhance Power Quality in Multi Machine System Under Disturbance*, vol. 554 (Springer LNEE, 2019)

Performance of Intelligent Controller-Based Bearingless Switched Reluctance Motor

E. Himabindu, D. Krishna, and Venu Madhav Gopala

1 Introduction

Bearingless switched reluctance motors (BRSMs)have been refined in contemporary eras which have doubly salient poles fabrication. BSRMs have bountiful improvements like tenacity, fault tolerance and cut rate. They develop into substitute explication under immense speed, hovering temperature surroundings [1]. The principle, numerical representations of motor drive, deferment is inspected copiously [2–4]. Established on these illustrations, a few control strategies, ingenious motor morphologies were also contemplated [5–7], and the superlative layout of windings was deliberated intensively [8, 9]. Investigators floated the study of the strategy for BSRMs, preeminently the measurement pattern, the intensification of torque, power density and machine efficiency, which are consistently supplication for the motor design. BSRM has an expanded scope of equipment in the industry, such as generator starters, electric vehicles (EV) and wind turbine fans (WTV). The need to support fewer motors has opened up an immense market and consolidated the new trend toward electric motors. The BSRM is dominated by magnetic bearing, SRM [10], which serves as a frictionless, lubrication-free high-speed efficacy. In the modernist era, various BSRM structures were understood; firstly, Bosch and Bichsel are considered advantages of the concepts of winding models without electric motor bearings [3–5]. After Takemoto, Suzuki, Akira Chiba imported an example from BSRM 8/6, in which the suspension forces and the net torque of the engine are conjugated to each other; for this reason, the tuning situation turns into perplexity during the conventional transmission transaction [3–6]. The attributes between suspension force, motor torque and single layer winding in a unipolar design were conditioned

E. Himabindu (✉)
Geethanjali College of Engineering and Technology, Hyderabad, India

D. Krishna · V. M. Gopala
Anurag University, Hyderabad, India

Sanjeevikumar et al. (eds.), *Advances in Renewable Energy and Electric Vehicles*, Lecture Notes in Electrical Engineering 767,
https://doi.org/10.1007/978-981-16-1642-6_28

by Wang et.al. In the abstract windings for the suspension force, the damaged torque produces premeditations at the isolated salient poles [5–15]. The similar was dedicated to specific BSRM 10/8, 12/14. A captivating centrifugal force can be increased among stator poles, rotor by virtue of its lower line of prominence among the stator poles and the rotor poles [16]. This originated net centrifugal force can be alienated into twin mechanisms, one is component of the suspension force in *xy* direction that is pragmatic to levitate the rotor to anticipated midpoint position (0, 0) in *xy* coordinate system, and another is module of the motor torque whichever is advantageous for the turning of rotor in circular direction [17–19]. In this article, the speed, position of the magnetic levitation rotor and its displacements are protected by the ANFIS controller, under different loads to diversify the parameters of the electric and mechanical motor. Complete driving is demonstrated in a distinctive loading environment in a precise way [20, 21]. By accepting the air-suspended rotor in center position proving the magnetic forces of suspension winding, the projected BSRM control scheme is divided into three stages.

- To suspend the rotor in a central location implementing the suspension currents to entire windings of the suspensions over the asymmetric converter together through the current hysteresis controller, the ANFIS controller.
- When rotor levitates, held in the midpoint position, it excites the torque currents from the motor to main stator winding to stimulate rotor at drawn speed.
- After confirmatory anticipated speed, then apply torque loads to rotor shaft.

The 12/14-BSRM winding scheme and its construction are shown in Fig. 1. The drive includes dyadic hysteresis control circuits for torque currents, four for suspension force. To achieve expected speed, the ANFIS rotor position controller with the variable structure entity is related to expected 12/14-BSRM [22–25]. Transmission

Fig. 1 Structure of 12/14-BSRM

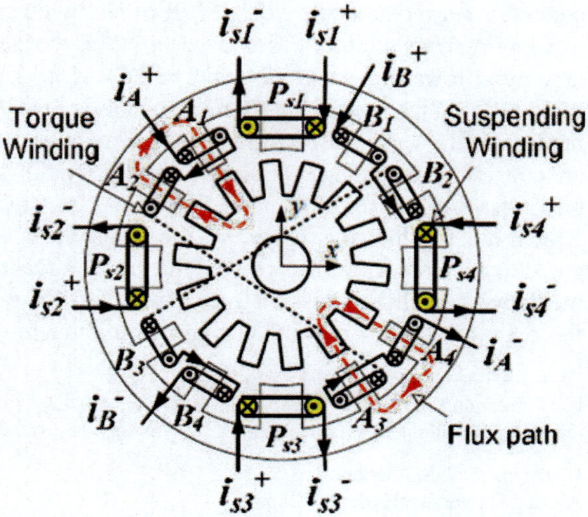

speed performance, rotor displacements and suspension forces in X–Y directions with PI are compared to FLC and ANFIS, and ANFIS shows the best performance in all conditions.

2 Modeling of 12/14-BSRM

By virtue of uncoupled nature exist amid torque, suspension forces equations are premeditated clearly. The state equations of motoring are proved by.

Force in x and y directions:

$$F_x = 4K_f(\varphi)N_m i_{ma} N_s i_{sa1} \tag{1}$$

$$F_y = 4K_f(\varphi)N_m i_{ma} N_s i_{sa2} \tag{2}$$

Inductance matrix L

$$[L] = \begin{bmatrix} L_{ma} & M_{(ma,sa1)} & M_{(ma,sa2)} \\ M_{(ma,sa1)} & L_{sa1} & M_{(sa1,sa2)} \\ M_{(ma,sa2)} & M_{(sa1,sa2)} & L_{sa2} \end{bmatrix} \tag{3}$$

where L_{ma}, L_{sa1} and L_{sa2} are the self-inductances of N_m and N_s, respectively. $M_{(ma,sa1)}$ is the mutual inductance between N_m and N_s.

Energy stored in a phase winding:

$$W_a = \frac{1}{2}[i_{ma} i_{sa1} i_{sa2}][L]\begin{bmatrix} i_{ma} \\ i_{sa1} \\ i_{sa2} \end{bmatrix} \tag{4}$$

Radial forces in matrix form:

$$\begin{bmatrix} F_\alpha \\ F_\beta \end{bmatrix} = i_{ma}\begin{bmatrix} K_{f1} & -K_{f2} \\ K_{f2} & K_{f1} \end{bmatrix}\begin{bmatrix} i_{sa1} \\ i_{sa2} \end{bmatrix} \tag{5}$$

where

$$K_{f1} = N_m N_s\left(\frac{\mu_0 hr(\Pi - 12|\theta_e|)}{6l_0^2} + \frac{32\mu_0 hrc|\theta_e|}{(4rc|\theta_e|l_0 + \Pi 6l_0^2)}\right) \tag{6}$$

$$K_{f2} = N_m N_s\left(\frac{\mu_0 hr(\Pi - 12|\theta_e|)}{12l_0^2} - \frac{2\mu_0 h}{l_0} + \frac{16\mu_0 hrc|\theta_e|\left(r|\theta_e|^2 + 2l_0\right)}{\Pi\left(4rc|\theta_e|l_0 + \Pi 6l_0^2\right)}\right) \tag{7}$$

Switching sequence per phase based on rotor position:

1. $-15^0 \leq \theta < 0^0$ (a—phase excitation)

$$\begin{bmatrix} F_{a1} \\ F_{a2} \end{bmatrix} = \begin{bmatrix} F_{\alpha} \\ F_{\beta} \end{bmatrix} = i_{\text{ma}} \begin{bmatrix} K_{f1} & -K_{f2} \\ K_{f2} & K_{f1} \end{bmatrix} \begin{bmatrix} i_{\text{sa1}} \\ i_{\text{sa2}} \end{bmatrix} \quad (8)$$

2. $0^0 \leq \theta < 15^0 (c - \text{phase excitation})$

$$\begin{bmatrix} F_{c1} \\ F_{c2} \end{bmatrix} = \begin{bmatrix} \cos 240 & -\sin 240 \\ \sin 240 & \cos 240 \end{bmatrix} \begin{bmatrix} F_{\alpha} \\ F_{\beta} \end{bmatrix} = i_{\text{mc}} \begin{bmatrix} K_{f1} & -K_{f2} \\ K_{f2} & K_{f1} \end{bmatrix} \begin{bmatrix} i_{\text{sc1}} \\ i_{\text{sc2}} \end{bmatrix} \quad (9)$$

3. $0^0 \leq \theta < 15^0 (b - \text{phase excitation})$

$$\begin{bmatrix} F_{b1} \\ F_{b2} \end{bmatrix} = \begin{bmatrix} \cos 120 & -\sin 120 \\ \sin 120 & \cos 120 \end{bmatrix} \begin{bmatrix} F_{\alpha} \\ F_{\beta} \end{bmatrix} = i_{\text{mb}} \begin{bmatrix} K_{f1} & -K_{f2} \\ K_{f2} & K_{f1} \end{bmatrix} \begin{bmatrix} i_{\text{sb1}} \\ i_{\text{sb2}} \end{bmatrix} \quad (10)$$

3 Proposed Control Method

See Figs. 2 and 3, Table 1.

Figure 4 shows control block of projected control modus. The actual torque and flux connection are approximated by the diagram. Forced torque can be obtained by adjusting the speed error from an ANFIS controller. Stationed at the outputs of the hysteresis controllers, elementary spatial voltage vectors are tricked. For supremacy of levitation, required radial forces are invaded by adjusting displacement errors of ANFIS. The true radial forces are stationed in resulting radial force expressions [20]. By bringing together the basic three-dimensional voltage vectors, the outputs of the radial force hysteresis controllers are convolved to recover equivalent spatial voltage

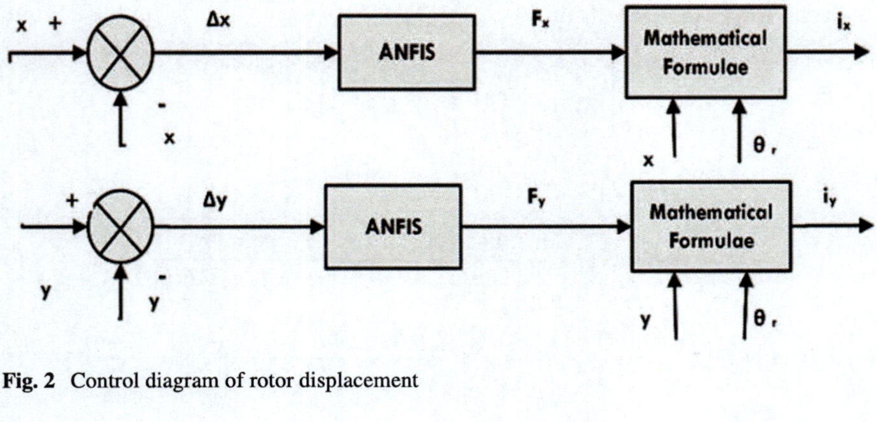

Fig. 2 Control diagram of rotor displacement

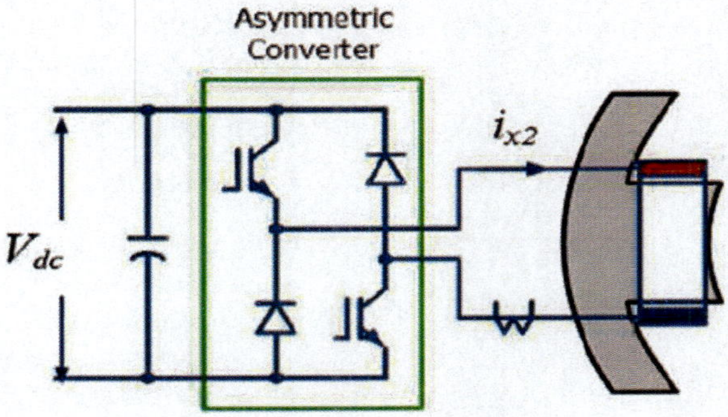

Fig. 3 Suspension winding asymmetric converter

Table 1 Suspension winding currents switching rules types

Desired force	Suspending force poles selection
If $F_x \geq 0, F_y \geq 0$	I_{s1} and I_{s2}
If $F_x \geq 0, F_y \leq 0$	I_{s1} and I_{s4}
If $F_x \leq 0, F_y \leq 0$	I_{s3} and I_{s4}
If $F_x \leq 0, F_y \geq 0$	I_{s3} and I_{s2}

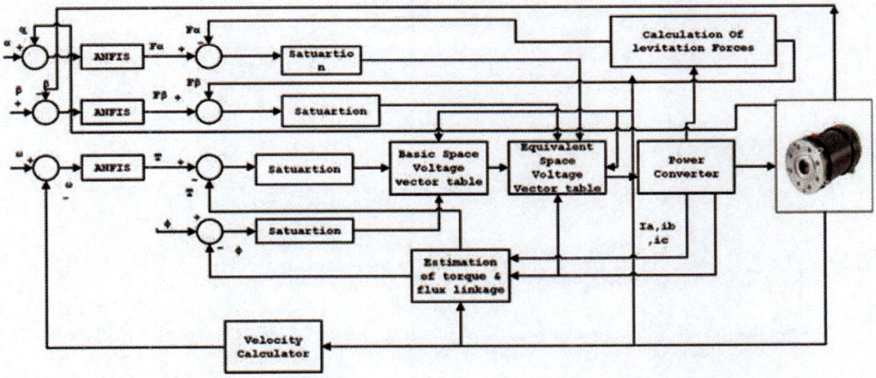

Fig. 4 Control diagram of proposed BSRM

vectors that are accustomed to resolve switching signals of the electronic power converter for BSRM [26]. The windings are stimulated by power to rotate rationally, levitate rotor synchronously. It must be emphasized that levitation forces will be inhibited straightly, under closed-loop control in the prospected control method, which can recuperate levitation pursuance of motors with fewer bearings (Figs. 5, 6 and 7).

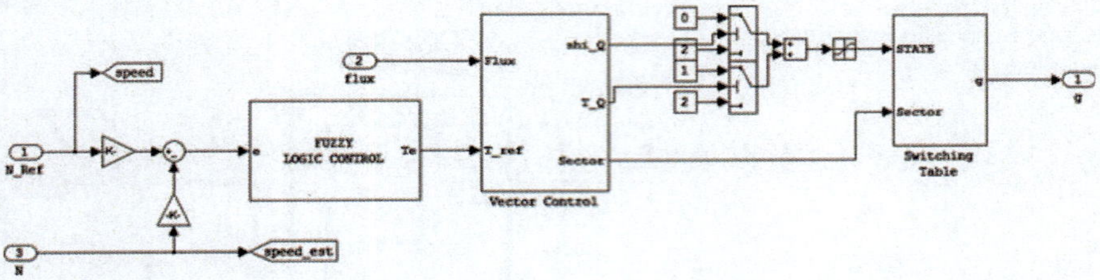

Fig. 5 Proposed ANFIS controller

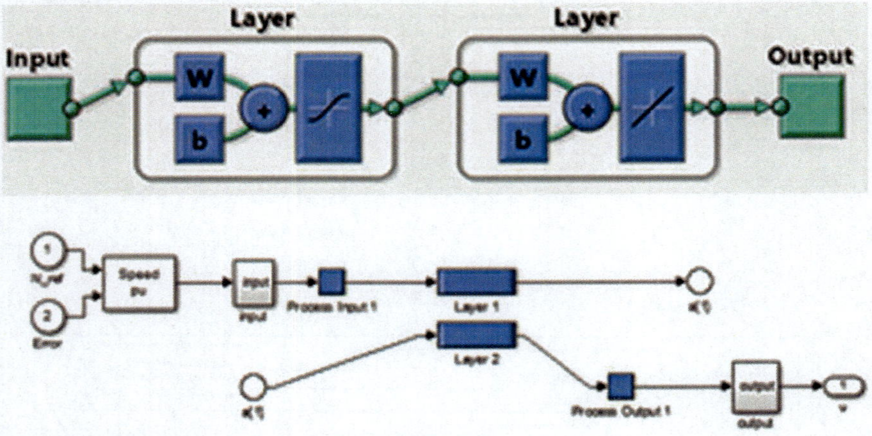

Fig. 6 Structure of ANN

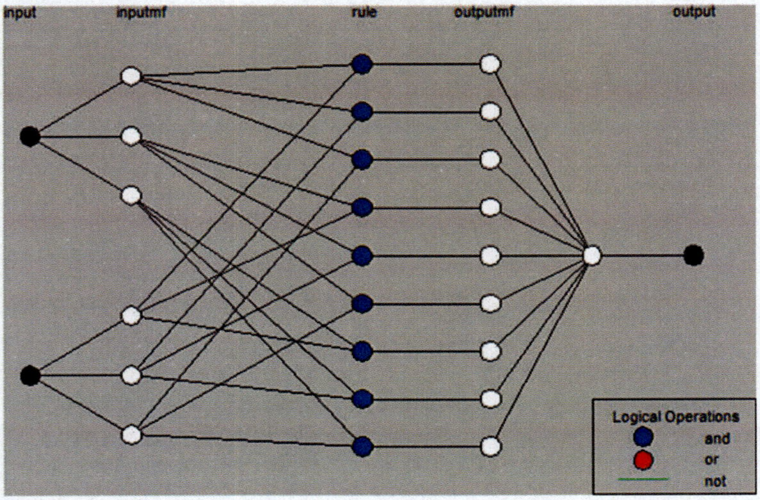

Fig. 7 Structure of ANFIS

ANFIS is a brand of adaptive networks (AN) that assimilates neural networks (NN), fuzzy logic principles (FLP). NNs are supervised learning algorithms that confer a set of chronological data for forecasting future values. The fuzzy logic control signal is caused by the activation of the rule base (RB). This rule base is used with chronological data, and it is random in nature. This indicates that the controller output is also random, which can prevent optimal results. The ANFIS effort can make the RB reduction more robust to the situation. In this approach, the RB is selected using NN techniques using the backpropagation (BP) algorithm. To reinforce its relevance, persuasion, the movable of fuzzy logic, that is, the approximation of a nonlinear system by fitting IF-THEN rules is connoted in this modeling run. This interval access makes ANFIS a universal estimator.

There are some important aspects of this method that need a better understanding. More specifically, they are as follows:

- There are no normal approaches for transfiguring human knowledge, involvement into rule base and the database of a FIS.
- Effective methods are needed to adjust membership functions in order to minimize measurement of output error or maximize the performance index.

ANFIS is a hybrid method that has distinct lineaments of interest. ANN can diagnose patterns in progress using the transformation to the encompassment with learning endowment. At the same time, the FLS can synthesize expert judgment and perform the culpable process. The capability to handle classical data as well as expert intelligence is being flexible to accustom to unconventional speculative precedence's are imperative the visages of the ANFIS model. ANFIS model has been hired in emergency management, which scrutinizes external environments and level of threats. The system is illuminated as shrewd because of the parameters and fuzzy rules, which are approximated by ANN intelligently.

4 Simulation Results

Figure 8 shows the current waveform of BSRM under dissimilar load conditions. It is observed that progress of projected BLSRM is identical w.r.t universal SRM. When motor rotates progressively at rated rpm, gradually suspended at medial position, source currents are stable.

Figure 9 shows the flux waveform of BSRM under incomparable load conditions. It is scrutinized that progress of projected BLSRM is identical w.r.t universal SRM. When motor rotates progressively at rated rpm progressively suspended at center position, source flux is stable.

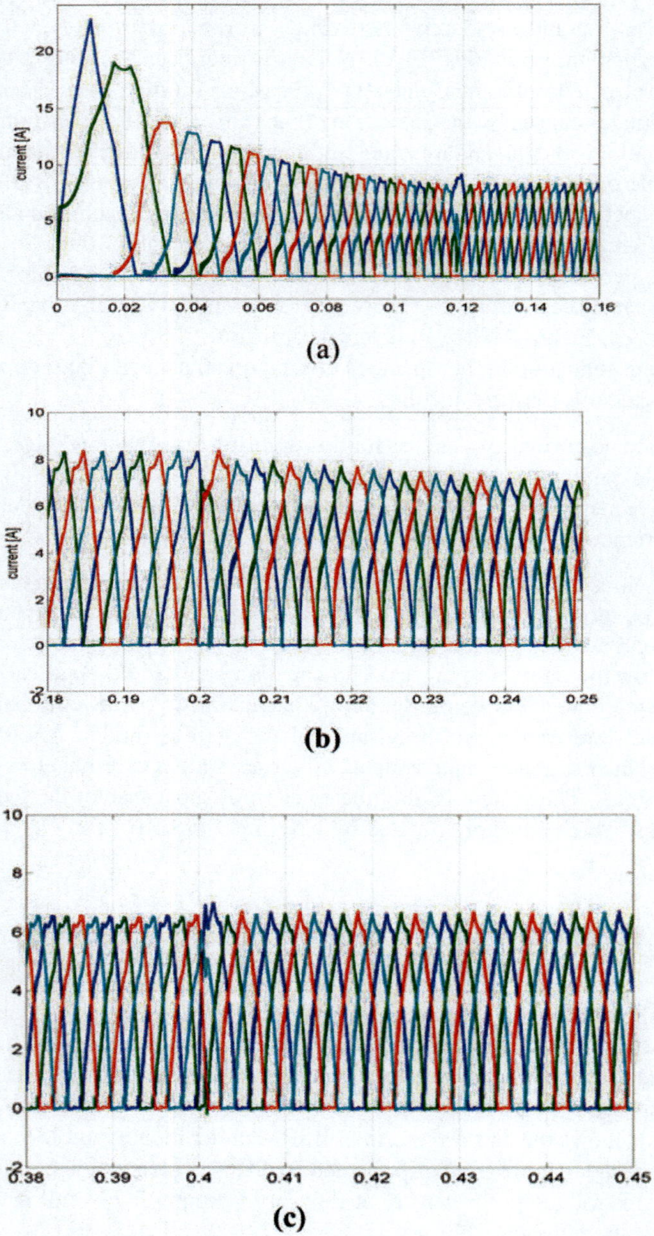

Fig. 8 Simulation result of current at **a** $s = 0$ s **b** $t = 0.18$ s and **c** $t = 0.38$ s

Fig. 9 Simulation result of flux at **a** $t = 0$ s **b** $t = 0.18$ s and **c** $t = 0.38$ s

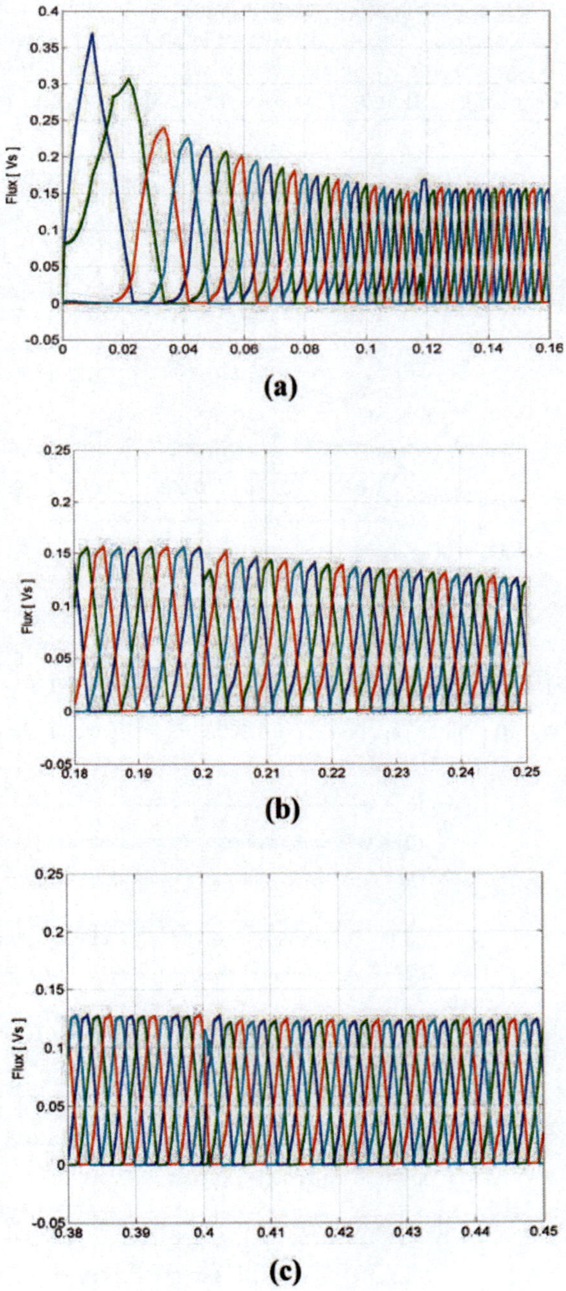

Figure 10 shows torque variations at $t = 0.082$ s. Once rotor is at midpoint position, torque currents are interpolated to main stages via asymmetric converter to expedite rotor to coveted speed with primary torque load of 0.082 Nm and 0.1 s. From Figs. 10 and 11, it is observed that progress of projected BLSRM is similar

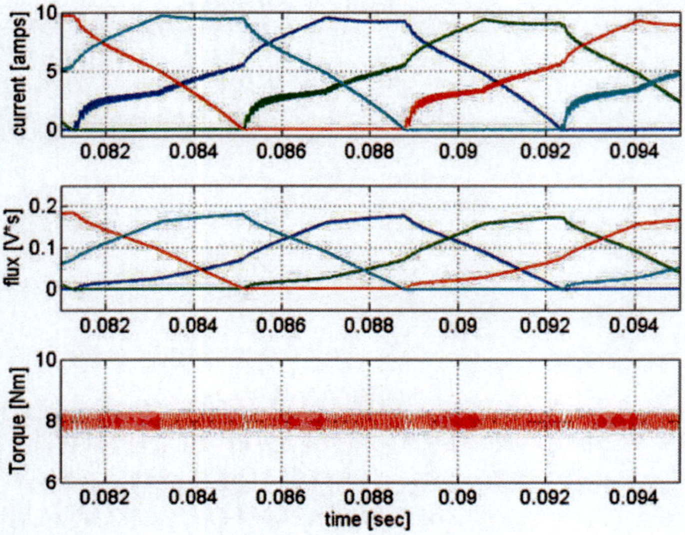

Fig. 10 Torque variation at $t = 0.082$ s of corresponding current and flux

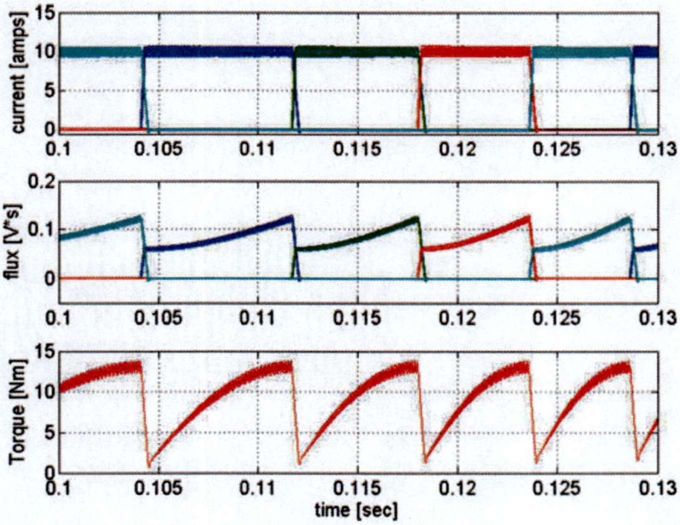

Fig. 11 Torque variation at $t = 0.1$ s of corresponding current and flux

with universal SRM. If motor rotates progressively at rated rpm, gradually deferred at center position, source currents and output torque are constant, and torque current has virtually no reflex on appending force. In order to inspect contemplated control strategies, the BSRM drive simulated various working circumstances (Figs. 12, 13 and 14).

Figure 15 shows the results of speed response with orientation speeds of 1000, 15,000 rpm, at different load torque speed results are confronted with FLC, ANN and ANFIS, and it observed that speed response is fairly good under ANFIS. Figure 16 shows process of drive beneath load torque disturbance. The load torque is altered at $t = 0$ s, 0.4 s and $t = 1$ s. When load torque is improved, here is a small dip in motor speed. In comparison with PI and FLC, ANFIS gives good speed path which is very nearer to orientation speed (Tables 2, 3, 4 and 5).

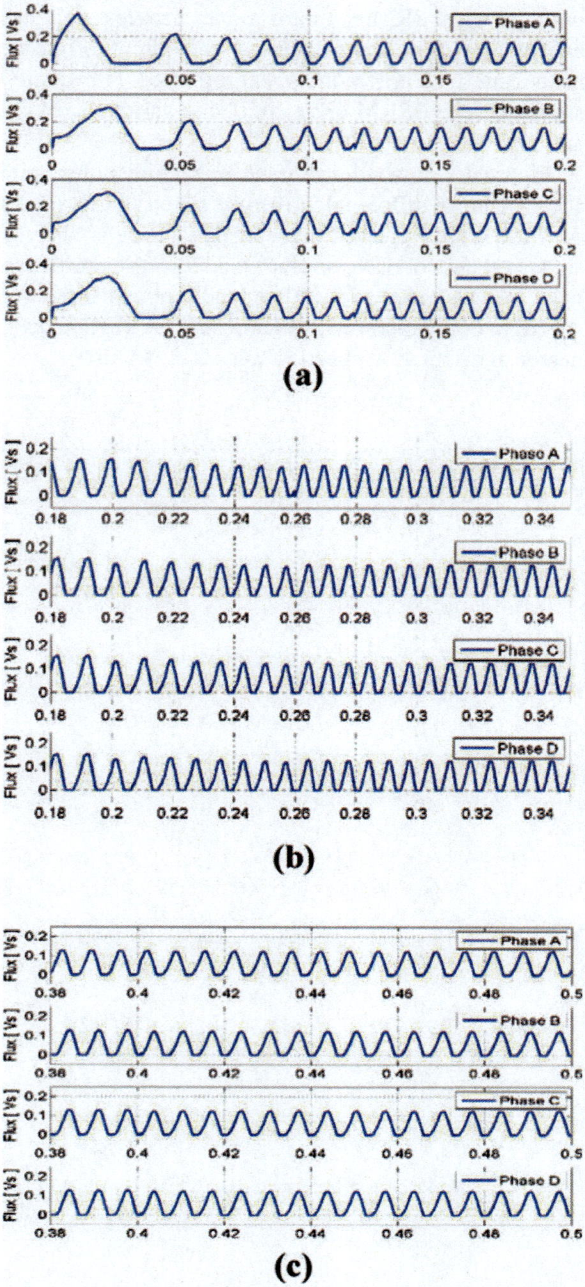

Fig. 12 Simulation results of flux at **a** $t = 0$ s **b** $t = 0.18$ s **c** $t = 0.38$ s for all phases

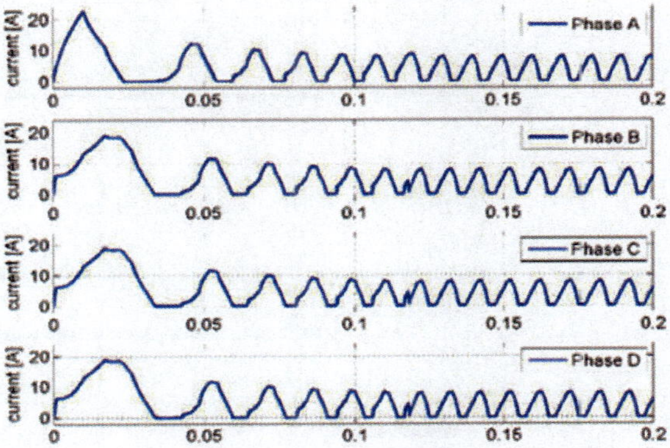

(a)

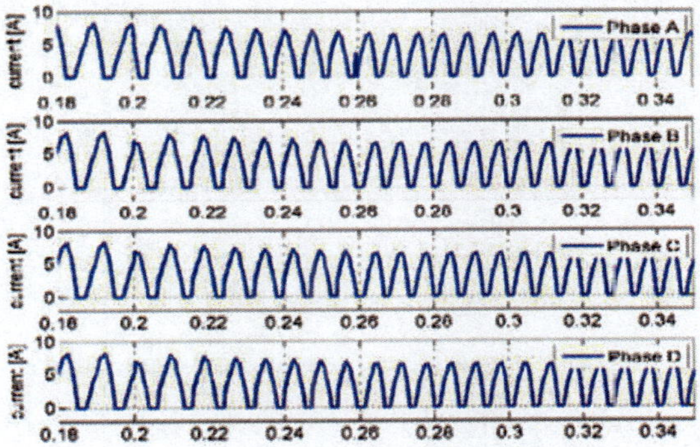

(b)

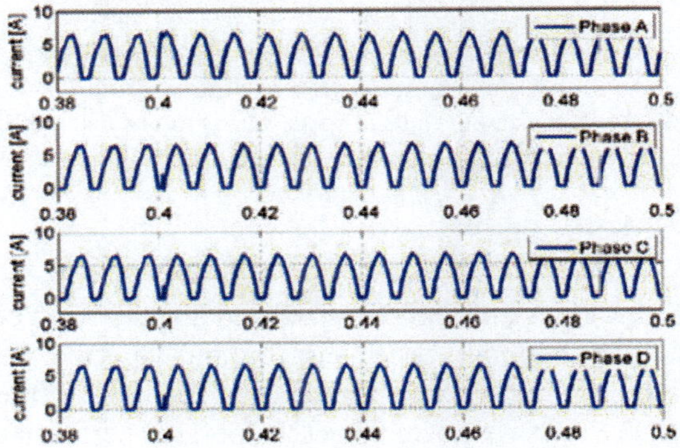

(c)

Fig. 13 Simulation results of current at **a** $t = 0$ s **b** $t = 0.18$ s and **c** $t = 0.38$ s

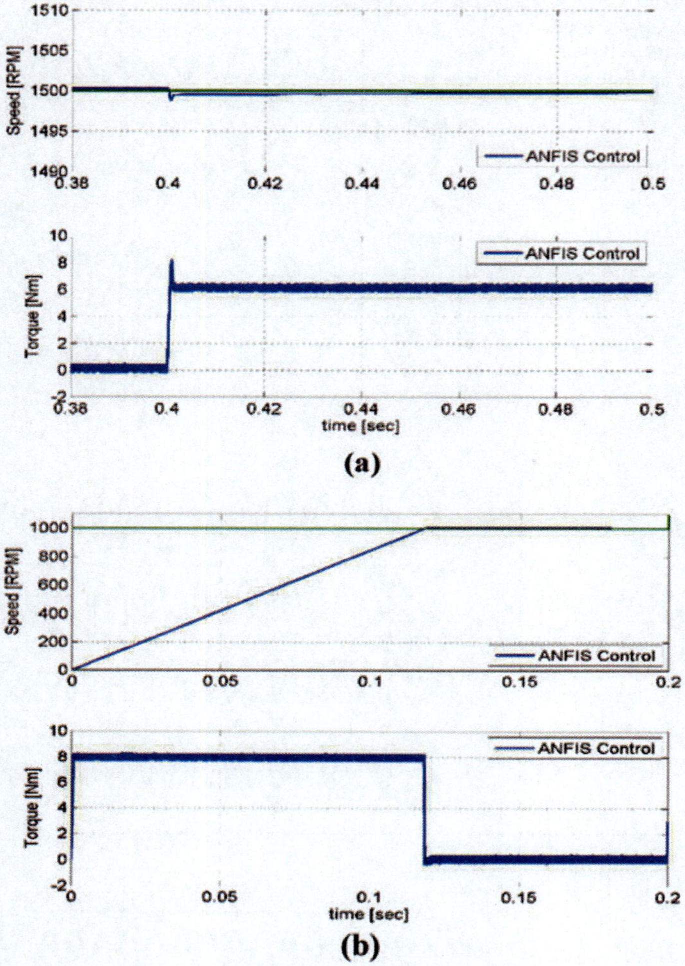

Fig. 14 Simulation results of speed versus torque using ANFIS at **a** $t = 0$ s **b** $t = 0.18$ s

5 Conclusion

In this article, the BSRM speed and displacement control were performed using the ANFIS controller. The robustness and expected speed of the BSRM were obtained under various load circumstances using the ANFIS controller and compared with PI, FLC and ANN to find out their higher performance.. The fast encounter rate, the strongest disturbance and rejection properties were reasonably examined.

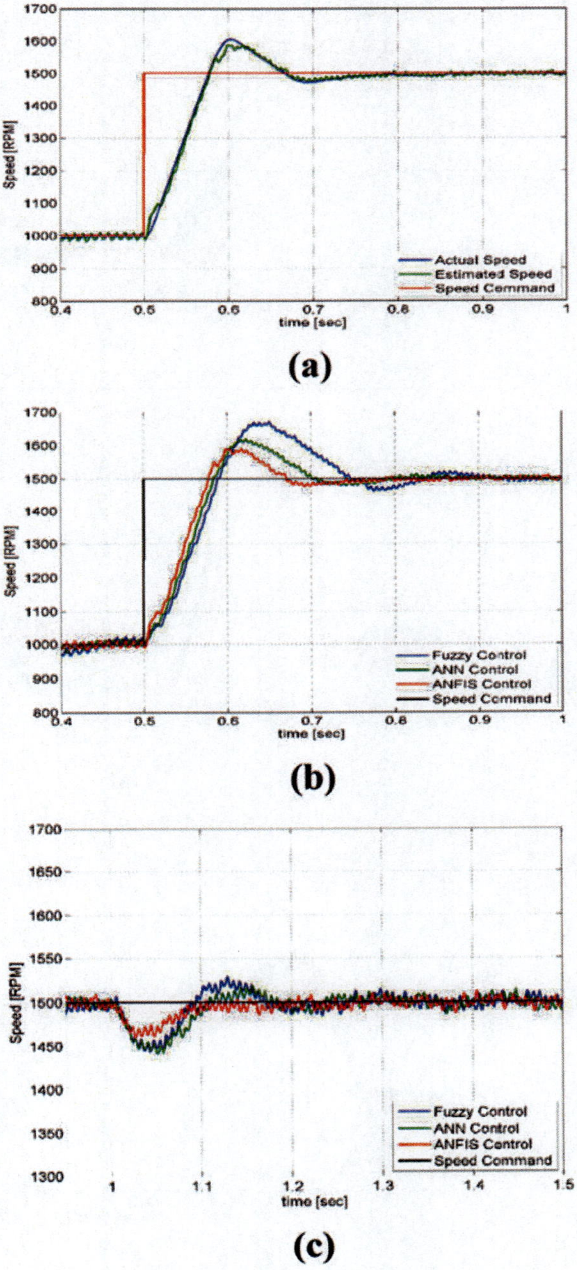

Fig. 15 Simulation results of speed using fuzzy, ANN, ANFIS control at **a** $t = 0.4$ s **b** $t = 0.5$ s **c** $t = 1$ s

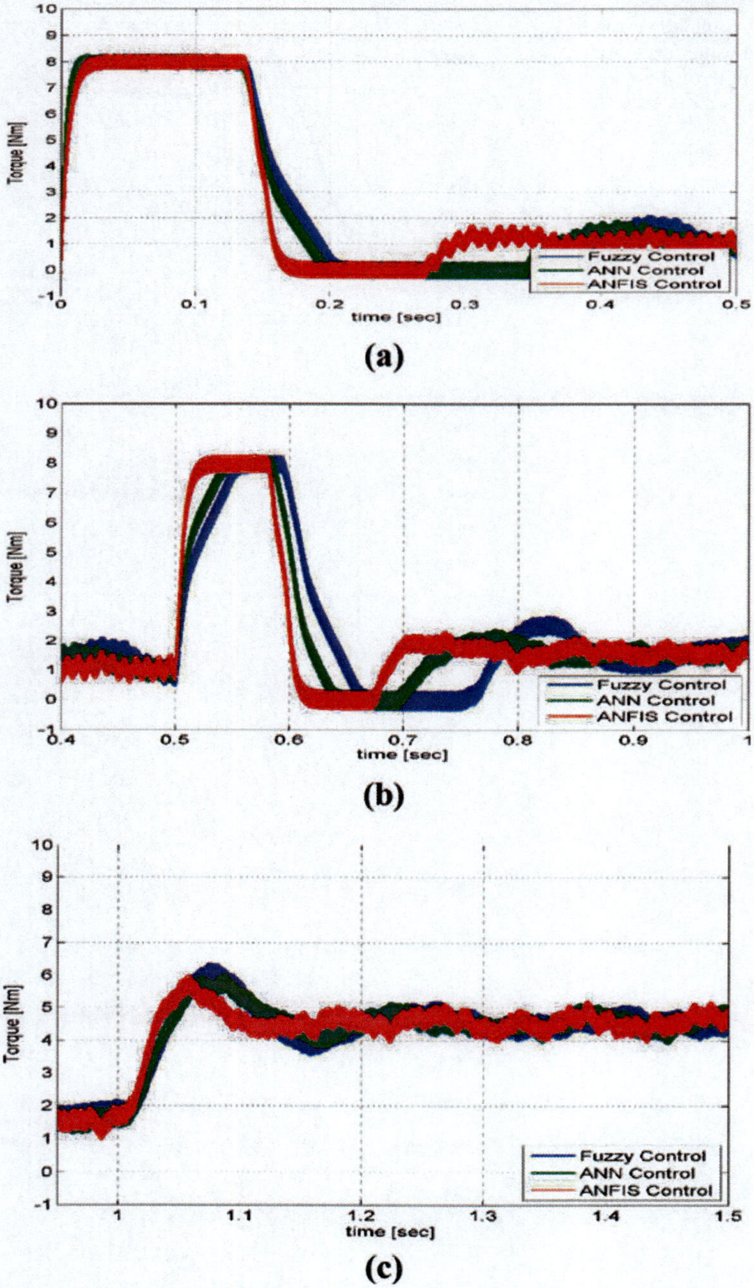

Fig. 16 Simulation results of speed using FUZZY, ANN, ANFIS control at **a** $t = 0.4$ s **b** $t = 0.5$ s **c** $t = 1$ s

Table 2 Performance comparisons by different controller at $t = 0$ s

Parameters	With PI	With fuzzy	With ANN	With ANFIS
Rise time (s)	0.06	0.04	0.02	0.01
Settling time (s)	0.7	0.65	0.64	0.6
Overshoot (%)	14	12	10.5	9

Table 3 Performance comparisons by different controller at $t = 0.4$ s

Parameters	With PI	With fuzzy	With ANN	With ANFIS
Rise time (s)	0.7	0.61	0.59	0.56
Settling time (s)	0.9	0.78	0.75	0.7
Overshoot (%)	12	11.5	9.5	8

Table 4 Performance comparisons by different controller at $t = 1$ s

Parameters	With PI	With fuzzy	With ANN	With ANFIS
Rise time (s)	1.08	1.06	1.04	1.01
Settling time (s)	1.1	1.09	1.085	1.08
Overshoot (%)	11	10.5	8.5	7

Table 5 Specifications of test motor

Parameters	Dimension
Rated power(motor)	1 kW
Maximum motor current/phase	4 A
Voltage\phase	250 V
Net torque	1 Nm
Speed	1500 rpm
Torque winding per phase resistance	0.8 Ω
Suspension winding per phase resistance	0.35 Ω
Pole arc of stator for radial force (deg)	32
Suspension voltage	250 V
Maximum suspension current	4 A
Pole arc of rotor (deg)	16

References

1. B. Liu, Survey of bearingless motor technologies and applications, in *Proceedings of the IEEE International Conference Mechatronics Automation (ICMA)* (Beijing, China, 2015), pp. 1983–1988

2. M. Takemoto, A. Chiba, T. Fukao, A new control method of bearingless switched reluctance motors using square-wave currents, in *Proceedings of the IEEE Power Engineering Society Winter Meeting Conference*, vol. 1. (2000), pp. 375–380
3. R. Bosch, Development of a bearingless electric motor, in *Proceedings of the International Conference Electric Machines (ICEM'88)* (Pisa, Italy, 1988), pp. 373–375
4. M. Takemoto, A. Chiba, H. Akagi, T. Fukao, Radial force and torque of a bearingless switched reluctance motor operating in a region of magnetic saturation. IEEE Trans. Ind. Appl. **40**(1), 103–112 (2004)
5. M. Takemoto, A. Chiba, T. Fukao, A method of determining the advanced angle of square-wave currents in a bearingless switched reluctance motor. IEEE Trans. Ind. Appl. **37**(6), 1702–1709 (2001)
6. J. Bichsel, The bearingless electrical machine, in *Proceedings of the International Symposium Magnetic Suspension Technology* (NASA Langley Research Center, Hampton, 1991), pp. 561–573
7. D. Krishna, M. Sasikala V. Ganesh, Mathematical modeling and simulation of UPQC in distributed power systems, in *2017 IEEE International Conference on Electrical, Instrumentation and Communication Engineering (ICEICE)* (Karur, 2017), pp. 1–5. doi: https://doi.org/10.1109/ICEICE.2017.8191886
8. Z. Xu, D.H. Lee, J.W. Ahn, Comparative analysis of bearingless switched reluctance motors with decoupled suspending force control. IEEE Trans. Ind. Appl. **51**(1), 733–743 (2015)
9. M. Takemoto, A. Chiba, H. Akagi, T. Fukao, Suspending force and torque of a bearingless switched reluctance motor operating in a region of magnetic saturation, in *Conference Record IEEE-IAS Annual Meeting* (2002), pp. 35–42
10. G. Yang, Z. Deng, X. Cao, X. Wang, Optimal winding arrangements of a bearingless switched reluctance motor. IEEE Trans. Power Electron. **23**(6), 3056–3066 (2008)
11. Z. Liu, Z. Deng, J. Cai, Y. Wu, S. Wang, Optimal design of a bearingless switched reluctance motor, in *Proceedings of the International Conference Applied Superconductivity Electromagnetic Devices* (Chengdu, China, 2009), pp. 241–245
12. M. Takemoto, K. Shimada, A. Chiba et al., A design and characteristics of switched reluctance type bearingless motors, 4th international symposium on magnetic suspension technology, NASA/CP-1998-207654 (1998), pp. 49–63
13. M. Takemoto, A. Chiba, T. Fukao, A new control method of bearingless switched reluctance motors using square-wave currents, in *Proceedings of the 2000 IEEE Power Engineering Society Winter Meeting* (Singapore, CD-ROM, 2000), pp. 375–380
14. H.J. Wang, D.H. Lee, J.W. Ahn, Novel bearingless switched reluctance motor with hybrid stator poles: concept, analysis, design and experimental verification, in *The Eleventh International Conference on Electrical Machines and Systems* (2008), pp. 3358–3363
15. D.-H. Lee, J.-W. Ahn, Design and analysis of hybrid stator bearingless SRM. J. Electr. Eng. Technol. **6**(1), 94–103 (2011)
16. D.H. Lee, Z.G. Lee, J. Liang, J.W. Ahn, Single-phase SRM drive with torque ripple reduction and power factor correction. IEEE Trans. Ind. Appl. **43**(6), 1578–1587 (2007)
17. H.J. Wang, Design and control of a novel bearingless switched reluctance motor, Industrial System Engineering of the Kyungsung University, Busan, Korea, PhD's thesis, 6 June 2009
18. Z. Xu, D.-H. Lee, J.-W. Ahn, Modeling and control of a bearingless switched reluctance motor with separated torque and suspending force poles
19. Z. Xu, F. Zhang, J.-W. Ahn, Design and analysis of a novel 12/14 hybrid pole type bearingless switched reluctance motor
20. Z. Xu, D.-H. Lee, J.-W. Ahn, Control characteristics of 8/10 and 12/14 bearingless switched reluctance motor, in *The 2014 International Power Electronics Conferences*
21. L. Chen, W. Hofman, Analytically computing winding currents to generate torque and levitation force of a new bearingless switched reluctance motor, in *Processing of the 12th International Power Electronics and Motion Control Conference* (Portoroz Slovenia, 2006), pp. 1058–1063
22. L. Chen, W. Hofman, Performance characteristics of one novel switched reluctance bearingless motor drive, in *Power Conversion Conference 2007 (PCC '07)* (Nagoya, 2007), pp. 608–613

23. L. Chen, W. Hofmann, Analysis of radial forces based on rotor eccentricity of bearingless switched reluctance motors, in *International Conference on Electrical Machines* (Rome, 2010)
24. X. Cao, H. Yang, L. Zhang, Z. Deng, Compensation strategy of levitation forces for single-winding bearingless switched reluctance motor with one winding total short circuited. IEEE Trans. Ind. Electron. **63**(9), 5534–5546 (2016)
25. D. Krishna, M. Sasikala, V. Ganesh, Adaptive FLC-based UPQC in distribution power systems for power quality problems, Int. J. Ambient Energy. http://doi.org/https://doi.org/10.1080/01430750.2020.1722232
26. E. Himabindu, M.G. Naik, Modular current cell topology of seven and fifteen level CSI with reduced count. Int. J. Innov. Technol. Explor. Eng. **8**(8), 2174–2178 (2019)

Review of Battery State-of-Charge Estimation Algorithms

Kaustubh Kaushik, Devang Sureka, and H. V. Gururaja Rao

1 Introduction

Historical dependence on fossil fuels and subsequent overuse of traditional fuels has pushed environmental concerns to an all-time high. The single largest source of pollution is the combustion engines used primarily in vehicles. The alternative to the old internal combustion-based vehicles is electric or hybrid vehicles that replace combustible fuel engines with battery-based power units, keeping the carbon emissions in check. The use of batteries in place of traditional fuels presents a different set of challenges. The state of charge estimation largely governs the safety, robustness, durability, and reliability of batteries. Over the years, various algorithms have been put forward, like open circuit voltage (OCVMs), Coulomb counting method (CCMs), model-based methods (MBMs), and ANN-based methods (ANNBMs). Few algorithms use complementary methods for even more accurate estimation. To improve the BMS productivity and guarantee the battery's private use, we calculate the battery's SOC at each second during the activity. SOC cannot be calculated directly because the lithium-ion battery forms a closed-loop system. Hence, we calculate the SOC for the entire battery framework [1, 34].

Battery management systems include parameters like state of health (SOH), state of charge (SOC), state of power (SOP), and state of life (SOL) in its control circuit and an analog sampling circuit. The control circuit calculates parameters based on analog signals' readings and directs the information through various communication ports to the central control unit. The BMS forms the backbone of electric vehicle technology, and we judge the performance based on parameters like range, power, and service life. A BMS comprises a variety of sensors, actuators, regulators, and

K. Kaushik · D. Sureka · H. V. G. Rao (✉)
Department of Electrical and Electronics, Manipal Institute of Technology, Manipal Academy of Higher Education, Manipal, India
e-mail: gururaj.rao@manipal.edu

Sanjeevikumar et al. (eds.), *Advances in Renewable Energy and Electric Vehicles*,
Lecture Notes in Electrical Engineering 767,
https://doi.org/10.1007/978-981-16-1642-6_29

signature lines. An implementable BMS aims to make sure the battery's energy dispensation is judicious, and the data provided to the vehicle's energy management system is accurate as possible.

Additionally, proper interventions to the battery structure on the off chance that it works in adulterated conditions are secured, and this is achieved by monitoring the charging and delivering pattern of batteries. The model circuit's chief endeavor evaluates the current, voltage, and temperature according to the control circuit's gating signal. Moreover, the control circuit keeps checking the charge (SOC) state and other related parameters like SOH and SOP through analog signals. Subsequently, pass the data to the vehicular power management system, and it gives critical decision components to the heads and power appointment of vehicular energy [2, 3, 35].

State of charge (SOC) is the base parameter based on which a BMS specifies or calculates other parameters, including SOL and SOP. SOC is the proportion of remaining battery capacity about the battery's maximum capacity represented in percentage. SOC accuracy and delivery in real-time help eliminate catastrophic consequences and better manage various other sub-systems. We can draw an analogy between the SOC and traditional fuel gauge in combustion-based vehicles. SOC measurement is a challenging task. Internal and external factors play an important role in accurate SOC measurements like battery-aging, charging-discharging cycles, and differential characteristics between cells connected parallelly [4, 35].

Since BMS stays one of the most fundamental variables, different techniques have been proposed to estimate the SOC precisely. From the 1960s, scholastics, specialists, and researchers have performed broad examination to do the battery SOC assessment. In [7–10], authors have introduced a point by point SOC assessment as far as in general examination progress, future improvement patterns, and the beginning of SOC assessment.

Regardless, there is no perfect method of the SOC assessment cycle and estimation assurance and how to negate the atmospheric changes. Thus, this paper will explore the gaps by emphasizing the pre-existing estimation methods and focusing on the battery pack rather than a single cell [35].

2 State of Charge Estimation Methods

2.1 Classification of Estimation Methods

SOC estimation methods are broadly classified in four different forms, namely (a) look-up-table-based examples of which are open circuit voltage (OCV) and AC impedance, (b) Coulomb counting method (CCM), (c) model-based estimation methods (MBM), and (d) data-driven methods which are further classified into data training and data model fusion method. For this paper, we will explore each type of broad estimation method and then further delve into an example of each of those general classifications.

2.1.1 Look-Up-Table-Based Method

The state of charge of batteries directly correlates with their external fixed parameters like the open circuit voltage (OCV) and impedance. This way, we shall calculate their limits, and after that, tally it against the provided table, which is prepared with the associations among SOC and at any rate one limits, we can predict the SOC [11–14, 35].

We shall look at an OCV example for looking up a table-based model of state of charge estimation. Open circuit voltage is estimated under the condition that the battery is separated from any heap and has enough unwinding opportunity to arrive at its interior balance. The relationship between the open circuit voltage and SOC is the most effective strategy for assessing SOC if a precise estimation of open circuit voltage is already given. Since the Li-ion battery's unwinding time might even exceed 10 h or considerably more, which influences this method's real-time applicability. The connection between this method and SOC is found to alter temperature and used-age [15–18, 34]. Broad works zero in on improving the open circuit voltage method with higher accuracy and precision by considering external factors proposed in [16, 20–22]. Also, the qualities of the OCV-SOC bend are firmly identified with battery science. For instance, the open circuit voltage and state of the charge curve is moderately level for lithium-iron-phosphate batteries, which implies a little mistake in OCV will cause a significant SOC assessment blunder. As figured, the distinction of open circuit voltage is only 72 mV in the SOC scope of 30–80%. Hence, the traditional open circuit voltage method is not precisely satisfactory for real-world online applications. Analysts take a quick OCV shot to improve its utility in short, unwinding time [23, 24, 34].

The method suggested that the open circuit voltage method has higher computational accuracy and is apt for online assessment. Even though this method stands up to numerous downsides, it is still being perfected for better relevance during online applications.

Figure 1 shows that the OCV of a LiPB cell shows a monotonically growing example with its SOC. Thus, if we know the OCV, we can gather battery SOC by looking into the table among OCV and SOC [34].

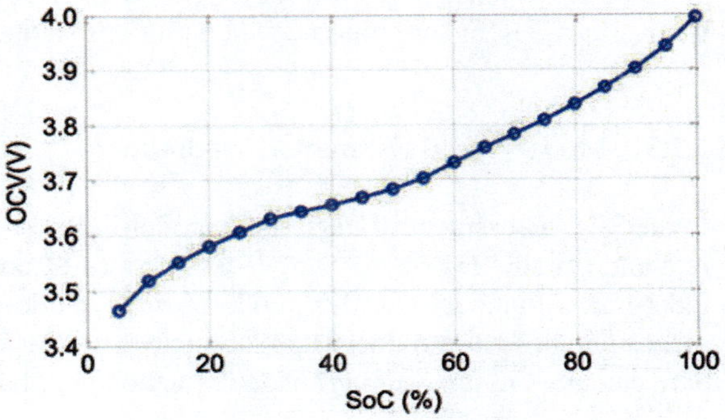

Fig. 1 OCV curve of Lip cell

2.1.2 Coulomb Counting Method

The coulomb counting method, also known as ampere-hour integral method, is another efficient method of calculation of SOC of a cell. It is defined as:

$$z(t) = z(0) - \int_{0}^{t} \frac{n_i I_l(t)}{Q} \mathrm{d}t \quad (1)$$

where $z(t)$ is the SOC at time t and $z(0)$ is the underlying SOC; n_i is the battery's Coulombic proficiency. I_t is current, which is positive and negative for charging and discharging conditions, respectively. From Eq. (1), we can characterize SOC as the limit-integration of current. Consequently, the Coulomb counting method is close to the perfect SOC calculation methodology in principle [34]. Be that as it may, in all actuality, the underlying state of charge of the battery is not known ahead of time due to self-release and irregular utilization. Mistakes from current sensors likewise collect in the computation cycle. To beat these downsides, improving the CCM is proposed in [15, 16, 34]. Since Coulombic accuracy influences the productivity of Coulomb counting method in Eq. (1), changing the productivity count during the releasing cycle will assist in improving calculation exactness [12, 13]. In any case, it is generally challenging to acquire its worth because the battery's test tests under various current rates are required [26, 27, 34]. Joining the OCV state-of-charge relationship is additionally an excellent method to make up for the deficiencies of the Coulomb counting method. In [27], the creators proposed to reset the underlying state of charge of this method by anticipating open circuit voltage in a shorter duration and naturally remunerating the assessing blunder. Contrasted and the regular CCM, the proposed technique increments by 2.07% the SOC assessment precision when a UDDS profile is utilized. The battery's underlying charge levels are obtained for this method by thinking about the open circuit voltage, resting time, and temperature impact [29, 34]. By adding the release proficiency, the error of SOC assessment is additionally diminished. Eliminating the commotion from the current sensor will further reduce the discrepancies in the state-of-charge calculation. For simple applications, if the underlying charge is calculated ahead of time, and more accurate sensors are employed for the power management system, and this method is advantageous and appropriate for continuous state-of-charge assessment [34].

2.1.3 Model-Based Estimation Methods.

In the SOC assessment strategies, the model-based one appears to be the most accurate and suitable for online SOC assessment as of now. Much work is defined and identified with model-based methods. Various online-based models for SOC assessment strategies are introduced and summed up in the accompanying section. The standard calculations are Kalman channel, Luenberger observer, PI (extent combination) observers, H∞, sliding–mode observer. The Kalman channel is the most preferred for nonlinear assessment and artificial intelligence-based applications. The authors in [35] talk about a PNGV model based on an improved SOC estimation method with Kalman filtering.

Reasoned in Eq. (1), the model-based SOC assessment strategies can likewise generally represented by [30]:

$$\{Z = \frac{n_i}{C_n}\dot{I}_t - L(\widehat{v_t} - v_t)v_t = h(Z, i_t, \ldots) \tag{2}$$

whereas the voltage at time t given by the voltage sensors, $\widehat{v_t}$ is defined as the voltage assessed by battery setup, $h(.)$, is the representation of the model of the battery [30, 34]. From Eq. (3), we can deduce that the feedback remuneration used in assessing the state of charge is given by the difference between the voltage estimated by the sensor and the one calculated by the model the battery. Considering the non-open loop structure, MBMs can manage obscure starting charge levels. In Eq. (2), the gain of remunerating the charge levels determined by Coulombic count is denoted by L. The model proposed in [31] consists of two equivalent RC circuits, depicts a straightforward structure and highly viable SOC assessment technique utilizing two free PI observers where one of them helps further improve the demonstrating exactness. Simultaneously, the second one calculates the open circuit voltage for SOC assessment all the while. Then again, H∞ observer is proposed for diminishing the impact of commotion and boundary vulnerability on the assessment precision [34]. A versatile H∞ channel is presented in [32] that improves the accuracy of charge levels assessment values opposing the sensor's clamor and incorrectness from the battery model. Using recursive least square for boundary refreshing, this strategy shows exact assessed charge levels in a piece of equipment tuned in the analysis [34]. Nonetheless, the Kalman channel is the most famous model-based assessment calculation because of its vigor to the commotion in the academic circles. Extended Kalman filter (EKF) is utilized in [33] to deduce battery internal temperature and charge levels simultaneously dependent on a novel thermoelectric model. He et al. [31] approve unscented Kalman filter (UKF)-based state-of-charge assessment on an embedded system [34]. As dangerously portrayed in this paper, MBMs depend on an exact battery model for acquiring precise SOC assessment. Nonetheless, the battery's internal characteristics alter while it charges and discharges and is a very complicated process. It is trying to develop a model that can be accurate and depicts all the battery attributes. Particularly, for non-offline applications, the battery model's computational intricacy should be limited to a sensible reach. Being harsh toward introductory SOC and influential to estimation clamor, MBMs are incredibly mainstream for various online SOC assessment applications [34].

2.1.4 Data-Driven Estimation Methods

Data-driven estimation methods are broadly classified into two sub-categories, namely data training (DT) and data model fusion method (DMF). The DT method is further based on two models first support vector machines and the second neural network-based estimation model [35]. Data-driven control techniques just utilize the info yield data of the framework to build up a regulator. Since these techniques do not need a precise plant model, the assessments and suppositions presented in the plant demonstrating step are excluded. Nonlinear measurable data displaying apparatuses are reasonable. They can show complex connections among data sources and yield or discover designs in the data. In [35], the neural organization is utilized to build up the SOC assessor, where the data layer is made up of parameters like current, temperature, and the charge level of the battery. The yield layer represents the voltage. Experiments have shown high levels of exactness with the given arrangement. Various types of fake neural organization (ANN) strategies and a few techniques like ANN are common in planning the nonlinear connection among data sources and yields. In SOC assessment, ANN can legitimately set up the relationship among state-of-charge-related values, including current, voltage, and temperature. Subsequently, designers can make an assessor with no prior data available about the battery [34]. The connection among the inputs (voltage, current, temperature) and charge levels is straightforwardly settled using a structure that is nonlinear and ANN-based. ANNBMs ought to be prepared to build up a nonlinear relationship and can operate continuously [34]. Two distinct ANN-based structures are applied to assess SOC in [32]. Using computed the limit blur, precise SOC is deduced from the ANN assessor during the battery's life expectancy. If suitable examples are chosen and improved boundaries are picked for the preparation cycle, the ANNBMs can introduce a precise SOC assessment for the preparation test [34].

Nonetheless, it is handily discovered that these techniques' practicability is firmly identified with the preparation cycle and the set of data, given that the test surrounding shift, the determination of ANN-based methods is limited to online forms. By and large, ANN-based processes are handily relocated for online usage in the wake of having been prepared disconnected [34]. A review of various model-based and data-driven methods for SOC estimation of batteries is discussed in [36]. In [37], various artificial intelligence and direct measurement techniques for SOC estimation are explored.

2.2 *Argument*

In the wake of presenting every one of the SOC assessment techniques' highlights, their appropriateness for online use is talked about in this part. As appeared by the past locale's evaluation, the fittingness for online utilization of the four essential SOC examination techniques. From Fig. 2, we can see that all these procedures have their central focuses and hindrances. Be that as it may, for non-offline applications

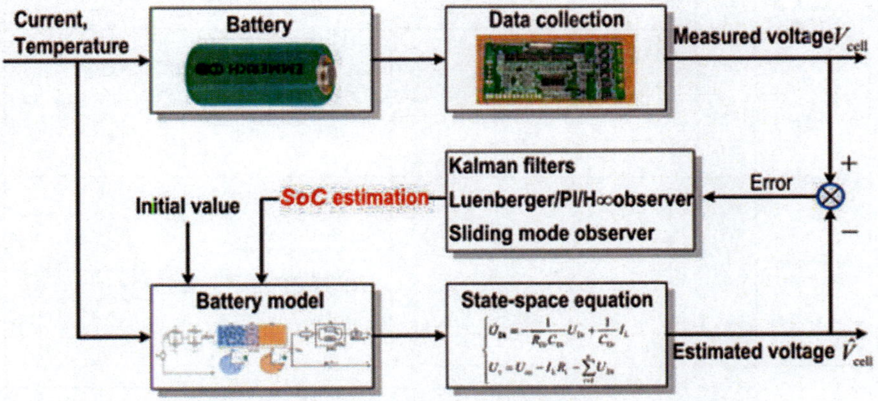

Fig. 2 General model-based estimation method

on battery management systems, exactness, stiffness, and computation expense are the three most significant elements to consider [34]. The electric vehicle application is based as an example to investigate and analyze these various techniques. From a clear perspective, every procedure can accomplish fantastic outcomes under exact circumstances. Since the Coulomb counting method is a non-closed-loop structure, starting charge level and estimation of current are, without a doubt, critical for its precision. Commonly, in hybrid applications, some current sensors' exact starting charge level and high exactness are deemed impractical [34] (Fig. 3).

OCVM depends on the exact OCV esteem for acquiring the assessed state of charge. The open circuit voltage could still be reached even after the EV left a while ago. Nonetheless, amidst the driving cycle, the current interference could likewise occur while the EV halts at the traffic signal. The current interference during the conditions is typically immensely small for battery unwinding. Subsequently, quick open circuit voltage assessment is pressing for the use of OCV-based method continuously [34]. Besides, the OCV and state of charge bend ought to be steep to ensure assessment precision. The exactness of MBM depends on building up an exact model of the battery in the assessment cycle. Choosing the apt model of the battery structure will improve assessment precision. Nonetheless, it is difficult to replicate the complex electrochemical process of battery by the equal circuit model, typically utilized in the MBM. Besides, the adjusted calculation's exhibition and combination are likewise firmly identified with an exact assessed SOC. This way, the precision of MBMs appears to be adequate for EV applications if the correct battery model and the appropriate assessment calculation are picked [34]. ANNBMs are amazingly precise if the whole EV driving cycle's present profile is like the preparation dataset. The viable application consistently experiences a wide range of working conditions, implying that power is an essential factor to mull over [35]. For EV applications, the battery pack ought to satisfy the extra force prerequisites of a continuous drive. The current, temperature, and age change frequently. A shut circle framework is generally

Categories of Methods	Principles	Advantages	Disadvantages
Coulomb counting method	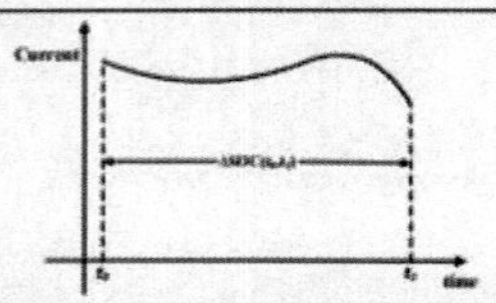	• Computational effectively; • Direct SOC calculation; • Easy to understand	• Accurate initial SOC is needed; • Current sensor error accumulated during the process.
Open circuit voltage method	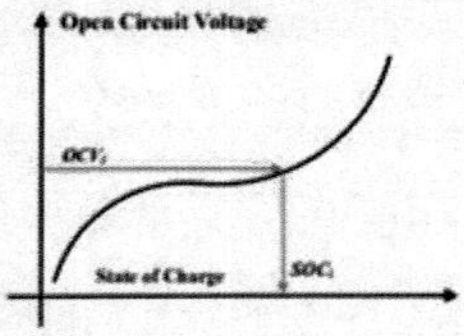	• One to one relationship between OCV and SOC; • Small amount of computation.	• Long relaxation time for OCV measurement; • Temperature, age and battery types affect the measurement result of OCV.
Model based method	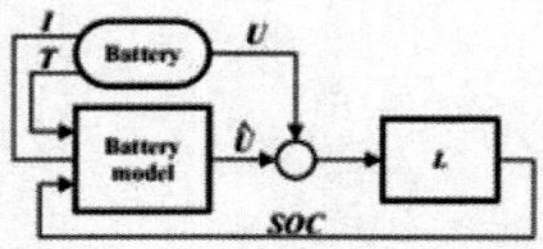	• Insensitive to initial SOC; • good robust; • High accuracy	• Rely on modeling accuracy; • High computation cost
ANN based method	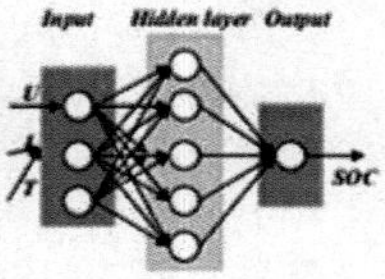	• Do not need previous knowledge about battery; • Easy transplant to hardware after offline training	• Large amount of training samples is needed; • Hard to generalize to different driving cycles.

Fig. 3 Comparison of various SOC estimation algorithms

heartier than an open-circle framework. Accordingly, MBMs have prevalent vigor contrasted and the other three. Notwithstanding, different strategies can likewise accomplish better strength by taking a few measures. It tends to be induced using Eq. (1) that the Coulomb count method strength under different operating cycles can be improved by thinking about the temperature and maturing impacts. Likewise, adding these impacts to the open circuit voltage state-of-charge bend helps change the OCV-based method under other working conditions. MBMs have better heartiness due to the input rectification. Since the battery model's exactness might be diminished during battery utilization, Internet refreshing of the battery model boundaries is essential for guaranteeing its vigor. Moreover, the assessment calculations ought to likewise be inhumane toward demonstrating and sensor blunders. Much preparing data under various working conditions ought to be gathered to improve the power of ANNBMs. The preparation cycle boundaries must be upgraded, and different approval cycles ought to be performed to keep away from the ideal nearby consequences of ANNBMs. Computational over-burden is always considered for equipment execution. Coulomb counting method and OCV-based methods are computationally effective because they include a straightforward count measure. MBMs are tedious, particularly, the Kalman channel containing framework activity in the assessment cycle. Ease applications will pick PI spectator or SM observer for lower calculation trouble. The ANN-based method is less complicated when disconnected before relocating to the installed framework. In outline, steps could be implemented to improve the exactness, vigor, and multifaceted computational nature of charge level's assessment strategies for online execution. For continuous applications, the most appropriate way is a decent compromise of all affecting components (e.g., the prerequisite of exactness, power, and computational exertion) [34].

3 Conclusion

The state of charge of batteries has a quick arranging relationship with their external fixed characteristics, like OCV and impedance. It might be gainfully used to change the inaccurate SOC. By the by, it is difficult to check the specific OCV always since the assessment of OCV of a battery is done by removing power and making sure the battery's rest for a widely inclusive period. Of course, the proportion of battery impedance relies upon the assessment contraption. Consequently, it cannot be executed for running electric vehicles. Such a SOC evaluation methodology is better for being applied to the laboratory atmosphere [35].

The CCM is otherwise called the Coulomb checking strategy. This methodology turns out for batteries because there are no essential outcomes during ordinary action. Nevertheless, for this strategy's evaluation by this strategy, three disadvantages ought to be overseen first. In any case, the initial state of charge must be known. Second, the battery current's assessment bumbles from sporadic aggravations, for instance, uproar and temperature drift, which are inevitable [35].

Lastly, Q should be recalibrated as the assortment of the battery's working conditions and developing levels. The mix of the recently referenced components would also decrease the steady nature of this method. Along these lines, the vital ampere-hour system can work with other supporting strategies, for example, model-based methods [35].

Model-based strategies have been the most robust assessment strategy. MBMs work best on blend methodology. It joins the CCM necessary procedure and battery open circuit voltage limits table-based investigating system by the batteries' state condition. It can be deduced that the system's charge levels probably go like some platform between the CCM and the investigating table-based procedures. An off-base state of charge check dictated by the ampere-hour vital procedure brings an off-base battery's open circuit voltage, and a while later, it grows the gauge bumble of the terminal voltage. The base estimate slip-up of the battery terminal voltage can be deduced accurately best when the charge's state has been deduced this way. The OCV can be used to address the evaluation botch [35].

Specifically, the DD approach can show critical points of interest in the accompanying cases:

1. The worldwide numerical model of the controlled framework is altogether obscure.
2. The vulnerabilities of the controlled framework model are huge.
3. The numerical model cannot be worked for characterizing the controlled framework with a questionable structure in its working cycle.
4. The component model of the controlled framework is excessively convoluted, or the quantity of the request is overly restrictive, or it is unrealistic to break down and plan.

3.1 Recommendation

Battery SOC assessment is vital for battery management system utilized in EV. This paper audits and looks at ordinary SOC assessment strategies, zeroing in their utilization in electric vehicles. Four types of state of charge assessment techniques for a pack of batteries have been efficiently assessed and summed up. Albeit multiple SOC estimation methods have been proposed and comparing progress and applications have been depicted, and the precise prediction and strategies for the correct administration of a pack of batteries cannot be resolved. The hypothetical exploration and innovative utilization of the SOC assessment are remaining difficulties [34, 35].

1. Multi-requirement, multi-scale, and multi-state joint/double assessment. Assessment of battery state of charge includes the exactness of beginning qualities and the estimation and comprises recognizing the way of limit debasement and the warm conduct of batteries. The current techniques predominantly work to amend the underlying mistake of state of charge or accomplish the joint/double assessment for battery limit and the state of charge. Nonetheless, they only occasionally think about the mechanical properties (exhaustion harm), electrical properties (corruption way of intensity), and warm properties (warm disappointment track) of batteries. The combination technique consolidates a DD control system, multi-scale multi-measurement improvement hypothesis, and ideal assessment hypothesis to give a viable answer for the multi-oblige multi-scale state joint assessment [35].
2. Usually, utilized models of the battery pack for EVs contain electrochemical models, equivalent circuit models, and electrochemical impedance models. EMs can display the unpredictable substance response cycle of batteries, yet they cannot give an extensive portrayal of limit corruption, warm disappointment, and batteries' mechanical exhaustion measure. The quality of the equivalent circuit models and electrochemical impedance models is that the models' structure and request are moderately straightforward. The restrictions are that they cannot outline the internal response energy and the limit corruption and maturing way of batteries. Every type of battery model has its pros and cons; hence, a combination of more than one model by brushing various sorts of battery models with calculated combination rule can accomplish excellent prescient execution

under questionable battery maturing levels, operating conditions, and materials used for fabrication of a battery [34, 35].

3. State of charge assessment for a hybrid association battery framework with solid time changing, nonlinear, and non-uniform qualities. The battery pack utilized in the electric vehicles comprises many battery cells. It is trying to guarantee the consistency of the boundary and state for all cells. What is awful, because of the unsettling influence of dubious working conditions, age levels, and the adjusting procedures, SOC assessment strategies intended for battery cells cannot guarantee the SOC assessment precision of the multi-cell battery pack. Therefore, this will, at last, prompt wasteful energy use. Hence, the battery pack's SOC assessment can be identical to a state assessment issue for a half-breed framework with solid time-differing, nonlinear, and non-uniform attributes. Hence, we can look for arrangements from the vulnerability displaying hypothesis, the framework ID hypothesis, and the data-driven control hypothesis [35].

SOC estimation algorithms based on both current and voltage values are ideal. The algorithm should be modeled such that its failure should accurately correspond to the battery/cell failure. The mixed SOC would have the highest applicability for almost all the cases combined, as it takes advantage of the complementary behavior of the other algorithms, i.e., it is more flexible and can take into account failure scenarios better.

References

1. J. Meng, M. Ricco, G. Luo, M. Swierczynski, D.-I. Stroe, A.-I. Stroe, R. Teodorescu, An overview of online implementable SOC estimation, in *A Treatise on Electricity and Magnetism*, vol. 2, 3rd ed., ed. by C. Maxwell. (Clarendon, Oxford, 1892, 2017), pp.68–73.
2. R. Yang, R. Xiong, H. He, H. Mu, C. Wang, A novel method on estimating the degradation and state of charge of lithium-ion batteries used for electrical vehicles. Appl. Energ. **207**, 331–340 (2017)
3. S. Peng, C. Chen, H. Shi, Z. Yao, State of charge estimation of battery energy storage systems based on adaptive unscented Kalman filter with a noise statistics estimator. IEEE Access **5**, 13202–13212 (2017)
4. X. Hu, R. Xiong, B. Egardt, Model-based dynamic power assessment of lithium-ion batteries considering different operating conditions. IEEE Trans. Ind. Inform. **10**(3), 1948–1959 (2014)
5. M. Lurie, H.N. Seier, R.C. Shair, State of charge indicators for nickel-cadmium batteries. State Charg. Indic Nickel Cadmium Batter. (1963)
6. C.W. Lillehei, A.B. Cruz, I. Johnsrude, R.D. Sellers, A new method of assessing the state of charge of implanted cardiac pacemaker batteries. Amer. J. Cardiol. **16**(5), 717 (1965)
7. X. Hu, C. Zou, C. Zhang, Y. Li, Technological developments in batteries: a survey of principal roles types and management needs. IEEE Power Energ. Mag. **15**(5), 20–31 (2017)
8. L. Lu, X. Han, J. Li, J. Hua, M. Ouyang, A review on the key issues for lithium-ion battery management in electric vehicles. J. Power Sour. **226**, 272–288 (2013)
9. M.A. Hannan, M.S.H. Lipu, A. Hussain, A. Mohamed, A review of lithium-ion battery state of charge estimation and management system in electric vehicle applications: challenges and recommendations. Renew. Sustain. Energ. Rev. **78**, 834–854 (2017)
10. Z. Li, J. Huang, B.Y. Liaw, J. Zhang, On state-of-charge determination for lithium-ion batteries. J. Power Sour. **348**, 281–301 (2017)

11. R. Xiong, J. Tian, H. Mu, C. Wang, A systematic model-based degradation behavior recognition and health monitoring method for lithium-ion batteries. Appl. Energ. **207**, 367–378 (2017)
12. V. Pop, H.J. Bergfeld, D. Danilov, P.P.L. Regtien, P.H.L. Notten, *Battery Management Systems: Accurate State-of-Charge Indication for Battery-Powered Applications*, vol. 9. (Springer 2008)
13. S. Rodrigues, N. Munichandraiah, A.K. Shukla, Review of state-of-charge indication of batteries by means of a.c. impedance measurements. J. Power Sour. **87**(1), 12–20 (2000)
14. X. Hu, F. Sun, Y. Zou, Comparison between two model-based algorithms for Li-ion battery SOC estimation in electric vehicles. Simul. Model. Pract. Theor. **34**, 1–11 (2013)
15. L. Lavigne, J. Sabatier, J.M. Francisco, F. Guillemard, A. Noury, Lithium-ion open-circuit voltage (OCV) curve modelling and its ageing adjustment. J. Power Sour. **324**, 694–703 (2016)
16. M.W. Cheng, Y.S. Lee, M. Liu, C.C. Sun, State-of-charge estimation with aging effect and correction for lithium-ion battery. IET Electr. Syst. Transp. **5**(2), 70–76 (2015)
17. S. Tong, M.P. Klein, J.W. Park, Online optimization of battery open circuit voltage for improved state-of-charge and state-of-health estimation. J. Power Sour. **293**, 416–428 (2015)
18. Y. Xing, W. He, M. Pecht, K.L. Tsui, State of charge estimation of lithium-ion batteries using the open-circuit voltage at various ambient temperatures. Appl. Energ. **113**, 106–115 (2014)
19. B. Pattipati, B. Balasingam, G.V. Avvari, K.R. Pattipati, Y. Bar-Shalom, Open circuit voltage characterization of lithium-ion batteries. J. Power Sour. **269**, 317–333 (2014)
20. W. Waag, D.U. Sauer, Adaptive estimation of the electromotive force of the lithium-ion battery after current interruption for an accurate state-of-charge and capacity determination. Appl. Energ. 416–427 (2013.)
21. N. Lin, S. Ci, D. Wu, A novel low-cost online state of charge estimation method for reconfigurable battery pack, in *2016 IEEE Applied Power Electronics Conference and Exposition (APEC)*, 20–24 March 2016, pp. 3189–3192
22. L. Pei, R. Lu, C. Zhu, Relaxation model of the open-circuit voltage for state-of-charge estimation in lithium-ion batteries. IET Electr. Syst. Transp. **3**(4), 112–117 (2013)
23. E. Leksono, I.N. Haq, M. Iqbal, F.X.N. Soelami, I.G.N. Merthayasa, State of charge (SOC) estimation on $LiFePO_4$ battery module using coulomb counting methods with modified peukert, in *2013 Joint International Conference on Rural Information and Communication Technology and Electric-Vehicle Technology (rICT & ICeV-T)*, 26–28 November 2013, pp. 1–4
24. K.S. Ng, C.-S. Moo, Y.-P. Chen, Y.-C. Hsieh, Enhanced coulomb counting method for estimating state-of-charge and state-of-health of lithium-ion batteries. Appl. Energ. **86**(9), 1506–1511 (2009)
25. A. Purwadi, A. Rizqiawan, A. Kevin, N. Heryana, State of charge estimation method for lithium battery using combination of coulomb counting and adaptive system with considering the effect of temperature, in *The 2nd IEEE Conference on Power Engineering and Renewable Energy (ICPERE) 2014*, 9–11 December 2014, pp. 91–95
26. Y. Zhang, W. Song, S. Lin, Z. Feng, A novel model of the initial state of charge estimation for $LiFePO_4$ batteries. J. Power Sour. **248**, 1028–1033 (2014)
27. Z. Caiping, W. Le. Yi, L. Xue, C. Wen, G.G. Yin, J. Jiuchun, Robust and adaptive estimation of the state of charge for lithium-ion batteries. IEEE Trans. Ind. Electron. **62**(8), 4948–4957 (2015)
28. M. Cacciato, G. Nobile, G. Scarcella, G. Scelba, Real-time model-based estimation of SOC and soh for energy storage systems. IEEE Trans. Power Electron. **32**(1), 794–803 (2017)
29. Y. Zhang, R. Xiong, H. He, W. Shen, A lithium-ion battery pack state of charge and state of energy estimation algorithms using a hardware-in-the-loop validation. IEEE Trans. Power Electron. **32**(99), 1–1 (2016)
30. C. Zhang, K. Li, J. Deng, S. Song, Improved real-time state-of-charge estimation of $LiFePO_4$ battery based on a novel thermoelectric model. IEEE Trans. Industr. Electron. **64**(1), 654–663 (2017)
31. H. He, R. Xiong, J. Peng, Real-time estimation of battery 579 state-of-charge with unscented kalman filter and rtos μcos-ii platform. Appl. Energ. **162**, 1410–1418 (2016)
32. Y. Shen, Adaptive online state-of-charge determination based on neuro-controller and neural network. Energ. Convers. Manag. **51**(5), 1093–1098 (2010)

33. M. Xu, J. Deng, S. Quan, SOC estimation of lithium battery based on improved kalman filtering algorithm, in *2016 International Conference on Industrial Informatics—Computing Technology, Intelligent Technology, Industrial Information Integration (ICIICII)* (2016)
34. J. Meng et al., An overview of online implementable SOC estimation methods for Lithium-ion batteries, in *2017 International Conference on Optimization of Electrical and Electronic Equipment (OPTIM) and 2017 Internationall Aegean Conference on Electrical Machines and Power Electronics (ACEMP)*, Brasov (2017), pp. 573–580. https://doi.org/10.1109/OPTIM.2017.7975030.
35. R. Xiong, J. Cao, Q. Yu, H. He, F. Sun, Critical review on the battery state of charge estimation methods for electric vehicles. IEEE Access **6**, 1832–1843 (2018). https://doi.org/10.1109/ACCESS.2017.2780258
36. D.N.T. How, M.A. Hannan, M.S. Hossain Lipu, P.J. Ker, State of charge estimation for lithium-ion batteries using model-based and data-driven methods: a review. IEEE Access **7**, 136116–136136 (2019). https://doi.org/10.1109/ACCESS.2019.2942213
37. P. Tejeswini, P. Sivraj, Artificial intelligence based state of charge estimation of li-ion battery for EV applications, in *2020 5th International Conference on Communication and Electronics Systems (ICCES)*, Coimbatore, India (2020), pp. 1356–1361. https://doi.org/10.1109/ICCES48766.2020.9137999

Self-Sustaining Community for a Green Future—A Case Study

H. C. Gururaj and Vasudha Hegde

1 Introduction

Energy demand in India during 2019–2020 stood at 1,290,247 Million units while the supply was 1,283,960 Million units accounting to a deficit of 0.5%. The fossil fuels help meet 64% of the energy demand, and the renewable sources account for about 24% [1]. With India being one of the fastest growing economies in the world, the energy demand is only going to increase, as witnessed in 2018; India's energy demand outpaced the growth globally [2]. Adding to the woes, India is the fourth largest auto market in the world [3]. Resulting in India being ranked third in greenhouse gases emissions globally [4]. In order to meet its pledge of reducing the carbon emissions by 35% around 2030, India has seen a massive acceleration of investment in renewable energy sector and also subsidizing the EVs enabling a switch over from conventional internal combustion engines. The EV technology has matured and the production costs have come down and now almost on par with conventional automobiles. Affordability of the EVs coupled with increased consciousness on carbon emissions is resulting in more and more people switching to EVs, doing their bit to curb carbon emissions.

Now, the flip side of it will be the sudden surge in energy demand to charge the ever-increasing number of EVs on the road. India cannot go back to fossil fuels or nuclear sources for its energy supply. The only way forward is the clean and green energy sources like wind and solar. India being a hot tropical county is blessed with

H. C. Gururaj (✉)
Department of Electrical and Electronics Engineering, DRR Government Polytechnic, Davangere, India

V. Hegde
Department of Electrical and Electronics Engineering, Nitte Meenakshi Institute of Technology, Bengaluru, India
e-mail: vasudha.hegde@nmit.ac.in

Sanjeevikumar et al. (eds.), *Advances in Renewable Energy and Electric Vehicles*, Lecture Notes in Electrical Engineering 767,
https://doi.org/10.1007/978-981-16-1642-6_30

abundant sunshine and wind energy which can be utilized at the community level, making it self-sufficient to meet both household and vehicular energy demands. Rest of the paper is divided into the following sections: Sect. 2 discusses on the related work, Sect. 3 describes the collection and analysis of the data for simulation, the proposed rooftop models for extracting wind and solar energy. Section 4 focuses on the simulation results and Sect. 5 gives the concluding remarks.

2 Related Work

In [5], the utilization of wind energy on top of high-rise building to generate electricity was proposed. Since the direction and velocity of wind is not constant, a geometric progress model was used to adapt power which is unstable in nature. The consumer load is satisfied by both power from the grid (standby source) and the power generated from rooftop wind turbines (primary source). The switch over is handled by automatic transfer switch. The consumer load is classified into subsystems based on geometric progress model. Rooftop wind turbines drive DC generators which power the lights in underground garages and roads in the residential areas. The proposed model would save 13,500 Kwh. In [6], VAWT coupled to permanent magnet synchronous generator (PMSG) was employed to generate electricity on top of a bus. The turbines are placed at the front end of the bus inside the chamber. As the vehicle moves on the highway with an average speed of 60–80 kms per hour, the wind speed will be sufficiently high to generate considerable amount of energy which can be stored in the battery for future consumption or utilized in real time to supply the local loads such as an air conditioner resulting in instant improvement in fuel economy of the vehicle. In [7], recent advancement in wind resource assessment pertaining to an urban habitat is put forth. The choice of wind turbine for exploiting the wind resource available at the site, and the role of urban aerodynamics in extraction of enhanced wind energy is presented. Prediction of wind speed in an urban area is carried out using wind tunnels and computational fluid dynamics (CFD). The paper discusses about building integrated wind turbines (BIWT) and diffuser augmented turbines for urban applications, placement of the turbines on different locations of the building and the effect of wind conditions on the placement. In [8], study pertaining to pattern of wind flow in built up areas of Singapore is carried out using measuring instruments like wind mast and light detection and ranging (LiDAR) profiler. The prevalent wind characteristic is comprehended by carrying out statistical analysis like wind rose and Weibull distribution. The research concluded that the Singapore's southern shore is ideal for wind energy generation. In [9], the installation of rooftop solar PV panels having a battery system with cost analysis and benefits were presented. Installation capacity was optimized and minimum rating of the battery required to store the energy was arrived at using charge/discharge algorithm. The results depend on load profile of the consumer, net metering scheme, and changes in charge/discharge algorithm. An enhanced algorithm could provide optimum results. In [10], an off-grid solar PV system's technical and economic analysis was carried out using a mathematical model. Taking into account, real load conditions and solar technology's cost of implementation for a hostel with 192 rooms spread over a total of 6 floors would be 1.33 Crore Indian Rupees (INR). The payback period was estimated to be around 10 years and reduction of carbon dioxide emissions would be 4490 tons and would yield a profit of 1.95 Crore INR in its total lifetime of 25 years is a great incentive for the consumer's shift from grid to solar PV system. In [11], the application of solar PV system of 10 KW capacity for charging EVs at work location in the Netherlands was proposed. The comparison is drawn between a conventional vehicle driven by gasoline and an EV charged using either utility grid or solar PV system. Also, a comparison was made between solar panels installed on rooftop and solar carport. Choosing an EV over conventional vehicle would result in a saving of 1520 Euros per annum and reduced carbon dioxide emissions of 38.5 g/km. In [12], the hybrid generation on top of the moving vehicle was presented. A chamber containing channels for airflow were placed on top of the bus, housing VAWT coupled to a PMSG. On top of the chamber flexible solar panels were placed. As India receives sunshine for more than 300 days in a year and there is no obstruction for wind flow on the highways. Both solar and wind energy can be utilized to generate electricity when the vehicle travels from source to destination. The energy generated can be stored in the batteries. Upon reaching the destination, the batteries can be used as plug and play in electric scooters or stored energy can be exported to the grid.

3 Design

This section discusses about the collection and processing of the data for simulation, systems for transforming wind and solar energy.

3.1 Collection and Processing of the Data for Simulation

In order to have realistic values for simulation, wind speed, solar irradiance, and temperature data were collected from National Aeronautics and Space Administration (NASA)'s—Goddard Space Flight Centre—MERRA-2 meteorological data [13] for the month of July-2019 pertaining to Chitradurga, Karnataka, India. The datasets were processed to extract the minimum, maximum, and average values of the quantities as depicted in Table 1. The data from Table 1 is input to the simulation models. The proposed model is comprised of two systems: system for wind energy transformation and system for solar energy transformation as shown in Fig. 1. The building is assumed to be three storied (about 35 feet tall).

Housing three families (one family in each floor) and every family is assumed to have a Bajaj Chetak e-Scooter and one of the families having a Tata Nexon electric car for commutation. The units/Kwh consumption and the electricity bills for a period

Table 1 Minimum, maximum, and average values of wind speed, solar irradiance, and temperature

Quantity	Minimum value	Maximum value	Average value
Wind speed in m/sec	3.13	12.5	7.17
Solar irradiance in w/m^2	0.86	947.64	333.22
Temperature in °C	20.37	33.7	25.04

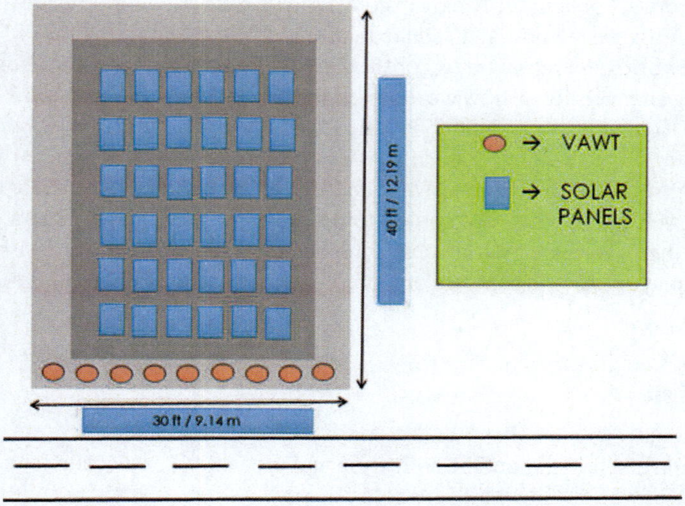

Fig. 1 Top view of the building showing VAWT and solar panels

of one year pertaining to a home are collected and average values were found out as shown in Table 2.

Bajaj Chetak has a 4 Kwh battery and a practical range of around 80 km [14]. And Tata Nexon electric car has a battery capacity of 30 Kwh and a range of around 250 km per recharge [15].

Assuming a daily city commute of 20 km for both scooters and car and a highway commute of 1000 km per month for the car. The Kwh consumption per month and the approximate bill to be paid for charging the batteries are as shown in Table 3. Table 4 shows the cumulative energy consumption per month for three homes in the building, three electric scooters, and one electric car used by the occupants.

Table 2 Units consumed in a month and the corresponding electricity bill for a family

Month	Units consumed in Kwh	Electricity bill in INR
Jun-19	170	1262
Jul-19	152	1027
Aug-19	205	1415
Sep-19	160	1171
Oct-19	143	967
Nov-19	151	1008
Dec-19	164	1090
Jan-20	168	1120
Feb-20	165	1097
Mar-20	180	1345
Apr-20	175	1302
May-20	168	1110
Average	167	1160

3.2 System for Wind Energy Transformation

The system comprises of vertical axis wind turbine (VAWT) coupled to a PMSG, a boost converter and the battery bank. Along with occupying less space and working with low wind speed, the VAWT does not need the wind to be blowing in a specific direction or with constant velocity. Hence, can do away with complicated yaw mechanism. All of which makes VAWT perfectly suit rooftop applications [16].

In order to keep the VAWT size practical, the blade dimensions are chosen as: height = 0.3 m and diameter = 0.25 m for an *H* type rotor [17] as shown in Fig. 2. Figure 3 shows the MATLAB and Simulink model used for wind energy transformation. 20 of such VAWTs are placed with sufficient spacing between them on the parapet wall of the building facing the road so that there is no obstruction for the wind flow.

The mechanical power, P_m obtainable from the wind is given by Eq. 1,

$$P_m = \frac{1}{2} \times C_p \times \rho \times A \times v^3 \text{ Watts} \tag{1}$$

where,

C_p →wind power co-efficient, ρ → density of air (kg/m^3), A → swept area (m^2), v → wind speed (m/sec)

The value of $C_p = 0.35$, $\rho = 1.225$ kg/m^3, $A = 0.3$ m$\times$0.25 m and the value of v varies from 3.13 to 12.5 m/sec based on which the value of P_m changes continuously. The rating of PMSG chosen is the one which is commercially available: 200 W, 24 V, and 600 rpm. Since the battery operates at 48 V, a boost converter is employed to step-up the generated voltage of 24 to 48 V.

Table 3 KWh consumption per month for electric scooter, car, and the corresponding approximate bill by the end of the month

Vehicle type	Vehicle name	Battery capacity in Kwh	Range in kilometers	Total kilometers in a month	Kwh consumed in a month to charge the batteries	Per unit cost in INR	Approximate bill by the end of the month in INR
Car	Tata Nexon	30.2	250	1440	180	7.8	1404
Scooters	Bajaj Chetak	4	80	600	32	7.8	250
	Bajaj Chetak	4	80	600	32	7.8	250
	Bajaj Chetak	4	80	600	32	7.8	250

Table 4 Cumulative units consumption for both household and vehicular energy demand

Vehicle/home	Kwh consumed in a month	Approximate electricity bill in INR
Car	180	1404
Scooter 01	32	250
Scooter 02	32	250
Scooter 03	32	250
Home 01	167	1160
Home 02	167	1160
Home 03	167	1160
Total	777	5634

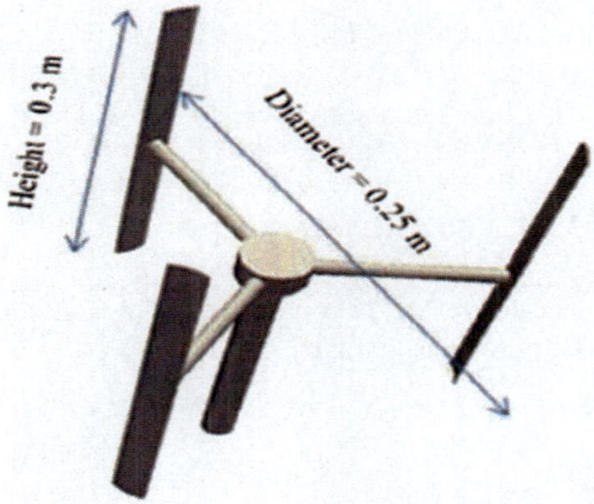

Fig. 2 *H* type VAWT rotor

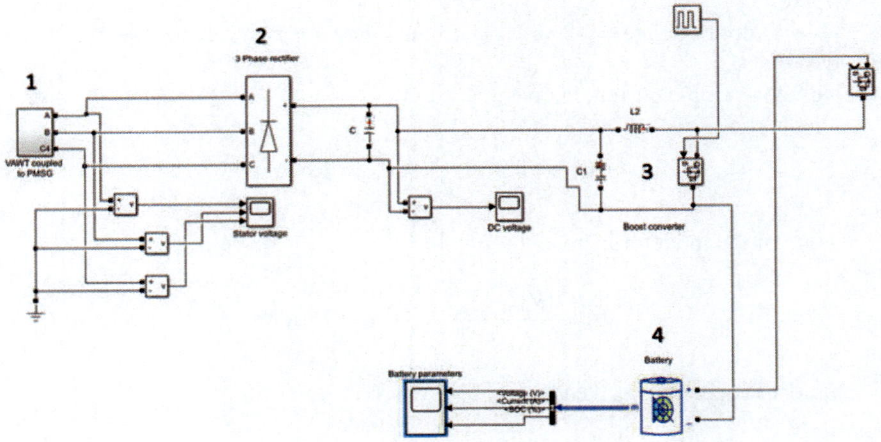

Fig. 3 MATLAB and Simulink model for wind energy transformation

MATLAB and Simulink model for wind energy transformation constitutes of four main parts.

3.2.1 Wind Generation

This block generates a random number between the minimum wind speed (3.13 m/sec) and maximum wind speed (12.15 m/sec) which holds constant for a specified time period before changing to another value, and the generated number is fed as an input to the wind turbine block. The output of which is given to the PSMG block. The output is a three-phase AC.

3.2.2 Three-Phase Rectifier

This block converts three-phase AC to DC.

3.2.3 Boost Converter

The converter [18] comprises of capacitor, inductor, pulse generator, and a MOSFET. Duty ratio is given by Eq. 2,

$$D = \frac{V_o - V_{\text{in}}}{V_o} \tag{2}$$

where,

$V_o \rightarrow$ Output voltage = 48 V, $V_{\text{in}} \rightarrow$ Input voltage = 24 V. Hence, the value of D is arrived at 0.5.

Inductor's ripple current is given by Eq. 3,

$$\Delta I_L = 0.2 \times \frac{V_o}{V_{\text{in}}} \times I_o = 1.52\text{ A} \tag{3}$$

Value of the inductor is given by Eq. 4,

$$L = \frac{(V_{\text{in}} - V_o) \times V_o}{\Delta I_L \times F_s \times V_{\text{in}}} = 875\ \mu H \tag{4}$$

Value of capacitor is given by Eq. 5,

$$C = \frac{I_o \times D}{\Delta V_c \times F_s} = 5800\ \mu F \tag{5}$$

3.2.4 Battery

The battery used is 48 V, 24 Ah Lithium-ion battery.

3.3 System for Solar Energy Transformation

Figure 4 shows the MATLAB and Simulink model for solar energy transformation. The system comprises of the following parts.

3.3.1 Solar Panels

The flexible solar panels [19] are chosen since they are light in weight and installation process is simple. The dimensions are: 1.245 m in length, 0.55 m in width, and 2.5 mm in thickness. Table 5 gives the solar panel specifications. The rooftop area of the building is 30ft/9.14 m wide and 40ft/12.19 m long. Considering the dimensions of rooftop and probability of shading from the parapet walls. 12 solar panels can be placed side by side in a row and 8 such rows can be accommodated. Hence, a total of 96 solar PV panels can be employed for generating electricity. But the total number of solar panels is limited to 90 in order to meet the voltage and current rating of the battery. The panels are placed south facing. Three panels can be connected in series

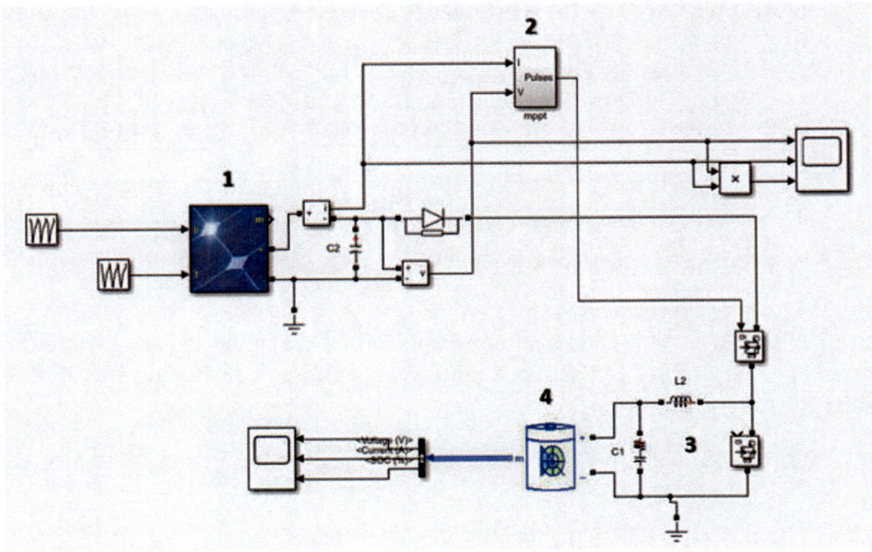

Fig. 4 MATLAB and Simulink model for solar energy transformation

Table 5 Solar panel specifications

Max-power: P_m (W)	100 W
Max-power voltage: V_{mp} (V)	18
Max-power current: I_{mp} (A)	5.56
Open-circuit voltage: V_{oc} (V)	21.24
Short circuit current: I_{sc} (A)	6.12
Max-system voltage: V_{dc} (V)	DC500V
Cell efficiency	18%
Operating temperature (C)	−20C to + 50C
Power tolerance (%) T	3%

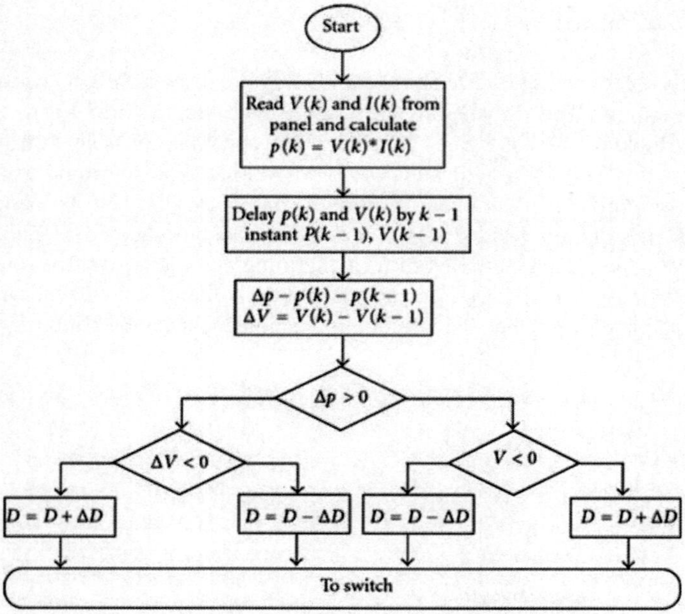

Fig. 5 Flowchart for P and O algorithm of MPPT

string (18 × 3 = 54 V) and five such combinations are connected in parallel string (5 × 5.56 = 27.8 A) and six such units are considered constituting a total of 90 solar PV panels.

3.3.2 Maximum Power Point Tracking

The flowchart for the Perturb and Observe algorithm of MPPT is shown in Fig. 5. Perturb and Observe algorithm [20] is adopted for tracking the maximum power from PV module.

3.3.3 Buck Converter

The output voltage from the series string is 54 V, whereas the battery operates at 48 V. Hence, calls for a buck converter [21]. The duty ratio is given by Eq. 6,

$$1 - D = \frac{V_O - V_{IN}}{V_O} \tag{6}$$

where,
$V_o \rightarrow$ Output voltage, $V_{in} \rightarrow$ Input voltage.
The value of D is arrived at 0.11.
The ripple current of the inductor is given by Eq. 7,

$$\Delta I_L = 0.2 \times \frac{V_o}{V_{in}} \times I_o = 4.44 \text{ Amps} \tag{7}$$

The value of inductance (L) is given by Eq. 8,

$$L = \frac{(V_{in} - V_o) \times V_o}{\Delta I_L \times F_s \times V_{in}} = 120\ \mu H \tag{8}$$

The value of capacitor C is given by Eq. 9,

$$C = \frac{I_o \times D}{\Delta V_c \times F_s} = 4635\ \mu F \tag{9}$$

3.3.4 Battery

The battery pack used is a 48 V, 24 Ah Lithium-ion batteries.

4 Simulation Results and Discussion

4.1 System for Wind Energy Transformation

The wind speed is varied between minimum and maximum values recorded in a month referring to Table 1. A random number between the specified limits of wind speed is generated and it holds constant for a particular amount of time, after which it changes to a new value. The simulation is run for 30 seconds and each value of wind speed is held constant for 10 seconds. Figure 6 shows the inputs to the block for generating random number to simulate real-life variations in wind speed. Figure 7

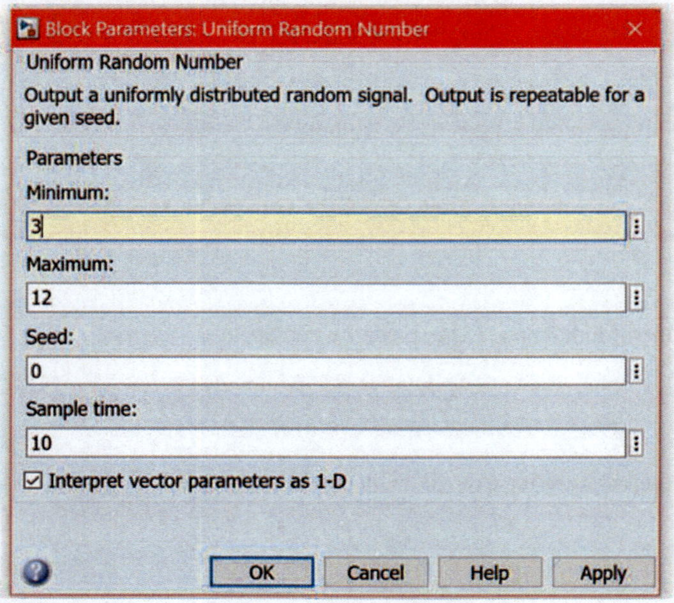

Fig. 6 Random number generator block for simulating real-life variations in wind speed

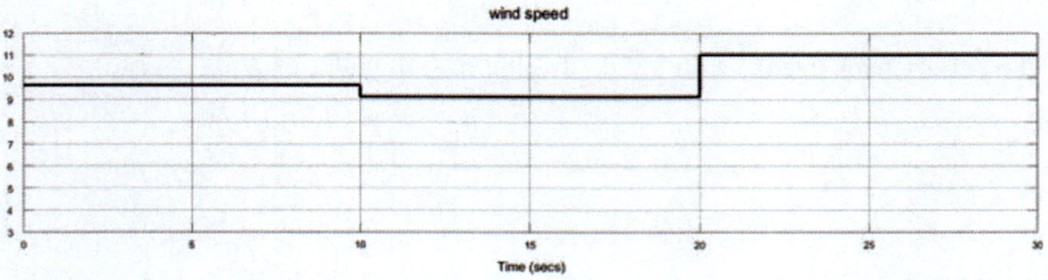

Fig. 7 Wind speed variation with time

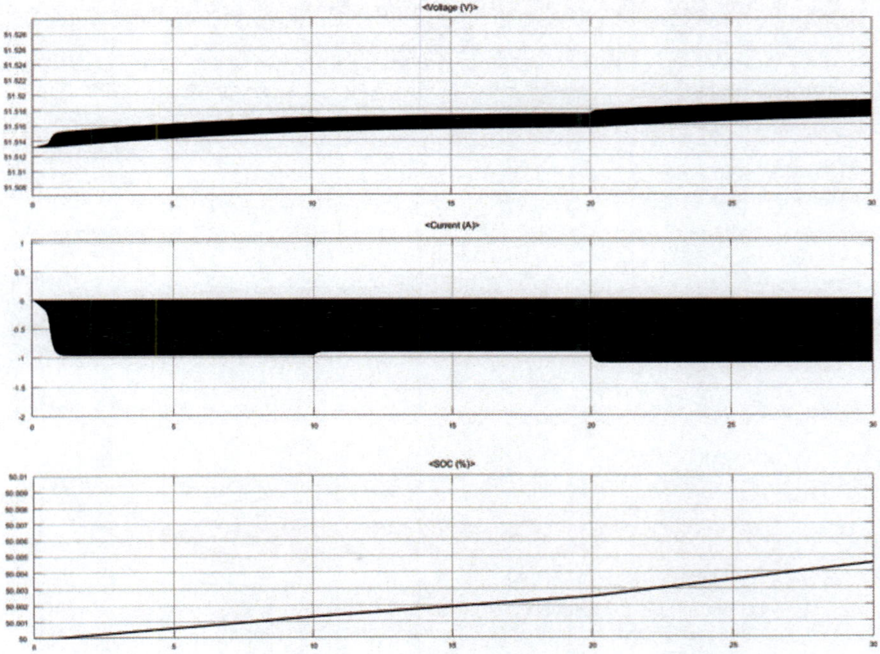

Fig. 8 Battery voltage, current, and SoC curves for wind energy transformation system

shows the variation in wind speed with respect to time. And as shown in Fig. 8 regardless of variations in wind speed, the battery is charged with constant voltage of around 51.5 V and the output current is around 1 A. Also, it is clear that SoC% is increasing which indicates charging of battery.

4.2 *System for Solar Energy Transformation*

The PV panel configuration for the specified series and parallel string combination is as shown in Fig. 9. And Fig. 10 shows the variation in irradiance and Fig. 11, the variation in temperature in accordance with values from Table 1 as input to the solar PV module to simulate real-life changes throughout the sunshine hours.

Figure 12 shows the voltage, current, and SoC curves. From the figure, the voltage is constant around 51 V and current is around 25 A and SoC has increased from 50 to 50.28% for a simulation time of 10 seconds indicating the charging of the battery.

From Fig. 8, the current is around 1 A, when 20 such VAWT + PMSG combinations are connected in parallel, total current will be $1 \times 20 = 20$ A. Hence, one 48 V, 24Ah battery will be charged in 72 min and in a day with 1440 min, 20 such batteries

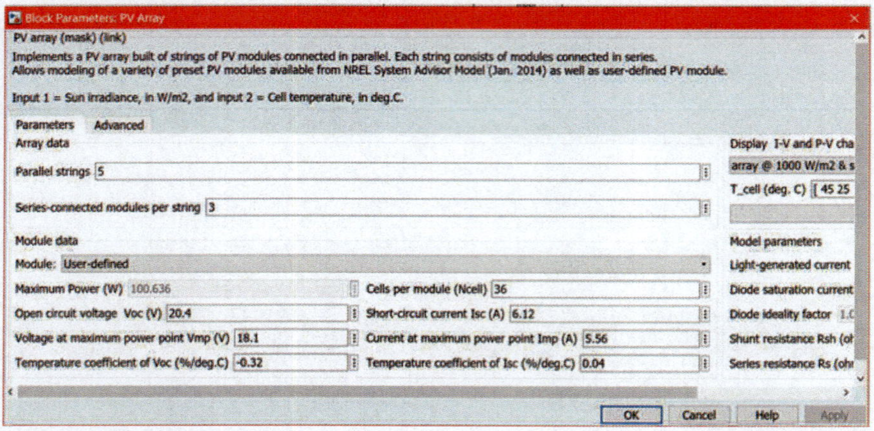

Fig. 9 PV array parameters

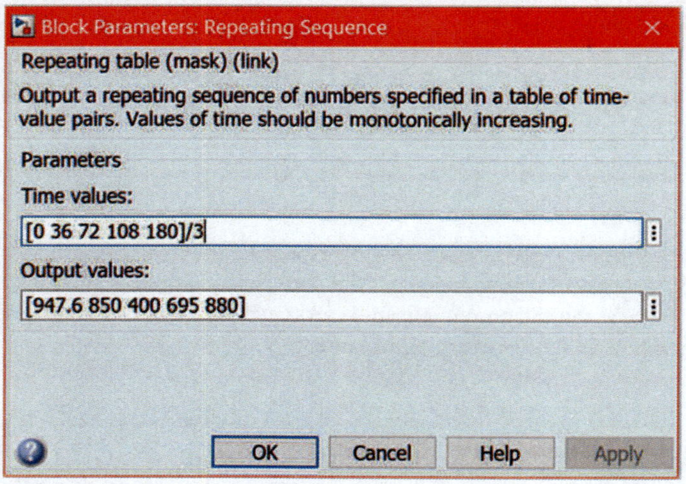

Fig. 10 Solar irradiance variation with time

can be charged or $20 \times 48 \times 24 = 23{,}040$ W or 23 Kwh of energy is generated utilizing the wind energy.

From Fig. 12, the rise in SoC is 0.28% in 10 seconds for 3 solar panels in series and 5 such parallel combinations. Hence, for 6 such units (having 3 solar panels in series and 5 such parallel combinations each), it will be $0.28\% \times 6 = 1.68\%$. So, for 60 seconds, it will be 10.08%. Converting it into number $= 0.1008$. For 1 hour the value is 6.048. Multiplying it with voltage and current rating, $6.048 \times 48 \times 24 = 6967.2$ watts or 6.967 Kw. Hence, 6 number of 48 V, 24Ah battery is charged by the end of an hour. Considering 06 sunshine hours in a day, 36 such batteries can

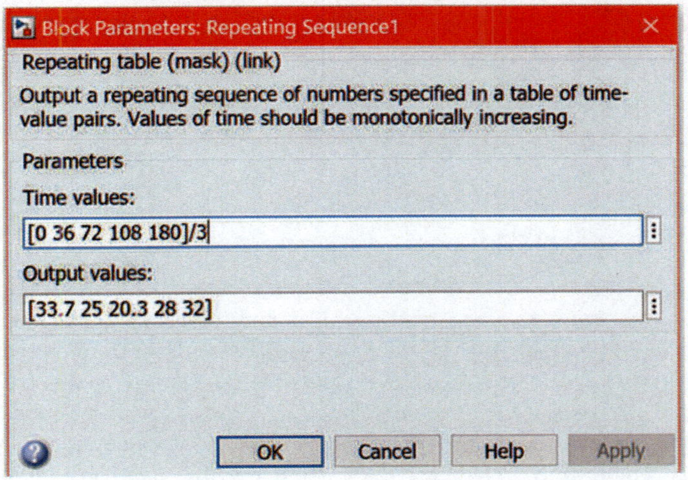

Fig. 11 Temperature variation with time

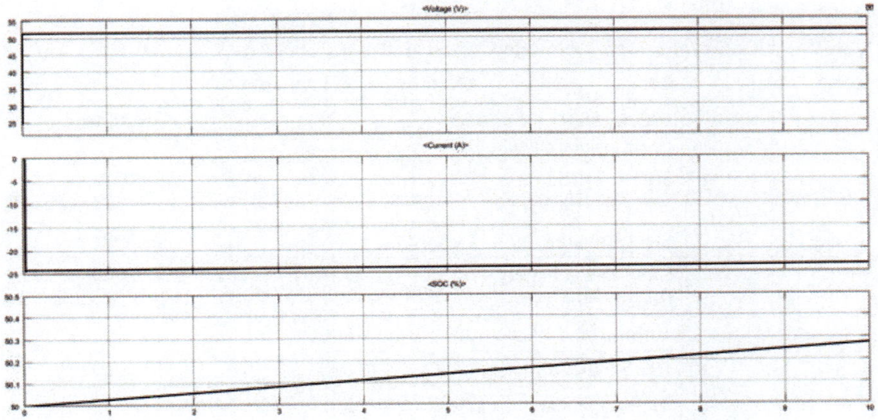

Fig. 12 Battery voltage, current, and SoC curves for solar energy transformation system

be charged or 6967.2 × 6 = 41803.2 Wh or 41.8 Kwh of energy is generated utilizing the solar energy. Put together a total of 56 number of 48 V, 24Ah batteries can be charged or 64.8 Kwh of energy is generated in day as depicted in Table 6. The total energy demand including household and vehicular needs is 777 units/Kwh in a month, whereas the generated units/Kwh is 1944, resulting in a surplus generation of 1167 units/Kwh. Out of which 167 units can be reserved as buffer for any unplanned increase in energy demand and remaining 1000 units can be exported to the utility grid generating income.

Table 6 Units/KWH generated in a month v/s the demand

Source	Approximate Kwh generated in a day	Approximate Kwh generated in a month	Approximate Kwh demand per month	Surplus Kwh generated	Buffer Kwh reserved	Approximate Kwh that can be exported to the utility grid
Solar	41.8	1254	777	1167	167	1000
Wind	23.0	690				
Total	64.8	1944				

Table 7 gives the breakup of the estimation of cost for installing the hybrid system on the rooftop of the building. All the prices shown are taken from online stores to keep the estimation of cost as accurate as possible. The approximate investment to be made is 1,386,000/- INR. If the community chooses to avail the subsidy by the government, the investment cost is reduced by 30% and the amount payable by the community would drop to 970,200/- INR. Table 8 shows the return on investment (ROI). When the subsidy is availed the ROI is 9.39 years and without subsidy it will take 12 years to get back the money invested.

Table 7 Estimation of cost for the rooftop hybrid system

S. No	Item	Quantity	Per unit cost in INR	Total cost in INR
1	100 W, 18 V flexible solar panel	90	3400	306,000
2	3 KVA hybrid on grid solar inverter	1	90,000	90,000
3	Steel structure	Lumpsum		30,000
4	Lithium-ion battery pack 48 V, 24Ah	56	13,000	728,000
5	Turbine blade setup	20	2000	40,000
6	200 W, 24 V PMSG	20	6000	120,000
7	24–48 V boost converter	20	600	12,000
8	Power cables	Lumpsum		20,000
9	Labor charges	Lumpsum		20,000
10	Miscellaneous charges	Lumpsum		20,000
Grand Total				1,386,000
30% Subsidy by the government				415,800
Amount payable by the community				970,200

Table 8 Return on investment

Return on Investment		
ROI	With government subsidy	Without government subsidy
Total investment in INR	970,200	1,386,000
Savings in INR/month in terms of electricity bill	5634	5634
Revenue generated/month by selling the excess units	$1000 \times 2.97 = 2970$	$1000 \times 3.99 = 3990$
Total benefit/month in INR	8604	9624
Number of months needed for ROI	113	144
Number of years needed for ROI	9	12

5 Conclusion

The main objective of proposing Lithium-ion batteries for the model is that, if the EV manufacturers devise a method where the depleted battery can be taken out of the vehicle and replaced by a fully charged one, then it eliminates the need to spend hours on charging the EVs.

For the case study considered, the proposed model has the potential to make the community completely self-sustained helping meet their household and vehicular energy demands, exporting the excess units generated, thereby increasing the share of clean energy on the grid. All this is done by utilizing freely available renewable energy sources on the rooftop of the building.

The model along with saving on the electricity bill helps generate revenue there by reducing the time needed for payback on investment. After the breakeven point, acts as a constant source of income. The model not only has monetary benefits but also is environmentally friendly as it makes an EV "truly non-polluting" by providing the energy to charge the batteries from renewable energy sources, which was not possible earlier as majority of the energy in the grid was generated using fossil fuels. Thus, zooming the community, the country, and the world toward a clean, green, and a sustainable future.

References

1. Power Sector at a Glance ALL INDIA [online]. (09 Nov 2020), Available: https://powermin.nic.in/en/content/power-sector-glance-all-india. Accessed on 23 Nov 2020
2. India's energy demand outpaces the global growth [online]. (26 Mar 2019), Available: https://www.thehindubusinessline.com/economy/india-outpace-global-energy-demand-growth-in-2018-iea/article26643852.ece. Accessed on 23 Nov 2020
3. India pips Germany, ranks 4th largest auto market now [online]. (24 Mar 2018), Available : https://economictimes.indiatimes.com/industry/auto/india-pips-germany-ranks-4th-largest-auto-market-now/articleshow/63438236.cms?from=mdr. Accessed on 23 Nov 2020

4. The carbon brief profile: India [online]. (14 Mar 2019), Available: https://www.carbonbrief.org/the-carbon-brief-profile-india. Accessed on Nov 23,2020
5. Z. Zhenglin, W. Hongyan, B. Xiaolong, X. Hongxiang, Study and application of the wind power on the tall building rooftops in large cities. 2011 International Conference on Materials for Renewable Energy & Environment, (Shanghai, 2011), pp. 566–569. doi: https://doi.org/10.1109/ICMREE.2011.5930876
6. H.C. Gururaj, H. Vasudha, Wind energy conversion system for a moving vehicle. Int. J Recent Technol. Eng. (IJRTE) ISSN: 2277–3878, **8**(3) (September 2019)
7. T. Stathopoulos, H. Alrawashdeh, A. Al-Quraan, B. Blocken, A. Dilimulati, M. Paraschivoiu, P. Pilay, Urban wind energy: some views on potential and challenges. J. Wind Eng. Ind. Aerodyn. **179**, 146–157, ISSN 0167–6105 (2018)
8. B.R. Karthikeya, S.P. Negi, N. Srikanth, Wind resource assessment for urban renewable energy application in Singapore. Renew. Energy **87**(1), 403–414, ISSN 0960–1481 (2016)
9. B. Tangwiwat, K. Audomvongseree, Benefit and cost analysis of the installation of rooftop solar PV with battery system. 2018 15th International Conference on Electrical Engineering/Electronics, Computer, Telecommunications and Information Technology (ECTI-CON), (Chiang Rai, Thailand, 2018), pp. 505–508. doi: https://doi.org/10.1109/ECTICon.2018.8619990
10. P. Sharma, H. Bojja, P. Yemula, Techno-economic analysis of off-grid rooftop solar PV system. 2016 IEEE 6th International Conference on Power Systems (ICPS), (New Delhi, 2016), pp. 1–5. doi: https://doi.org/10.1109/ICPES.2016.7584208
11. G.R.C. Mouli et al., Economic and CO2 emission benefits of a solar powered electric vehicle charging station for workplaces in the Netherlands. 2016 IEEE Transportation Electrification Conference and Expo (ITEC), (Dearborn, MI, 2016), pp. 1-7. doi: https://doi.org/10.1109/ITEC.2016.7520273
12. H.C. Gururaj, V.M. Shivareddy, H.Vasudha, M.J. Nagaraj, Hybrid Power Generation on top of the Vehicle. 2019 Global Conference for Advancement in Technology (GCAT), (BANGALURU, India, 2019), pp. 1-7. https://doi.org/10.1109/GCAT47503.2019.8978331
13. Global Modeling and Assimilation Office (GMAO), MERRA-2 tavg1_2d_slv_Nx: 2d,1-Hourly,Time-Averaged,Single-Level,Assimilation,Single-Level Diagnostics V5.12.4. Goddard Earth Sciences Data and Information Services Center (GES DISC), (Greenbelt, MD, USA, 2015). https://doi.org/10.5067/VJAFPLI1CSIV. Accessed 15 July 2020
14. Bajaj Chetak electric scooter launch today; price, features, specifications, other important details (14 Jan 2020) [online]. Available :https://www.indiatoday.in/auto/latest-auto-news/story/bajaj-chetak-electric-scooter-launch-today-price-features-specifications-other-important-details-revealed-1636656-2020-01-14. Accessed on 23 Nov 2020
15. Tata Nexon EV launch today; price, features, specifications, battery, range, other details explained (28 Jan 2020) [online]. Available :https://www.indiatoday.in/auto/latest-auto-news/story/tata-nexon-ev-launch-today-price-features-specifications-battery-range-warranty-all-other-details-explained-1640744-2020-01-28. Accessed on 23 Nov 2020
16. B.H. Khan, *Non-Conventional ENERGY Resources* 2nd edn
17. H. Dai, Z. Yang, L. Song, Mathematical modeling for H-type vertical axis wind turbine, Proceedings of the 11th IEEE International Conference on Networking, Sensing and Control, (Miami, FL, 2014), pp. 695–700. https://doi.org/10.1109/ICNSC.2014.6819710
18. A. Causo, G. Dall'Aglio, A. Sala, A. Salati, Power converter for vertical axis micro-wind generators. 2011 International Conference on Clean Electrical Power (ICCEP), (Ischia, 2011), pp. 304–307. https://doi.org/10.1109/ICCEP.2011.6036303
19. K. Trautz et al.,High efficiency flexible solar panels. 2013 IEEE 39th Photovoltaic Specialists Conference (PVSC), (Tampa, FL, 2013), pp. 0115-0119. https://doi.org/10.1109/PVSC.2013.6744111

20. M. Kumar, M. Kachhwaya, B. Kumar, Development of MATLAB/Simulink based model of PV system with MPPT. 2016 7th India International Conference on Power Electronics (IICPE), (Patiala, 2016), pp. 1–4. https://doi.org/10.1109/IICPE.2016.8079336
21. B.R. Jogi, Y.K. Bharath, A.A. Kadle, A DC-DC buck converter for standalone solar DC micro-grid system using perturb and observe MPPT algorithm. 2019 Global Conference for Advancement in Technology (GCAT), (BANGALURU, India, 2019), pp. 1-5. https://doi.org/10.1109/GCAT47503.2019.8978278

Simulation-Based Design for an Energy-Efficient Building

Veena Mathew, Ciji Pearl Kurian, and Aravind Babu

1 Introduction

The building information modeling is a tool that could virtually model a building which helps the stakeholders to understand and visualize how the building would look like and analyze the energy performance in the early design stage [1]. There are various software tools for analyzing the energy performance of a building, such as eQUEST, Energy Plus, TRNSYS, and Autodesk Revit. Multiple parameters which affect the energy consumption of the building includes window to wall ratio, building orientation, shading coefficient, visible light transmittance, the heights of window and windowsill, and the control method of lighting. By choosing an optimal value of these parameters, the energy usage can be reduced [2–6]. The incapability of static glazing in fulfilling the variations in heat transmittance due to its fixed properties, filtering an optimum glazing technique for window, has got a crucial role in energy management, daylight harnessing, and inside comfort. Comparison of energy management methods in office buildings with electrochromic (EC) windows and other static glazing materials for three different climates; hot and dry, cold, mixed cold and hot, provide with a significant amount of energy savings of 45% for all the locations [7, 8]. This paper simulates various building parameters that influence building energy consumption and the impact of static and dynamic types of glazing to ensure the building sustainability.

V. Mathew · C. P. Kurian (✉) · A. Babu
Department of Electrical and Electronics, Manipal Institute of Technology, MAHE Manipal, Udupi, India
e-mail: ciji.pearl@manipal.edu

Sanjeevikumar et al. (eds.), *Advances in Renewable Energy and Electric Vehicles*, Lecture Notes in Electrical Engineering 767,
https://doi.org/10.1007/978-981-16-1642-6_31

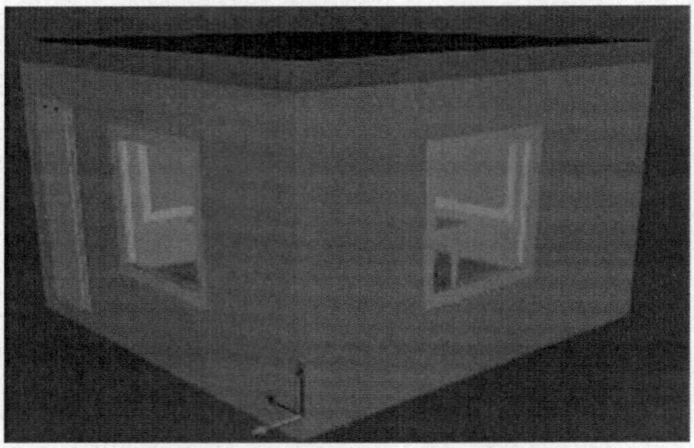

Fig. 1 3-D view of the test room

Table 1 Existing input parameters of test room selected for study

Parameter	Specification
Roof	4-inch lightweight concrete ($U = 0.2245$ BTU/h.ft^2.°F)
Walls	8-inch lightweight concrete block ($U = 0.1428$ BTU/h.ft^2.°F)
Ceiling	8-inch lightweight concrete ceiling ($U = 0.2397$ BTU/h.ft^2.°F)
Floor	Passive floor, no insulation, tile or vinyl ($U = 0.5210$ BTU/h.ft^2.°F)
HVAC	Residential 14 SEER/0.9 AFUE split/packaged gas < 5.5 ton

2 Simulation

2.1 Existing Condition

A test room with dimensions 4.17 m × 4.17 m × 2.31 m located in MIT, Manipal, was considered for analysis. For the simulation purpose, among the four windows, one window was found at a time. Figure 1 shows the 3-D view of the test room and Table 1 listed the existing input parameters for the simulation.

2.2 Effect of Static Glazing on Energy Consumption

The mean energy use intensity (EUI) of the building for the different glazing types, for east, west, north, south window orientations are shown in Fig. 2. Table 2 shows the different glazing considered for the analysis. The highest energy consumption was observed with the windows having clear glass (Type 1) with all solar gain transmitted. So, the choice of glazing has highly impacted by the emissivity constant of the same.

The low value of SHGC constitutes less heat gain in the interior space. The insulation offered by the glass material will be more if U value decreases. The visible

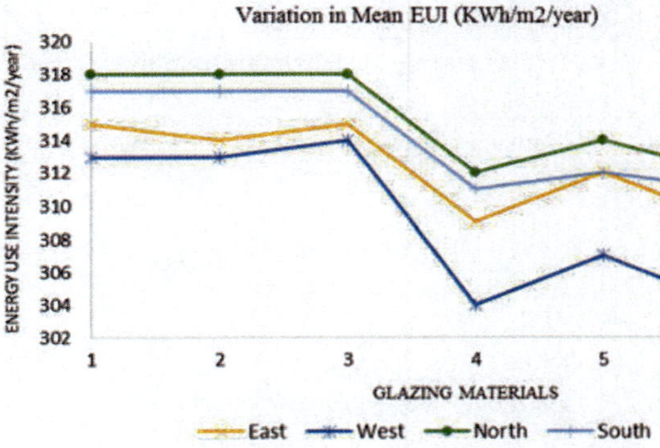

Fig. 2 Mean EUI for the location Udupi

Table 2 Specifications for glazing selected for the study

Glazing type	Characteristics	U (W/m^2/K)	SHGC	VLT
1	Clear element (all solar gain transmitted)	5.689	0.86	0.9
2	Uncoated single glazing 1/4-inch-thick-clear glass	5.90	0.80	0.54
3	1/8-inch Pilkington glazing	5.543	0.86	0.9
4	Double glazing ¼-inch-thick-clear/low-E $e = 0.05$ glass	1.986	0.26	0.42
5	Double glazing ¼-inch-thick-clear/low-E $e = 0.1$ glass	1.986	0.39	0.45
6	Triple glazing ¼-inch-thick-clear/low-E $e = 0.05$ glass	1.532	0.26	0.55
7	Triple glazing ¼-inch-thick-clear/low-E $e = 0.1$ glass	1.532	0.36	0.59

light transmittance of glass is its ability to transmit visible part of the spectrum of the solar radiation, expressed as a value between 0 and 1. Hence, the glazing having higher VLT should be preferred for effective utilization of the daylight for visual tasks.

2.3 *Effect of Building Materials on Energy Consumption.*

The building orientation, window to wall ratio (WWR), window shades, and window glasses are some of the factors which decide the total energy consumption for a building. The wall, roof materials, infiltration, and occupancy are also key factors in determining the energy consumption of a building. Table 3 shows the building parameters that account for least EUI. Figure 3 shows the EUI for the test room with current and best building parameters.

Table 3 Factors accounting for the least EUI

Glazing	Double glazing ¼-inch-thick-clear/low E ($e = 0.05$ glass)-type D
Window shades	2/3 window height
WWR	15%
Roof material	R60
Infiltration	0.17 ACH
Daylight and occupancy controls	Yes
Solar surface coverage	90%
HVAC loads	High-efficiency package terminal AC

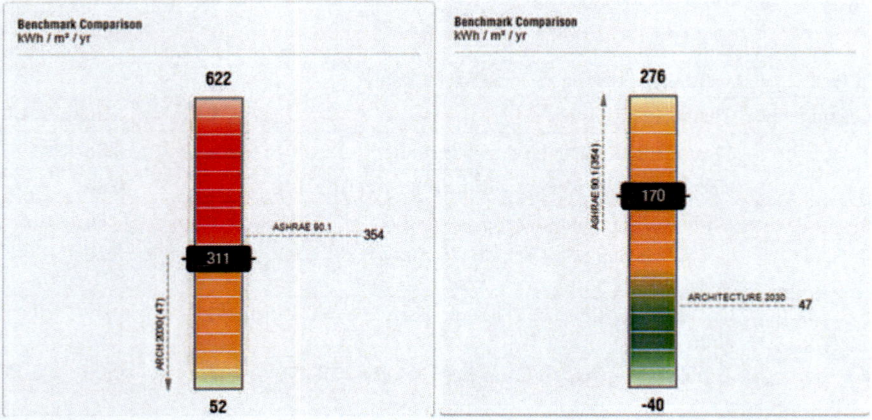

Fig. 3 EUI for the test room with existing and best parameters

2.4 *Effect of Surrounding Building on Energy Consumption*

The effect of surrounding building on the energy usage of the present building was evaluated in this study. In Fig. 4, a building mass is modeled near the window of the present building. The estimation of energy consumption of the building was then done for the seven glazing types, for different cases, where a window was placed in east, west, north, south directions, separately. A reduction of about 1–2 kWh/m^2/year in mean EUI was observed (i.e., around 17–34 KWh/year) in comparison with the case when the building was modeled as a single entity. The maximum EUI decreases by around 4–5 KWh/m^2/year.

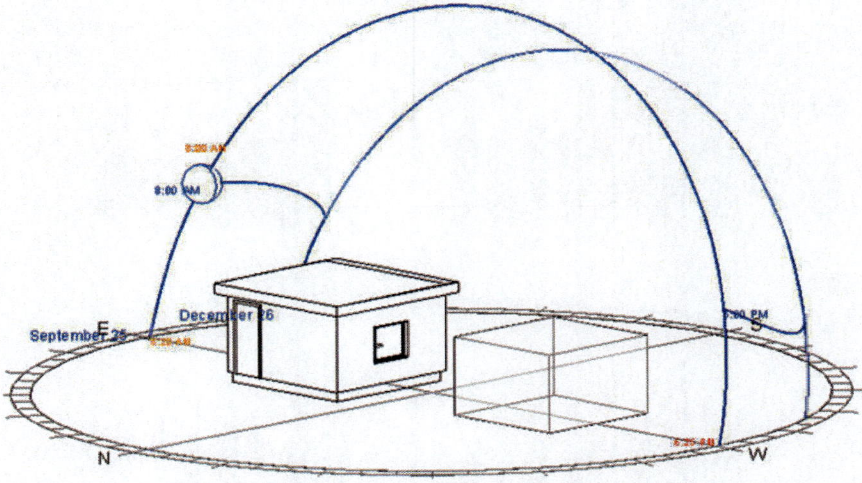

Fig. 4 Building model with an obstruction placed near the window

2.5 *Cooling Load Consumption for the Test Room with Occupancy and Glazing Types.*

The variation in the cooling load requirement for all window orientations with the number of occupants from 1 to 4 is shown in Fig. 5 for glazing '4'. The cooling set point was taken at 25 °C. The building model with the west window was observed to have more cooling load requirement compared to the windows in other directions due to the high solar gain that occurs after the mid-day. The peak cooling load for most of the cases was observed at around 2–3 pm in May. It could be inferred from the results that the cooling load of HVAC systems increases with the increase in occupancy. An increase of around 1160 Btu/hr (0.34 KW), i.e., more than 15% increase in cooling energy demand can be observed in most of the cases when the number of occupants increases from 1 to 4. The variation in cooling load in all the window orientations with occupancy was found to be similar for all the climates with different cooling set point temperatures.

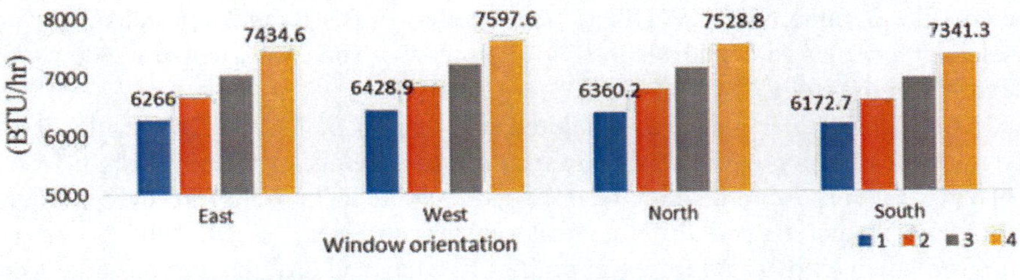

Fig. 5 Relationship between cooling load requirement and occupancy

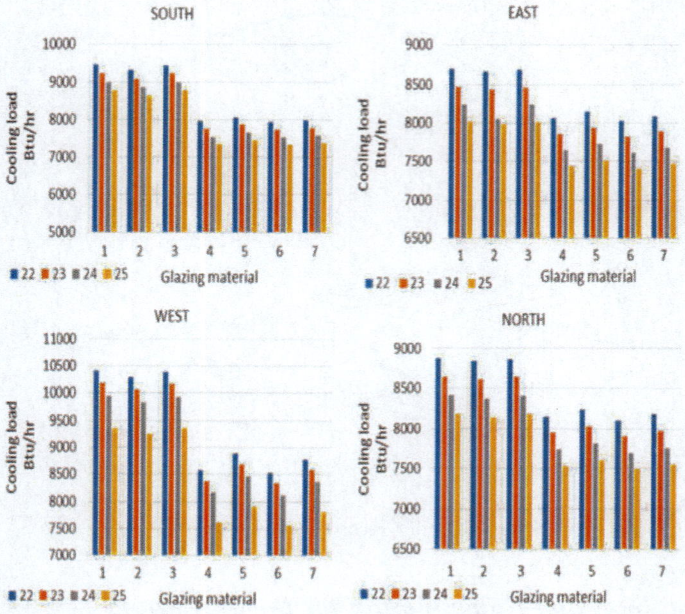

Fig. 6 Relationship between cooling load requirement and set point temperature

For better analysis regarding the variation in energy consumption in the Udupi region, during changes in temperature, the simulations were carried out with different set point temperatures for all types of glazing in each direction. The set points taken are 22, 23, 24, and 25 °C. Figure 6 shows the effect of temperature set point on cooling load in each direction for the highest occupancy, 4. As the set point ascended by 1 °C, the results depict about a 3% reduction in cooling capacity, which points toward the relevance of set point temperature.

2.6 *Analysis Using Dynamic Glazing*

The building modeling parameters and glazing materials with nearly similar parameters (like *U* value, SHGC, VLT) as mentioned in Tables 1 and 2, respectively, were selected for solar gain analysis in IES VE software. The east-oriented window was chosen for the study.

The highest solar gain was exhibited on the 21st of March. The results show significant variation in the solar gain for different timings of the day with different glazing. Figures 7 and 8 show the solar gain variation (in Watts) for the test room for glazing materials with different heat transfer coefficient '*U*' (in $W/m^2.K$) for the east-oriented window: for different timings of a day and different months of a year.

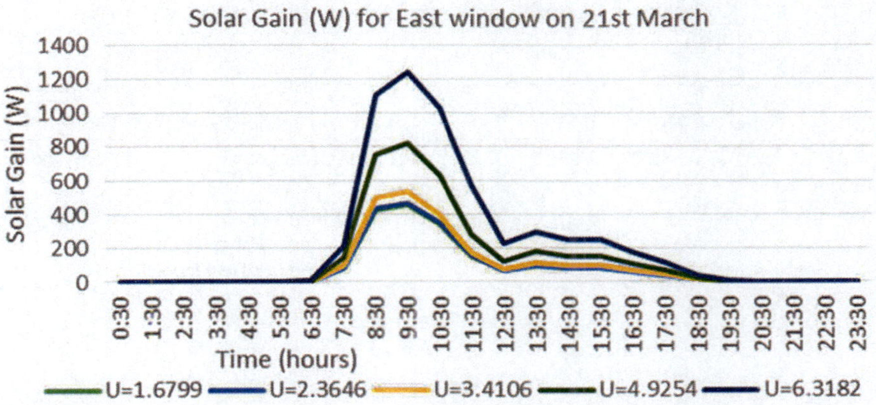

Fig. 7 Solar gain for different timings of a day for the east-oriented window

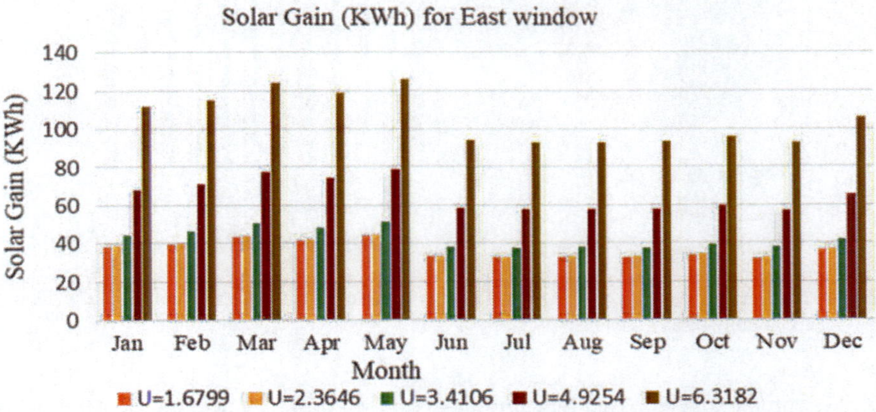

Fig. 8 Solar gain for different months of a year for east-oriented window

To analyze the energy performance of an electrochromic window, 'eQUEST,' a US Department of Energy (DOE) dynamic energy interpretation program, was used. Electrochromic Absorbing Bleached (2844)/Colored (2845) and double glazing had compared for which with the properties as listed in Table 4 below:

The same test room building in Udupi was modeled using eQUEST software for warm and humid location, to check for its performance using electrochromic glazing. Two different operation schedules were chosen for the simulation. The building was modeled for the same parameters mentioned in Table 1. Figure 9 shows the cooling load consumption for both glazings in east orientation with Udupi climate (warm and humid) for March on an hourly basis operation schedule and the whole year every month schedule. Since March shows the highest solar gain, it suits the best for the simulation purpose. From Fig. 8, it is clear that solar gain is more in the morning time

Table 4 Specifications of glazing and operation schedule for the simulation in eQUEST

Type	*U* value	SHGC	VLT	Operation schedule	
				Annual	For March
Double glazing low-e(0.04)	0.23	0.28	0.4	Throughout the year	Through out
EC (Colored) 6 + Ar 12.7 + Low-e 6	1.27	0.12	0.10	Summer season	8 am-Noon
EC (Bleached) 6 + Ar 12.7 + Low-e6	1.27	0.6	0.66	Winter season (January/February and November/December)	Noon-5 pm

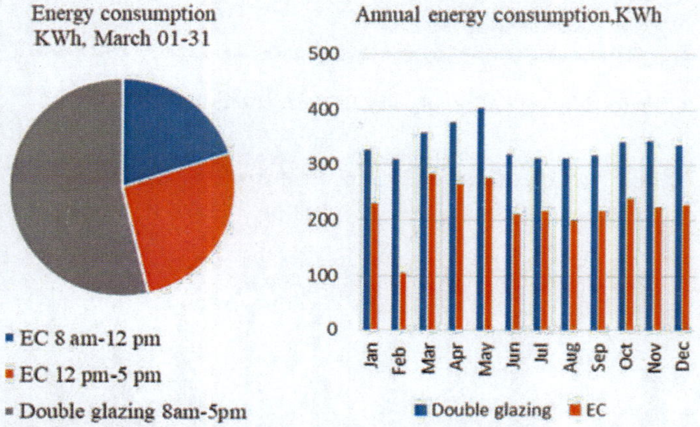

Fig. 9 Cooling load consumption for both glazings in east orientation with Udupi climate on hourly basis and monthly basis schedule

and is comparatively lesser after mid-day (east window). Hence, the EC window is scheduled to operate in the colored state during morning time and in the bleached state after mid-day. During the winter season, for more solar transmittance ECW at bleached state is used, and for the other days of the year, the set colored to minimize the cooling load.

The results obtained from simulations show that the energy usage could be reduced to 2840 KWh annually from 4066KWh with an energy demand of 850.3 KWh for winter seasons and 1989 KWh for the other months (February to October). Around 30% of energy savings could be achieved if double glazing with low E (0.04) is replaced by the electrochromic window. With monthly operation schedule on an hourly basis, for the month 10% reduction in cooling load demand for EC glazing could be achieved.

3 Conclusion

Evaluation of building energy provides an overview of conclusions to improve the building design in early stages.

Here, the simulation was conducted to study the overall energy performance of building under consideration with warm and humid climate. A comparison of energy consumption between double glazing and electrochromic window was also performed. The following results are extracted from the analysis.

i) The optimal building parameters bring an annual energy savings of 45% was attained, which is more prominent in green building design.
ii) For a given glazing type, the main factors in a building that contributes toward higher EUI are roofing material, HVAC systems, and glazing material. With a highly efficient HVAC system and roofing material with higher R-value, the energy savings could improve in an interior space.
iii) Among the analyzed static window technologies, multi-pane glazing leads to a significant reduction in the heat transfer on windows (double and triple glazing). In all climate scenarios, compared to the base case (clear glazing), double glazing can achieve more than a 10% hike in cooling load savings.
iv) Apart from the glazing, occupancy, and set point temperature majorly affects the cooling load in an interior space. As occupancy varies from 1 to 4, more than a 15% increase in cooling energy demand was observed in all cases with different glazing materials. Reduction in cooling demand of 3% followed when set point temperature was ascended by 1°C.
v) Solar gains were found to be the major contributor to cooling load demand in the interior space. Also, the simulation shows that in a building, overall heat gain is mostly affected by glazing type and roofing material.
vi) Energy savings of 30% could be achieved if a window with double glazing having permanent optical and heat properties is replaced by an electrochromic window glazing for monthly basis annual scheduled operations and 10% on an hourly basis for March in monthly operation schedule.

With the above results, it could be inferred that either a dynamic glazing or a proper window shading control like blinds can provide more cooling load savings compared to the traditional static ones. Also, with the derived energy-efficient measures and the optimum shading control strategy, in an interior space, an economical, energy-efficient scheme for daylight and artificial light systems can be designed. This paper provides a methodology for the designers and experts in the field of building construction to optimize the comfort level and energy performance of the interior spaces in different climatic zones.

References

1. P. Singh, A. Sadhu, Multicomponent energy assessment of buildings using building information modeling. Sustain. Urban Areas **1**(49), 101603 (2019)
2. J. Yu, Y. Liu, C. Xiong, H.J. Chao, Study on daylighting and energy conservation design of transparent envelope for an office building in hot summer and cold winter zone. Procedia Eng. **1**(121), 1642–1649 (2015)
3. Q.G. Deng, G.Y. Cao, Z.C. Liu, Z.S. Wang, Y. Yang, X.Y. He, J.J. Yu, Annual daylight glare evaluation: Impact of weather file selection. Light. Res. Technol. **50**(3), 446–455 (2018)
4. R. Debnath, R. Bardhan, Daylight performance of a naturally ventilated building as a parameter for energy management. Energy Procedia. **1**(90), 382–394 (2016)
5. C.P. Kurian, S. Milhoutra, V.I. George, Sustainable building design based on building information modeling (BIM), in 2016 IEEE International Conference on Power System Technology (POWERCON) (IEEE, 22 Nov 2016), (pp. 1–6)
6. D.W. Perera, D. Winkler, N.O. Skeie, Multi-floor building heating models in MATLAB and Modelica environments. Appl. Energy **1**(171), 46–57 (2016)
7. N.L. Sbar, L. Podbelski, H.M. Yang, B. Pease, Electrochromic dynamic windows for office buildings. Int. J. Sustain. Built Environ. (2012)
8. B.R. Park, et al., Improvement in energy performance of building envelope incorporating electrochromic windows (ECWs). Energies. **12**(6), 1181 (2019)

Study and Optimization of Piezoelectric Materials for MEMS Biochemical Sensor Applications

M. J. Nagaraj, V. Shantha, N. Nishanth, and V. Parthsarathy

1 Introduction

Micro-electro-mechanical system (MEMS) is combination of electrical and mechanical components and embedded on a chip [1, 2], which will be used to generate a small dimension system. MEMS-based sensors are being used in various applications such as, automotive detection electronics, biomedical equipment, clinical diagnosis, drug screening and chemical sensors for warfare and pathogen detection. Micromachined microcantilevers are still in the field of research and are mainly used as biochemical sensors. In biochemical sensors, the binding of biochemical molecules would change the mechanical properties of the microcantilever. The biochemical molecules are immobilized onto the microcantilever, hence free end of the microcantilever would bend and induce small voltage by piezoelectric device, and the same can be used to assay the biochemical concentration on the microcantilever. In static detection, a stress gradient is induced across the microcantilever due to the molecular accumulation on the surface of the microcantilever. The displacement due

M. J. Nagaraj (✉)
Department of Electrical and Electronics Engineering, Nitte Meenakshi Institute of Technology, Bengaluru, India
e-mail: nagaraj.mj@nmit.ac.in

V. Shantha
Department of Mechanical Engineering, Sir M. Visvesvaraya Institute of Technology, Bengaluru, India
e-mail: shantha_mech@sirmvit.edu

N. Nishanth
National Institute of Technology Karnataka, Bengaluru, India

V. Parthsarathy
Department of Electronics and Communication Engineering, Cambridge Institute of Technology (North Campus), Bengaluru, India

Sanjeevikumar et al. (eds.), *Advances in Renewable Energy and Electric Vehicles*,
Lecture Notes in Electrical Engineering 767,
https://doi.org/10.1007/978-981-16-1642-6_32

to the bending moment can be determined by using optical methods such as position sensitive detector (PSD). When studied dynamically, the targeted biochemical molecules that would bind onto the cantilever surface will result in the change of resonance frequency of microcantilever [3, 4].

The relationship between minimum measurable input force gradient and resulting deflection of piezoelectric microcantilever using scanning force microscopy was developed by Zhang et. al. [5, 6]. Electro-mechanical characteristics of piezoelectric was investigated by Smits and Choi [7]. Methods of closed-loop control to reduce effective error to measure deflection of multilayer piezoelectric cantilever were developed by Cheng et. al. [8, 9].

Piezoelectric device embedded on the microcantilever (Fig. 1) replaces the conventional method of PSD measurement. Piezoelectric biochemical sensors are devices that are used for the determination of analytes by affine interactions without the application of any further reagents. Piezoelectricity or piezoelectric (Fig. 2) effect is a phenomenon of inducing voltage when a material is stressed mechanically, i.e. mechanical energy is getting converted to electrical energy. Similarly, we can convert electrical energy to mechanical energy by giving voltage to a piezoelectric material's surface, which would cause mechanical stress or oscillation. Piezoelectric biochemical sensor is a group of analytical devices which can be used for conventional label-free detection of an analyte [10, 11].

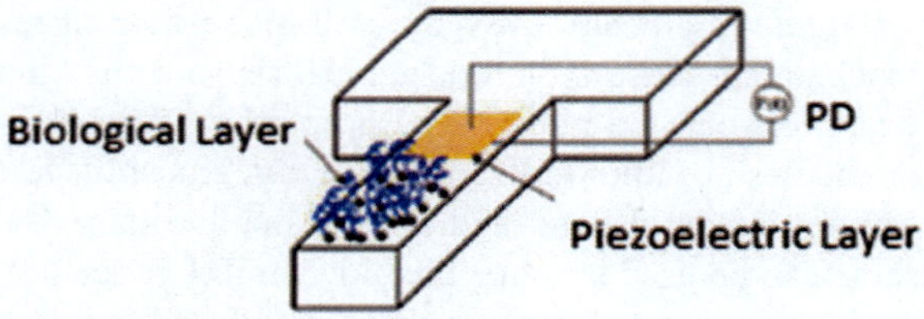

Fig. 1 Piezoelectric method

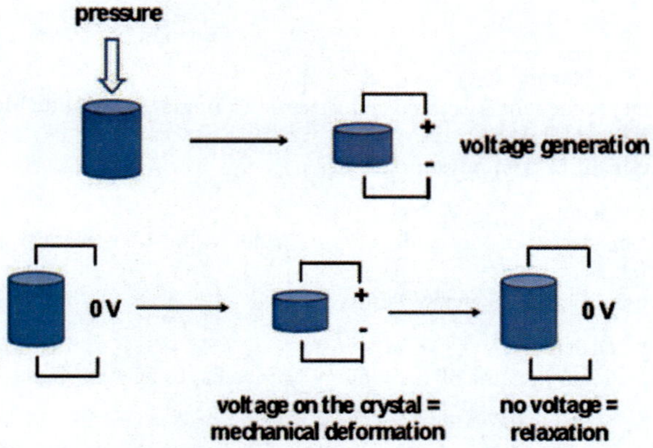

Fig. 2 Concept of Piezoelectric material

2 Experimental

The lateral view model for a multilayer piezoelectric microcantilever (Fig. 3) is based on 3D solid elements. The ratio of breadth to length for this simulation is arbitrarily taken as 1:2.5 to get more displacement. Different chemical compounds mentioned in Table 1 were used as the piezoelectric layer in the microcantilever. The particle mass used for the experiment is silicon rectangular block and it weighs about 0.285 pg. The dimension of the particle is 1 μm × 1 μm × 0.1 μm. The particle (Fig. 4) would be placed at 5 μm from the free end of the microcantilever during the experiment, since deflection is more at the free end of the cantilever beam. The weight applied by the simulated biomolecules on the plane of the piezoelectric material will be in the Z-direction. Once the force due to weight of the biomolecule is accumulated on the piezoelectric material, there is a change of position of atoms in the crystalline structure, hence electric charge would vary on the piezoelectric material surface. The vice versa process also causes the atoms to displace, i.e. if an external voltage is applied across the two ends of the piezoelectric material, there is displacement of atoms in the crystalline structure due to the change in the positions of holes and electrons. Due to resultant bipolar torque, the microcantilever displaces. The bending of the microcantilever is dependent on the chemical reaction that takes place on the surface [12].

The deflection of microcantilever (Z) is computed by using Eq. (1) [12],

$$Z = \frac{3(1 - \nu)L^2 \delta s}{T^2 E} \tag{1}$$

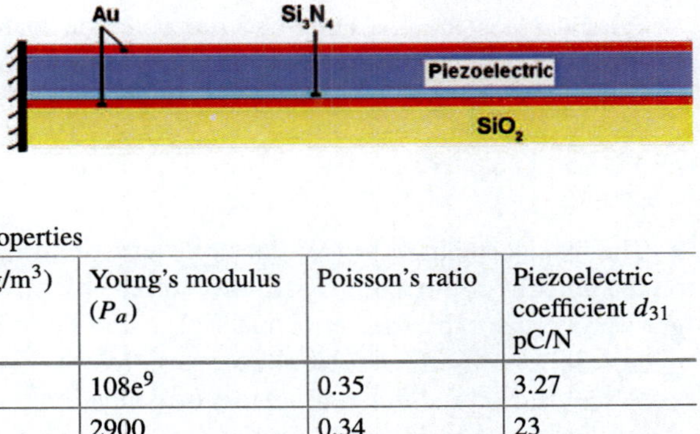

Fig. 3 Lateral view of the designed microcantilever

Table 1 Piezoelectric material properties

Material	Density(kg/m^3)	Young's modulus (P_a)	Poisson's ratio	Piezoelectric coefficient d_{31} pC/N
ZnO	5680	108e^9	0.35	3.27
PVDF	1780	2900	0.34	23
PZT-5H	7500	63e^9	0.31	283
PZT-8	7600	63e^9	0.33	105
Aluminium Nitride	3300	345e^9	0.24	1.9
Quartz	2651	71.7e^9	0.17	2.31(d_{11})

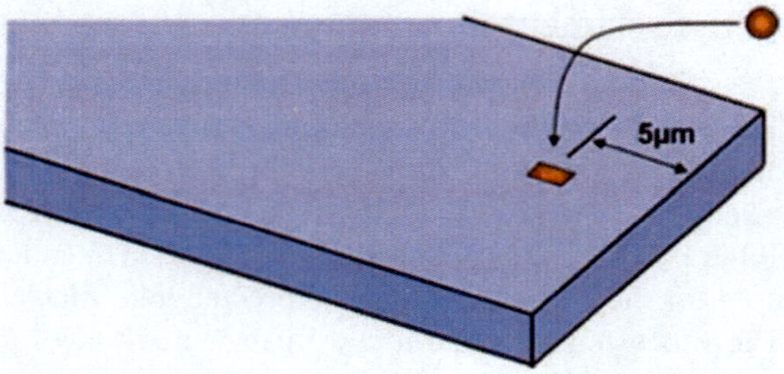

Fig. 4 Simulating cell on microcantilever

Where ν is the Poisson's ratio, L is length of the microcantilever, δ_s is the differential surface stress, T is thickness of the microcantilever and E is the Young's modulus of the piezoelectric material. A microcantilever is designed by placing a skeletal layer of piezoelectric material on a bulky material having high elasticity. The elastic material used will be in static equilibrium and will not have electrical charge. The relationship between deflection of the tip of microcantilever and voltage is computed using Eq. (2):

$$Z = \frac{d_{31} 3L^2 E_p V}{T^2 E_e} \tag{2}$$

From the above equation, the equation for V will be:

$$V = \frac{T^2 E_e Z}{3 d_{31} L^2 E_p} \tag{3}$$

Where d_{31} is the coefficient of piezoelectric material, E_e and E_p are Young's moduli for elastic and piezoelectric material, respectively.

$$V = \frac{E_e (1 - \nu) \delta s}{T^2 E_p E} \tag{4}$$

The displacement of the microcantilever is in the range of 10^{-9} and is correlated on the amount of biochemical reaction on the plane of the microcantilever. Table 1 gives properties of piezoelectric material, and in Table 2, specification of piezoelectric microcantilever used for the studies, respectively.

Figure 5 shows simulation results of deflection and generated voltage of piezoelectric microcantilever for different piezoelectric materials. Figure 6 shows graph of electric potential obtained for different piezoelectric material.

Table 2 Specification of piezoelectric microcantilever

Materials	Function						
Gold-Au	Electrodes of piezoelectric layer						
ZnO, PVDF, PZT-5H, PZT-8, Aluminium Nitride, Quartz	Piezoelectric layer						
Si_3N_4	Used to insulate and cancel the initial charges in the microcantilever						
SiO_2	Flexible basic layer of the microcantilever						
Structure	Length	Width	Thickness	Gold	Piezoelectric materials	Si_3N_4	SiO_2
Unit (μm)	50	20	1.6	0.2	0.5	0.2	0.5

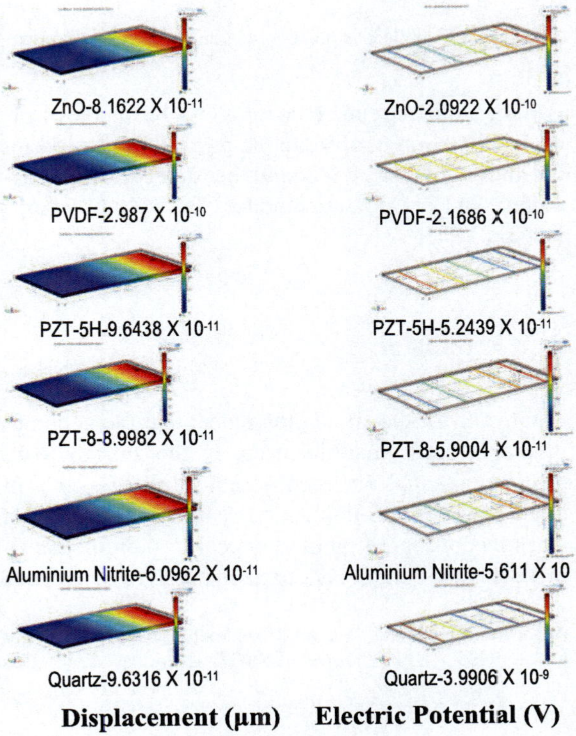

Fig. 5 Simulation results of deflection and generated voltage of piezoelectric microcantilever for different piezoelectric materials

3 Results and Discussion

Figure 6 shows the electric potential induced in different piezoelectric materials when a mass of the 0.285 pg is immobilized on the surface of piezoelectric microcantilever at 5 μm from the free end of the microcantilever during the analysis; since deflection

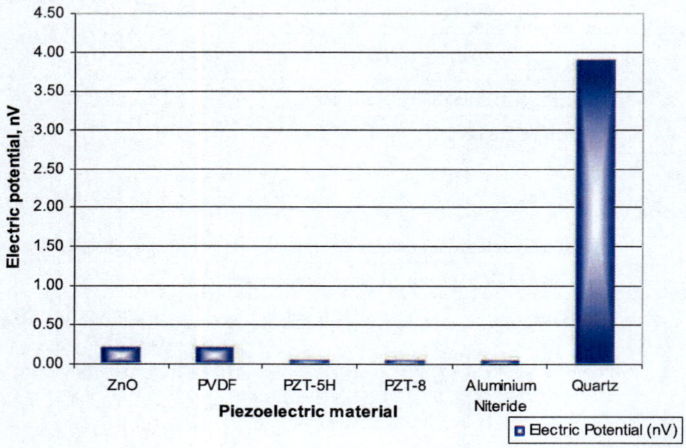

Fig. 6 Graph of electric potential obtained for different piezoelectric material

is more at the free end of the cantilever beam, the piezoelectric layer is changed during each simulation to observe the maximum displacement and induced voltage. From the above analysis, it is found that quartz material displaces more and induces more voltage and ZnO material induces less voltage when compared to other piezoelectric materials [13].

4 Conclusion

The results of the simulation reveal that the generated voltage depends on the type of piezoelectric material used. In the analysis carried, it was observed that the electric potential obtained is more, i.e. 3.999 nV in quartz material compared to other materials. In this analysis, it is concluded that to improve the sensitivity of (for low concentration of chemical or biomolecules) biochemical sensor, preferred piezoelectric material is quartz material to be embedded on microcantilever.

Acknowledgements This work has been carried out in Centre for Nano Materials and MEMS under VGST—KFIST project at NMIT—National MEMS Design Centre, Bengaluru, India.

References

1. J. Fritz, M.K. Baller, H.P. Lang, H.J. Guntherodt, C. Gerber, J.K. Gimzewski, Translating biomolecular recognition into nanomechanics. Science **288**(5464), 316–318 (2000)
2. G. Wu, R.H. Dater, K. Hansen, T. Thundat, R. Cote, A. Majumdar, Bioassay of prostate specific antigen (PSA) using microcantilevers. Nat. Biotechnol. **19**, 856–860 (2001)

3. A. Gupta, R. Bashir, Single virus particle mass detection using micro resonators with nanoscale thickness. Appl. Phys. Lett. **84**(11), 1976–1978 (2004)
4. B. Ilic, Y. Yang, H.G. Craighead, Virus detection using nanoelectromechanical devices. Appl. Phys. Lett. **85**(13), 2604–2606 (2004)
5. X.D. Zhang, C.T. Sun, Minimum detectable forcegradients of piezoelectric microcantilever. J. Micromech. Microeng. **5**, 231–236 (1995)
6. V. Tabard-Cossa, M. Godin, L.Y. Beaulieu1, P. Grutter, A differential microcantilever-based system for measuring surface stress changes induced by electrochemical reactions. Sens. Actuators B **107**(in press, 2005), 233–241
7. J.G. Smits, W. Choi, The constituent equations of piezoelectric heterogeneous bimorph. IEEE Trans. Ultrason. Ferroelectr. Freq. control **38**(3), 256–270 (1991)
8. H.-M. Cheng, M.T. Ewe, R. Bashir, G.T.-C. Chiu, Modeling and control of piezoelectric cantilever beam micro-mirror and micro-laser arrays to reduce imagebanding in electro photographic processes. J. Micromech. Microeng. 11487–498 (2001)
9. D.M. Welda, A. Kapitulnik, Feedback control and characterization of a microcantilever using optical radiation pressure. Appl. Phys. Lett. **89**, 164102 (Oct 2006)
10. M. Pohanka, The piezoelectric biosensors: principles and applications, a review. Int. J. Chem. Sci. Int. J. Electrochem. Sci. **12**, 496–506 (2017). https://doi.org/10.20964/2017.01.44
11. M. Pohanka, Overview of piezoelectric biosensors, immunosensors and DNA sensors and their applications. Materials **11**, 448 (2018)
12. M. Norouzi, A. Kashaninia, Design of piezoelectric microcantilever chemical sensors in COMSOL multiphysics area. J. Electr. Electron. Eng. (2009)
13. M.J. Nagaraj, Design and analysis of MEMS based microcantilever for biosensors, M.Tech, Thesis (2014)

Techno-Socio-Economic Sizing of Solar–Diesel Generator-Based Autonomous Power System Using Butterfly-PSO

Priyanka Paliwal

1 Introduction

In recent years, the focus of electric power industry has seen a shift towards increased utilization of renewable energy sources [1–3]. As per IEA, present solar installation is able to meet about 3% of electricity demand worldwide [4]. India has immense solar potential of around 750 GWp [5]. However, one significant concern associated with solar generation is their highly intermittent nature. The merits and demerits of solar generation are as follows [3]:

1.1 Merits

- Solar generation does not pose any fuel costs. Thus, operating costs are quite low.
- Reduction in harmful environmental emissions offers long-term benefits on human health and lifestyle.
- Increased reliance on locally available resources.

1.2 Demerits

- Capital cost is high. Thus, initial investment is significantly increased.
- Reliability concerns due to uncertainties associated with solar irradiance.

P. Paliwal (✉)
Department of Electrical Engineering, Maulana Azad National Institute of Technology, Bhopal, India
e-mail: priyankapaliwal@manit.ac.in

Sanjeevikumar et al. (eds.), *Advances in Renewable Energy and Electric Vehicles*,
Lecture Notes in Electrical Engineering 767,
https://doi.org/10.1007/978-981-16-1642-6_33

The primary concern in utilization of solar power is the impact of variability of solar irradiance on reliability aspects. This is more relevant if solar power generation is used for APS. The reliability issues can be handled by integrating solar generation with storage systems or conventional generators such as diesel. The integration of diesel generation can provide reliability benefits; albeit they have harmful impact on environment. Thus, an off-grid system has to be planned using an optimum combination of solar and diesel generation so as to strike a balance between techno-socio-economic parameters.

Optimal sizing problem using a mix of generation technologies has been discussed in several Refs. [6–12]. The sizing problem has been primarily focused on system reliability, cost and environmental emissions. Paliwal et.al. have proposed optimal sizing based on reliability evaluation for autonomous [6, 13] as well as grid connected systems [9]. Akram et al. [7] have solved two-level optimization for renewable energy-based generators and storage. Saif et al. [14] have formulated a linear programming-based bi-objective optimization problem aiming at minimizing cost and CO_2 emission using time series analysis. Gavanidou et al. [15] have used probabilistic method for evaluating the performance of wind–diesel energy systems. Khatod et al. [10] have carried out optimal planning of autonomous system comprising of diesel generator, solar and wind energy sources. Katsigiannis et al. [11] have optimized the component size of PV–wind–diesel–biodiesel–fuel cell battery-based hybrid power system. Ajlan et al. [16] have directed a planning study for off-grid systems. Optimal resource planning with different generation combination has also been done in [17, 18]. A variety of optimization techniques have been applied for solving component sizing problem such as genetic algorithm [19, 20], PSO [21], simulated annealing–Tabu search [11] and linear programming [14].

This paper proposes a techno-socio-economic assessment-based component sizing for a solar–diesel generator-based APS. The solar irradiance has been modelled using beta probability density function. The diesel generator is integrated along with solar to counterbalance the effect of variability of solar irradiance. A proficient meta-heuristic called as Butterfly-particle swarm optimization (BF-PSO) is used to obtain the optimum sizing problem. The obtained results have been evaluated to provide an enhanced insight into planning formulation. A case study has been carried out for APS sited in Jaisalmer, India.

The organization of paper is as follows: In Sect. 2, the modelling of solar irradiance and output power has been explained. Section 3 presents the problem formulation. In Sect. 4, the methodology involving implementation of Butterfly-PSO has been discussed. Section 5 presents case study and discussion on results. In Sect. 6, conclusion from work has been discussed.

2 Solar Power Modelling

The modelling of solar irradiance is carried out using beta probability distribution as follows [6]:

Fig. 1 I-V characteristics of PV module

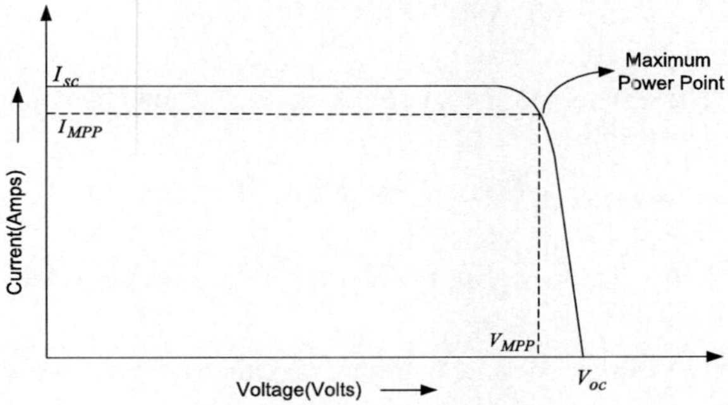

$$f_b(s) = \begin{cases} \dfrac{\Gamma(\alpha+\beta)}{\Gamma(\alpha)\Gamma(\beta)} s^{(\alpha-1)}(1-s)^{\beta-1} \\ \text{for} \quad 0 \le s \le 1, \quad \alpha \ge 0, \quad \beta \ge 0 \\ 0, \quad \text{otherwise} \end{cases} \tag{1}$$

where s = Solar irradiance in kW/m^2, $f_b(s)$ = Beta distribution function for s, α, β = Parameters of beta distribution function, Γ = Gamma function.

The α, β parameters can be calculated from mean and standard deviation of beta distribution. Figure 1 represents the curve at a particular level of solar irradiance and temperature corresponding to standard test condition (STC).

The power from solar generators can be calculated as [6, 13]

$$T_C = T_A + s \cdot \frac{N_{OT} - 20}{0.8} \tag{2}$$

$$I = s[I_{SC} + K_i(T_C - 25)] \tag{3}$$

$$V = V_{OC} - K_v T_C \tag{4}$$

where T_c = cell temperature, °C, T_A = ambient temperature, °C, N_{OT} = nominal operating temperature of cell, °C, I = PV module short-circuit current at conditions other than standard test condition (STC), A, K_i = short-circuit current temperature coefficient, A/°c, V = open-circuit voltage at conditions other than STC, V, K_v = open-circuit voltage temperature coefficient, V/°c. The maximum power from a solar generator comprising of N modules can be obtained as:

$$P_{solar}(s) = N \cdot FF \cdot V \cdot I \tag{5}$$

where N = number of modules, FF = fill factor, V = open-circuit voltage, volts, I = short-circuit current, A.

3 Problem Formulation

The techno-socio-economic criteria and objective function are explained in following subsections.

3.1 *Techno-Socio-Economic Assessment*

The techno-socio-economic assessment is carried out as follows:

(i) Technical assessment

The technical assessment is done by calculating expected energy not served over planning years. In present study, it is incorporated as a constraint on system reliability as follows:

$$\text{EENS} \leq \text{EENS}_{\text{defined}} \tag{6}$$

where EENS = expected energy not served, kWh, $\text{EENS}_{\text{defined}}$ = defining reliability standard, kWh.

(ii) Social assessment

The social assessment is carried out by calculating social cost of carbon. This is done by calculating cost of emissions from diesel generators as:

$$\text{Cost}_{\text{emission}} = P_{\text{diesel}} \times \text{SCC} \tag{7}$$

where P_{diesel} = power output from diesel generator, kW, SCC = social cost of carbon, /kWh.

(iii) Economic assessment

The economic assessment is carried out by calculating total life cycle cost (TLCC) over planning years. TLCC is used as objective function.

3.2 *Objective Function and Constraints*

$$\text{Minimize} \quad \text{TLCC} \tag{8}$$

The different cost components associated in determination of TLCC are calculated as explained in [6].

Subjected to following constraints:

(i) Reliability constraint [defined by Eq. (6)].

(ii) Power balance constraint: For any instant of time, the power balance between generator and load must be maintained.
(iii) Component capacity constraint as follows:

$$C_{\text{solar}_{\min}} \leq C_{\text{solar}} \leq C_{\text{solar}_{\max}} \tag{9}$$

$$C_{\text{diesel}_{\min}} \leq C_{\text{diesel}} \leq C_{\text{diesel}_{\max}} \tag{10}$$

where $C_{\text{solar}_{\min}}$ and $C_{\text{solar}_{\max}}$ is minimum and maximum capacity, respectively, of solar generators, kW.

$C_{\text{diesel}_{\min}}$ and $C_{\text{diesel}_{\max}}$ are minimum and maximum capacity, respectively, of diesel generators, kW.

4 Methodology: Implementation of BF-PSO

The problem of obtaining component sizes for a solar–diesel generator APS is a constrained, discrete combinatorial optimization problem. For solving optimization problem, a modified version of classic PSO called as Butterfly-PSO [22] has been used. The BF-PSO impersonates the natural intelligence and information sharing mechanism of butterflies during nectar search process. The velocity and position of each particle are calculated as follows:

$$\begin{aligned} v_{id}(k+1) &= w_k v_{id}(k) \\ &+ S_k(1-P_k)C_1 r_1[\text{Pbest}_{id}(k) - x_{id}(k)] \\ &+ P_{kg} C_2 r_2[\text{gbest}_d(k) - x_{id}(k)] \end{aligned} \tag{11}$$

$$x_{id}(k+1) = x_{id}(k) + \alpha_k v_{id}(k+1) \tag{12}$$

where x_{id} = position, v_{id} = velocity, Pbest_{id} = personal best and gbest_d = global best position of dth dimension of ith particle, w_k = inertia weight for kth iteration, C_1 and C_2 = acceleration coefficients and r_1 and r_2 are random variable (0–1), S_k = butterfly sensitivity towards flower and P_k = probability of food for kth iteration, α = time varying probability coefficient.

The values of sensitivity and probability are considered varying between 0.0 and 1.0 and are expressed as a function of iteration as follows:

$$S_k = e^{-(\text{iteration}_{\max} - \text{iteration}_k)/\text{iteration}_{\max}} \tag{13}$$

where $\text{iteration}_{\max}$ = maximum number of iterations and iteration_k = kth iteration count.

$$P_k = \text{Fitness}_{\text{gbest},k} / \sum \text{Fitness}_{\text{pbest},k} \tag{14}$$

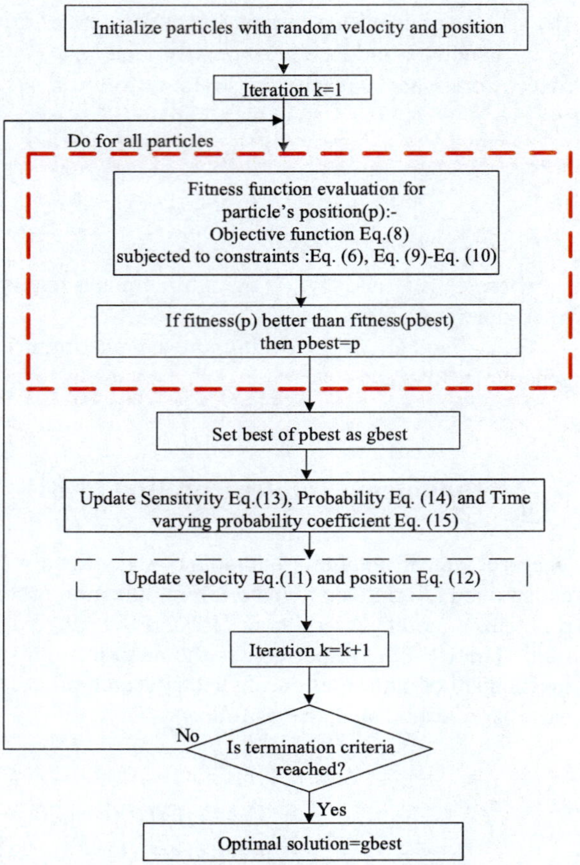

Fig. 2 Implementation of BF-PSO for optimization problem

where $\text{Fitness}_{\text{pbest},k}$ and $\text{Fitness}_{\text{gbest},k}$ = Fitness of pbest and gbest solutions, respectively, in kth iteration, P_{kg} = probability of global best (for global solution, P_{kg} = 1)

$$\alpha_k = \text{rand} * P_k \tag{15}$$

where rand = random number in the interval [0, 1].

The flowchart describing application of BF-PSO is presented in Fig. 2.

5 Case Study: Results and Discussions

The techno-socio-economic-based assessment of APS is carried out for the site of Jaisalmer, India. The peak load of system is 200 kW, and the load data has been

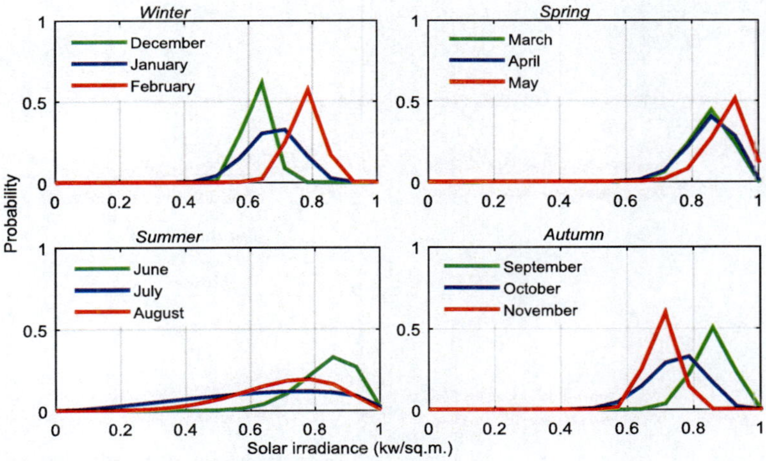

Fig. 3 Probability distribution of solar irradiance

obtained from [23]. The data for solar irradiance has been derived from [24]. The planning horizon is considered as 20 years. As discussed in Sect. 2, solar irradiance has been modelled using beta pdf. Figure 3 demonstrates the probability distribution of solar irradiance using beta pdf. The pdfs have been obtained for all the four seasons for time segment 1:00 to 2:00 p.m. Such pdfs are obtained for all time segments for period under study.

The technical and economic parameters considered in the analysis have been obtained from [6]. For incorporating technical criteria which is embodied as a constraint on system reliability, two cases are considered as follows:

Case-1 EENS $\leq 0.2\%$ of total load demand.
Case-2 EENS $\leq 0.1\%$. of total load demand.

The optimal solving problem is solved using Butterfly-PSO explained in Sect. 4. The convergence characteristics for both the cases are presented in Fig. 4. Table 1 presents the component sizes obtained using techno-socio-economic assessment.

It can be observed from Table 1 that when increased reliability standards are desired, the system cost increases. In comparison with Case-1, Case-2 sets higher reliability standard wherein the percentage of expected energy not served should be less than 0.1% of total load demand. Thus, TLCC increases for Case-2. Table 2 presents different cost components pertaining to both the cases.

Figure 5 presents the power output available from solar generators over a period of one year. It can be observed that it is highly intermittent in nature and cannot be relied upon for maintain system reliability. Thus, integration of a firm generation such as diesel is essential in order to satisfy requisite reliability criteria. The importance of integrating diesel generator can be further explained with reference to Fig. 6. It

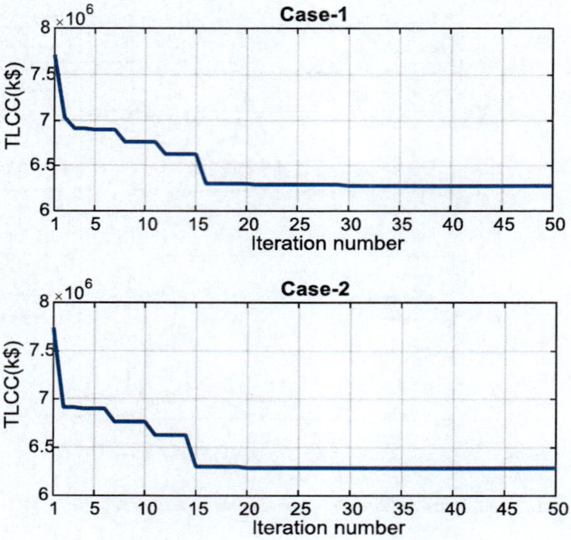

Fig. 4 Convergence characteristics of BF-PSO

Table 1 Results of techno-socio-economic assessment

Case	Capacity of solar generators (kW)	Capacity of diesel generators (kW)	TLCC (k$)	LCOE ($/kWh)	EENS (kWh)
1	180	180	6275.6	0.5952	1369.3
2	180	200	6287.6	0.5958	243.98

Table 2 Cost components for two cases

Case	Capital cost (k$)	Operating cost	$Cost_{emission}$
1	896.18	5319.256	77.75
2	901.736	5327.313	77.872

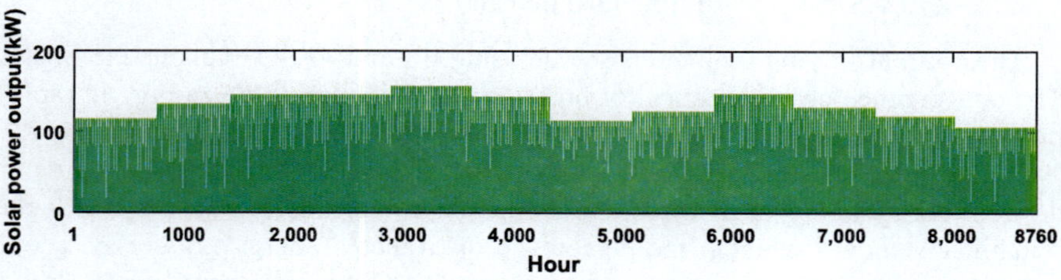

Fig. 5 Power output from solar generator

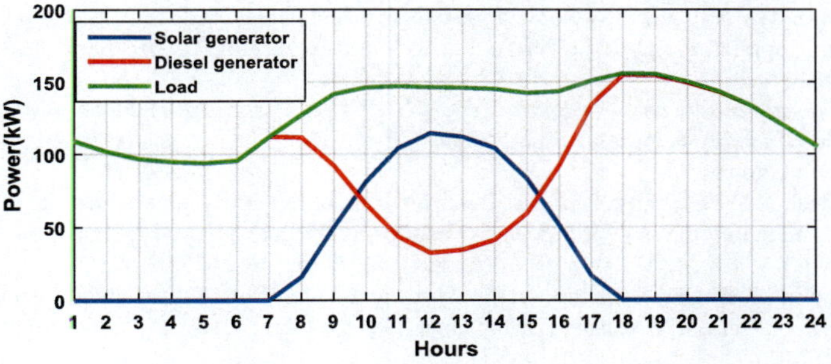

Fig. 6 Power distribution from solar and diesel generator for a typical day

can be clearly observed that during non-sunshine hours, diesel generators maintain uninterrupted supply to load.

6 Conclusion

In this paper, the component sizing for an APS comprising of solar and diesel generators has been presented. Techno-socio-economic assessment of different possible combinations has been carried out in order to come up with component sizes which can yield adequate reliability, reduce emissions and are economically viable. Diesel generators alone can ensure system reliability. However, there social impact is high due to emissions. On the other hand, solar power presents a clean and environment-friendly alternative but is highly stochastic. An integration of these two complementary technologies can counterbalance the drawbacks associated with both. In order to capture the uncertainties of solar irradiance, modelling has been done using beta pdf. The optimization problem is framed within the concept of techno-socio-economic assessment. Butterfly-PSO has been employed to obtain the component sizes. In order to visualize system planning from a broader perspective, two cases ascertaining different reliability standards have been considered. It has been observed that additional costs are to be incurred if reliability standards are raised. The results and analysis presented in this paper can provide insight into different aspects of planning of APS.

References

1. S.K. Wankhede, P. Paliwal, M.K. Kirar, Increasing penetration of DERs in smart grid framework: a state-of-the-art review on challenges, mitigation techniques and role of smart inverters. J. Circuits Syst Comput. (2020). https://doi.org/10.1142/S0218126620300147
2. S.K. Bhargava, S.S. Das, P. Paliwal, Multi-objective optimization for sizing of solar-wind based hybrid power system: a review. Int. J. Innovative Res. Sci. Eng. Technol. **3**(4), 195–201 (2014)

3. P. Paliwal, N.P. Patidar, R.K. Nema, Fuzzy logic based model for determination of battery state of charge applied to hybrid power system. J. World Acad. Eng. Sci. Technol. **171**, 1164–1168 (2012)
4. IEA, *Global Energy and CO_2Status Report 2019*. IEA. Paris. (2019). https://www.iea.org/reports/global-energy-co2-status-report-2019
5. www.mnre.gov.in
6. P. Paliwal, N.P. Patidar, R.K. Nema, Determination of reliability constrained optimal resource mix for an autonomous hybrid power system using Particle Swarm Optimization. Renewable Energy **63**, 194–204 (2014). https://doi.org/10.1016/j.renene.2013.09.003
7. U. Akram, M. Khalid, S. Shafiq, Optimal sizing of a wind/solar/battery hybrid grid-connected microgrid system. IET Renew. Power Gener. **12**(1), 72–80 (2018). https://doi.org/10.1049/iet-rpg.2017.0010
8. A. Kumar, A.R. Singh, Y. Deng, X. He, P. Kumar, R.C. Bansal, A novel methodological framework for the design of sustainable rural microgrid for developing nations. IEEE Access **6**, 24925–24951 (2018). https://doi.org/10.1109/ACCESS.2018.2832460
9. P. Paliwal, N.P. Patidar, R.K. Nema, Probabilistic indices for analyzing the impact of DER penetration on system reliability. IET Renew. Power Gener. (2020). https://doi.org/10.1049/iet-rpg.2019.1214
10. D.K. Khatod, V. Pant, J. Sharma, Evolutionary programming based optimal placement of renewable distributed generators. IEEE Trans. Power Syst. **99**, 1–13 (2010)
11. Y.A. Katsigiannis, P.S. Georgilakis, E.S. Karapidakis, Hybrid simulated annealing–tabu search method for optimal sizing of autonomous power systems with renewables. IEEE Trans. Sustain. Energy. **3**(3), 330–338 (2012)
12. P. Paliwal, N.P. Patidar, R.K. Nema, A novel method for reliability assessment of autonomous PV-wind-storage system using probabilistic storage model. Int. J. Electr. Power Energy Syst. **55**, 692–703 (2014)
13. P. Paliwal, Reliability constrained planning and sensitivity analysis for solar-wind-battery based isolated power system. Int. J. Sustain. Energy Plann. Manag. (2020). https://doi.org/10.5278/ijsepm.4599
14. A. Saif, K.G. Elrab, H.H. Zeineldin, S. Kennedy, J.L. Jr. Kirtley, Multi-objective capacity planning of a PV-wind-diesel-battery hybrid power system, in *IEEE International Energy Conference and Exhibition*, pp. 217–222 (2010)
15. E.S. Gavanidou, A.G. Bakirtzis, P.S. Dokopoulos, A probabilistic method for the evaluation of the performance and the reliability of wind-diesel energy systems. IEEE Trans. Energy Convers. **8**(2), 197–206 (1993)
16. A. Ajlan, C.W. Tan, A. Mohamed Abdilahi, Assessment of environmental and economic perspectives for renewable-based hybrid power system in Yemen. Renew. Sustain. Energy Rev. **75**, 559–570 (2017). https://doi.org/10.1016/j.rser.2016.11.024
17. N. Huanna, Y. Lu, Z. Jingxiang, W. Yuzhu, W. Weizhou, L. Fuchao, Flexible-regulation resources planning for distribution networks with a high penetration of renewable energy. IET Gener. Transm. Distrib. **12**(18), 4099–4107 (2018). https://doi.org/10.1049/iet-gtd.2018.5531
18. L. Wang, C. Singh, Multicriteria design of hybrid power generation systems based on a modified particle swarm optimization algorithm. IEEE Trans. Energy Convers. **24**(1), 163–172 (2009). https://doi.org/10.1109/TEC.2008.2005280
19. H. Yang, L. Lu, Z. Fang, Optimal sizing method for stand-alone hybrid solar–wind system with LPSP technology by using genetic algorithm. Sol. Energy **82**(4), 354–367 (2008)
20. Y.A. Katsigiannis, P.S. Georgilakis, E.S. Karapidakis, Multiobjective genetic algorithm solution to the optimum economic and environmental performance problem of small autonomous hybrid power systems with renewables. IET Renew. Power Gener. **4**(5), 404–419 (2010)
21. L. Wang, C. Singh, Compromise between cost and reliability in optimum design of an autonomous hybrid power system using mixed-integer PSO algorithm, in *International Conference on Clean Electrical Power*, pp. 682–689 (2007)

22. R. Kamdar, P. Paliwal, Y. Kumar, A hybrid multi-agent based BFPSO algorithm for optimization of benchmark functions. J. Circ. Syst. Comput. (2019). doi: https://doi.org/10.1142/S0218126620501121
23. Reliability test system task force of the application of probability methods subcommittee, The IEEE reliability test system-1996. IEEE Trans. Power Syst. **14**(3), 1010–1020 (1999)
24. India Meteorological Department. Solar Radiant Energy over India. Ministry of Earth Sciences. (2009)

Total Harmonic Reduction for a Series H-Bridge Multistage Inverter with Different Switching Methods

S Anand, Sujata Shivashimpiger, Shreeram V. Kulkarni, and C. H. Venkata Ramesh

1 Introduction

Inverter is devices which as the ability to convert DC into AC. Inverters were used to drive the load when the grid is off. However in recent time due to increased advancement in technology, inverters are used in motor drives, UPS, power system utilization [1].

Multilevel inverters have emerged as advanced research in the field of industrial applications. Multilevel inverter can be easily connected for high-end applications to obtain green energy. Gird-connected PV system [2] shows a VSIs used to convert DC voltage into AC voltage and feeds the energy to the load and grid through LC filter circuit. The inverter has to be controlled in order to obtain harmonic less voltage to obtain effective power quality.

Multilevel inverter using phase disposition (PO) modulation is introduced in [3]. The main disadvantage here is voltage is not balanced in capacitors which is sorted out in [4] using series multistage inverters. The advantage of cascaded multilevel inverters are minimum number of components, reliably, and modularity. The conventional multilevel inverters will need more components which increase the cost and complexity as the number of bridges is increased. Several filter methods are used to overcome harmonics [6] by using different PWM methods. Power semiconductors are widely used to achieve higher switching voltage levels and for reduction of harmonics, several selective harmonic elimination methods are used to reduce the

S. Anand (✉) · S. Shivashimpiger · S. V. Kulkarni · C. H. V. Ramesh
Department of EEE, Nitte Meenakshi Institute of Technology, Bengaluru, India
e-mail: anand.s@nmit.ac.in

S. V. Kulkarni
e-mail: shreeram.kulkarni@nmit.ac.in

C. H. V. Ramesh
e-mail: venkataramesh.ch@nmit.ac.in

Sanjeevikumar et al. (eds.), *Advances in Renewable Energy and Electric Vehicles*,
Lecture Notes in Electrical Engineering 767,
https://doi.org/10.1007/978-981-16-1642-6_34

harmonics [7]. The concept of switching angle enhances the switching schemes and analyzes the previous methods like space vector modulation, pulse width modulation, multiple pulse width modulation, nearest modulation used in reduction of harmonics [8–12].

The cascaded H-bridge multilevel inverter is used for high-voltage applications since it carries independent voltage sources; the voltage sharing adeptness is greatly increased [13].

In this paper, a nine-level series H-bridge multilevel inverter is added which includes switching angle and time calculation for switching the inverters on/off by applying triggering pulses using EP and HH method and by comparing the conventional and proposed inverter for reduction n total harmonic distortion (THD).

2 Methodology

Figure 1 shows the block diagram structure with input fed by DC source and from DC source to the auxiliary switches, usually IGBT is used as an auxiliary switch and firing gate pulses are given to operate the IGBT switches.

The circuit in Fig. 2 shows that there are two components: one is the switching component and the other is to generate positive and negative half cycle component. Each IGBT switch is triggered by providing switching gate pulses from the pulse generator block, and pulse with appropriate control is taken care to generate a nine-step voltage so that the circuit is used to obtain a nine-step multistage output. The switching operation for on/off of IGBT switches is taken care by following the switching pattern as shown in Table 1. The current flows from switch S_1 to R_1 to load to R_2 to D_4 to D_3 to D_2 to S_1. The diode D_1 is reverse biased, and remaining diodes D_2, D_3, D_4 are forward biased. During next interval, the switches S_1 and S_2 are turned ON by switching gate pulses. The current flows from load to R_2 to D_4 to D_3 to S_2 to S_1 to R_1 to load. In each interval, the switches turn ON and OFF simultaneously to generate multilevel output.

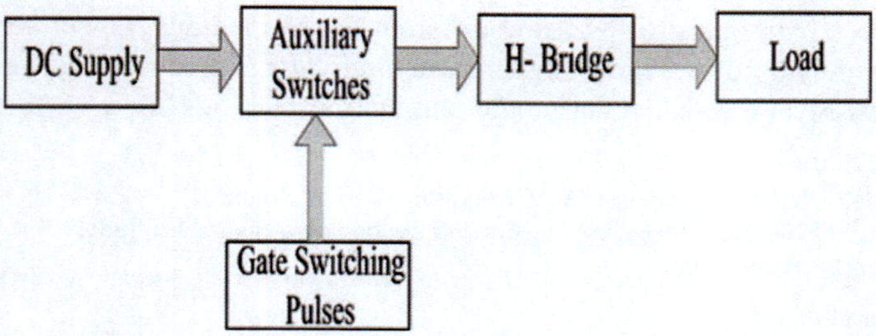

Fig. 1 Basic block diagram

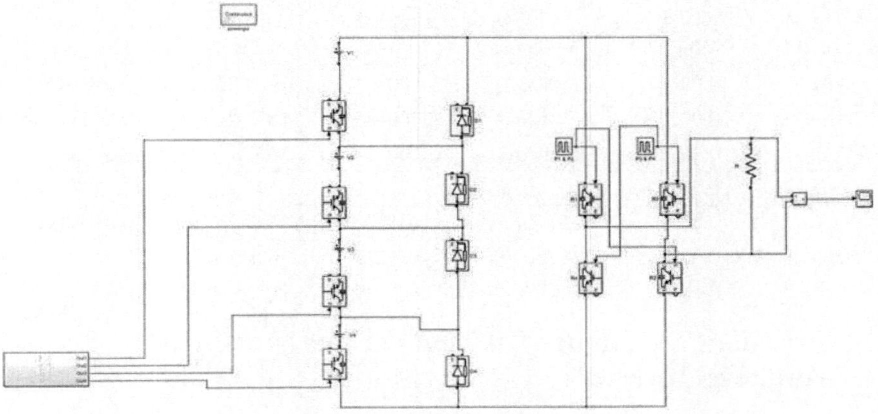

Fig. 2 Simulink model for proposed nine-stage inverter

Table 1 Switching table for proposed nine-multilevel inverter

Level	S_1	S_2	S_3	S_4	R_1	R_2	R_3	R_4
0	OF	OFF	OFF	OFF	OFF	OFF	OFF	OFF
V	ON	OFF	OFF	OFF	ON	ON	OFF	OFF
2 V	ON	ON	OFF	OFF	ON	ON	OFF	OFF
3 V	ON	ON	ON	OFF	ON	ON	OFF	OFF
4 V	ON	ON	ON	ON	ON	ON	OFF	OFF
3 V	ON	ON	ON	OFF	ON	ON	OFF	OFF
2 V	ON	ON	OFF	OFF	ON	ON	OFF	OFF
V	ON	OFF	OFF	OFF	ON	ON	OFF	OFF
0	OFF	OFF	OFF	OFF	OFF	OFF	OFF	OFF
−V	ON	OFF	OFF	OFF	OFF	OFF	ON	ON
−2 V	ON	ON	OFF	OFF	OFF	OFF	ON	ON
−3 V	ON	ON	ON	OFF	OFF	OFF	ON	ON
−4 V	ON	ON	ON	ON	OFF	OFF	ON	ON
−3 V	ON	ON	ON	OFF	OFF	OFF	ON	ON
−2 V	ON	ON	OFF	OFF	OFF	OFF	ON	ON
−V	ON	OFF	OFF	OFF	OFF	OFF	ON	ON
0	OFF	OFF	OFF	OFF	OFF	OFF	OFF	OFF

Table 2 Switching angles by EP method

Angle (α)	Time (s)	Angle (α)	Time (s)	Angle (α)	Time (s)
20° (α_1)	0.00111	140° (α_7)	0.00777	260° (α_{13})	0.01444
40° (α_2)	0.00222	160° (α_8)	0.00888	280° (α_{14})	0.01555
60° (α_3)	0.00333	180° (α_9)	0.00999	300° (α_{15})	0.01666
80° (α_4)	0.00444	200° (α_{10})	0.01111	320° (α_{16})	0.17777
100° (α_5)	0.00555	220° (α_{11})	0.01222	340° (α_{17})	0.018888
120° (α_6)	0.00666	240° (α_{12})	0.01333	360° (α_{18})	0.02

3 Switching Techniques Used for Various Nine-Step Multilevel Inverter

For the nine-steps multilevel inverter, the switching angles are obtained from the following methods.

3.1 *Equal Phase (EP) Method*

In the equal phase method, the main switching angles are obtained from formula below

$$\alpha * j = j^{*}180^\circ/n \text{ where } j = 1, 2, \ldots, (n-1)/2 \quad (1)$$

The main switching angles by EP method for the nine-level inverter are obtained as follows:

$$\alpha_1 = 20^\circ$$
$$\alpha_2 = 40^\circ$$
$$\alpha_3 = 60^\circ$$

The total switching angles is obtained by EP method and calculated as shown in Table 2.

3.2 *Half Height (HH) Method*

The half height method is used to obtain the switching angles from the below formula.

$$\alpha * j = \sin -1((2j-1)(s-1)) \quad (2)$$

Table 3 Switching angles by HH method

Angle (α)	Time (s)
7.18 (α_1)	0.000398
22.02° (α_2)	0.001223
38.68° (α_3)	0.00214
61.044° (α_4)	0.00339
UPTO $\alpha_i = 360°$	

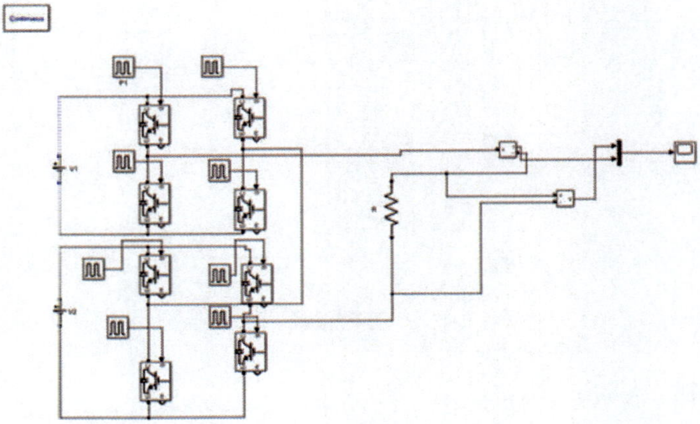

Fig. 3 Five-stage multilevel inverter

where $j = 1, 2, \ldots (s - 1)/2$ and $s =$ stages of switching levels.

In this method [5], the switching angles of the nine-level inverter are obtained from Table 3.

4 Conventional Multi-Level Inverter

Figure 3 shows the conventional five-level inverter circuit, and Fig. 4 shows a seven-level inverter circuit; these inverters are compared, that as the level of stages in an multilevel inverter is elevated and then the THD level is decreased.

5 Results

See Figs. 5, 6, 7, 8, 9, 10, and 11.

From the comparison Table 4, we obtain that HH method is efficient in reducing the THD compared to EP method.

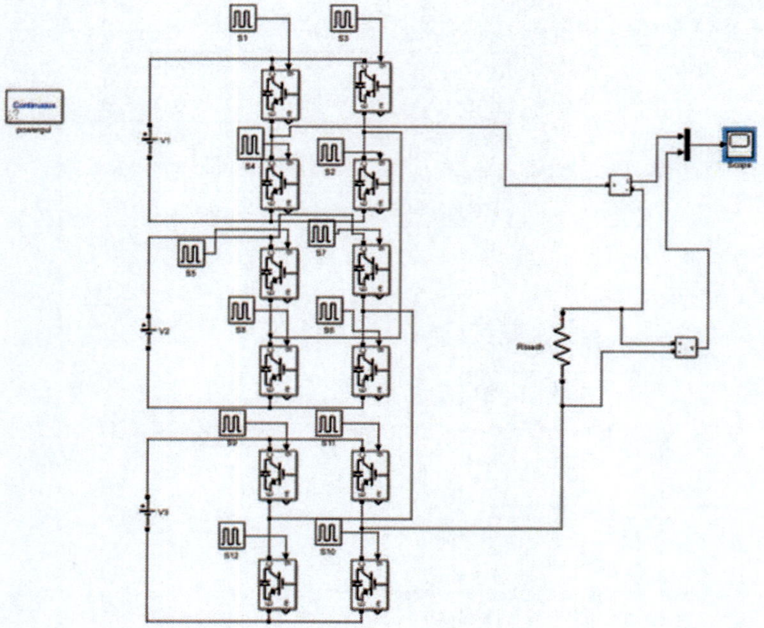

Fig. 4 Seven-stage multilevel inverter

6 Conclusion

In this paper, we conclude that the total harmonic distortion is obtained by using different switching methods and conclude that half height method is effective and efficient in reducing the harmonics, in turn reducing the number of switches for a proposed inverter in distinction to conventional multilevel inverter, since as the number of levels are increased in a conventional multilevel inverter, the switching losses increases in turn reducing the Total harmonic distortion but in the new proposed nine-level inverter the usage of number of switches is reduced in turn reducing the switching losses and obtaining better harmonic spectrum by using different switching angle methods. Hence, we can conclude that this proposed method requires less switching devices and the harmonics can be eliminated by elevating the number of stages and reducing the switches and switching losses in the new proposed method.

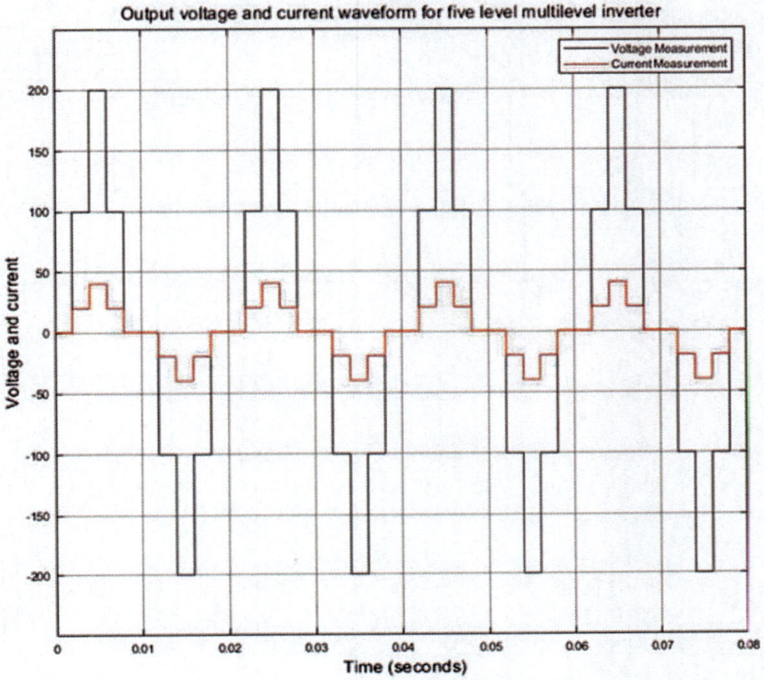

Fig. 5 Simulink output for five-stage conventional inverter

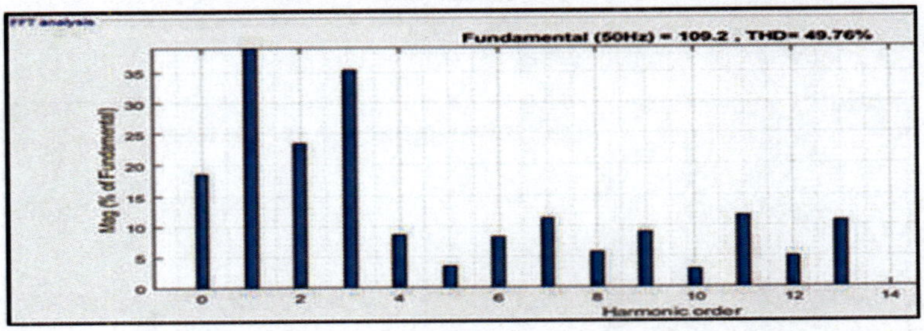

Fig. 6 THD output for five-stage inverter

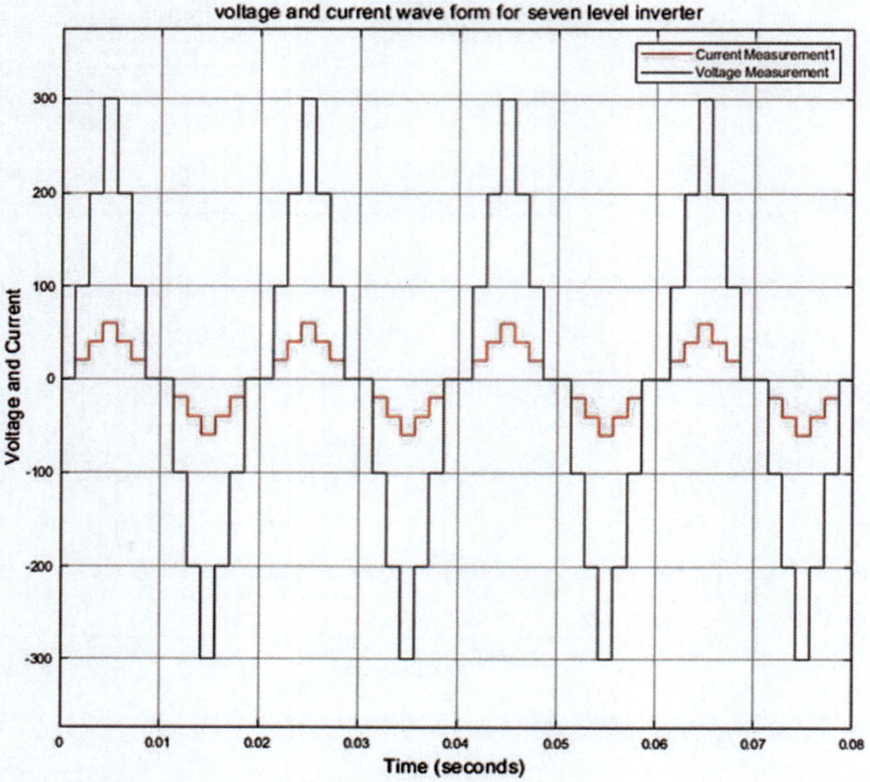

Fig. 7 Seven-stage conventional inverter output voltage

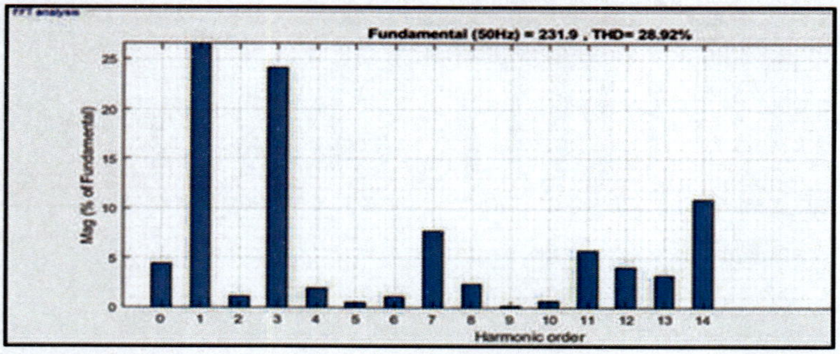

Fig. 8 THD for seven-stage multilevel inverter

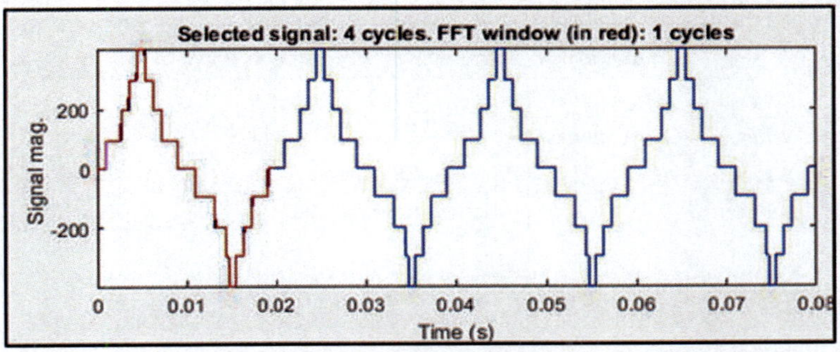

Fig. 9 Nine-stage output voltage of proposed inverter by EP method

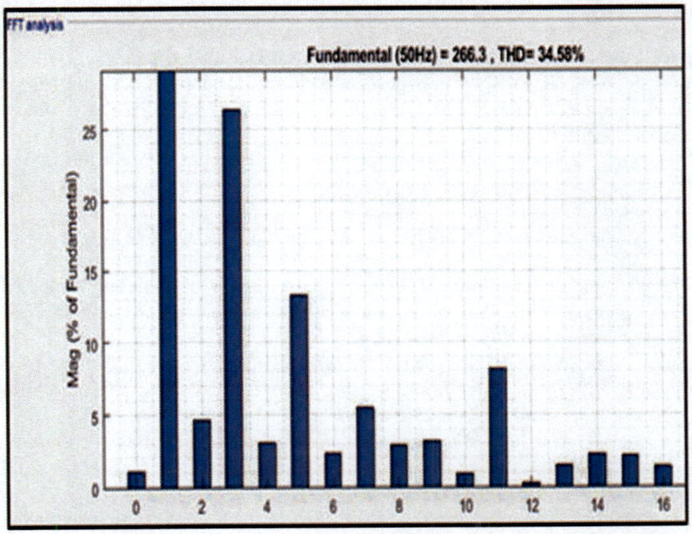

Fig. 10 THD for nine-stage proposed inverter by EP method

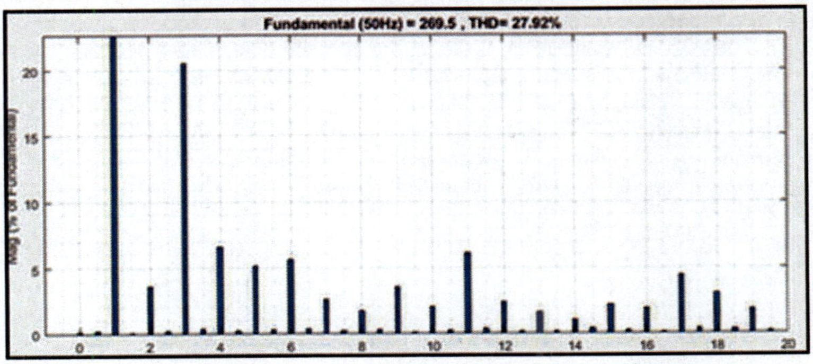

Fig. 11 THD for nine-stage proposed inverter by HH method

Table 4 THD analysis: EP method, HH method, and conventional method

S. No	m Level	Conventional	Proposed	Number of switches	%THD
1	5 Level	Conventional	–	8	49.76%
2	7 Level	Conventional	–	12	28.92%
3	9 Level	–	Proposed EP method	8	34.58%
4	9 Level	–	Proposed H–H method	8	27.92%

References

1. A. Nabae, I. Takahashi, H. Akagi, A new neutral-point-clamped PWM inverter. IEEE Trans. Ind. Appl. **5**, 518–523 (1981)
2. A. Ajami, A. Mokhberdoran, M.R.J. Oskuee, A new topology of multilevel voltage source inverter to minimize the number of circuit devices and maximize the number of output voltage levels. J. Electr. Eng. Technol. **8**(6), 1321–1329 (2013)
3. E. Babaei, S. Alilu, S. Laali, A new general topology for cascaded multilevel inverters with reduced number of components based on developed H-bridge. IEEE Trans. Industr. Electron. **61**(8), 3932–3939 (2013)
4. M.R.J. Oskuee, M. Karimi, S.N. Ravadanegh, G.B. Gharehpetian, An innovative scheme of symmetric multilevel voltage source inverter with lower number of circuit devices. IEEE Trans. Industr. Electron. **62**(11), 6965–6973 (2015)
5. F.L. Luo, H. Ye, Advanced DC/AC inverters: applications in renewable energy. CRC Press (2017)
6. M.S. Dahidah, G. Konstantinou, V.G. Agelidis, A review of multilevel selective harmonic elimination pwm: formulations, solving algorithms, implementation and applications. IEEE Trans. Power Electron. **30**(8), 4091–4106 (2015)
7. S. Sezen, A. Aktaş, M. Ucar, E. Özdemir, Design and operation of a multifunction photovoltaic power system with shunt active filtering using a single-stage three-phase multilevel inverter. Turk. J. Electr. Eng. Comput. Sci. **25**, 1412–1425 (2017)
8. Y. Liu, B. Ge, H. Abu-Rub, F.Z. Peng, An effective control method for three-phase Quasi-Z-source cascaded multilevel inverter based grid-tie photovoltaic power system. IEEE Trans. Indus. Electron. **61**(12), 6794–6802 (2014)
9. "An effective control method for quasi-Z-source cascade multilevel inverter-based grid-tie single-phase photovoltaic power system. IEEE Trans. Indus. Inf. **10**(1), 399–407 (2014)
10. M. Narimani, B. Wu, Z. Cheng, N.R. Zargari, A new nested neutral point-clamped (NNPC) converter for medium-voltage (MV) power conversion. IEEE Trans. Power Electron. **29**(12), 6375–6382 (2014)
11. H. Lou, C. Mao, D. Wang, J. Lu, L. Wang, Fundamental modulation strategy with selective harmonic elimination for multilevel inverters. IET Power Electronics **7**(8), 2173–2181 (2014)
12. A. Salem, E. Ahmed, M. Orabi, M. Ahmed, Study and analysis of new three-phase modular multilevel inverter. IEEE Trans. Industr. Electron. **63**(12), 7804–7813 (2016)
13. A. Noman, A.A. Al-Shamma, K. Addoweesh, A. Alabduljabbar, A. Alolah, Cascaded multilevel inverter topology based on cascaded H-bridge multilevel inverter. Energies **11**(4), 895(2018)

Variable Frequency and Voltage Control of Induction Motor for Electric Vehicles

Anup Shetty and Suryanarayana K.

1 Introduction

Impacts on the environment by fossil fuel powered vehicles are driving automobile industry toward battery-powered vehicles [1]. Toxic gases excreted by combustion engine based vehicles are one of the major contributors to environmental pollution [2]. Electric motor-driven vehicle development is growing rapidly because of features like operation with less noise, full-rated torque utilization at starting and reliability.

Induction motors are preferred in industries in view of its robustness in operation and control [3]. Optimal performance of the motor could be attained only at the rated speed. However, most of the applications require variable speed operation of motors. Traditional speed control driving techniques with mechanical moving parts and large transformers are simple and economic at the cost of poor efficiency and derated performance [5].

Controlling the induction motor with variable frequency and variable voltage with constant voltage-to-frequency ratio is one among the various speed control drive techniques which are in practice [7]. V/f control method is a simple, cost-effective, and efficient for driving three-phase induction motor. Open-loop V/f algorithm is adopted where there is no need of accurate speed control [5].

Digital implementation of the proposed control algorithm is done using NXP makes MC56F84789 controller which has 100 MHz core frequency. This digital signal controller has two on-chip eFlexPWM modules containing four submodules which are capable of generating complementary pair of PWMs, making them preferred for three-phase motor control applications [6]. High-speed ADC module of

A. Shetty (✉) · Suryanarayana K.
Department of Electrical and Electronics Engineering, NMAM Institute of Technology, Karkala, India

Suryanarayana K.
e-mail: suryanarayana@nitte.edu.in

Sanjeevikumar et al. (eds.), *Advances in Renewable Energy and Electric Vehicles*,
Lecture Notes in Electrical Engineering 767,
https://doi.org/10.1007/978-981-16-1642-6_35

12-bit resolution is used for control signal measurement. PWM modules incorporate fault signal feature which is used to disable PWM signal generation on undesired conditions [8].

Section 1 presents a brief introduction to the digital implementation of induction motor drives. Section 2 discusses the operation of an induction motor. V/f control theory is discussed in Sect. 3. Section 4 presents the system overview of the proposed idea. Section 5 presents software implementation aspects. Section 6 gives the simulation results and observations, and Section 7 presents the conclusion.

2 Induction Motor Operation

When the rated three-phase AC supply is applied to the stator windings of motor, a magnetic flux of fixed magnitude, rotating at synchronous speed is generated. The magnetic flux travels through the air gap between stator and rotor of the motor, links the rotor surface through the rotor conductors which are stationary during startup [5]. Relative speed difference between the rotating magnetic flux and stationary rotor flux induces an electro-motive force (EMF). The induced EMF frequency is as same as the power supply frequency under standstill condition. The induced EMF generates a current in rotor as the rotor bars of the motor are short-circuited at the ends. EMF magnitude is proportional to the relative velocity between the rotating flux and the rotor conductors. As current carrying conductor under the magnetic field experiences force, rotor starts rotating. Rotor will rotate in a direction such that it opposes the cause inducing EMF in the rotor conductor as per Lenz's findings. Rotor tries to catch up with the speed of rotating flux to reduce the relative velocity [5]. Practically, rotor rotates with a speed which is always less than that of rotating flux [4]. The speed difference between stator flux and rotor is called as slip speed. Slip speed depends upon the amount of mechanical load applied on the rotor shaft.

Speed of the stator flux is called the synchronous speed denoted as N_s could be written as:

$$N_s = \frac{120f}{P} \tag{1}$$

where P is the number of poles which is always an even integer number, f is the supply frequency.

The speed at which rotor flux rotates is called as base speed (N).

Percentage slip could be written as:

$$\%\text{Slip} = \frac{N_s - N}{N_s} * 100 \tag{2}$$

Induction motor fed from three-phase supply is a self-starting motor. Typical speed–torque characteristics of an induction motor when the stator winding is excited

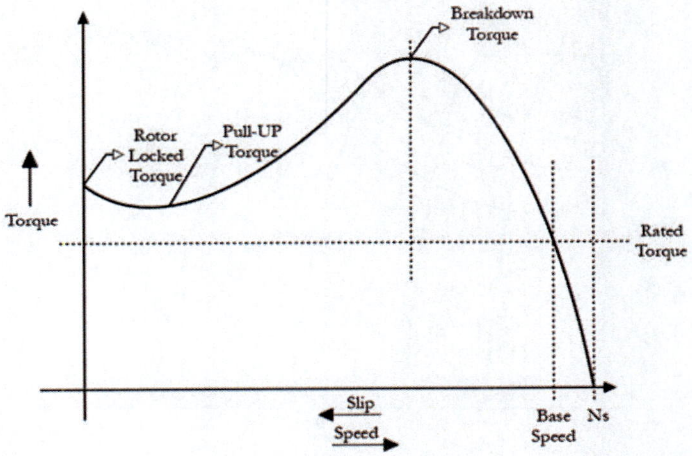

Fig. 1 Induction motor speed–torque characteristics

with rated voltage and frequency is as in Fig. 1. During starting, the motor is capable of generating a torque which is 1.5 times the rated torque [5]. When the rotor is in standstill condition, induction motor will draw current which is nearly 2–3 times the rated motor current [5]. This high initial current helps the motor to generate the toque which is called as rotor locked torque to overcome rotor inertia and frictional opposition. Rotor locked torque provides momentum for the rotor, and the rotor starts rotating.

When motor speed increases, the current starts to fall in the windings. It draws the rated current when operating at rated speed and delivering rated torque [4]. The motor can drive the rated load at this operating point. Motor could be overloaded based on the manufacturer maximum overloading data. Loading motor beyond its rated value causes motor to slow down and motor delivers torque up to a maximum value as of breakdown torque. Further, loading motor results in motor stalling. Even running the motor in overloaded condition is allowed only till the manufacturer overloading limit.

Motor overloading is not favorable as the flow of current above its rated value causes excessive resistive losses in motor winding leading to a temperature rise of motor windings and poor motor performance [4]. Based on the class of insulation and cooling facility available, some motors are capable of running effectively in overloaded conditions [5].

Induction motor supplied directly from the grid performs best at rated speed and load conditions. As in the speed–torque characteristics curve, torque is nonlinear over the entire speed range. To obtain the rated torque and performance at any desired speed, operation of the induction motor is controlled using simple and cost-effective variable frequency with variable voltage driving methods. These techniques shift the motor performance curve to the left as in Fig. 2, so that the rated performances

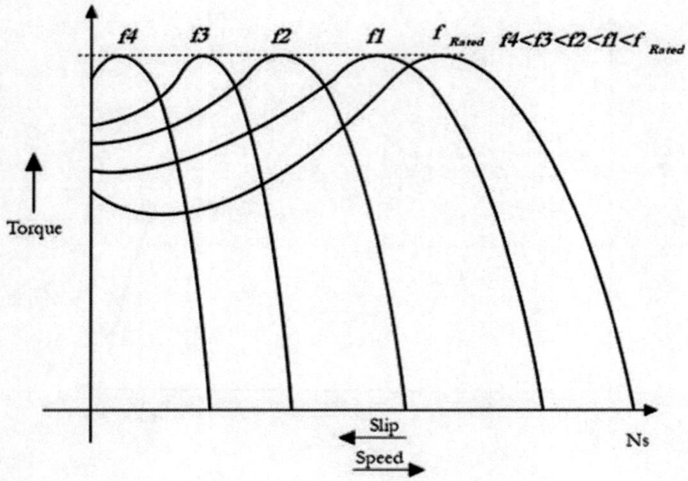

Fig. 2 Induction motor speed–torque characteristics with V/f control

are obtained at the lower speed conditions which are desirable in electric vehicle operation.

3 V/F Control Theory

Motor torque generated is directly proportional to the stator magnetic flux developed [3]. Stator voltage is responsible for producing the flux and could be written as:

$$V\alpha\Phi * \omega \tag{3}$$

stator voltage (V) is proportional to the product of magnetic flux (Φ) and angular velocity(ω).

$$V\alpha\Phi * 2\pi f \tag{4}$$

$$\Phi\alpha \mathrm{V/f} \tag{5}$$

As per (5), flux developed in the stator is directly proportional to the ratio of stator voltage and the frequency of the applied voltage. Constant torque can be obtained by keeping the V/f ratio constant throughout the speed range.

The voltage-to-frequency ratio applied to the motor is kept constant from minimum frequency to rated frequency as in Fig. 3. Torque generated by the motor

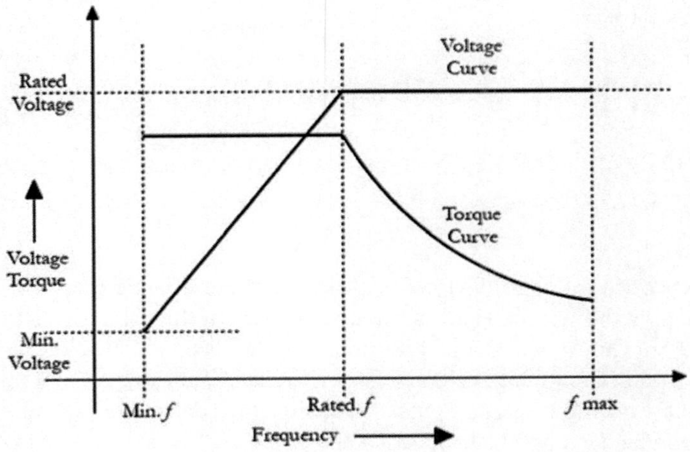

Fig. 3 Stator voltage versus frequency profile under V/f control

is kept constant with V/f control until the operating point reaches rated motor conditions. Frequency of operation can be further increased causing field weakening, which will lead to increasing speed at the cost of decreasing motor torque. Since stator voltage applied is limited by the winding insulation and source voltage constraints, stator voltage is maintained at rated voltage in the operating region above base frequency.

Driving the induction motor with V/f control is desirable with advantages such as lower starting current, chance of driving motor at speeds greater than rated speeds with smooth acceleration and deceleration. V/f control is most suited in applications with less accurate speed requirements. Advancements in microcontroller performance and computation speeds have enabled modern motor drive development with minimal time and efficient control algorithm.

4 System Overview

The power source for the system is a battery array of 25 cells with total capacity of 325 V 40 Ah. In-house developed galvanic isolated SMPS is used to generate suitable voltage levels for control circuit and gate driver circuits of the inverter module. DC bus of the battery pack is connected to the SEMIKRON inverter module consisting of IGBTs with system ratings as in Table 1. In the connection path, startup resistor of suitable rating has been included to limit the inrush current drawn by the DC bus capacitor C, when the start switch is closed. The bypass switch is used to bypass the startup resistor R when the capacitor gets charged to a certain voltage. Speed control reference signal is sensed from a Hall effect based foot pedal. Control panel

Table 1 Inverter module specifications

S. No	Description	Value
1	Input voltage	650 V DC
2	Power rating	14 Kw
3	Rated output voltage	440 V AC
4	IGBT DESAT protection	Enabled
5	Rated current	30 A

with reverse/forward control switch and reverse driving mode indicator buzzer is incorporated in the system. Hardware block diagram of the system is as in Fig. 4.

Control card housing MC56F84789 controller has been used to control induction motor hardware. Control card incorporates necessary features like the generation of PWM signals, analog-to-digital signal conversion, fault detection for disabling PWM signals upon fault event, LCD and LEDs for necessary visual indications.

MC56F84789 is an NXP make 32-bit 56800EX core-based powerful controller incorporating MCU and DSP functionalities. Controller can handle up to 100 MIPS at 100 MHz core frequency. Two high-speed 8-channel 12-bit on-chip ADC modules with programmable amplifier can operate at 5 MHz rate. It has two eFlexPWM modules with up to 24 PWM outputs. Rich set of control and programming flexibility added to the PWM modules makes this controller most suited for motor control and power electronic converter design applications [8]. Controller chip includes modules which are mainly used in power electronic circuit operations such as 8 channel 16-bit DAC, two 16-bit quad timers, two periodic interrupt timers, programmable comparator with integrated 6-bit DAC, one 12-bit DAC and inter-peripheral crossbar module. Communication modules integrated with UART, SPI, I2C, and CAN protocols are available as inbuilt peripherals.

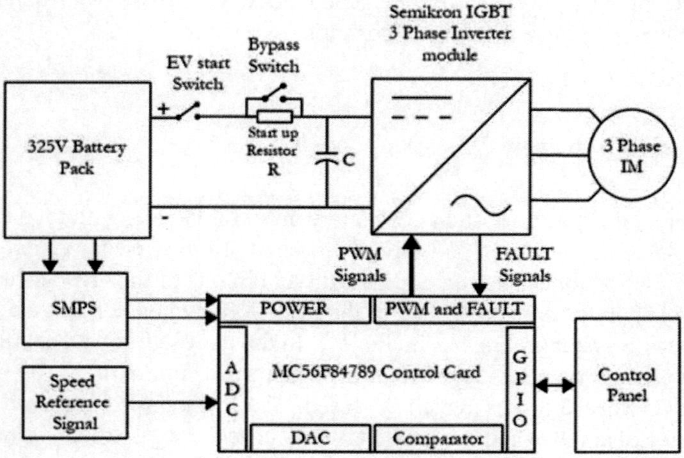

Fig. 4 System block diagram

Table 2 Induction motor specifications

S. No	Description	Value
1	Input voltage	3 Phase, 220 V AC
2	Power rating	7.5 HP
3	Frequency	50 Hz
4	Rated current	23 A
5	Rated speed	1457 RPM
6	Winding connection	Delta

Custom-made three-phase squirrel cage induction motor with ratings as in Table 2 is used to realize the proposed vehicle driving system. Inverter module consisting of IGBTs is used to generate variable frequency and voltage for induction motor. IGBTs with antiparallel diodes are used to freewheel the motor winding energy.

5 Software Implementation

Induction motor V/f control is implemented with sinusoidal pulse width modulation technique. SPWM signals are generated with lookup table-based approach. Sine wave look-up table is generated with the help of MATLAB/Simulink. Unipolar SPWM-based lookup table is used to store samples which are fed as duty cycle value for PWM modules. Phase shift required for three-phase sine wave generation is achieved by indexed lookup table sample feeding. ADC module is triggered by PWM module for each PWM reload cycle at 2 kHz rate. Quad timer interrupt subroutine is called at 500 Hz rate to read ADC to which control signal is connected. Modulation index parameter is used to control the output voltage of the inverter. Frequency and modulation index calculation based on the V/f profile is done within the timer interrupt subroutine. Inverter IGBT PWM switching frequency is set at 2 kHz.

The software flow diagram is as in Fig. 5.

6 Simulation, Results, and Observations

Simulink model is developed to generate the desired SPWM signals to drive the three-phase inverter model. Sine waves with frequency of 50 Hz are generated which are used as reference fundamental signals. The three-phase sine waves which are 120 degrees apart are as in Fig. 6. Inverter output voltage magnitude is controlled by controlling modulation index of the reference signals generated. Likewise, inverter output frequency is controlled by varying the frequency of reference signals.

Modulation is carried out with the help of carrier saw tooth signal with the frequency 2 kHz, which is the switching frequency of inverter module. The generated carrier signal is as in Fig. 7.

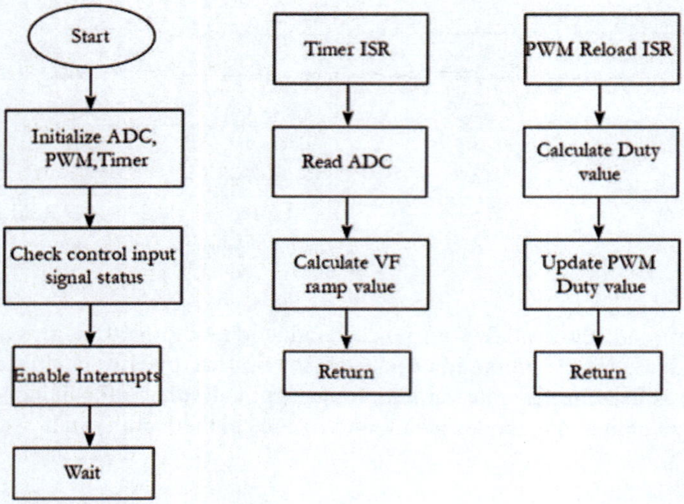

Fig. 5 Software flow diagram

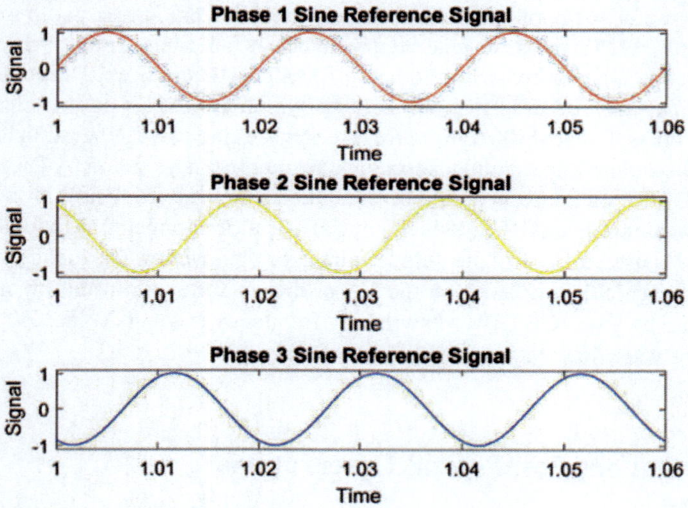

Fig. 6 Reference signals at frequency 50 Hz

Fundamental reference signal and carrier signals are passed through a comparator to generate sinusoidal pulse width modulated signals. SPWM signals as in Fig. 8 are used as gate drive signals of the inverter module.

Inverter module output is connected to induction motor windings. When SPWM gate driving signals are applied to inverter module, the motor voltage and current readings are as in Fig. 9. Unfiltered voltage is applied to the motor winding. Due to

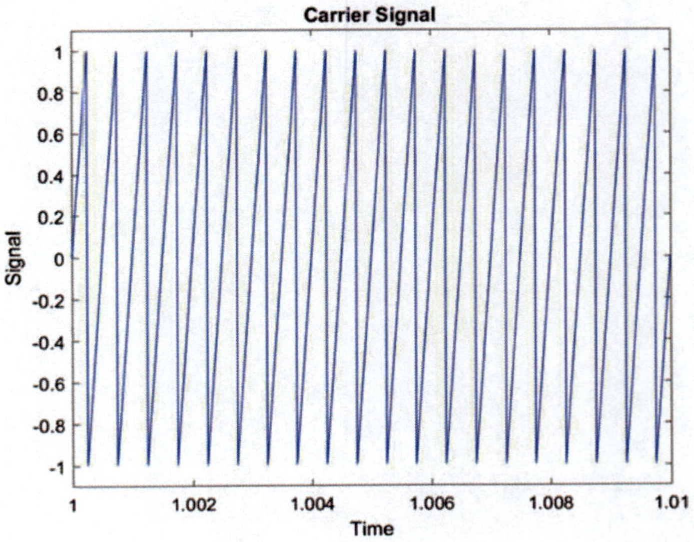

Fig. 7 Carrier signal at frequency 2 kHz

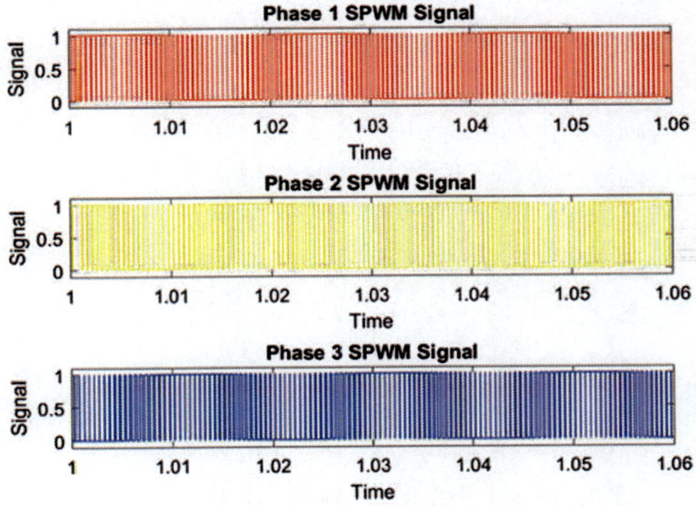

Fig. 8 SPWM signals generated $f = 2$ kHz

the large inductance of the motor windings, carrier signal gets filtered as in motor current waveform.

Simulated motor speed is 1470 RPM when the reference speed is 1500 RPM is as in Fig. 10.

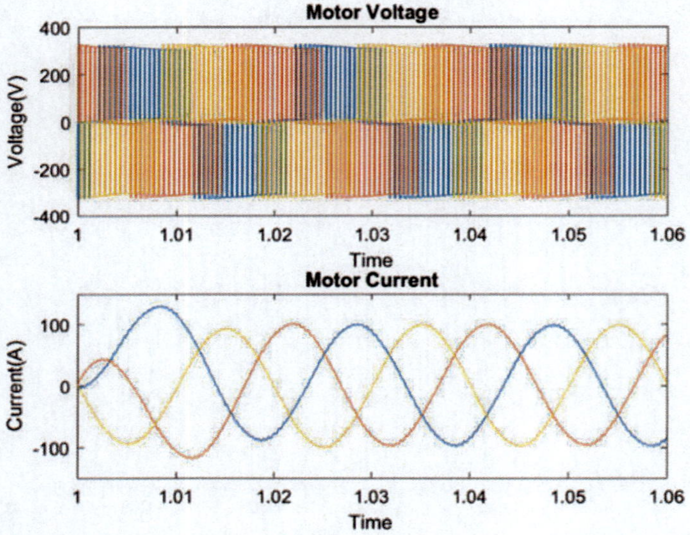

Fig. 9 Motor voltage and current waveforms

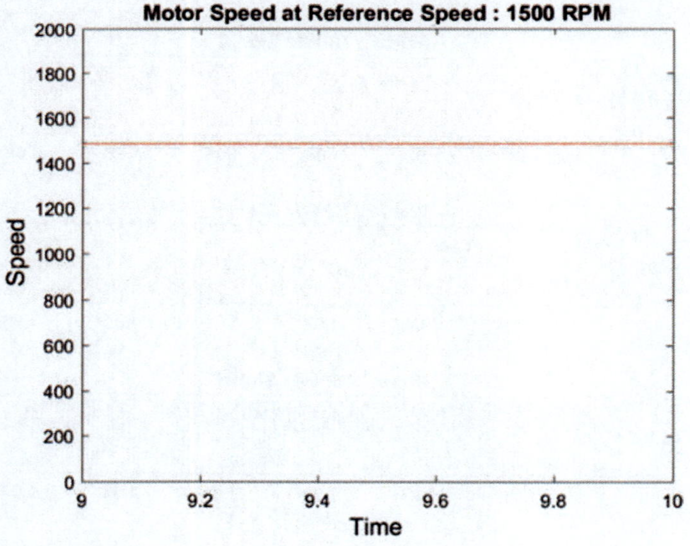

Fig. 10 Motor speed

Simulink model consists of SPWM signal generation section as in Fig. 11a, inverter model, DC source, and three-phase induction motor as in Fig. 11b.

SPWM signals generated with the help of microcontroller are as in Fig. 12. These control signals are given to the driver modules of the respective phase of the inverter module.

SPWM signals generated are passed through low-pass filter with cutoff frequency 200 Hz. Filtered PWM signals are as in Fig. 13. Carrier of frequency 2 kHz has been eliminated, and fundamental frequency reference signal has been reproduced.

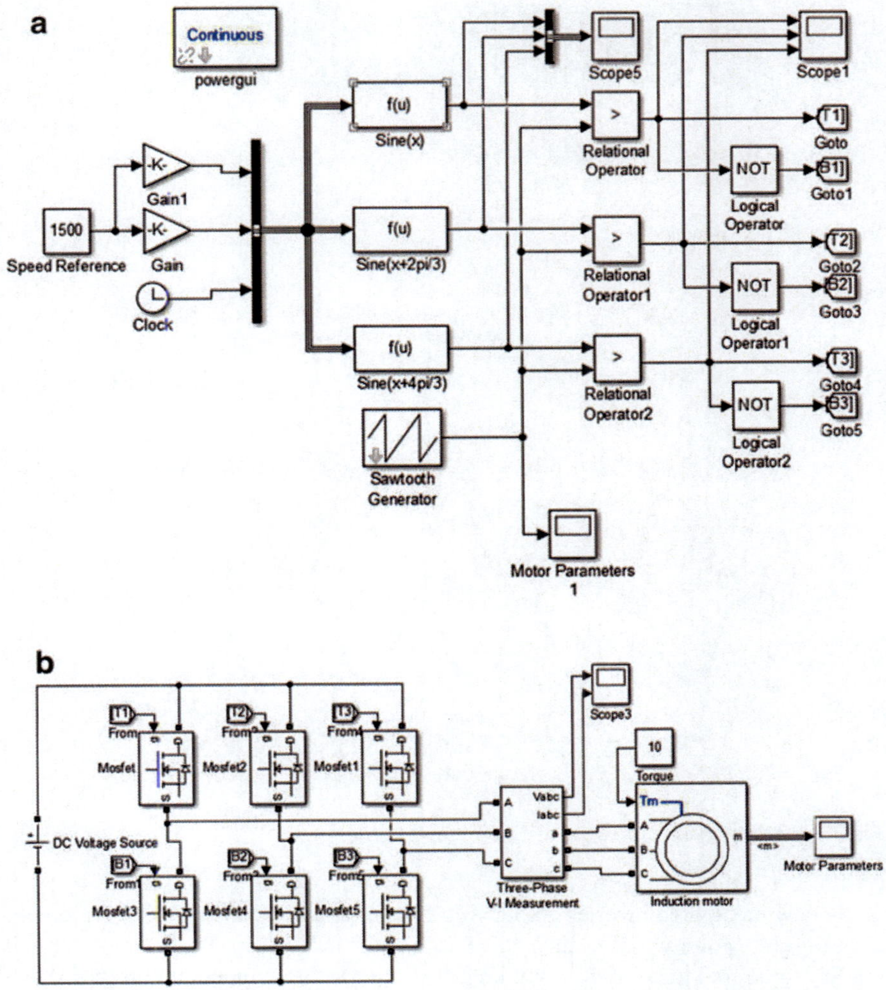

Fig. 11 **a** Simulink model of V/F driving system **b** Simulink model of V/F driving system

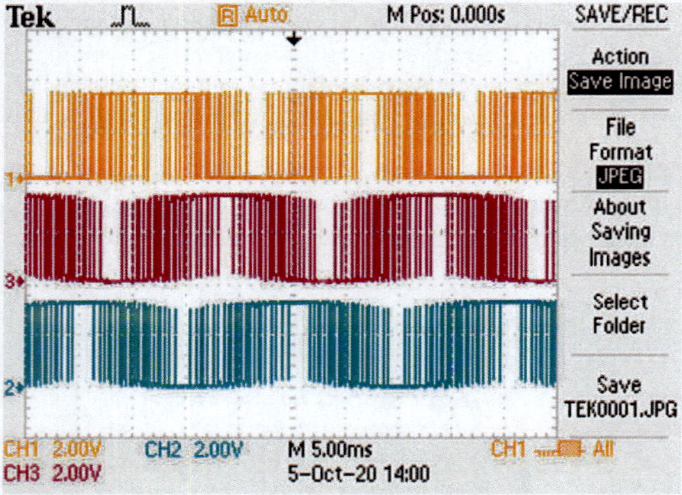

Fig. 12 SPWM signals

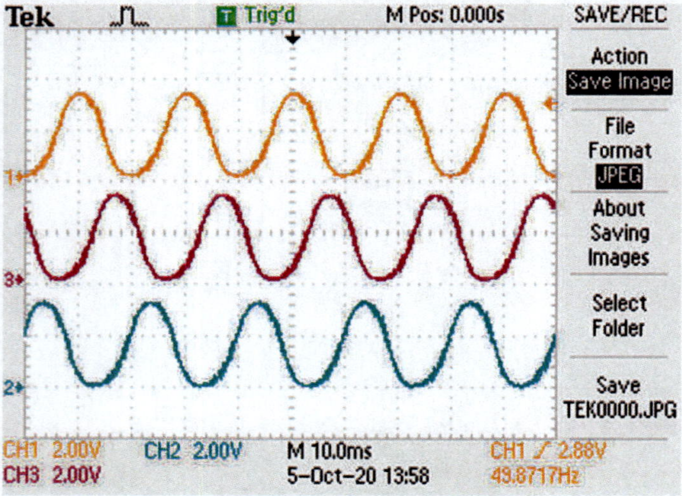

Fig. 13 SPWM signals filtered

SPWM modulated inverter output voltage connected across the motor winding is as in Fig. 14. Motor winding currents of each phase is as in Fig. 15. Inverter output is applied unfiltered to the motor winding. It is observed that inductor current follows reference sinusoidal signal pattern as in Fig. 15.

The proposed drive system has been implemented on a four-wheeler vehicle where the induction motor was able to drive the vehicle with two people including vehicle

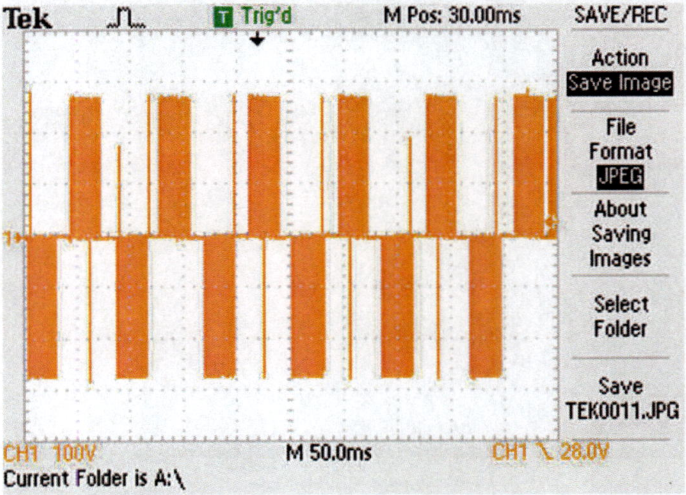

Fig. 14 Motor winding voltage

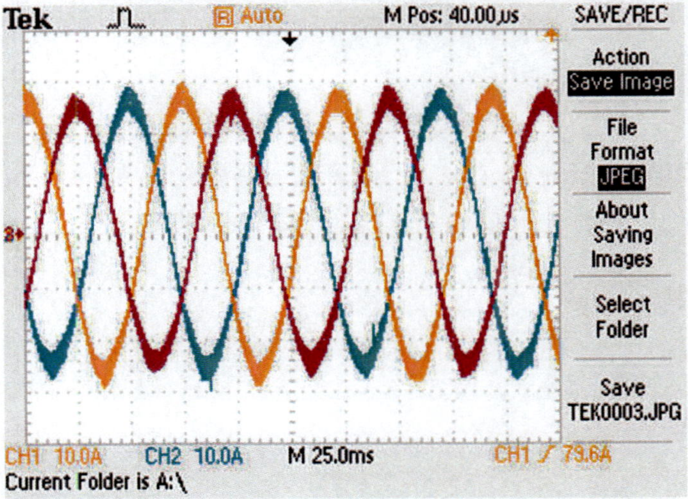

Fig. 15 Motor winding currents

body weight of 800 kg (Approx.) and battery weight of 400 Kgs (Approx.). Electric drive system implemented on the 4-wheeler vehicle is shown in Fig. 16. Drive circuitry assembly is as in Fig. 17.

Induction motor is coupled to the rare wheel drive mechanism of the four-wheeler as in Fig. 18.

Fig. 16 Electric vehicle

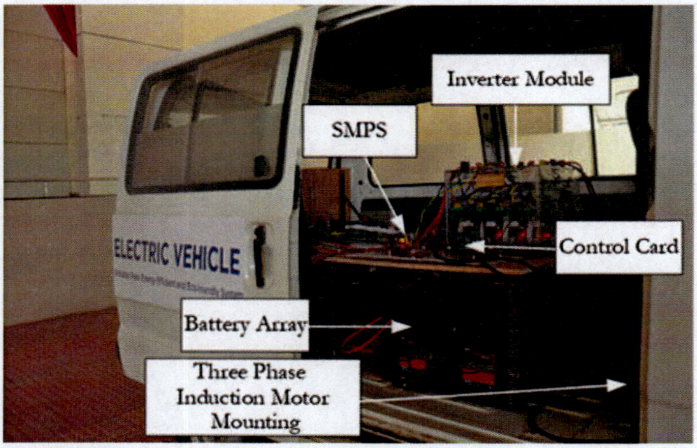

Fig. 17 EV drive circuit

7 Conclusion

Open-loop V/f control of an induction motor is simulated using MATLAB/Simulink. Software implementation of the same is done with the help of MC56F84789 microcontroller. Desired performance of the motor is obtained by software fine-tuning. Smooth acceleration and regenerative braking techniques are implemented.

Fig. 18 Motor wheel coupling

References

1. C.C. Chan, The state of the art of electric and hybrid vehicles. Proc. IEEE **90**(2), 247–275 (2002). https://doi.org/10.1109/5.989873
2. G. Graditi, G. Langella, C. Laterza, M. Valenti, Conventional and electric vehicles: a complete economic and environmental comparison, in *2015 International Conference on Clean Electrical Power (ICCEP)* (Taormina, 2015), pp. 660–665. https://doi.org/10.1109/ICCEP.2015.7177590.
3. K. Singh, A. Dalal, P. Kumar, Analysis of induction motor for electric vehicle application based on drive cycle analysis, in *2014 IEEE International Conference on Power Electronics, Drives and Energy Systems (PEDES)* (Mumbai, 2014), pp. 1–6. https://doi.org/10.1109/PEDES.2014.7042134.
4. A. Hughes, *Electric Motors and Drives, Fundamentals, Types and applications* (Elsevier publications Pvt Ltd.)
5. P. Yadamale, AN843, *Speed Control of 3-Phase Induction Motor Using PIC18 Microcontrollers* (Microchip Technology Inc., 2002)
6. P. Ulhr, Z. Kubiczek, AN1911, Phase AC motor control with V/Hz open loop using DSP56F80xx, App Note, NXP Semiconductor Inc. (2001)
7. S.M. Goda, Y.S. Elkoteshy, A.N. Ouda, A.E. Elawa, Scalar control technique for three phase induction motor for electric vehicle application, in *2018 Twentieth International Middle East Power Systems Conference (MEPCON)* (Cairo, Egypt, 2018), pp. 705–711. https://doi.org/10.1109/MEPCON.2018.8635187
8. MC56F847xx Reference Manual with Addendum, MC56F84789, MC56F847XXRM Rev. 2.0, 03/2016, Freescale Semiconductor, Inc.
9. Hardware Design Document: MC56F84789 Based Control Card,V0C/2017, Department of EEE, NMAMIT, Nitte
10. User Manual: SEMIKRON power electronics teaching system, Rev 0:2016, Semikron